GNSS, Space Weather and TEC Special Features

GNSS, Space Weather and TEC Special Features

Editors

Serdjo Kos
José Fernández
Juan F. Prieto

MDPI • Basel • Beijing • Wuhan • Barcelona • Belgrade • Manchester • Tokyo • Cluj • Tianjin

Editors

Serdjo Kos
University of Rijeka
Croatia

José Fernández
Instituto de Geociencias IGEO (CSIC-UCM)
Spain

Juan F. Prieto
Universidad Politécnica de Madrid
Spain

Editorial Office
MDPI
St. Alban-Anlage 66
4052 Basel, Switzerland

This is a reprint of articles from the Special Issue published online in the open access journal *Remote Sensing* (ISSN 2072-4292) (available at: https://www.mdpi.com/journal/remotesensing/special_issues/GNSS_Space_Weather_TEC).

For citation purposes, cite each article independently as indicated on the article page online and as indicated below:

LastName, A.A.; LastName, B.B.; LastName, C.C. Article Title. *Journal Name* **Year**, *Volume Number*, Page Range.

ISBN 978-3-0365-7594-0 (Hbk)
ISBN 978-3-0365-7595-7 (PDF)

Contents

About the Editors

Serdjo Kos

Prof. Dr. Serdjo Kos is a Full Professor (tenured). For two terms, he was the Dean of the Faculty of Maritime Studies, University of Rijeka, and for four terms he was the Head of the Department of Nautical Science. Some of his scientific research interests include: the theories of loxodromic and orthodromic navigation; terrestrial, astronomical, and electronic navigation; satellite navigation, satellite and inertial navigation systems (ionospheric/tropospheric delay errors in GNSS systems, positional dispersion of accuracy in GNSS systems, satellite positioning errors due to extreme ionospheric and geomagnetic disturbances, GeoRSS systems and technologies, errors of inertial navigation systems, etc.); multimodal transport networks; intermodal/multimodal transport systems; optimization and simulation in maritime traffic; and risk assessment on the use of electronic navigation devices. Due to his significant scientific contribution to the theory of loxodromic navigation at the Royal Geographical Society in London, he received an international award from the British Royal Institute of Navigation; it should be emphasized that he is the only member of the Council of the Royal Institute of Navigation in London who is not a citizen of the crown countries of the Commonwealth.

José Fernández

Prof. Dr. José Fernández was born in 1963 and received a degree (1986) and a PhD (1992) in Geodesy from the Complutense University of Madrid (UCM). He was a professor at the UCM until 2002, when he became a member of research staff of the Spanish Council for Scientific Research (CSIC). He is currently a Full Research Professor at the Institute of Geosciences (IGEO). His research focuses on the field of Geodesy and its application to natural and anthropogenic hazards, both in observation (using terrestrial and space techniques) and modeling aspects. His developed models have been applied, in combination with inversion methods, for the interpretation of displacements and gravity changes in volcanic, seismic, landslides, and subsidence areas. He has participated in more than 85 research projects (national, European Union, ESA, NASA, and other space agencies and international organizations), acting as a Coordinator, Responsible Investigator, or Responsible Co-researcher in about 60 of them. He is currently an editor of Scientific Reports (Nature Group), Remote Sensing, and Contributions to Geophysics and Geodesy, and a member of the Editorial Advisory Board of the Journal of Geodetic Science. He has been a part of different working groups of the International Association of Geodesy.

Juan F. Prieto

Juan F. Prieto (PhD Geodesy & Geomatics, Associate Professor at Universidad Politécnica de Madrid, Spain). Juan F. Prieto holds a PhD in geodesy and geomatics (Universidad Politécnica de Madrid, Spain). Since 2000, he has been teaching in the technical engineering programs at the School of Surveying, Geodesy, and Cartography Engineering (UPM). His research is developed within the framework of geodesy and geomatics, and its application to the shape and dimensions of the Earth, natural risks (earthquakes, volcanoes, terrain instabilities), and anthropogenics, using terrestrial and space techniques, and data processing. His research also involves the development of national GNSS geodetic networks in general (in Europe, Asia, and Africa), as well as of crustal displacements and gravity variation fields in volcanic and seismic areas (Spain). All these works are fundamental tools for studying the crises or the crustal properties of the Earth and the mantle in general. Concerning geodetic monitoring, he has jointly applied new observational and data processing methodologies in Radar Synthetic Aperture Interferometry (InSAR), gravimetry and GNSS, in collaboration with other Spanish national and international research groups. He has participated in more than 50 research projects (Spanish national, EU, ESA, NASA, and other space agencies, the World Bank, and international organizations), serving as the coordinator or responsible investigator in seven of them. He has directed, or co-directed, more than 50 B.S. and M.S theses.

remote sensing

MDPI

Editorial

Editorial for the Special Issue "GNSS, Space Weather and TEC Special Features"

Serdjo Kos [1,*], José Fernández [2] and Juan F. Prieto [3]

1 Faculty of Maritime Studies, University of Rijeka, 51000 Rijeka, Croatia
2 Institute of Geosciences (IGEO), CSIC-UCM, Calle del Doctor Severo Ochoa, 7, Ciudad Universitaria, 28040 Madrid, Spain
3 ETS de Ingenieros en Topografía, Geodesia y Cartografía, Universidad Politécnica de Madrid, 28031 Madrid, Spain
* Correspondence: serdjo.kos@pfri.uniri.hr

Citation: Kos, S.; Fernández, J.; Prieto, J.F. Editorial for the Special Issue "GNSS, Space Weather and TEC Special Features". *Remote Sens.* **2023**, *15*, 1182. https://doi.org/10.3390/rs15051182

Received: 28 January 2023
Revised: 13 February 2023
Accepted: 13 February 2023
Published: 21 February 2023

For high-quality scientific communication in the field of technical and natural sciences, it is of utmost importance to ensure clarity of the text, logical mathematical argumentation, and the possibility of verifying the obtained theoretical results using appropriate experiments.

The publication of research results requires the skill of scientifically communicating relevant data and their mutual logical connection into a purposeful and comprehensible whole.

In the domain of electronic navigation, satellite navigation (GNSS) is one of the most important modern complex systems. GNSS is a key infrastructure for supporting the development and improvement of not only navigation and civil engineering infrastructures but also power grid systems, banking operations, global transportation systems, and global communication systems. Today, GNSS requires the use of several positioning networks and sensors, such as radio networks and micro electromechanical systems (MEMS), among others. Earth's atmosphere, especially the ionosphere, troposphere, etc., is a huge laboratory where multiple processes and phenomena occur that directly affect the propagation of electromagnetic waves. Like all complex systems, GNSS technology also undergoes certain evolutionary stages. Some factors affecting the future evolution of GNSS technology include the appearance of new signals and frequencies and the use of complementary technologies, but in the domain of GNSS technologies, it is essential to study the impact of space weather on GNSS systems. Another area of research related to GNSS technologies is vertical Total Electron Content (TEC) distribution and anomalies related to earthquakes and volcanic eruptions on Earth.

There are many challenges that must be addressed, because they affect the reliability, accuracy, and all other essential parameters of GNSS systems. This Special Issue seeks to address some of these issues by publishing manuscripts on topics such as GNSS risk assessment, different effects of space weather disturbances on the operation of GNSS systems, environmental impacts on the operation of GNSS systems, GNSS positioning error budgets, TEC special features in volcano eruptions. A total of 17 scientific papers are published. Some specific updates and improvements presented in this Special Issue include the following:

– Contribution to the research of the effects of Etna volcano activity on the features of the Ionospheric Total Electron Content behaviour—In this paper [1], volcanic activity was modeled using volcanic radiative power (VRP) data obtained using the Middle InfraRed Observation of Volcanic Activity (MIROVA) system. The estimated minimal night TEC values were averaged over defined index days of the VRP increase. During the analyzed period of 19 years, volcano activity was categorized according to pre-defined criteria. The influence of current space weather and short-term solar activity on TEC near the volcano was systematically minimized. The results showed mean/median TEC increases of approximately +3 standard deviations from the overall mean values, with peak values placed approximately 5 days before the VRP increase and followed by general TEC depletion around the time of the actual volcanic activity increase. Additionally, a TEC oscillation pattern was found over the volcano site with a half-period of 6.25 days. The results mainly indicate that the volcanic activity modified the ionospheric dynamics within the nearby ionospheric region before the actual VRP increase, and that the residual impact in the volcano's surrounding area could be attributed to terrestrial endogenous processes and air–Earth currents. These changes could be detected according to criteria predefined in the research: during quiet space weather conditions, while observing night-time TEC values, and within the limits of low short-term solar influence.

- Lithosphere Ionosphere Coupling Associated with Seismic Swarm in the Balkan Peninsula from ROB-TEC and GPS—The authors of [2] detected and analyzed pre-earthquake ionospheric anomalies (PEIAs) using TEC data from the Royal Observatory of Belgium (ROB), and analyzed coseismic ionospheric disturbance (CID) using vertical TEC (VTEC) from the GPS stations in earthquake preparation areas. The results showed that PEIAs appeared to increase continuously from 08:00–12:00 UT in the 3 days before a seismic swarm of Mw > 5.0. The ionosphere over the seismogenic zones exhibited large-scale anomalies when multiple seismogenic zones of the Balkan Peninsula spatially and temporally overlapped. Moreover, the TEC around the earthquake centers showed a positive anomaly lasting for 7 h. In a single seismogenic zone in Greece, the TEC around the earthquake center reached over +3.42 TECu. In addition, the CID observed from GPS stations showed that with the increase in the number of earthquakes, the ionosphere over the seismogenic area was more obviously disturbed, and after three strong earthquakes, the TEC suddenly decreased over the seismogenic area and formed a phenomenon similar to an ionospheric hole. The authors suggested that a lithosphere–atmosphere–ionosphere coupling mechanism existed before the seismic swarm appeared in the Balkan Peninsula, and earthquake-induced VTEC anomalies occurred more frequently within a 3–10-day window before the earthquake; this phenomenon was particularly evident when multiple seismogenic zones overlapped spatiotemporally.
- Multi-Station and Multi-Instrument Observations of F-Region Irregularities in the Taiwan–Philippines Sector—This paper [3] presents a multi-station and multi-instrument system, organized and proposed for ionospheric scintillation and equatorial spread-F (ESF) specification and their associated motions in the Taiwan–Philippines sector. The issues related to the scintillation and ESF event observed on 26 October 2021, under magnetically quiet conditions, are presented and discussed. The authors first indicated the existence of a plasma bubble in the Taiwan–Philippines sector using FormoSat-7/Constellation Observing System for Meteorology, Ionosphere, and Climate-2 (FS7/COSMIC2) GPS/GLONASS radio occultation observations. The authors verified the latitudinal extent of the tracked plasma bubble using recorded ionograms from the Vertical Incidence Pulsed Ionospheric Radar located at Hualien, Taiwan. Thye also discussed the spatial and temporal variabilities of two-dimensional vertical scintillation index VS_4 maps based on simultaneous GPS L1-band signal measurements from 133 ground-based receivers located in Taiwan and the surrounding islands. They used two high-sampling, software-defined GPS receivers and characterized targeted plasma irregularities by carrying out spectrum analyses of the received signal. They found that the derived plasma irregularities moved eastward and northward, and the smaller the irregularity scale, the higher the spectral index and the stronger the scintillation intensity at lower latitudes in the target irregularity feature.
- Landslide Deformation Prediction Based on a GNSS Time Series Analysis and Recurrent Neural Network Model—The authors of this paper [4] developed a novel Attention Mechanism with a Long Short-Term Memory Neural Network (AMLSTM NN) model based on Complete Ensemble Empirical Mode Decomposition with Adap-

tive Noise (CEEMDAN) landslide displacement prediction. The CEEMDAN method was implemented to ingest a landslide Global Navigation Satellite System (GNSS) time series. The AMLSTM algorithm was then used to realize prediction work, in conjunction with multiple impact factors. The Baishuihe landslide was adopted to illustrate the capabilities of the model. The results showed that the CEEMDAN-AMLSTM model achieved competitive accuracy and has significant potential for landslide displacement prediction.

- CubeSat Observation of the Radiation Field of the South Atlantic Anomaly—This paper [5] presents the results of one-and-a-half years of observations of the South Atlantic Anomaly radiation field, measured using a CubeSat in polar orbit with an elevation of 540 km. The position was calculated using an improved centroid method that took into account the area of the grid. The dataset consisted of eight campaigns measured at different times, each with a length of 22 orbits (~2000 min). The radiation data were combined with GPS position data. Westward movement was detected at 0.33°/year and southward movement at 0.25°/year. The position of the fluence maximum featured higher scatter than the centroid position.
- Fractal Nature of Advanced Ni-Based Superalloys Solidified on Board the International Space Station—In this paper [6] presents advanced analytical techniques for describing experimentally obtained microstructures analyzed on cross-sectional images of different samples. The samples were processed on the International Space Station using a device from the Materials Science Laboratory-Electromagnetic Levitator (MSL-EML) based on electromagnetic levitation principles. The authors also applied several aspects of fractal analysis and obtained important results regarding fractals and Hausdorff dimensions related to the surface and structural characteristics of CMSX-10 alloy samples. Using scanning electron microscopy (SEM) (Zeiss LEO 1550), the authors analyzed the microstructures of samples solidified in space and successfully performed fractal reconstruction of the samples' morphologies. The fractal analysis was extended to microscopic images based on samples solidified on Earth, establishing new developments in knowledge of their advanced structures.
- GNSS-IR Snow Depth Retrieval from Multi-GNSS and Multi-Frequency Data—In this paper [7], the authors analyzed SNR data of the Global Positioning System (GPS), Global Orbit Navigation Satellite System (GLONASS), Galileo satellite navigation system (Galileo), and BeiDou navigation satellite system (BDS) from the P387 station of the U.S. Plate Boundary Observatory (PBO). Lomb–Scargle periodogram (LSP) spectrum analysis was used to compare the difference in reflector height between the snow-free and snowy surfaces in order to determine the snow depth, which was compared with the PBO snow depth. First, the different frequency results of the multi-GNSS system were analyzed. The retrieval accuracy of the different GNSS systems was analyzed through multi-frequency mean fusion. The joint retrieval accuracy of the multi-GNSS system was analyzed through mean fusion. The results showed that the different frequencies of the multi-GNSS system had a strong correlation with the PBO snow depth, and that the accuracy was better than 10 cm. The multi-frequency mean fusion of different GNSS systems could effectively improve the retrieval accuracy, which was better than 7 cm. The joint retrieval accuracy of the multi-GNSS system was further improved, with a correlation coefficient (R) of 0.99 between the retrieval snow depth and the PBO snow depth, and the accuracy was better than 3 cm.
- Determination of Navigation System Positioning Accuracy Using the Reliability Method Based on Real Measurements—In this paper [8], a new method was presented for determining navigation system positioning accuracy based on a reliability model where the system's operation and failure statistics were referred to as life and failure times. Based on real measurements, the method proposed in this article was compared with the classical method (based on the 2DRMS measure). In the analyses, real (empirical) measurements were made using the following principal modern navigation positioning systems: the Global Positioning System (GPS) (168′286 fixes),

the Differential Global Positioning System (DGPS) (900′000 fixes), and the European Geostationary Navigation Overlay Service (EGNOS) (900′000 fixes). Research performed on real data showed that the reliability method provided a better estimate of navigation system positioning accuracy compared to the 2DRMS measure.

– Evaluating Total Electron Content (TEC) Detrending Techniques in Determining Ionospheric Disturbances during Lightning Events in A Low Latitude Region—In this paper [9], TEC was detrended using several methods to show this impact. Information from the detrended TEC may or may not necessarily represent a geophysical parameter. Two commonly used detrending methods, the Savitzky–Golay filter and polynomial fitting, were evaluated during thunderstorm events in Hong Kong. A two-step detection and distinguishing approach was introduced alongside linear correlation in order to determine the best detrending model. The Savitzky–Golay filter on order six and with a time window of 120 min performed best in detecting lightning events, and had the highest moderate positive correlation of 0.4. Moreover, the best time frame was 120 min, which suggests that the observed disturbances could be travelling ionospheric disturbance (TID), with lightning as the potential source.
– Seasonal and Interhemispheric Effects on the Diurnal Evolution of EIA: Assessed by IGS TEC and IRI-2016 over Peruvian and Indian Sectors—In this paper [10], the global total electron content (TEC) map in 2013, retrieved from the International Global Navigation Satellite Systems (GNSS) Service (IGS), and the International Reference Ionosphere (IRI-2016) model were used to monitor the diurnal evolution of the equatorial ionization anomaly (EIA). The statistical analyses were conducted during geomagnetically quiet periods in the Peruvian and Indian sectors, where equatorial electrojet (EEJ) data and reliable TEC were available. The EEJ was used as a proxy to determine whether the EIA structure was fully developed. To characterize dynamics accounting for the full development of EIAs, the authors defined and statistically analyzed the onset, first emergence, and peaks of the northern and southern crests based on the proposed crest-to-trough difference (CTD) profiles. These time points extracted from IGS TEC showed typical annual cycles in the Indian sector, which can be summarized as being of winter hemispheric priority, i.e., the development of EIAs in the winter hemisphere was ahead of that in the summer hemisphere. Additionally, the same time points showed abnormal semiannual cycles in the Peruvian sector, that is, EIAs develops earlier during two equinoxes/solstices in the northern/southern hemisphere. The authors suggested that the onset of EIAs is a consequence of the equilibrium between sunlight ionization and ambipolar diffusion. The latter term was not considered in modeling the topside ionosphere in IRI-2016, which resulted in a poor capacity of IRI to describe the diurnal evolution of EIAs. The meridional neutral wind's modulation of the ambipolar diffusion can explain the annual cycle observed in the Indian sector, while the semiannual variation seen in the Peruvian sector might be due to additional competing effects induced by the F-region height changes.
– A Graph Convolutional Incorporating GRU Network for Landslide Displacement Forecasting Based on Spatiotemporal Analysis of GNSS Observations—In this paper [11], a novel graph convolutional incorporating GRU network (GC-GRU-N) was proposed and applied to landslide displacement forecasts. The model conducts attribute-augmented graph convolution (GC) operations on GNSS displacement data with weighted adjacency matrices and an attribute-augmented unit to combine features, including the displacement, the distance, and other external influence factors, to capture spatial dependence. The output of multi-weight graph convolution is then applied to the gated recurrent unit (GRU) network to learn temporal dependencies. The related optimal hyper-parameters were determined via comparison experiments. When applied to two typical landslide sites in the Three Gorge Reservoir (TGR), China, GC-GRU-N outperformed the comparative models in both cases. The ablation experiment results showed that the attribute augmentation, which considers external factors of landslide displacement, can further improve the model's prediction performance.

- Regional Ionospheric Corrections for High Accuracy GNSS Positioning—This paper [12] aimed to evaluate the accuracy of a local/regional ionospheric delay model across Australia using a linear interpolation method. The accuracy of the ionospheric corrections was assessed as a function of both different latitudinal regions and the number and spatial density of GNSS Continuously Operating Reference Stations (CORSs). This research showed that, for a local region of 5° latitude × 10° longitude in mid-latitude regions of Australia (~30° to 40°S) with approximately 15 CORS stations, ionospheric corrections with an accuracy of 5 cm can be obtained. In Victoria and New South Wales, where dense CORS networks exist (nominal spacing of ~100 km), the average ionospheric correction accuracy can reach 2 cm. For sparse networks (nominal spacing of >200 km) at lower latitudes, the average accuracy of ionospheric corrections was within the range of 8 to 15 cm; significant variations in the ionospheric errors of some specific satellite observations during certain periods were also found. In some regions such as Central Australia, where there are a limited number of CORSs, this model was impossible to use. On average, centimeter-level-accuracy ionospheric corrections can be achieved if there are sufficiently dense (i.e., nominal spacing of approximately 200 km) GNSS CORS networks in the region of interest. Based on the current availability of GNSS stations across Australia the authors proposed a set of 15 regions of different ionospheric delay accuracies with extents of 5° latitude × 10° longitude across continental Australia.
- Retrieval of Soil Moisture Content Based on Multisatellite Dual-Frequency Combination Multipath Errors—In this paper [13], the authors defined a soil moisture inversion method based on a multisatellite dual-frequency combined multipath error: a multipath error calculation model of dual-frequency carrier phase (L4 Ionosphere Free, L4_IF) and dual-frequency pseudorange (DFP) without an ionospheric effect was constructed. The data of five epochs were selected before and after the time point of the effective satellite period to define the multipath error model and error equation, and the delay phase for soil moisture retrieval was solved. The proposed method was verified using Plate Boundary Observatory (PBO) P041 site data. The results showed that the Pearson correlation coefficients (R) of the L4_IF and DFP methods at the P041 station were 0.97 and 0.91, respectively. To better verify the results' reliability and the proposed method's effectiveness, the soil moisture data of the MFLE station, about 210 m away from P041 station, were used as the verification data in this paper. The results showed that the delay phase solved using the multipath error and soil moisture were strongly correlated. The Pearson correlation coefficients (R) of the L4_IF and DFP methods at the MFLE station were 0.93 and 0.86, respectively. In order to improve the inversion accuracy of GNSS-IR soil moisture, this paper constructed the prediction model of soil moisture using linear regression (ULR), a back propagation neural network (BPNN), and a radial basis function neural network (RBFNN), and evaluated the accuracy of each model. The results showed that the soil moisture retrieval method based on a multisatellite dual-frequency combined multipath error can replace the traditional retrieval method and effectively improve the time resolution of GNSS-IR soil moisture estimation.
- Comparative Study of Predominantly Daytime and Nighttime Lightning Occurrences and Their Impact on Ionospheric Disturbances—In this paper [14], the hourly occurrence of lightning and its impact on ionospheric disturbances, quantified using the rate of total electron content index (ROTI), were assessed. The linear correlation between diurnal lightning activity and ROTI in the coastal region of southern China, where lightning predominates in the daytime, was initially negative, contrary to a positive correlation in southern Africa, where lighting predominates in the evening. After appreciating and applying the physical processes of gravity waves, electromagnetic waves, and the Trimpi effect arising from lightning activity, and the time delay impact they have on the ionosphere, the negative correlation was overturned to a positive one using cross-correlation.

- Phase Centre Corrections of GNSS Antennas and Their Consistency with ATX Catalogues—In this paper [15], the results of research on the adequacy of antenna phase centre correction (PCC) variations are presented following an analysis of its component—the antennas' phase centre offset (PCO). Height differences were determined using different independent methods: EUREF Permanent Network (EPN) combined solutions, Precise Point Positioning (PPP), and the single baseline solution. The results of GNSS processing were attributed to direct geometric levelling outputs. The research was conducted within the Global Positioning System (GPS) only, and the experiment was based on a comparison of the height differences between four GNSS antennas located on the roof of a building: two permanent station antennas and two auxiliary points. The antennas were located at similar heights; precise height differences were determined via geometric levelling, both at the beginning and the end of the session. Post-processing was conducted with the use of a GPS system, precise ephemeris, the adopted antenna correction model, and a zero-elevation mask. For one of the antennas, a change in the antenna characteristic model from IGS08 to IGS14 led to an 8 mm difference in height. Older antennas used in the national (or transnational) permanent network need individual PCCs.
- Assessment of the Water Vapor Tomography Based on Four Navigation Satellite Systems and Their Various Combinations—In this paper [16], experiments in Hong Kong were conducted to analyze and assess the performance of GPS, BDS, GLONASS, Galileo, and their combinations in water vapor tomography. The numerical results showed that the number of available signal rays varied widely in the four satellite systems, and the value could be increased by the combination of satellite systems. The combinations also increased the number of voxels crossed by the signal rays, but this value was not directly related to the number of available signal rays; the number and distribution of the voxels with sufficient signal rays, which were most closely related to the structure of the tomographic model, showed no obvious differences in the four satellite systems and their combinations. Comparative results of slant water vapor (SWV), estimated using GNSS data and water vapor density derived from radiosonde data, revealed that the differences in the water vapor tomography of the four satellite systems were small, and their combinations resulted in limited improvement in the tomographic results.
- A New Approach for Improving GNSS Geodetic Position by Reducing Residual Tropospheric Error (RTE) Based on Surface Meteorological Data—In this paper [17], a study was performed based on GLONASS positioning solutions and the accompanying meteorological parameters in a defined and harmonized temporal-spatial frame of three locations in the Republic of Croatia. A multidisciplinary approach-based analysis from a navigational science point of view was applied. The residual amount of satellite positioning signal tropospheric delay was quantitatively reduced by employing statistical analysis methods. The result of the statistical regression was a model that correlates surface meteorological parameters with RTE. Considering the input data, the model has a regional character, and it is based on the Saastamoinen model of zenith tropospheric delay. The verification results showed that the model reduced the RTE, and thus, increases the geodetic accuracy of the observed GNSS stations (with horizontal components of position accuracy of up to 3.8% and vertical components of up to 4.37%, respectively). To obtain these results, the root mean square error (RMSE) was used as the fundamental parameter for position accuracy evaluation. Although it was developed based on GLONASS data, the proposed model also showed a considerable degree of success in the verification of geodetic positions based on the Global Positioning System (GPS). The purpose of the research, and one of its scientific contributions, was that the proposed method can be used to quantitatively monitor the dynamics of changes in the deviations of X, Y, and Z coordinate values along coordinate axes. The results showed that there is was distinct interdependence of the dynamics of Y and Z coordinate changes (with almost mirror symmetry), a result which had not

previously been investigated or published. The resultant positions of the coordinates were created by deviations in the coordinates along the *Y* and *Z* axes; in the vertical plane of space, the deviations of the coordinate X (horizontal plane) were mostly uniform and independent of deviations along the *Y* and *Z* axes. The proposed model showed the realized state of the statistical position equilibrium of the selected GNSS stations, which were observed using RTE values. Although it is of regional character, the model is suitable for application in larger areas with similar climatological profiles and for users who do not require the achievement of a maximum level of geodetic accuracy using Satellite-Based Augmentation Systems (SBAS) or other more advanced, time-consuming, and equipment-consuming positioning techniques.

A series of scientific papers were published in this Special Issue that have made a significant scientific contribution to its thematic domain (more detailed scientific contributions are presented for each published article); however, there are many questions and unknowns to be solved using appropriate scientific methods and research, for example, investigating and establishing the regularity of the dynamics of changes in the user coordinates X, Y, Z along the coordinate axes x,y,z as a function of the ionospheric delay of the satellite signal, or the effects of the influence of volcanic eruptions on satellite determination of the user's position.

Just as science itself has no end or completion, the scientific research conducted in the domain of this Special Issue is neither finished nor completed; however, it will continue to be conducted in the future, with the aim of developing a deeper understanding of and scientific explanations for a series of processes related to this topic.

Funding: This research was funded by the University of Rijeka grant number—uniri—ethnic—18–66 and by grant G2HOTSPOTS (PID2021-122142OB-I00) from the MCIN/AEI/10.13039/501100011033/FEDER, UE. This work represents a contribution to CSIC Thematic Interdisciplinary Platform TELEDETEC.

Data Availability Statement: Data Availability are presented in each mentioned paper described in this Editorial article.

Conflicts of Interest: The author declares no conflict of interest.

References

1. Toman, I.; Brčić, D.; Kos, S. Contribution to the Research of the Effects of Etna Volcano Activity on the Features of the Ionospheric Total Electron Content Behaviour. *Remote Sens.* **2021**, *13*, 1006. [CrossRef]
2. Wei, L.; Li, J.; Liu, L.; Huang, L.; Zheng, D.; Tian, X.; Huang, L.; Zhou, L.; Ren, C.; He, H. Lithosphere Ionosphere Coupling Associated with Seismic Swarm in the Balkan Peninsula from ROB-TEC and GPS. *Remote Sens.* **2022**, *14*, 4759. [CrossRef]
3. Tsai, L.-C.; Su, S.-Y.; Lv, J.-X.; Bullett, T.; Liu, C.-H. Multi-Station and Multi-Instrument Observations of F-Region Irregularities in the Taiwan–Philippines Sector. *Remote Sens.* **2022**, *14*, 2293. [CrossRef]
4. Wang, J.; Nie, G.; Gao, S.; Wu, S.; Li, H.; Ren, X. Landslide Deformation Prediction Based on a GNSS Time Series Analysis and Recurrent Neural Network Model. *Remote Sens.* **2021**, *13*, 1055. [CrossRef]
5. Kovář, P.; Sommer, M. CubeSat Observation of the Radiation Field of the South Atlantic Anomaly. *Remote Sens.* **2021**, *13*, 1274. [CrossRef]
6. Mitić, V.; Serpa, C.; Ilić, I.; Mohr, M.; Fecht, H.-J. Fractal Nature of Advanced Ni-Based Superalloys Solidified on Board the International Space Station. *Remote Sens.* **2021**, *13*, 1724. [CrossRef]
7. Tu, J.; Wei, H.; Zhang, R.; Yang, L.; Lv, J.; Li, X.; Nie, S.; Li, P.; Wang, Y.; Li, N. GNSS-IR Snow Depth Retrieval from Multi-GNSS and Multi-Frequency Data. *Remote Sens.* **2021**, *13*, 4311. [CrossRef]
8. Specht, M. Determination of Navigation System Positioning Accuracy Using the Reliability Method Based on Real Measurements. *Remote Sens.* **2021**, *13*, 4424. [CrossRef]
9. Osei-Poku, L.; Tang, L.; Chen, W.; Mingli, C. Evaluating Total Electron Content (TEC) Detrending Techniques in Determining Ionospheric Disturbances during Lightning Events in A Low Latitude Region. *Remote Sens.* **2021**, *13*, 4753. [CrossRef]
10. Wan, X.; Zhong, J.; Xiong, C.; Wang, H.; Liu, Y.; Li, Q.; Kuai, J.; Cui, J. Seasonal and Interhemispheric Effects on the Diurnal Evolution of EIA: Assessed by IGS TEC and IRI-2016 over Peruvian and Indian Sectors. *Remote Sens.* **2021**, *14*, 107. [CrossRef]
11. Jiang, Y.; Luo, H.; Xu, Q.; Lu, Z.; Liao, L.; Li, H.; Hao, L. A Graph Convolutional Incorporating GRU Network for Landslide Displacement Forecasting Based on Spatiotemporal Analysis of GNSS Observations. *Remote Sens.* **2022**, *14*, 1016. [CrossRef]
12. Dao, T.; Harima, K.; Carter, B.; Currie, J.; McClusky, S.; Brown, R.; Rubinov, E.; Choy, S. Regional Ionospheric Corrections for High Accuracy GNSS Positioning. *Remote Sens.* **2022**, *14*, 2463. [CrossRef]

13. Nie, S.; Wang, Y.; Tu, J.; Li, P.; Xu, J.; Li, N.; Wang, M.; Huang, D.; Song, J. Retrieval of Soil Moisture Content Based on Multisatellite Dual-Frequency Combination Multipath Errors. *Remote Sens.* **2022**, *14*, 3193. [CrossRef]
14. Osei-Poku, L.; Tang, L.; Chen, W.; Chen, M.; Acheampong, A.A. Comparative Study of Predominantly Daytime and Nighttime Lightning Occurrences and Their Impact on Ionospheric Disturbances. *Remote Sens.* **2022**, *14*, 3209. [CrossRef]
15. Borowski, L.; Kudrys, J.; Kubicki, B.; Slámová, M.; Maciuk, K. Phase Centre Corrections of GNSS Antennas and Their Consistency with ATX Catalogues. *Remote Sens.* **2022**, *14*, 3226. [CrossRef]
16. Yang, F.; Wang, J.; Wang, H.; Gong, X.; Wang, L.; Huang, B. Assessment of the Water Vapor Tomography Based on Four Navigation Satellite Systems and Their Various Combinations. *Remote Sens.* **2022**, *14*, 3552. [CrossRef]
17. Bakota, M.; Kos, S.; Mrak, Z.; Brčić, D. A New Approach for Improving GNSS Geodetic Position by Reducing Residual Tropospheric Error (RTE) Based on Surface Meteorological Data. *Remote Sens.* **2022**, *15*, 162. [CrossRef]

remote sensing

MDPI

Article

A New Approach for Improving GNSS Geodetic Position by Reducing Residual Tropospheric Error (RTE) Based on Surface Meteorological Data

Mario Bakota [1], Serdjo Kos [2], Zoran Mrak [2] and David Brčić [2,*]

[1] Faculty of Maritime Studies, University of Split, Ruđera Boškovića 37, 21000 Split, Croatia
[2] Faculty of Maritime Studies, University of Rijeka, Studentska 2, 51000 Rijeka, Croatia
* Correspondence: david.brcic@pfri.uniri.hr

Abstract: Positioning error components related to tropospheric and ionospheric delays are caused by the atmosphere in positioning determined by global navigation satellite systems (GNSS). Depending on the user's requirements, the position error caused by tropospheric influences, which is commonly referred to as zenith tropospheric delay (ZTD), must be estimated during position determination or determined later by external tropospheric corrections. In this study, a new approach was adopted based on the reduction of residual tropospheric error (RTE), i.e., the unmodeled part of the tropospheric error that remains included in the total geodetic position error, along with other unmodeled systematic and random errors. The study was performed based on Global Navigation Satellite System (GLONASS) positioning solutions and accompanying meteorological parameters in a defined and harmonized temporal-spatial frame of three locations in the Republic of Croatia. A multidisciplinary approach-based analysis from a navigational science aspect was applied. The residual amount of satellite positioning signal tropospheric delay was quantitatively reduced by employing statistical analysis methods. The result of statistical regression is a model which correlates surface meteorological parameters with RTE. Considering the input data, the model has a regional character, and it is based on the Saastamoinen model of zenith tropospheric delay. The verification results show that the model reduces the RTE and thus increases the geodetic accuracy of the observed GNSS stations (with horizontal components of position accuracy of up to 3.8% and vertical components of position of up to 4.37%, respectively). To obtain these results, the Root Mean Square Error (RMSE) was used as the fundamental parameter for position accuracy evaluation. Although developed based on GLONASS data, the proposed model also shows a considerable degree of success in the verification of geodetic positions based on Global Positioning System (GPS). The purpose of the research, and one of its scientific contributions, is that the proposed method can be used to quantitatively monitor the dynamics of changes in deviations of X, Y, and Z coordinate values along coordinate axes. The results show that there is a distinct interdependence of the dynamics of Y and Z coordinate changes (with almost mirror symmetry), which has not been investigated and published so far. The resultant position of the coordinates is created by deviations of the coordinates along the *Y* and *Z* axes—in the vertical plane of space, the deviations of the coordinate X (horizontal plane) are mostly uniform and independent of deviations along the *Y* and *Z* axes. The proposed model shows the realized state of the statistical position equilibrium of the selected GNSS stations which were observed using RTE values. Although of regional character, the model is suitable for application in larger areas with similar climatological profiles and for users who do not require a maximum level of geodetic accuracy achieved by using Satellite-Based Augmentation Systems (SBAS) or other more advanced, time-consuming, and equipment-consuming positioning techniques.

Keywords: GNSS; tropospheric error; surface meteorological data; statistical position equilibrium; Saastamoinen model of zenith tropospheric delay

Citation: Bakota, M.; Kos, S.; Mrak, Z.; Brčić, D. A New Approach for Improving GNSS Geodetic Position by Reducing Residual Tropospheric Error (RTE) Based on Surface Meteorological Data. *Remote Sens.* **2023**, *15*, 162. https://doi.org/10.3390/rs15010162

Academic Editor: Yunbin Yuan

Received: 1 December 2022
Revised: 16 December 2022
Accepted: 23 December 2022
Published: 27 December 2022

1. Introduction and Background

A total satellite positioning error budget can be decomposed in satellite, receiver, and propagation medium components. The latter refers to atmospheric layers that affect the path of satellite navigation signals, namely the ionosphere and the troposphere. In addition, other present errors with a measurable and significant impact on GNSS positioning include signal multipath errors, receiver noise error, satellite and receiver clock errors, and satellite orbital error [1–3].

Tropospheric error is caused by the propagation of a radio navigation signal through the lowest layer of the atmosphere. In general, the troposphere is divided into two layers: the wet (non-hydrostatic) layer (with a height of up to about 10 km above the Earth's surface) and the dry (hydrostatic) layer (which is 10–40 km above the Earth's surface) [4,5]. These layers cause delays in satellite navigation signals. The troposphere is a non-dispersive medium; therefore, the magnitude of this error component does not depend on signal frequency and cannot be determined as is the case with ionospheric delay. Tropospheric error (which is the reduction of radio navigation signal propagation speed and its deviation from the geometric path and is also commonly called tropospheric delay) can be modeled based on the fundamental meteorological parameters of the troposphere: temperature, humidity, and atmospheric pressure [6–10]. Various models have been developed to predict and reduce tropospheric delay with different scopes of application, including the following: the Two-Quartic Hopfield model (n/a) ((n/a)—not specified or valid for any elevation angle.) [11], the Saastamoinen model (10° and above) [12], the Modified Hopfield model (n/a) [13], the Marini model (10° and above) [14], the Davis et al. model (C*f*A) (5° and above) [15], the Ifadis model (2° and above) [16], and the Askne and Nordius model (n/a) [17], etc. The above models allow the value of the tropospheric delay to be estimated with varying degrees of accuracy depending on the input components of the model (including dry, wet, or total delay and with or without mapping function); therefore, they have different elevation angles of mapping function (given in the brackets). Tropospheric delay caused by the wet component is a consequence of the presence of water vapor in all its forms in the upper layer of the troposphere (up to 10 km). Tropospheric delay is partly caused by non-hydrostatic causes; as such, the zenith wet delay (ZWD) is much smaller in absolute terms (several millimeters) than the zenith hydrostatic delay (ZHD—which reaches several tens of centimeters) [4]. The ZWD cannot be accurately modeled using surface meteorological data due to the extreme space and time volatile features of water vapor. With existing tropospheric models, the error caused by ZWD can be reduced by up to 10–20% of its actual value [18]. The causes of the wet component are not in hydrostatic equilibrium; therefore, models based on the partial pressure of water vapor or relative surface humidity do not provide sufficient accuracy. They require empirical constants that vary spatially and temporally [19], although approaches and models for ZWD estimation are being developed based on surface meteorological parameters [20].

The causes of the hydrostatic component of the zenith delay (or atmospheric dry gasses and the non-dipole component of water vapor refraction [18]) are in hydrostatic equilibrium [19] and are therefore determined relatively simply and precisely using the Saastamoinen model. Both components form the ZTD. Approaches of existing tropospheric delay models for ZHD and ZWD differ, and certain models [21–26] of ZTD usually include a non-hydrostatic zenith component which depends primarily on the temporal-spatial distribution of water vapor and the height of the distribution, with the influence of water vapor being the most important. Considering that the insufficient modeling capabilities of the non-hydrostatic causes of tropospheric delay in the zenith direction limit the accuracy of the mapping function for other signal elevation angles, the separation of the mapping function from the zenith delay allowed the development of a number of new models of mapping functions [27–32]. Such models can combine hydrostatic and non-hydrostatic causes and can be combined with tropospheric delay in the zenith direction. This results in hybrid models separated by hydrostatic and non-hydrostatic causes. In general, tropospheric errors in GNSS measurement range from 2–2.4 m in the zenith direction and 25 m at horizontal elevation angles [33,34].

The new approach used in this study investigates the overall effect of tropospheric delay on the accuracy of the GNSS geodetic position of the selected area. Several studies [10,35] show the relationship between tropospheric delays based on radiosonde signal measurements and deviations from position accuracy where the main effect was found to be the atmospheric refraction expressed by the number of N units, which is a value that varies greatly in time and space [36–38]. In this way, it is possible to combine the influence of the tropospheric delays in radio signals (usually with peak values up to several tens of millimeters) with the amounts of slant tropospheric errors in the GNSS system which can reach several tens of meters.

A model of empirical character is proposed which combines the value of the non-modeled geodetic position RTE GNSS with the meteorological surface parameters of the observed positions. Rather than determining the appropriateness of a particular model of tropospheric delay (either in the zenith or slant direction), the goal of the presented research was to determine the existence of structural dynamics of deviations of x, y, and z coordinates based on the observed GNSS stations in the function of reducing the non-modeled tropospheric error based on relevant and real surface meteorological data.

The basis for determining tropospheric errors within GNSS position accuracy errors was the Saastamoinen model. The proposed model, together with the previously modeled part of the tropospheric error, reduces the tropospheric error of the GNSS position by correlating the unmodeled portion of the tropospheric error with real meteorological parameters, thereby increasing the overall geodetic accuracy of the determined position.

The following section presents the methodology used in the development of the proposed model, including data collection, statistical regression, and the proposed model validation. The obtained verification results for each location and time period are presented in the third section. The performance of the proposed model and the periodicity characteristics of RTE are presented and discussed in the fourth section, including its potential suitability with GPS. Concluding remarks and possible directions for further research are given in the final section.

2. Methodology

The development of the presented model is based on combined GLONASS positioning solutions and meteorological data from GNSS reference stations in the mid-latitudes of the Republic of Croatia. Positioning solutions were calculated based on position data in the Receiver Independent Exchange (RINEX) 2.0 compressed format [39]. Initial theories regarding the development of the model included the assumption that the propagation medium error, the user segment, and the microenvironment errors have a constant time–space character. Initial limitations of the proposed model include:

- The presented methodology implies the creation of the most mutual and frequent alignment of positional and meteorological parameters, and thus the choice of used regional GNSS stations is conditioned (the availability of regional meteorological records is within a 10-min frequency).
- In accordance with the initial spatial limitation, the selection of available GNSS positional data measurements was limited to the GPS and GLONASS, and there was no possibility of using records from other GNSS systems. As a full-fledged part of the GNSS, GLONASS data were used due to their relative underrepresentation in similar research.
- The main goal of the research was to determine a possible statistically significant correlation between realistic surface meteorological parameters and the geodetic accuracy of GNSS position deviations. Therefore, the model was developed based on positional and meteorological data with a geographical resolution of 3.7° × 2.9° (geographical grid) which declared as of regional character.

- The format of the available GNSS data limited the possibilities of their processing. Therefore, single-frequency (L1) positioning and the Klobuchar model for ionospheric delay were used. Input clock parameters and ephemerides are contained in the navigation message (including the broadcast ephemerides and clock parameters). Processing of solid tides and multipath corrections was not accessible.
- Other more accurate positioning techniques, such as PPP (Precise Point Positioning) or RTK (Real-time Kinematics), were not supported.

2.1. Time Frame of the Study

The time frame used for the creation and validation of the proposed model covers the year 2019. The model verification was performed using data from 2014 and 2015. Data from the same locations were analyzed since the model verification required the mutual compatibility of meteorological and GLONASS data at all stages of the study.

2.2. Meteorological Data Collection

The set of meteorological input data was determined based on the relationship between the propagation of radio signals through the neutral atmosphere and tropospheric dynamic processes. Hydrostatic and non-hydrostatic causes of tropospheric delay were identified. The independent input variables for the model were: pressure P (hPa), temperature T (°C), precipitation water PWV (mm), precipitation Pr (mm), and relative humidity Rh (percent). Meteorological data can be interpolated from existing numerical weather models (NWM), which is acceptable for analyzing existing tropospheric models, but not for developing a new model according to the selected regional model development. Therefore, in this study, meteorological data collected using automatic meteorological instruments at selected GNSS locations were used by the State Hydrometeorological Institute of the Republic of Croatia (DHMZ) [40] as a source of meteorological data. The time resolution of the meteorological data was 10 min.

Given the current limitations of available data sources, the study was based on the well-known approach of determining the tropospheric error by determining pseudoranges [10]. The used data delimit the area from 42.60° to 46.38° North latitude and from 15.22° to 18.11° East longitude with an altitude range from 64.3 m to 457.9 m above sea level (Figure 1).

Figure 1. The wide geographic area of selected GNSS measuring stations.

The climatological profiles of the observed locations differ. According to the Köppen climate classification, which is determined based on the average annual course of air temperature and precipitation, Čakovec is classified as *Cb* (which is a moderately warm, rainy climate with an average monthly temperature of the coldest month between −3 °C and 18 °C) [41]. At the same time, the warmest month of the year has an average temperature of less than 22 °C [40]. There are no particularly dry months in the year, and it is the area with the least precipitation in the cold season. Čakovec has a humid climate according to Thornthwaite's climate classification which is based on the ratio of the amount of water needed for potential evapotranspiration and expressed by the humidity coefficient IP/E [41]. Zadar and Dubrovnik belong to areas with a temperate climate with long and hot summers (*Ca*-mark according to the Köppen climate classification), while Zadar belongs to an area with a subhumid (semi-humid) climate in terms of the humidity coefficient IP/E [42]. Dubrovnik has a humid climate due to heavier precipitation. Water vapor and precipitation amounts are the main meteorological input parameters for determining the non-hydrostatic zenith component of the GNSS position. The above measurement locations do not have significantly pronounced differences in their humidity coefficients; as such, they represent a suitable choice for creating the proposed model, which is also applicable to larger geographical areas with similar climate profiles.

2.3. Geographic GNSS Data Collection

The study was carried out based on data from locations equipped with GNSS stations that provided predicted meteorological parameters with adequate time resolution. Therefore, measuring stations in the Regional Reference Frame Sub-Commission for Europe (EUREF) in Čakovec, Zadar, and Dubrovnik were selected. The basic stations' data are listed in Table 1 [43].

Table 1. General data from the EUREF measuring stations used (made by authors according to [43]).

Station Log	City	Latitude φ (° N)	Longitude λ (° E)	Elevation h (m)	ECEF (ETRS 89) Coordinates [1] (m)		
					x	y	z
CAKO00 HRV	Čakovec	46.387	16.439	222.1	4,227,250.7	1,247,280.6	4,595,193.3
DUB200 HRV	Dubrovnik	42.650	18.110	457.9	4,465,932.8	1,460,581.6	4,299,308.5
ZADA00 HRV	Zadar	44.113	15.227	64.3	4,425,737.1	1,204,734.5	4,417,173.4

[1] Earth centered, earth fixed coordinate system (The European Terrestrial Reference System, 1989).

2.4. Model Development

The development of the model involves the harmonization of two parallel input components: the value of the geodetic deviation of the user's position caused by the tropospheric error and the temporal-spatial harmonization of the corresponding meteorological input data.

2.4.1. Determination of the Size of the Tropospheric Error

Basically, the determination of the tropospheric error by measuring the pseudorange can be conducted in two ways:

(i) by determining and removing all systematic and random errors; and

(ii) by determining the deviation from the known position using a preselected tropospheric model and estimating the deviation residuals as unknown parameters [25].

In the first method, the pseudorange is calculated according to the general formula [10]:

$$R^{i}_{A} = \rho^{i}_{A} + \delta\rho_{mul} + \delta\rho_{rel} + c\delta_{A} - c\delta^{i} + I^{i}_{A} + T^{i}_{A} + e^{i}_{A} \quad (1)$$

where R^{i}_{A} is the pseudorange from position A to satellite i; ρ^{i}_{A} is the geometric distance; c is l speed of light; I^{i}_{A}, T^{i}_{A} is ionospheric and tropospheric delay; $c\delta_{A}, c\delta^{i}$ is the satellite orbit and clock error; $\delta\rho_{mul}$ is the multipath trajectories error; $\delta\rho_{rel}$ is relativistic error; and e^{i}_{A} is the random error.

The multipath error $\delta\rho_{mul}$ can be determined programmatically, usually based on the receiver and antenna equipment manufacturer. Sources of ionospheric delay, relativistic errors, satellite orbit errors, and clock errors are included in the navigation message or, in the case of post-processing, from the appropriate ground truth data. In any case, it is necessary to perform software processing to ensure appropriate data sources are used for additional corrections. For additional verification, the values of the isolated and determined tropospheric delays of a known position were compared with the values determined using the radiosonde signals or another system [10].

The model development was based on the determination of the user's position in accordance with the selected modeling approach and its initial limitations. It was obtained by determining the pseudoranges and isolating the accuracy deviation from the user's position caused by the tropospheric error within the total positioning error. A position determined in this way can be defined in a simpler form as the difference between the signal reception time which is determined based on its clock and time (determined by the satellite clock), which can be represented with the following expression [4,44]:

$$P_{r,i}^{S} = c\left(\bar{t}_r - \bar{t}^{s}\right) \tag{2}$$

where $P_{r,i}^{S}$ is the pseudorange of the i-th satellite; $\bar{t}_r$ is the time of the signal reception determined by the receiver clock (s); and $\bar{t}^{s}$ is the time of signal transmission determined by the satellite clock (s).

The misalignment of the satellite and receiver clocks (with input data contained in the navigation message), the ionospheric, tropospheric, and measurement error, and the expression (2) takes the form (3) by introducing the parameters of the geometric distance between the antennas of the satellite and the receiver:

$$\begin{aligned} P_{r,i}^{S} &= c\left((t_r + dt_r(t_r)) - \left(t^S + dT^S\left(t^S\right)\right)\right) + \varepsilon_P \\ &= c\left(t_r - t^S\right) + c\left(dt_r(t_r) - dT^S\left(t^S\right)\right) + \varepsilon_P \\ &= \left(\rho_r^S + I_{r,i}^S + T_r^S + c\left(dt_r(t_r) - dT^S\left(t^S\right)\right) + \varepsilon_P\right. \\ &= \rho_r^S + c\left(dt_r(t_r) - dT^S\left(t^S\right)\right) + I_{r,i}^S + T_r^S + \varepsilon_P \end{aligned} \tag{3}$$

where $I_{r,i}^{S}$ is the geometric distance between the satellite and receiver antenna; dt_r, dT^S is the receiver and satellite clock offset; $I_{r,i}^{S}$ is the ionospheric error; T_r^S is the tropospheric error; and ε_P is the measurement error.

The conversion of the geodetic position into a data set in ECEF ETRS89 format with values expressed in meters is expressed according to the following expressions:

$$el^2 = f(2-f) \tag{4}$$

$$\nu = \frac{a}{\sqrt{1 - el^2 \sin\phi_r^2}} \tag{5}$$

$$r_r = \begin{pmatrix} (\nu + h)\cos\phi_r \cos\lambda_r \\ (\nu + h)\cos\phi_r \sin\lambda_r \\ \nu\left(1 - el^2\right)\sin\phi_f \end{pmatrix} \tag{6}$$

where λ_r is the geodetic longitude; ϕ_r is the geodetic latitude; h is the ellipsoidal height; a is the semi-major axis of the reference Earth ellipsoid (6,378,137 m); el^2 is the first numerical eccentricity of the ellipsoid; and f is the flattening coefficient of the reference Earth ellipsoid.

The Saastamoinen model of tropospheric zenith correction was used, and the present parameters were transmitted in the navigation message as a common source of input program corrections for all measuring stations and periods and for standardizing other program settings. The input program parameters for pressure, absolute temperature, and partial pressure are determined by the expressions of the standard atmosphere model [44]:

$$P = 1013.25 \cdot \left(1 - 2.2557 \cdot 10^{-5} h\right)^{5.2568} \quad (7)$$

$$T = 15 - 6.5 \cdot 10^{-3} h + 273.15 \quad (8)$$

$$e = 6.108 \cdot exp\left(\frac{17.15\, T - 4684}{T - 38.45}\right) \cdot \frac{h_{rel}}{100} \quad (9)$$

where P is the total air pressure (hPa); T is the temperature in degrees Kelvin; e is the partial atmospheric pressure (hPa); h is the geodetic altitude above sea level; and h_{rel} is the relative humidity. The applied Saastamoinen model uses a constant relative humidity value of 70%.

The tropospheric correction in this configuration was calculated using the following algorithms [45,46]. For the mapping function in the selected program setting, the Niell model is calculated according to the expression [47–49]:

$$m(\varepsilon) = m_w(\varepsilon)\{1 + \cot \varepsilon \cdot (G_N \cos z + G_E \sin z)\} \quad (10)$$

where ε is the signal elevation angle; z is the signal zenith angle; G_N is the tropospheric gradient in the north direction; G_E is the tropospheric gradient in the east direction; and m_w is the non-hydrostatic mapping function of the non-hydrostatic Niell (New) Mapping Functions (NMF).

The total tropospheric delay was calculated according to the following expression:

$$d_{tro} = m_d(\varepsilon) d_d^z + m(\varepsilon)\left(d_{tot}^z - d_d^z\right) \quad (11)$$

where d_{tro} is the total tropospheric delay; d_d^z is the hydrostatic component of zenith tropospheric delay (in meters and determined by the Saastamoinen model); d_{tot}^z is the tropospheric total zenithal retardation; and m_d is the (NMF) hydrostatic mapping function. The parameters of tropospheric gradients and total tropospheric delay in zenith direction were estimated using the extended Kalman filter (EKF) [45]. The parameters for the correction of the ionospheric delay (A—the numerical coefficient of the maximum total free electron content in the ionospheric layer F2; F—the index of solar activity; and Ap—the daily index of geomagnetic activity) included in the GLONASS navigation message (broadcast ionosphere model) [47] have the following form:

$$P_{ion} = (\alpha_0, \alpha_1, \alpha_2, \alpha_3, \beta_0, \beta_1, \beta_2, \beta_3)^T \quad (12)$$

The determination of ionospheric delay $I_r^{S\wedge}$ (m) was performed using Klobuchar's model due to single-frequency computing. Despite the efficiency of Klobuchar's model, more effective reducing of the ionospheric delay would be achieved using an iono-free combination or dual-frequency receiver. However, considering the format of available input data, a single-frequency receiver with the Klobuchar model was used. Input data regarding clock parameters and ephemerides were sent in the navigation message (broadcast ephemerides and clock parameters) and used in calculations in the following form [44,46,50]:

$$P_{eph}(t_b) = \left(x, y, z, v_x, v_y, v_z, a_y, a_y, a_z, \tau_n, \gamma_n\right) \quad (13)$$

In addition, the initial software setup included a single positioning mode and a 3° elevation mask value (due to the scope of applicability of the NMF mapping function) to isolate the deviation in the geodetic accuracy of the user's position caused by the tropospheric error component.

2.4.2. Statistical Analysis and Model Development

The remaining value of the tropospheric error was determined based on the difference in the geodetic accuracy of the position determined both with and without tropospheric correction (using the Saastamoinen model). It is important to emphasize that, in addition to the tropospheric component, such determined deviations still contain several unmodeled systematic and random errors, including the residual ionospheric error, the satellite position and clock error, errors related to multipath, and solid tides. However, the only input difference when comparing the accuracy of the geodetic position was the application of the Saastamoinen model of tropospheric correction (as the other applied algorithms were identical); therefore, the effect of the resulting final difference between the two models can be deterministically attributed to the tropospheric component within the total geodetic position error.

The difference in geodetic accuracy of the user's position can be analyzed as a function of the influence of the applied Saastamoinen model of tropospheric delay correction since all other parameters were set identically in both cases. This provided the theoretical basis for quantifying the effect of the Saastamoinen model on improving the geodetic accuracy of the observed GNSS positions.

The resulting RTE value of the known position caused by the non-modeled part of the tropospheric delay was subjected to a statistical regression procedure to determine the correlation with real meteorological surface parameters. Statistical correlation with its positive parameters are the basis for developing a mathematical model to reduce RTE as a function of surface meteorological parameters (as independent input variables).

The stages in the creation of the proposed model are shown in Figure 2.

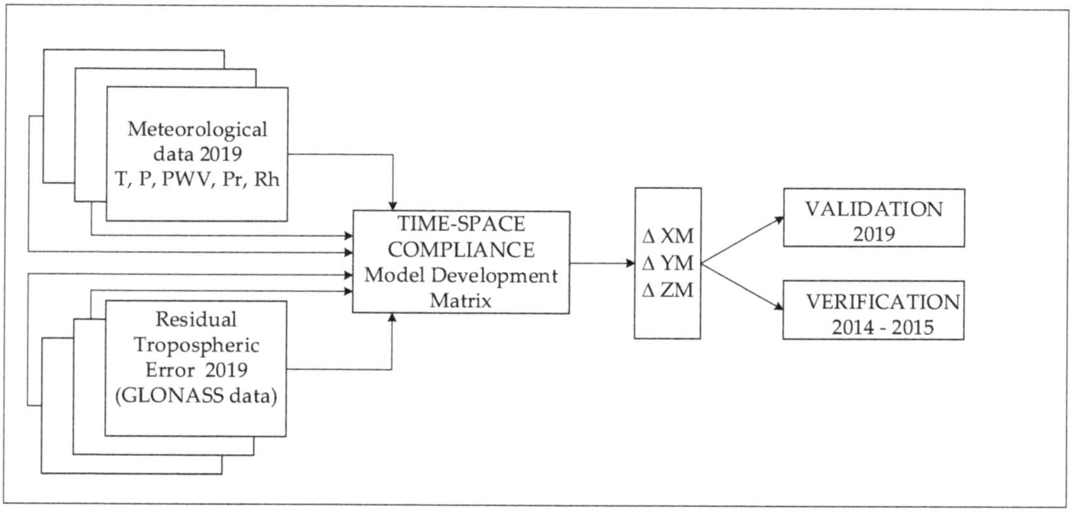

Figure 2. The proposed model development.

Meteorological independent input variables are as follows: T is the temperature (°C); Pr is the precipitation (mm); Rh is the relative humidity (percentage); PWV is the precipitation water (mm); P is the pressure (hPa); and ΔX_M, ΔY_M, and ΔZ_M are deviations along the x, y, and z axes in the proposed model (m).

The proposed model contains a mathematical expression for each axis as the observed deviation from the exact geodetic position followed the ECEF coordinate system. The final form of the proposed model is shown as follows. The amounts of the coefficients of the input predictors are the result of the regression analysis that describes the form and intensity of the mutual connection between the meteorological input predictors and the output variable RTE. For the X axis, the model component ΔX_M has the form:

$$\Delta X_M = \sqrt{-4.66541 - 0.01108Rh + 0.07062P + 0.11806Pr - 0.05187T + 0.1246PWV} \tag{14}$$

For the Y axis, the model component ΔY_M has the form:

$$\Delta Y_M = -\left(\sqrt{0.06656 - 0.00004Rh - 0.00003P + 0.00041Pr + 0.000047T - 0.00345PWV}\right) \tag{15}$$

For the Z axis, the model component ΔZ_M has the form:

$$\Delta Z_M = -\frac{1}{7.17152 - 0.01049R_h - 0.00407P + 0.12603P_r - 0.03837T + 0.01937PWV} \tag{16}$$

The final form of the proposed model (P_{Si}) is based on an extension of the existing Saastamoinen model and represents the sum of the corrections made by the Saastamoinen model and the proposed model for each axis:

$$P_{Si} = M_S + \begin{cases} \Delta X_M \\ \Delta Y_M \\ \Delta Z_M \end{cases} \tag{17}$$

where M_S is the correction value realized by the Saastamoinen model (in m).

2.4.3. Validation of the Proposed Model

Validation was performed using the cross-validation method with a 50:50 train-test ratio. Common statistical indicators of significance of the model components for a single coordinate axis include p value, multiple correlation coefficients, multiple determination, and adjusted determination coefficient. The results are shown in Table 2.

Table 2. Statistical indicators of the proposed model validation.

	Multiple Correlation	Multiple Determination	Adjusted Determination Coefficient	p-Value
X axis	0.115	0.013	0.013	0.00037
Y axis	0.034	0.001	0.001	0.00273
Z axis	0.092	0.008	0.008	0.00019

The obtained results show statistically significant correlations (upper limit $p = 0.05$), i.e., there is a statistically significant correlation between the independent input variables (the meteorological parameters) and the dependent output variable (RTE) for each coordinate axis, expressed by the Equations (14)–(16). The obtained correlation coefficients show: (i) the presence of a statistically significant correlation between the input predictors and the dependent variable in all components of the proposed model; (ii) a statistically significant prediction of the criterion variable by the input predictors in all components of the proposed model; and (iii) a statistically significant relationship between the obtained correlation, the number of samples (the input records), and the number of input variables (the predictors) of all components of the proposed model. Although the existence of a statistically significant connection between the input predictors and the output RTE variable is shown, the realized adjusted determination coefficients show different connection intensities, which is evident considering the associated low values of the coefficients of the input predictors in the model component for the y axis.

3. Results and Findings

The verification of the proposed model was performed using independent input data and meteorological data from the same locations for the period of 2014–2015. The coordinates of the geodetic positions for the period of 2014–2015 were calculated in an identical way and with the same input algorithms as the geodetic positions for 2019. The proposed model was created on this basis. The fundamental parameter that highlights the success of the proposed model is the reduction of the RMSE deviation of the positional accuracy compared to the positional accuracy achieved exclusively by the Saastamoinen model. The verification process was performed separately for each location and period. The movement of the coordinates along the coordinate axes in the horizontal and vertical planes and the inversion of the deviation of individual coordinates were observed using the proposed validated and verified empirical model; for example, y and z are complexly and graphically shown on a series of diagrams provided in Appendix A of the paper.

3.1. Verification for Čakovec Location

The correction values generated by the proposed model are shown in Table 3.

Table 3. Values of RTE corrections by the proposed model for Čakovec GNSS location (in m).

City	Year	Range	x	y	z
Čakovec	2014	Max	1.69	−0.13	−0.31
		Min	−0.64	−0.17	−0.71
	2015	Max	1.61	−0.12	−0.35
		Min	−0.85	−0.18	−0.75

The proposed model generates negative RTE correction values for coordinate axes y and z for 2014 and 2015. For the x axis, corrections were made in the range of −0.65 m to 1.69 m (for both years). The absolute amounts of the corrections range from 0.04 m (y-axis in 2014) to 2.46 m (x-axis in 2015).

The values of the RMSE, the standard deviation (STD), and the associated statistical parameters for the Čakovec GNSS position are shown in Table 4.

Table 4. Comparison of RMSE, STD, range, and median values for the Čakovec GNSS location obtained using the Saastamoinen model and the proposed model (in m).

	Saastamoinen Model			Proposed Model		
2014	x	y	z	x	y	z
Max	27.731	15.087	26.671	27.110	15.257	27.168
Min	−21.003	−17.843	−20.919	−21.186	−17.690	−20.328
RMSE	3.666	2.246	4.676	3.565	2.245	4.518
STD	3.593	2.244	4.400	3.553	2.244	4.396
Median	0.753	−0.113	−1.517	0.355	0.045	-0.969
2015	x	y	z	x	y	z
Max	23.429	12.394	20.610	22.653	12.549	21.036
Min	−15.533	−15.782	−23.515	−15.842	−15.621	−22.958
RMSE	3.681	2.355	4.111	3.539	2.362	4.036
STD	3.464	2.355	4.027	3.436	2.355	4.025
Median	1.318	−0.024	−0.815	0.925	0.136	−0.280

Application of the proposed model shows that values of the minimum and maximum deviations do not change significantly with ranges remaining approximately within the usual scale. There was a noticeable tendency for one deviation value to decrease while the opposite value increased and their absolute range values remained the same. Moreover, both minimum and maximum deviation values of the z axis decreased in the proposed model for the observed year of 2014.

The 2014 results showed a simultaneous improvement in the accuracy of the position RMSE value, which is the basic parameter for the success of the proposed model. Improvements in accuracy were obtained for the x-axis (10.04 cm/2.73%), y-axis (0.09 cm/0.04%), and z-axis (15.85 cm/3.38%). At the same time, the parameter STD improved for the x- and z-axis, and the value STD decreased for the y-axis (−0.03 cm/−0.01%). An improvement of the measurement frequency distribution was also observed. In 2014, the median value obtained with the proposed model for all axes decreased and approached the initial value of the deviation (zero), and in 2015, an improvement of the median value of the x- and z-axes and a deterioration of the y-axis was observed. An improvement in position accuracy was seen when comparing the RMSE values over the indicated period. For 2015, the proposed model simultaneously improved the x-axis position accuracy (14.26 cm/3.87%) and the z-axis accuracy (7.56 cm/1.84%), but it simultaneously worsened the y-axis position accuracy (−0.75 cm/−0.32%). It is clear that an overall improvement in position accuracy was achieved when observing the ratio of improvement and the deterioration of the overall position accuracy at the same time.

3.2. Verification for Zadar Location

The correction values generated by the proposed model are shown in Table 5.

Table 5. Values of RTE corrections based on the proposed model for Zadar GNSS location (in m).

City	Year	Range	x	y	z
Zadar	2014	Max	−0.67	−0.17	−0.80
		Min	0.90	−0.12	−0.29
	2015	Max	−0.85	−0.18	−0.86
		Min	0.88	−0.11	−0.28

It can also be seen that the values of generated RTE corrections for the GNSS station in Zadar were negative for the y and z axes in the observed period and ranged from −0.85 m to 0.9 m for the x axis. The absolute corrections' values ranged from 0.055 m (y-axis in 2014) to 1.74 m (x-axis in 2015). The statistical parameters RMSE, STD, range, and median for the Zadar GNNSS position are shown in Table 6.

Table 6. Comparison of the achieved values of RMSE, STD, minimum, maximum, and median values for the Zadar GNSS position obtained using the Saastamoinen model and the proposed model (in m).

	Saastamoinen Model			Proposed Model		
2014	x	y	z	x	y	z
Max	24.007	14.924	43.551	23.987	15.070	43.995
Min	−68.814	−28.262	−69.978	−69.145	−28.093	−69.437
RMSE	4.266	2.565	5.235	4.233	2.556	5.008
STD	4.257	2.554	4.737	4.232	2.554	4.737
Median	0.284	−0.279	−2.129	0.076	−0.124	−1.520
2015	x	y	z	x	y	z
Max	31.536	16.820	25.550	31.410	16.971	25.978
Min	−50.777	−20.134	−37.990	−50.937	−19.988	−37.365
RMSE	4.312	2.505	4.147	4.233	2.521	4.133
STD	3.836	2.498	4.130	3.809	2.498	4.127
Median	2.051	0.105	−0.372	1.859	0.260	0.219

Comparison of the parameters showed that the proposed model did not have a significant effect on the reduction of the final values of the accuracy deviations, and the range of the minimum and maximum values remained approximately the same. However, there was a significant change in the grouping values of the measurement frequencies for 2014 in all coordinate axes. For 2015, a positive shift of the median was obtained for the x and z axes, and an additional deviation from the zero value was registered for the y axis (0.26 m versus 0.10 m). The 2014 results showed an improvement in the accuracy of the RMSE value of the position at the same time for all axes. Accuracy improvements were obtained for the x-axis (3.33 cm/0.78%), the y-axis (0.9 cm/0.37%), and the z-axis (22.71 cm/4.33%). The parameter STD improved for the x-axis, and the value STD decreased for the y-axis (−0.03 cm/−0.01%)

and the z-axis (−0.006 cm/−0.001%). For 2015, the results of the predictive model showed the RMSE values of the x-axis (11.21 cm/2.6%) and the z-axis (1.45 cm/0.35%), while the y-axis had a decrease in the RMSE value (−1.65 cm/−0.65%). The same tendency was seen in the parameter STD: the x-axis and z-axis obtained improvements, and a deterioration of the y-axis STD value (−0.03cm/−0.01%) was recorded.

3.3. Verification for Dubrovnik Location

The correction values generated by the proposed model for the Dubrovnik GNSS location are shown in Table 7.

Table 7. Values of the RTE corrections based on the proposed model for the Dubrovnik GNSS location (in m).

City	Year	Range	x	y	z
Dubrovnik	2014	Max	1.69	−0.13	−0.36
		Min	−0.65	−0.18	−0.71
	2015	Max	0.80	−0.11	−0.29
		Min	−0.96	−0.18	−0.82

The values shown indicated that the model generated negative correction amounts of RTE for the y- and z-axes over the observed two-year period and values within the limits of −0.96 m to 1.69 m for the x-axis. The values of the absolute correction amounts ranged from 0.047 m (y-axis in 2014) to 2.33 m (x-axis in 2014). The values of the RMSE, standard deviation (STD) and associated statistical parameters for Dubrovnik GNSS position are shown in Table 8.

Table 8. Comparison of the realized values of RMSE, STD, range, and medians for the Dubrovnik GNNSS position obtained using the Saastamoinen model and the proposed model (in m).

	Saastamoinen Model			Proposed Model		
2014	x	y	z	x	y	z
Max	56.471	43.660	66.975	56.238	43.820	67.519
Min	−40.470	−34.888	−49.833	−41.077	−34.733	−49.264
RMSE	4.519	2.736	5.457	4.532	2.711	5.218
STD	4.514	2.687	4.796	4.497	2.687	4.798
Median	−0.176	−0.455	−2.424	−0.542	−0.297	1.867
2015	x	y	z	x	y	z
Max	47.847	13.006	24.051	47.525	13.184	24.494
Min	−28.606	−21.630	−21.456	−29.167	−21.458	−20.922
RMSE	4.313	2.602	4.337	4.291	2.598	4.226
STD	4.180	2.598	4.185	4.175	2.598	4.186
Median	1.134	−0.120	1.096	1.047	0.038	−0.538

Analysis of the presented accuracy parameters showed that the proposed model contributed to a deterioration of the RMSE value in 2014 for the x-axis (−1.35 cm/−0.3%), while it improved the accuracy of the other two axes, especially the z-axis (23.87 cm/4.37%), and also the y-axis (2.5 cm/0.92%). As for the two previously observed positions, there was no significant change in the value for the deviations range. An improvement in accuracy was observed for the median parameter in 2014 in the y and z axes, and an increase in the deviation was recorded for the x axis (from -0.17 m to 0.54 m). The parameter STD improved in the x-axis (1.78 cm/0.39%) and worsened in the y-axis (−0.004 cm/−0.001%) and z-axis (−0.12 cm/−0.02%).

For 2015, the proposed model improved the accuracy (parameter RMSE) of the x-axis (2.2 cm/0.57%), y-axis (0.3 cm/0.13%), and z-axis (11.15 cm /2.52%). The value of STD also improved for the x-axis, while it decreased for the y-axis (−0.03 cm/−0.01%) and the z-axis (−0.14 cm/−0.03%). At the same time, there were shifts in the median value toward the central value for all axes, with shifts on the y-axis (from -1.679 to 0.03 m) and the z-axis (from -3.84 to 0.53 m).

4. Discussion

Summary results of the verification procedure for the observed locations are presented in Table 9. The table contains data obtained by comparing the RMSE and STD parameters of the position accuracy obtained using the Saastamoinen model and the proposed model for all observed locations within the observed time frame.

Table 9. Total results of RMSE and STD values achieved using the proposed model.

City & Year	RMSE & STD Tendency	x		y		z	
		[%]	[cm]	[%]	[cm]	[%]	[cm]
Čakovec, 2014	RMSE	−2.739	−10.043	−0.042	−0.094	−3.389	−15.851
	STD	−1.118	−4.020	0.014	0.033	−0.075	−0.331
Čakovec, 2015	RMSE	−3.875	−14.267	0.322	0.758	−1.840	−7.569
	STD	−0.827	−2.867	0.007	0.016	−0.047	−0.192
Zadar, 2014	RMSE	−0.781	−3.335	−0.374	−0.962	−4.338	−22.717
	STD	−0.572	−2.438	0.012	0.031	0.001	0.006
Zadar, 2015	RMSE	−2.600	−11.214	0.659	1.652	−0.355	−1.475
	STD	−0.725	−2.783	0.013	0.033	−0.055	−0.229
Dubrovnik, 2014	RMSE	0.300	1.356	−0.920	−2.518	−4.374	−23.873
	STD	−0.395	−1.783	0.001	0.004	0.026	0.127
Dubrovnik, 2015	RMSE	−0.517	−2.232	−0.132	−0.343	−2.572	−11.156
	STD	−0.119	−0.500	0.014	0.036	0.0346	0.144

Positive values represented an increase in RMSE and STD values, i.e., a decrease in position accuracy, while negative values represented a decrease in the aforementioned statistical parameters, i.e., an increase in geodetic position accuracy. The graphical results of the performance of the proposed model are shown in Figures 3 and 4 where the curve represents the relative value (in percentages) and the columns represent the absolute amount (in cm).

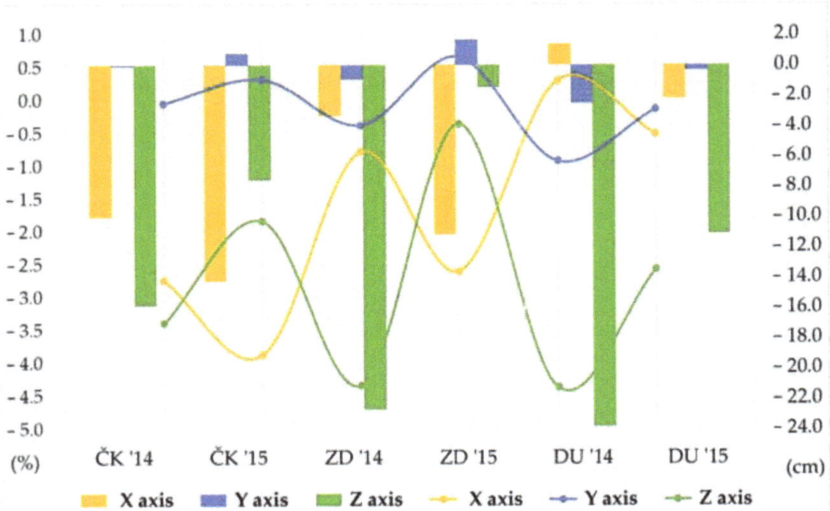

Figure 3. Movement of RMSE deviation values using the proposed model from the exact geodetic position of the observed locations expressed in absolute (right y-axis) and relative (left y-axis) values.

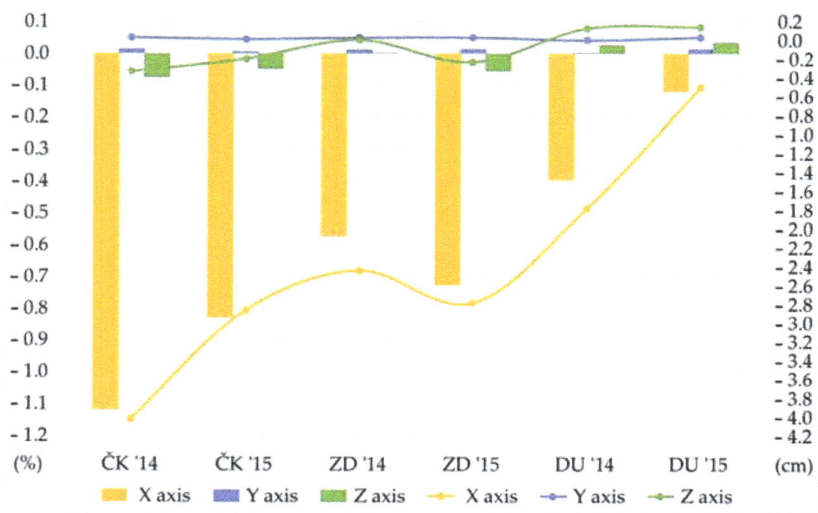

Figure 4. Movement of the parameter STD using the proposed model of observed locations expressed in absolute (right y-axis) and relative (left y-axis) values.

The presented results of RMSE as the main performance parameter showed that the proposed model improved the overall geodetic accuracy of the observed positions. The x-axis position deviations decreased for five measurements, while the deviations in the y-axis and especially in the z-axis decreased simultaneously (Dubrovnik, 2014).

The model for the y-axis achieved a decrease in the deviation in four measurements, while where there was an increase in the deviation in the remaining measurements (Čakovec, 2014 and Zadar, 2015), and there improvement in the accuracy of the position along the second horizontal x-axis was achieved in addition to the improvement achieved along the vertical z-axis.

The results of the obtained STD values as measures of the variability of the obtained results were half-hearted (partial). Improvement was obtained in nine measurements and worse results were obtained in the other nine measurements, as shown in Figure 4. It is important to point out that the absolute amounts of the obtained worse results were within the limits of 0.0014 cm to 0.1448 cm, although they were nominally worse than the STD value obtained using the Saastamoinen model. Considering the mentioned amounts, the obtained worse results can be ignored in further interpretation of the results and evaluation of the study's success.

4.1. Model Suitability for Application within the GPS

The proposed model shows a certain degree of success in the application within the GPS, although it was developed on the basis of GLONASS position records. The results showing the movement of the RMS parameter are shown in Figure 5.

The proposed model generally increases geodetic accuracy throughout the verification period. The RMS value of the x-axis deviation for all cities in the entire verification period reduced with absolute reduction amounts ranging from 3.29 cm to 32.26 cm (1.32–11.08%). At the same time, the geodetic accuracy decreased along the y-axis for the absolute values from 1.64 cm to 4.77 cm (1.09% to 5.54%). In the z axis, the model showed a variable result: it achieved an improvement in four of the six measurements (in absolute amounts from 0.65 cm to 2.49 cm, i.e., 0.34% to 1.23%). In the remaining two measurements, the proposed model degraded geodetic accuracy (varying from 3.55 cm to 16.32 cm, i.e., 1.78% to 7.95%). Differences in the model's ability to reduce the residual tropospheric delay of the satellite signal when compared to GPS are due to a number of reasons, including the different spatial constellation of systems, differences in modulations of the satellite navigation signals, etc. For adequate results within the GPS, a regression analysis should be performed as a function of the interdependence of meteorological input predictors and RTE output variables based on GPS positional data, which further observes and analyzes the dynamics and variations of RTE movements.

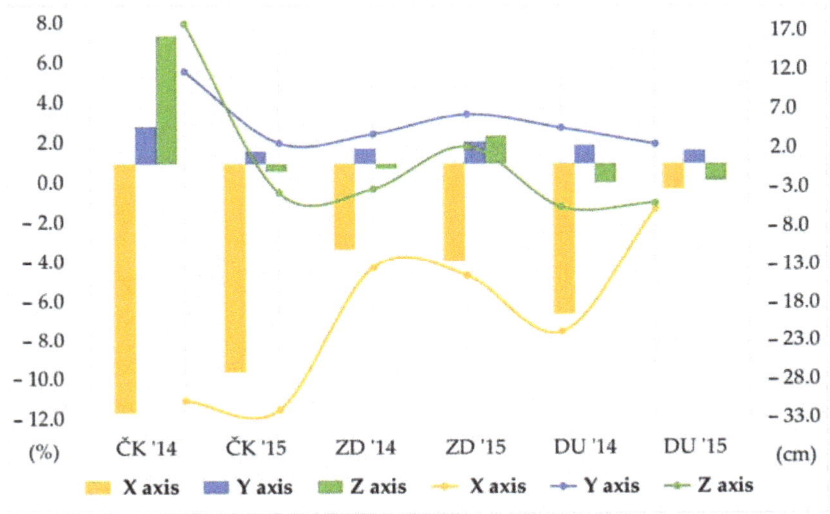

Figure 5. Movement of the RMSE deviation value using the proposed model from the exact geodetic position of the observed locations obtained using a GPS system and expressed in absolute (right *y*-axis) and relative (left *y*-axis) values.

As expected, the results do not achieve an improvement as with the GLONASS system. Given the time and space limitations of the research, GLONASS was chosen as an affirmed part of the GNSS system since the main goal of the research was the development and adoption of a methodology based on statistical analysis. However, the verification of the proposed model using GPS data showed the potential of the adopted approach and the possibility of further development based on the input of multi-GNSS positioning data for a wider GNSS application.

4.2. *Periodic Effect of the Proposed Model on the Positioning Accuracy*

The proposed model showed the possibility of a quantitative influence on the reduction of the tropospheric error which increases the accuracy of GLONASS positions. The tropospheric error is a stochastic phenomenon due to natural causes; however, the conducted research proves the possibility of developing a statistically significant correlation between tropospheric dynamics (expressed as a set of meteorological parameters) and the improvement of GNSS position accuracy. The regularity of the movement of the absolute differences of the values of RTE according to the proposed and Saastamoinen models can be observed; thus, indirectly, the movement of the geodetic accuracy achieved via the GLONASS system can also be observed. The annual motions shown in Figures 6–8 (see Appendix A for additional Figures A1–A7) show the annual oscillation for each position axis. The value for the *x*-axis is shown in the upper part of the graph, while the middle part of the graph shows the values for the *z*-axis, and the lower part of the graph shows the realized values for the *z*-axis. The property of periodicity of the motion manifests itself at smaller temporal resolutions—for example, at quarterly or monthly time frames—regardless of the selected GNSS position (see Appendix A: GNSS Position Zadar 2015).

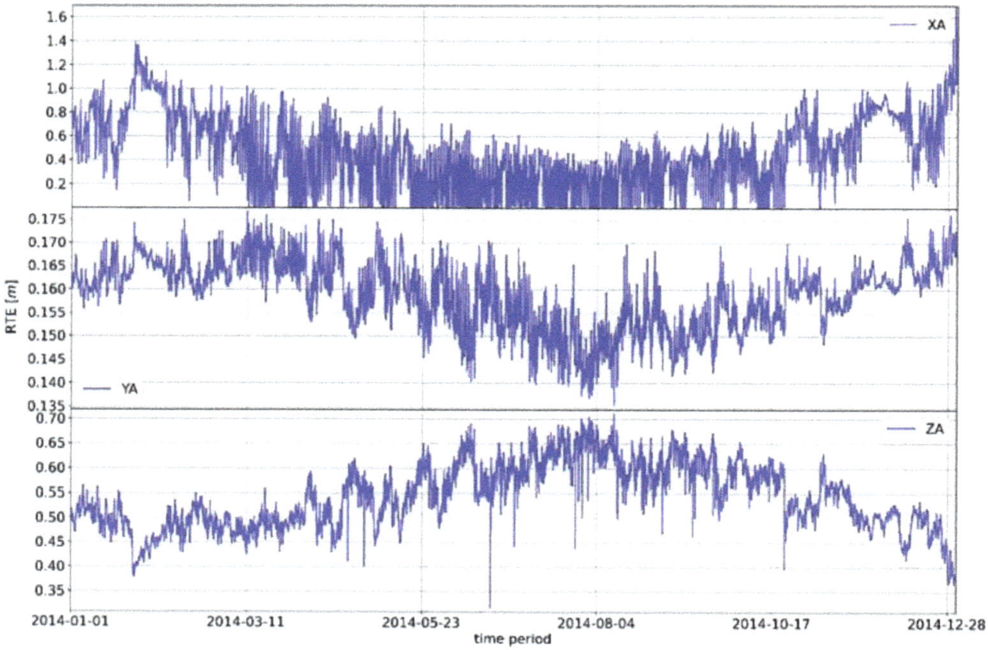

Figure 6. Movement of the absolute difference of the parameter RTE at the GNSS position Čakovec in 2014.

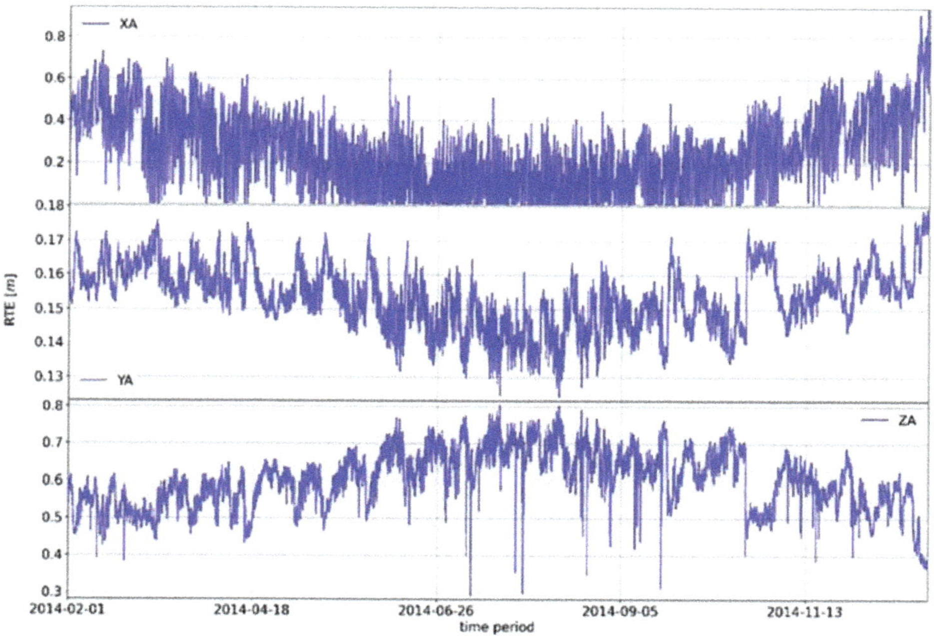

Figure 7. Movement of the absolute difference of the parameter RTE at the GNSS position Zadar in 2014.

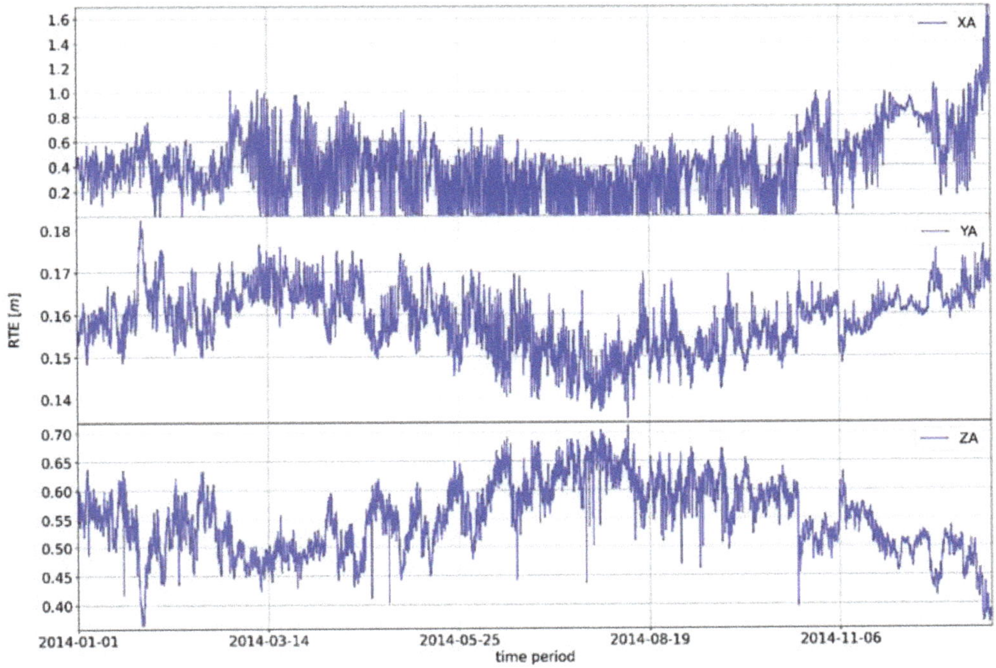

Figure 8. Movement of the absolute difference of the parameter RTE at the GNSS position Dubrovnik in 2014.

The annual trends in the difference of shown absolute RTE deviations clearly indicate the seasonal nature of the tropospheric error. On the *x*-axis, the difference in absolute RTE deviations typically reaches a maximum in the winter months, although the upper values of the deviations may also occur in the later periods of spring or autumn, i.e., in summer in the case of climatological deviations for the mentioned seasons.

The same pattern of correlation of the differences in absolute deviations from RTE along the y and z axes is evident in the other two observed GNSS positions. There are differences in the absolute values (amounts) of the RTE deviation difference, but there are no differences in the general distribution of the measured RTE deviation difference.

4.3. Observations on the Specifics of the Conducted Research

The dynamic specificities of the presented relationship between the positioning accuracy and the absolute values of the RTE difference are shown in the vertical level in the direction of the *z*-axis (regarding the relationship with the zenith angle of the incoming radio navigation signal) and in the horizontal plane in the direction of the *y*-axis. The vertical plane is defined by the *y*- and *z*-axes; therefore, improvement in the positional accuracy is expressed in such a way that the increase in deviation along the *y*-axis is followed by a decrease in deviation along the *z*-axis and vice versa. At the horizontal level, the deviation along the *x*-axis is somewhat proportionally related to the value achieved along the *y*-axis. Simultaneous deviations along the *y*- and *z*-axis show the mutual relationship achieved at the vertical level which they jointly define in the observed space as the starting point.

It is possible to define the RTE vector using the origin in the center O of the coordinate space (*Oxyz*) based on analytical interpretation of the accuracy dynamics of GNSS positions (observed based on the movement of the difference of the absolute values of RTE realized by the Saastamoinen model and the proposed model where the reduction of the parameter RTE improves the geodetic accuracy of the GNSS position). The motion of the RTE vector within the stationary *Oxyz* system describes the translational motion of the RTE parameter/point within the available (six) degrees of freedom. The proposed model shows improvement of the RTE parameter within the observed *Oxyz* system of taking the position (or the reduction of the RTE parameter) independent of time where the specified property of the *Oxyz* system is also independent of time [51]. Therefore, it is justified to conclude that the observed GNSS system has reached a state of statistical position equilibrium. The original assumption was that a decrease in the RTE parameter leads to a decrease in the tropospheric error which increases the geodetic accuracy of the position. The observed GNSS GLONASS system also reaches this state due to the realized movement of the RTE parameter.

5. Conclusions

The proposed model showed its success by reducing deviations from exact GNSS geodetic positions by acting on the non-modeled part of the tropospheric error. The basis of the proposed model was the existing Saastamoinen model, and correction values (obtained using the proposed model) for each axis should have been added to this model. The proposed model needs meteorological input parameters which can be interpolated from a given standard weather model depending on availability; alternatively, standard predicted input values can be used (as is the case of the absence of meteorological parameters in the Saastamoinen model).

The proposed model was verified at the same GNSS stations within a two-year period. The proposed model showed an improvement in position accuracy achieved by reducing the residual tropospheric error when compared to the Saastamoinen model. The model did not achieve a simultaneous improvement in all axes for all locations during the entire verification period, but it demonstrated superiority over the Saastamoinen model. Improvements in the horizontal axes of the position up to a maximum of 3.87% were achieved (14.26 cm), while the accuracy of the second horizontal axis reduced by 0.65% (1.65 cm) for two measurements. At the same time, the accuracy of the height component of the position improved in all measurements to a maximum of 4.37% (23.87 cm).

The proposed model must be used programmatically (software) as a complement to the Saastamoinen model, although there are certain limitations in terms of the geographical area of application, i.e., the possibility of application in areas with similar climate profiles. At the same time, the optimal application areas of the proposed model are found in stationary and dynamic systems to determine the position of the user in real time with lower accuracy. Therefore, it is clear that the model can be used within the existing application areas of the Saastamoinen model with all existing limitations and advantages.

An additional result of the applied methodology and the use of the RTE parameter is the statistical position equilibrium of the observed GNSS GLONASS positions defined by the oscillation around the central position values within the *Oxyz*-space. The model was developed based on GLONASS position data; however, it also shows a certain level of success in the verification of geodetic positions based on GPS position. The possible continuation of this research includes applying this suggested approach to developing a model based on multi-GNSS positioning data and verifying its effectiveness with other available GNSS systems. Future research of this type will be focused on determining the positional statistical balance of the X, Y, and Z coordinates along the x, y, and z coordinate axes as a function of various satellite navigation arguments.

Author Contributions: Conceptualization, S.K. and M.B.; methodology, S.K. and M.B.; software, M.B.; validation, S.K., Z.M. and D.B.; formal analysis, M.B. and Z.M.; investigation, M.B.; resources, D.B.; data curation, M.B.; writing—original draft preparation, S.K. and M.B.; writing—review and editing, D.B.; visualization, M.B.; supervision, S.K. and D.B.; project administration, D.B.; funding acquisition, S.K. All authors have read and agreed to the published version of the manuscript.

Funding: This research received no external funding.

Data Availability Statement: The GNSS observation data are available at http://www.epncb.oma.be/index.php (accessed 29 August 2021). Meteorological data are available at https://meteo.hr/proizvodi.php?section=katalog_zahtjevi&param=zahtjev_podaci_usluge upon request (accessed 29 August 2021).

Acknowledgments: This work was fully supported by the University of Rijeka under the project uniri-tehnic 18-66: *Research of environmental impacts on satellite navigation systems.*

Conflicts of Interest: The authors declare no conflict of interest.

Abbreviations

The following abbreviations were used in the manuscript:

DHMZ	Croatian Meteorological and Hydrological Service
ECEF	Earth Centered, Earth Fixed
EKF	Extended Kalman Filter
EUREF	Regional Reference Frame Sub-Commission for Europe
GLONASS	Global Navigation Satellite System
GNSS	Global Navigation Satellite System
GPS	Global Positioning System
NMF	(New) Mapping Functions
NWM	Numerical Weather Models
PPP	Precise Point Positioning
PWV	Precipitable Water Vapor
RINEX	Receiver Independent Exchange
RMSE	Root Mean Square Error
RTE	Residual Tropospheric Error
RTK	Real-time Kinematics
SBAS	Satellite-based Augmentation Systems
STD	Standard Deviation
ZHD	Zenith Hydrostatic Delay
ZTD	Zenith Tropospheric Delay
ZWD	Zenith Wet Delay

Appendix A

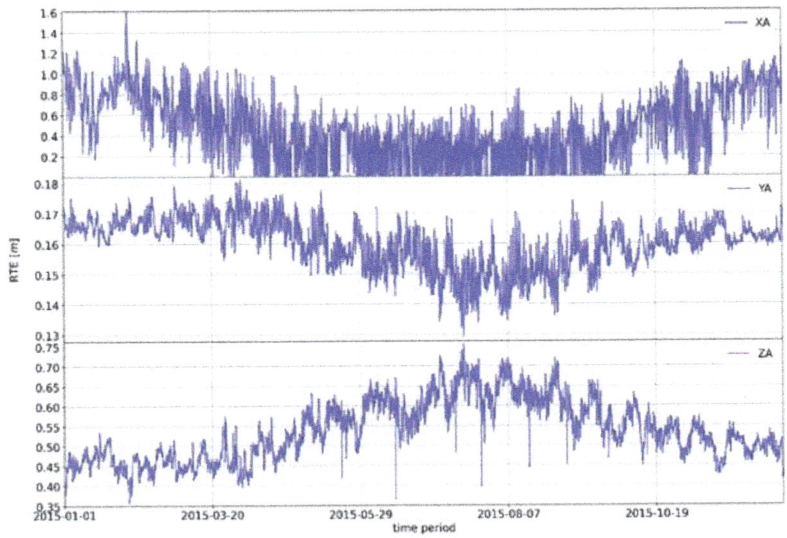

Figure A1. Movement of the absolute difference of the RTE GNSS Čakovec position parameter in 2015.

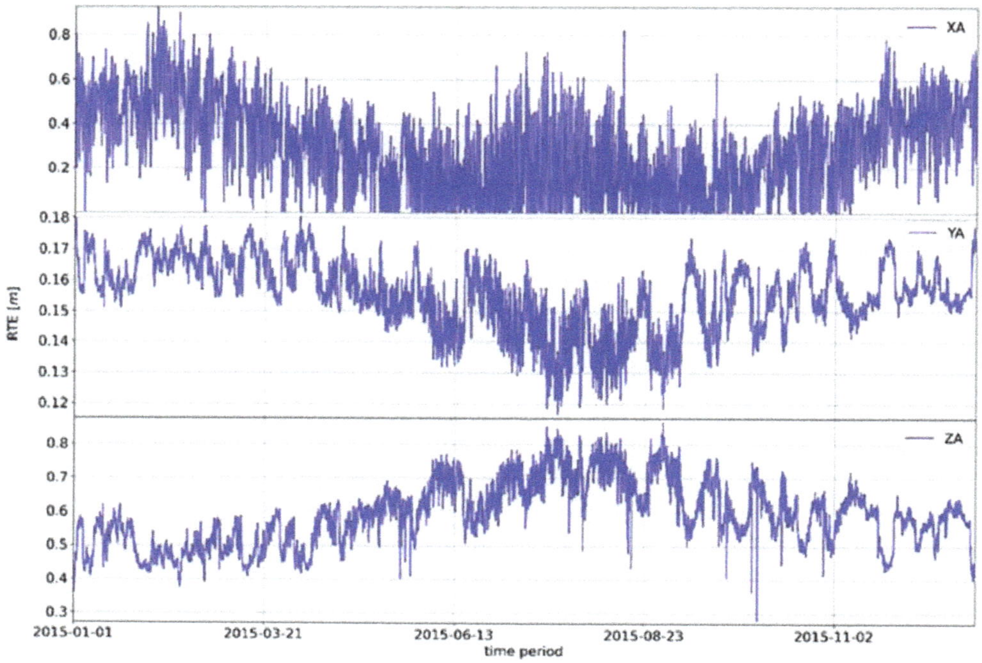

Figure A2. Movement of the absolute difference of the RTE GNSS Zadar position parameter in 2015.

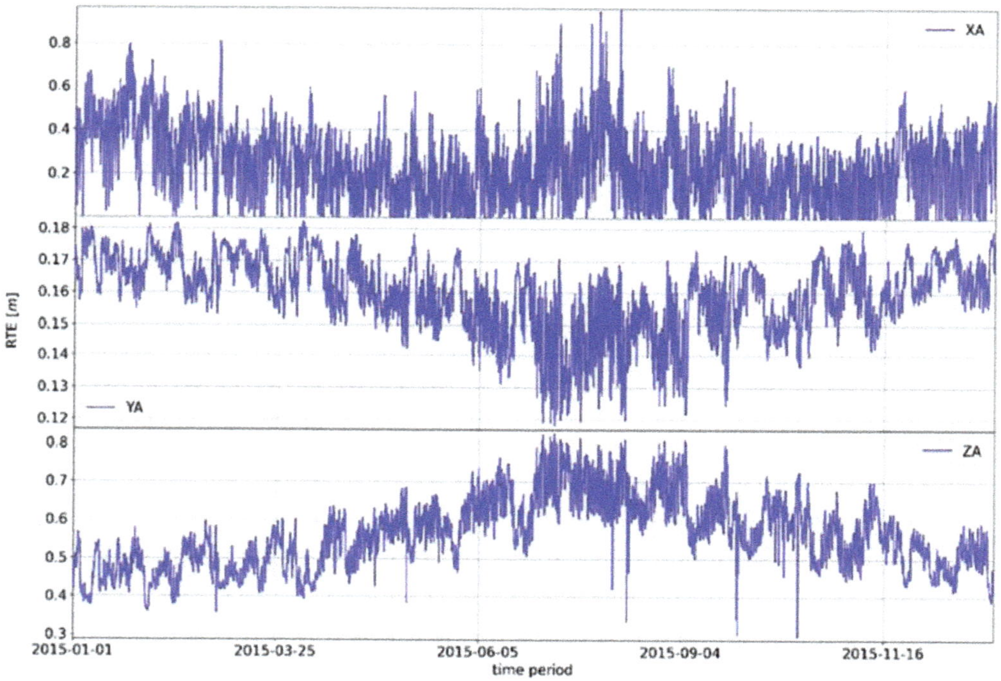

Figure A3. Movement of the absolute difference of the RTE GNSS Zadar position parameter in 2015.

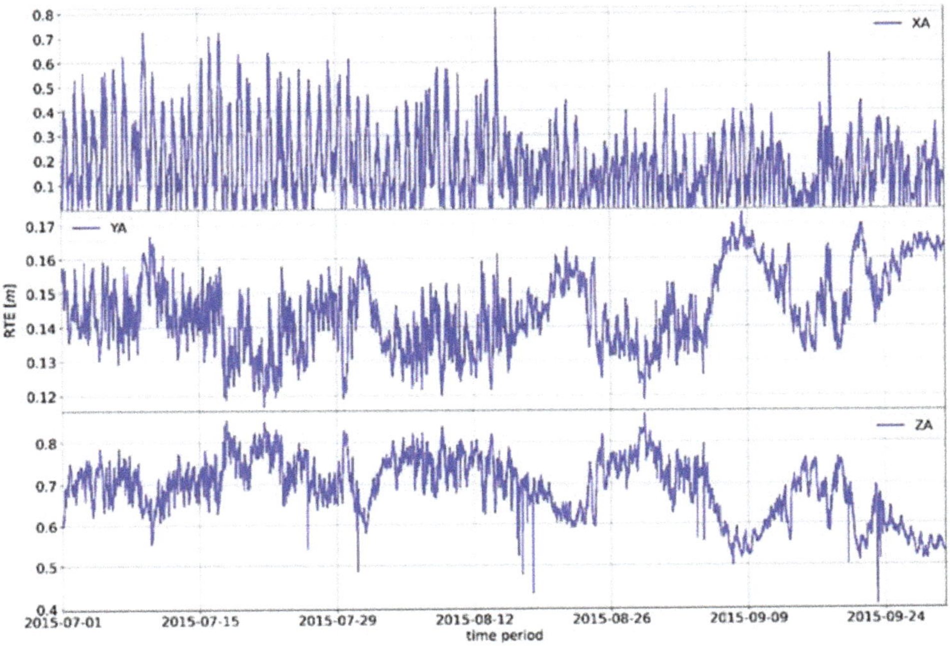

Figure A4. Movement of the absolute difference of the RTE GNSS Zadar position parameter in the period July–September 2015.

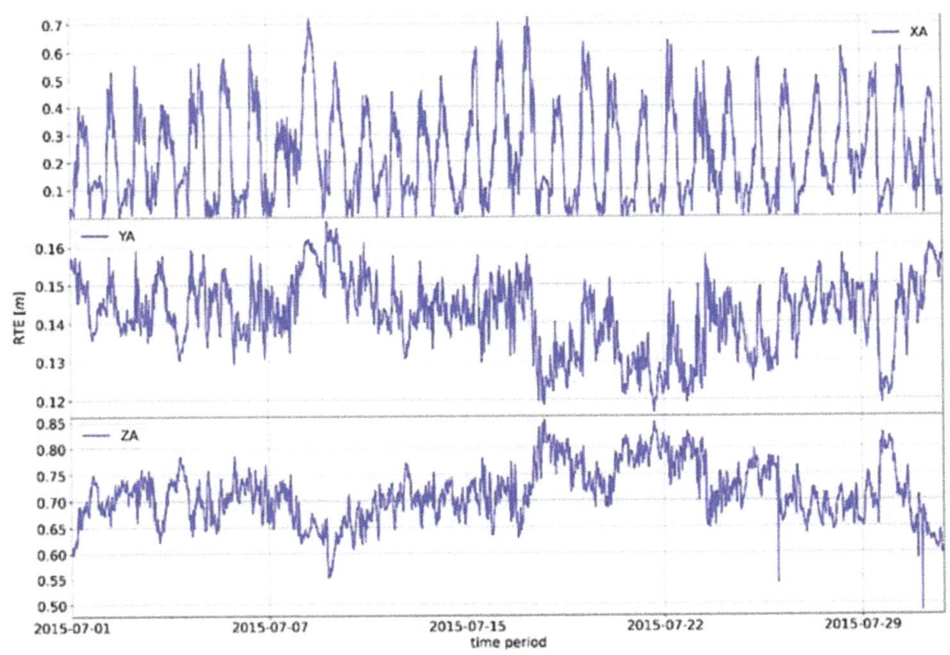

Figure A5. Movement of the absolute difference of the RTE GNSS position parameter in Zadar, July 2015.

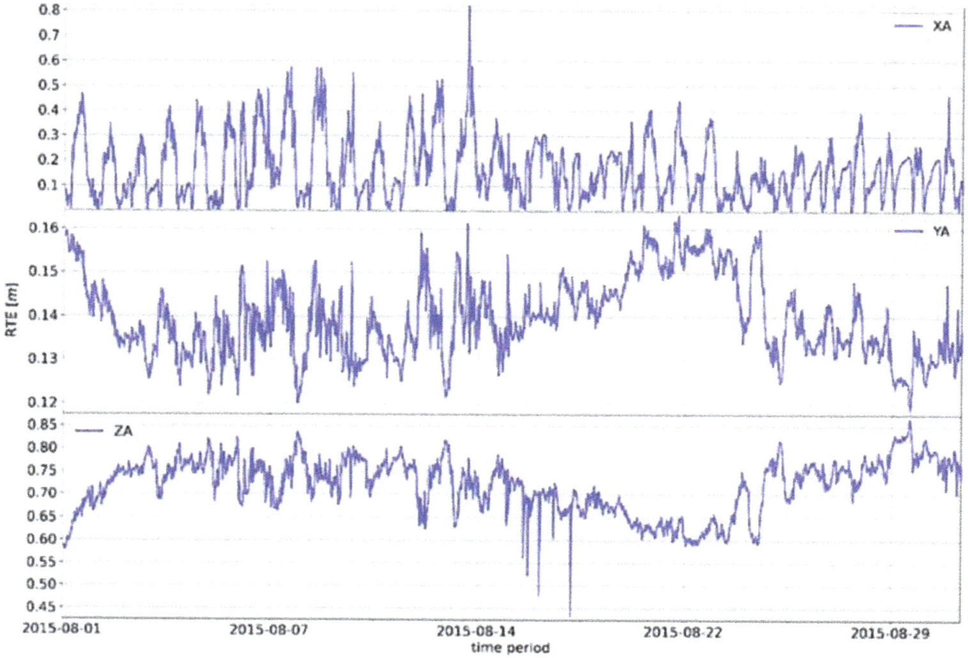

Figure A6. Movement of the absolute difference of the RTE GNSS position in Zadar, August 2015.

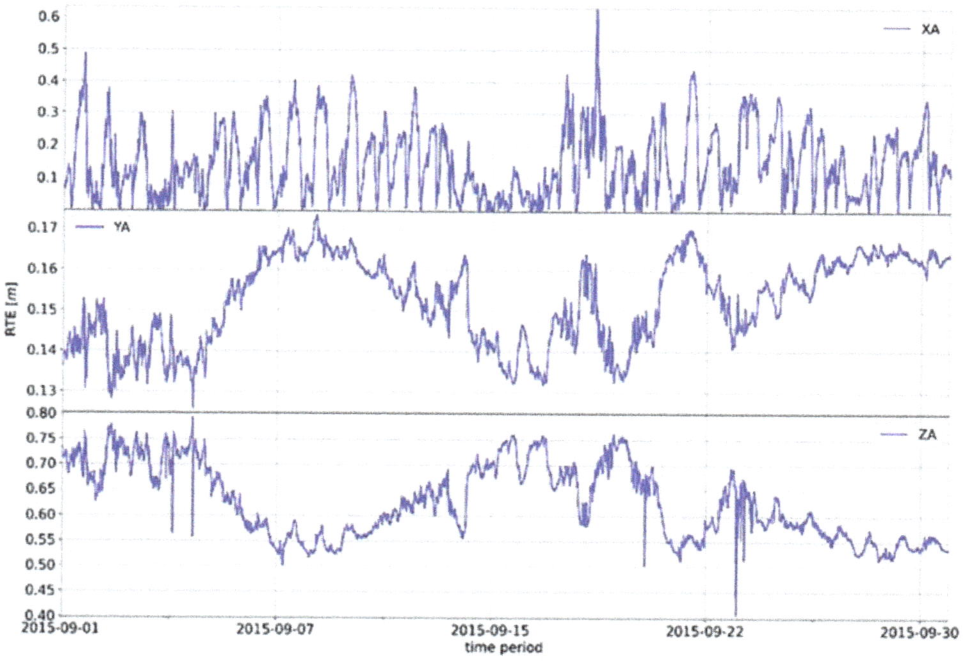

Figure A7. Movement of the absolute difference of the RTE GNSS position parameter in Zadar, September 2015.

References

1. Ansari, K.; Jamjareegulgarn, P. Effect of Weighted PDOP on Performance of Linear Kalman Filter for RTK Drone Data. *IEEE Geosci. Remote Sens. Lett.* **2022**, *19*, 3513504. [CrossRef]
2. Kaplan, E.D.; Hegarty, C.J. *Understanding GPS: Principles and Application*, 2nd ed.; Artech House: Boston, MA, USA, 2006.
3. Brčić, D.; Ćelić, J.; Valčić, S. Reconstruction of Geomagnetic Event as Observed in Northern Adriatic Region and Its Correlation with GPS Single-frequency Positioning Deviations. *Int. J. Mar. Navig. Saf. Sea Transp.* **2020**, *14*, 349–357. [CrossRef]
4. Subirana, J.S.; Zornoza, J.M.J.; Hernández-Pajares, M. *GNSS DATA PROCESSING Volume I: Fundamentals and Algorithms Acknowledgements*; ESA Communications ESTE: Leiden, The Netherlands, 2013.
5. Elektronika, Uređaji. Širenje Valova. In *Tehnička Enciklopedija*; LZMIK: Zagreb, Croatia, 1998; pp. 623–627.
6. Ustinov, U.M.; Afanasev, V.V.; Pripotnyuk, A.V.; Marinich, A.N. *Судовые Радиолокационные Системы*, 3rd ed.; Веленара: Sankt-Petersburg, Russia, 2009.
7. Jin, S.P.; Cardellach, E.; Xie, F. GNSS Atmospheric and Multipath Delays. In *GNSS Remote Sensing*; Springer: Dordrecht, The Netherlands, 2014; Volume 19, pp. 17–30.
8. El-Arini, M.B. Tropospheric Effects on GNSS Bertram Arbesser-Rastburg. In Proceedings of the Atmosphere and its Effect on GNSS Systems, Santiago, Chile, 14–16 April 2018; pp. 146–165.
9. Dodo, S.Y.; Ojigi, J.D.; Tsebeje, L.M. Determination of the best-fit Tropospheric Delay Model on the Nigerian Permanent GNSS Network. *J. Geosci. Geomat.* **2015**, *3*, 88–95.
10. Katsougiannopoulos, S.; Pikridas, C.; Rossikopoulos, D.; Ifadis, I.M.; Fotiou, A. Tropospheric Refraction Estimation Using Various Models, Radiosonde Measurements and Permanent GPS Data. In Proceedings of the XXIII FIG Congress, Munich, Germany, 8–13 October 2006.
11. Hopfield, H.S. Two-quartic tropospheric refractivity profile for correcting satellite data. *J. Geophys. Res.* **1969**, *74*, 4487–4499. [CrossRef]
12. Saastamoinen, J. Contributions to the theory of atmospheric refraction. *Bull. Géodésique* **1972**, *105*, 279–298. [CrossRef]
13. Pershin, D.Y. Comparative analysis of tropospheric delay models in precise point positioning in satellite navigation systems GLONASS/GPS. *Вестник Новосибирского Государственного Университета. Серия: Информационные Технологии* **2009**, *7*, 84–91.
14. Marini, J.W. Correction of Satellite Tracking Data for an Arbitrary Tropospheric Profile. *Radio Sci.* **1972**, *7*, 223–231. [CrossRef]
15. Davis, J.L.; Herring, T.A.; Shapiro, I.I.; Rogers, A.E.E. Geodesy by radio interferometry: Effects of atmospheric modeling errors on estimates of baseline length. *Radio Sci.* **1985**, *20*, 1593–1607. [CrossRef]
16. Ifadis, I.M. The excess propagation path of radio waves: Study of the influence of the atmospheric parameters on its elevation dependence. *Surv. Rev.* **2013**, *31*, 289–298. [CrossRef]
17. Askne, J.; Nordius, H. Estimation of tropospheric delay for microwaves from surface weather data. *Radio Sci.* **1987**, *22*, 379–386. [CrossRef]
18. Shrestha, S.M. *Investigations into the Estimation of Tropospheric Delay*; University of Calgary: Calgary, AB, Canada, 2003.
19. Kleijer, F. *Troposphere Modeling and Filtering for Precise GPS Leveling*; Delft University of Technology: Delft, The Netherlands, 2004.
20. Xia, P.; Xia, J.; Ye, S.; Xu, C. A new method for estimating tropospheric zenith wet-component delay of gnss signals from surface meteorology data. *Remote Sens.* **2020**, *12*, 3497. [CrossRef]
21. Dousa, J.; Elias, M. An improved model for calculating tropospheric wet delay. *Geophys. Res. Lett.* **2014**, *41*, 4389–4397. [CrossRef]
22. Parameswaran, K.; Saha, K.; Raju, S.C. Development of a regional tropospheric delay model for GPS-based navigation with emphasis to the Indian Region. *Radio Sci.* **2008**, *43*, 1–10. [CrossRef]
23. Parameswaran, K.; Raju, C.S.; Saha, K. Region-specific Tropospheric Delay Model for the Indian Subcontinent. In Proceedings of the ICG-Meeting, Bangalore, India, 5–7 September 2007.
24. Isioye, O.A.; Combrinck, L.; Botai, J. Performance Evaluation of Blind Tropospheric Delay correction Models over Africa. *S. Afr. J. Geomat.* **2015**, *4*, 502. [CrossRef]
25. Hu, Y.; Yao, Y. A new method for vertical stratification of zenith tropospheric delay. *Adv. Sp. Res.* **2019**, *63*, 2857–2866. [CrossRef]
26. Yang, F.; Guo, J.; Zhang, C.; Li, Y.; Li, J. A regional zenith tropospheric delay (Ztd) model based on Gpt3 and Ann. *Remote Sens.* **2021**, *13*, 838. [CrossRef]
27. Morel, L.; Pottiaux, E.; Durand, F.; Fund, F.; Boniface, K.; Junior, P.S.d.; van Baelen, J. Validity and behaviour of tropospheric gradients estimated by GPS in Corsica. *Adv. Sp. Res.* **2015**, *55*, 135–149. [CrossRef]
28. Deng, Z.; Bender, M.; Zus, F.; Ge, M.; Dick, G.; Ramatschi, M.; Wickert, J.; Löhnert, U.; Schön, S. Validation of tropospheric slant path delays derived from single and dual frequency GPS receivers. *Radio Sci.* **2011**, *46*, 6007. [CrossRef]
29. Mohammed, A.N.; Abdelfatah, M.A.; Mousa, A.E.K.; El-Fiky, G.S. Regional Egypt tropospheric mapping function model. *Alex. Eng. J.* **2018**, *57*, 4157–4165. [CrossRef]
30. Younes, S.A.M.; Afify, H.A. Accuracy improvement of tropospheric delay correction models in space geodetic data. Case study: Egypt. *Geod. Cartogr.* **2014**, *40*, 148–155. [CrossRef]
31. Yang, Y.F.; Chen, X.P.; Yao, M.H.; Zhou, C.L.; Liao, C.M. Research on zenith tropospheric delay modeling of regional cors network. *Int. Arch. Photogramm. Remote Sens. Spat. Inf. Sci.* **2020**, *XLII-3/W10*, 1197–1200. [CrossRef]
32. Dodo, J.D.; Ekeanyanwu, U.O.; Ono, M.N. Evaluation of Five Tropospheric Delay Models on Global Navigation Satellite System Measurements in Southern Nigeria. *J. Geosci. Geomat.* **2019**, *7*, 201–211. [CrossRef]

33. Li, F.; Zhang, Q.; Zhang, S.; Lei, J.; Li, W. Evaluation of Spatio-Temporal Characteristics of Different Zenith Tropospheric Delay Models in Antarctica. *Radio Sci.* **2020**, *55*, e2019RS006909. [CrossRef]
34. Krueger, E.; Schueler, T.; Arbesser-rastburg, B. The Standard Tropospheric Correction Model for the European Satellite Navi−Gation System Galileo. In Proceedings of the General As-sembly URSI, New Delhi, India, 2005; Available online: https://www.researchgate.net/profile/Bertram_Arbesser$-$Rastburg/publication/252717445_THE_STANDARD_TROPOSPHER-IC_CORRECTION_MODEL_FOR_THE_EUROPEAN_SATELLITE_NAVIGATION_SYSTEM_GALILEO/links/00b4952c318413b26d000000/THE$-$STANDARD$-$TROPOSPHERIC$-$CORRECTION$-$MODEL$-$FOR$-$THE$-$EUROPEAN$-$SATELLITE$-$NAVIGATION$-$SYSTEM$-$GALILEO.pdf (accessed on 30 November 2021).
35. Snajdrova, K.; Boehm, J.; Willis, P.; Haas, R.; Schuh, H. Multi-technique comparison of tropospheric zenith delays derived during the CONT02 campaign. *J. Geod.* **2006**, *79*, 613. [CrossRef]
36. Ajayi, G. *Physics of the Tropospheric Radiopropagation*; International Centre for Theoretical Physic: Trieste, Italy, 1989.
37. Grabner, M.; Kvicera, V. Atmospheric Refraction and Propagation in Lower Troposphere. In *Electromagnetic Waves*; Prentice Hall: Hoboken, NJ, USA, 2011. [CrossRef]
38. Coleman, C.J. A direction-sensitive model of atmospheric noise and its application to the analysis of HF receiving antennas. *Radio Sci.* **2002**, *37*, 3-1–3-10. [CrossRef]
39. Hatanaka, Y. A Compression Format and Tools for GNSS Observation Data. *Bull. Geogr. Surv. Inst.* **2008**, *55*, 21–30.
40. DHMZ—Državni Hidrometeorološki Zavod. Available online: http://meteo.hr/index.php (accessed on 28 November 2021).
41. Volarić, B.; Lisac, I. Klimatska podjela Hrvatske prema značajkama godišnjeg hoda temperature zraka. *Acta Geogr. Croat.* **1984**, *19*, 3–11.
42. Tadić, M.P.; Čapka, G.M.; Zaninović, K.; Cindrić, K. Drought Vulnerability in Croatia. *Agric. Conspec. Sci.* **2014**, *79*, 31–38.
43. EUREF Permanent GNSS Network. Available online: http://www.epncb.oma.be/index.php (accessed on 16 August 2021).
44. RTKLIB: Documents. Available online: http://www.rtklib.com/rtklib_document.htm (accessed on 10 November 2021).
45. Astudillo, M.J.; Lau, L.; Tang, Y.T.; Moore, T. Analysing the Zenith Tropospheric Delay Estimates in On-line Precise Point Positioning (PPP) Services and PPP Software Packages. *Sensors* **2018**, *18*, 580. [CrossRef]
46. RTKLIB: An Open Source Program Package for GNSS Positioning. Available online: http://www.rtklib.com (accessed on 29 November 2021).
47. Niell, A.E. Global mapping functions for the atmosphere delay at radio wavelengths. *J. Geophys. Res. Solid Earth* **1996**, *101*, 3227–3246. [CrossRef]
48. Niell, A.E. Improved atmospheric mapping functions for VLBI and GPS. *Earth Planets Space* **2000**, *52*, 699–702. [CrossRef]
49. Niell, A.E. Preliminary evaluation of atmospheric mapping functions based on numerical weather models. *Phys. Chem. Earth Part A Solid Earth Geod.* **2001**, *26*, 475–480. [CrossRef]
50. ICD GLONASS CDMA. Информационно-Аналитический Центр Контроля ГЛОНАСС и GPS. Available online: https://www.glonass-iac.ru/ (accessed on 16 August 2021).
51. Plischke, M.; Bergersen, B. *Equilibrium Statistical Physics*, 3rd ed.; World Scientific Publishing Co. Pte. Ltd.: Singapore, 2006.

remote sensing

MDPI

Article

Assessment of the Water Vapor Tomography Based on Four Navigation Satellite Systems and Their Various Combinations

Fei Yang [1,2,3,4,*], Jinyang Wang [1], Hongsen Wang [1], Xu Gong [1], Lei Wang [5] and Bin Huang [2]

1 College of Geoscience and Surveying Engineering, China University of Mining and Technology-Beijing, Beijing 100083, China; sqt2000204095@student.cumtb.edu.cn (J.W.); zqt2000204104@student.cumtb.edu.cn (H.W.); 1810250229@student.cumtb.edu.cn (X.G.)
2 Key Laboratory of South China Sea Meteorological Disaster Prevention and Mitigation of Hainan Province, Haikou 570203, China; bhuang_hms@163.com
3 State Key Laboratory of Geodesy and Earth's Dynamics, Innovation Academy for Precision Measurement Science and Technology, Chinese Academy of Sciences, Wuhan 430077, China
4 Beijing Key Laboratory of Urban Spatial Information Engineering, Beijing 100038, China
5 State Key Laboratory of Information Engineering in Surveying, Mapping and Remote Sensing, Wuhan University, Wuhan 430079, China; lei.wang@whu.edu.cn
* Correspondence: yangfei@cumtb.edu.cn

Abstract: With the gradual improvement of Galileo and the opening of BDS-3 services, water vapor tomography based on multi-GNSS can be effectively carried out to reconstruct three-dimensional water distribution. In this paper, experiments in Hong Kong were conducted to analyze and assess the performances of GPS, BDS, GLONASS, and Galileo and their combinations in water vapor tomography. Numerical results show that the number of available signal rays varies widely in the four satellite systems, and the value can be increased by the combination of satellite systems; the combinations also increase the number of voxels crossed by signal rays, but this value is not directly related to the number of available signal rays; the number and distribution of the voxels with sufficient signal rays, which most closely related to the structure of the tomographic model, show no obvious differences in the four satellite systems and their combinations. Comparative results of slant water vapor (SWV) estimated by GNSS data and water vapor density derived from radiosonde data reveal that the differences in the water vapor tomography of the four satellite systems are small, and their combinations have limited improvement in the tomographic results.

Keywords: GNSS meteorology; water vapor; tomography; multi-GNSS

Citation: Yang, F.; Wang, J.; Wang, H.; Gong, X.; Wang, L.; Huang, B. Assessment of the Water Vapor Tomography Based on Four Navigation Satellite Systems and Their Various Combinations. *Remote Sens.* **2022**, *14*, 3552. https://doi.org/10.3390/rs14153552

Academic Editors: Serdjo Kos, José Fernández and Juan F. Prieto

Received: 21 June 2022
Accepted: 22 July 2022
Published: 24 July 2022

1. Introduction

Since the concept of GNSS meteorology was first proposed by Bevis et al., the water vapor information derived from GNSS has drawn increasing attention in the meteorological and GNSS communities [1]. The precipitable water vapor (PWV), which refers to the height of an equivalent column of water vapor [2], has been widely validated to achieve mm-level accuracy based on the conversion of GNSS zenith tropospheric delay [3]. Further, the three-dimensional water vapor information can also be inversed by using GPS signals as scanning rays in the research area, which is called water vapor tomography.

Braun et al. first proposed the concept of GPS water vapor tomography [4] and Flores et al. first realized it using the data from the Kilauea network in Hawaii [5]. The research region, covered by ground GPS receivers, is discretized into finite voxels according to its latitude, longitude, and altitude, and the unknown estimated parameter of the voxels are assumed to be constant during a given period. The GPS-derived slant water vapor is regarded as the observations for water vapor tomography.

In modeling the GPS water vapor tomography, it is found that the geometric distribution of the observed signals is an inverted cone due to the fixed structure of GPS sites and satellites [6]. The direct effect caused by this phenomenon is the presence of tomographic voxels without signal rays passing through, especially at the lower and edge layer of the area of interest. It also makes many voxels be penetrated by only a very small number of signals. From the perspective of the water vapor tomography model, it often leads to a large number of zero elements appearing in the coefficient matrix, which becomes a sparse matrix [7]. This is the fundamental cause of the ill-posed problem in GPS water vapor tomography, which seriously restricts its stable and high-accuracy solution. Obviously, enriching the observation equation of the GPS water vapor tomography is an effective way to overcome the above problem by introducing various observation information, and related research has been carried out.

Based on voxel horizontal boundary selection and non-uniform symmetrical division, Chen et al. and Yao et al. proposed an optimized approach of horizontal voxel division to introduce more signal rays penetrated from the top layer into the observation equation [8,9]. The similar effect can be obtained by the method of constructing the tomographic buffer area carried out by Trzcina et al. and Sa et al. [10,11]. These methods are limited to specific tomographic regions and certain experimental periods. Adavi et al. explored how to use the constructed virtual reference sites to augment location-specific GPS observations [12]. The virtual signals were also introduced to the tomography model using the calculated mapping function and ZWD/PWV of corresponding site and the elevation and azimuth of specified virtual satellite [6,13]. Studies have shown that it is a feasible method to incorporate the GPS signal rays passed through form the side face into the tomography model; for example, Zhao et al. constructed the unit scale factor for these signals using the radiosonde and reanalysis data [14], Zhang et al. and Hu et al. established the height factor models adapted to these signals from side face [15,16]. In addition, Zhao et al. tried to extend the observations of GPS sites outside the tomographic region into the tomography model based on the GPT2w and TMF models [17,18]. The above methods all rely on external data or models and tend to introduce new error for the observation information. On the other hand, some have attempted to add multi-source observation information from various sensors into the GPS tomography model, such as the COSMIC occultation data [19], the GNSS-R data [20], the InSAR data [21,22], WRF output data [23], LEO constellation-augmented data [24], PWV data derived from FY-3, and MODIS [25,26]. However, the spatiotemporal resolution, availability, and consistency with the tomographic region are the factors that seriously restrict the fusion of the above observations into the tomography model.

It is more reasonable and convenient to construct the GNSS water vapor tomography model together with the observations from GPS and the other three satellite navigation systems. Bender et al. simulated GPS, GLONASS, and Galileo data and introduced the method for obtaining three-dimensional water vapor information by tomography technique in multi-satellite systems [27]. Wang et al. compared the tomographic accuracy of BDS and GPS based on simulated data, and showed that the result using 9 satellites of BDS is basically comparable to that of GPS [28]. Xia et al. and Benevides et al. carried out the water vapor tomography experiments of GPS combined with GLONASS and GPS combined with Galileo in Hongkong and Lisbon regions, respectively [29,30]. Dong et al. and Zhao et al. utilized the measured data derived from different numbers of BDS2 satellites and combined it with GPS and GLONASS data to construct the tomography model in Wuhan and Guiyang, respectively [31,32]. With the gradual improvement of Galileo and the opening of BDS-3 services, the above experiments based on simulated data or incomplete satellite data cannot fully reflect the current status of water vapor tomography based on multi-GNSS. Therefore, this paper aims to explore the differences between the four satellite navigation systems and their combination in water vapor tomography, including the modeling process and the reconstructed results.

2. Materials and Methods

The observations in GNSS water vapor tomography are the slant water vapor (SWV) which can be converted from slant wet delay (SWD) as follows [33]:

$$\mathrm{SWV} = \frac{10^6}{\rho_w \times \frac{R}{m_w}\left(\frac{k_3}{T_m} + k_2 - \frac{m_w}{m_d} \times k_1\right)} \times \mathrm{SWD} \tag{1}$$

where ρ_w refers to the liquid water density with the unit of $\mathrm{g{\cdot}m^{-3}}$; $R = 8314\ \mathrm{Pa{\cdot}m^3{\cdot}K^{-1}{\cdot}kmol^{-1}}$ denotes the universal gas constant; m_w and m_d represent the molar mass of water and the dry atmosphere and their values are 18.02 $\mathrm{kg{\cdot}kmol^{-1}}$ and 28.96 $\mathrm{kg{\cdot}kmol^{-1}}$, respectively; T_m is the weighted mean temperature, which can be calculated by using surface temperature [34,35]; k_1, k_2, and k_3 are the empirical physical constants, which are equal to 77.60 $\mathrm{K{\cdot}hPa^{-1}}$, 70.4 $\mathrm{K{\cdot}hPa^{-1}}$, and $3.739 \times 10^5\ \mathrm{K{\cdot}hPa^{-1}}$, respectively [36]. After mapping the zenith wet delay (ZWD) and the wet delay gradients into the elevation direction, the SWD can be obtained as follows [37]:

$$\begin{aligned} \mathrm{SWD} &= m_w(ele) \times \mathrm{ZWD} \\ &+ m_w(ele) \times \cot(ele) \times \left(G_{NS}^w \times \cos(azi) + G_{WE}^w \times \sin(azi)\right) \end{aligned} \tag{2}$$

where m_w indicates the wet mapping function and the global mapping function (GMF) was used in this paper; *ele* and *azi* denote the satellite elevation and azimuth angles, respectively. G_{WE}^w and G_{NS}^w represent the wet delay gradient parameters in the east-west and north-south directions, respectively. Affected by water vapor along the signal ray, ZWD is the wet component of zenith total delay (ZTD) which is the primary parameter retrieved from GNSS observation. To obtain ZWD, the zenith hydrostatic delay (ZHD) should be subtracted from ZTD [38]. In this paper, the Saastamoinen model is used to calculate the accurate ZHD using the pressure measurements as follows [39]:

$$\mathrm{ZHD} = \frac{0.002277 \times P_s}{1 - 0.00266 \times \cos(2\varphi) - 0.00028 \times H} \tag{3}$$

where φ and H denote the latitude and geodetic height of the GNSS site, respectively. P_s is the measured surface pressure.

In the water vapor tomography, the SWV value is also an integral expression of water vapor along the slant path from the ground receiver and GNSS satellite, given by the following:

$$\mathrm{SWV} = 10^{-6} \cdot \int \rho(s) ds \tag{4}$$

where $\rho(s)$ in $\mathrm{g{\cdot}m^{-3}}$ denotes the water vapor density and ds refers to the path traveled by a satellite signal ray. After discretizing the tomographic region into finite voxels, the observation equation of GNSS water vapor tomography can be established based on the distances of GNSS signal rays crossing the divided voxel and the unknown estimated water vapor density with each voxel. It can be expressed as follows:

$$\mathrm{SWV} = \sum_{i=1}^{n} d_i \cdot x_i \tag{5}$$

where n represents the total number of divided voxels in the research region. d_i denotes the distance of signal rays inside voxel i, which can be calculated by using the coordinates of GNSS sites and satellites. x_i refers to the water vapor density of voxel i, which is the unknown estimated parameter.

In water vapor tomography, two types of constraints are widely used in tomographic modeling along with the observation equation, one is the horizontal constraint and the other is vertical constraint, since a spatial relation exists between water vapor in a specific voxel and its surrounding ones. For the horizontal constraint, it assumes that the distribution of water vapor density is relatively stable in the horizontal direction within a small region, and represents the relationship between the water vapor density of a certain voxel and those of its adjacent voxels in the same layer. For the vertical constraint, it refers to the exponential relationship between the water vapor density of voxels for two consecutive layers. These two constraints can be expressed as follows:

$$w_1^h x_1 + w_2^h x_2 + \cdots w_{i-1}^h x_{i-1} - x_i + w_{i+1}^h x_{i+1} + \cdots w_m^h x_m = 0 \tag{6}$$

$$x_i - w_{i+m}^v x_{i+m} = 0 \tag{7}$$

where m is the total number of voxels in the same layer. w^h and w^v denote the horizontal weighted coefficient and the vertical weighted coefficient, respectively. The horizontal weighted coefficient is constructed based on the Gaussian inverse distance weighted function as the following equation:

$$w_{i-1}^h = -\frac{e^{-\frac{d_{i,i-1}^2}{2\sigma^2}}}{\sum\limits_{j=1}^{m} e^{-\frac{d_{j,i}^2}{2\sigma^2}}} \tag{8}$$

where d is the distance between the two voxels. j is a number from 1 to m, represented the voxel ordering of the same horizontal layer. σ denotes the smoothing factor. The vertical weighted coefficient is constructed based on exponential function as follows:

$$w_{i+m}^v = e^{(h_{i+m} - h_i)/H_s} \tag{9}$$

where h represents the height of the corresponding voxel and H_s refers to the water vapor scale height with an empirical value of 1.5 km [40]. Note that each voxel has corresponding equations for horizontal and vertical constraints.

Thus, the tomography model for water vapor reconstruction can be established by combining the observation equation of multi-GNSS and the two types of constraint equations.

$$\begin{pmatrix} y_{swv}^G \\ y_{swv}^C \\ y_{swv}^R \\ y_{swv}^E \\ 0 \\ 0 \end{pmatrix} = \begin{pmatrix} A_G \\ A_C \\ A_R \\ A_E \\ H \\ V \end{pmatrix} \cdot x \tag{10}$$

where y_{swv} denotes the vector with SWV values derived from these four satellite systems; A represents the coefficient matrices of the observation equation for different types of satellite systems; H and V are the coefficient matrices of the horizontal and vertical constraints, respectively. The tomography solution of the unknown water vapor density vector x can be obtained as follows:

$$\begin{aligned} \hat{x} = &\left(A_G^T P_G A_G + A_C^T P_C A_C + A_R^T P_R A_R + A_E^T P_E A_E + H^T P_T H + V^T P_V V\right)^{-1} \\ &\cdot\left(A_G^T P_G y_{swv}^G + A_C^T P_C y_{swv}^C + A_R^T P_R y_{swv}^R + A_E^T P_E y_{swv}^E\right) \end{aligned} \tag{11}$$

where P represents the weighting matrices of different equations, which are determined by an optimal weighting method using the variance components estimation and homogeneity test [41]. Note that the number of satellite systems in Equations (8) and (9) can be adjusted in the experiment.

3. Results

3.1. Experimental Description

In this paper, the Hong Kong satellite Positioning Reference Station Network (SatRef) was selected to conduct the water vapor tomography experiment. We divided this research region into 560 voxels ranging from 113.87° to 114.35°, from 22.19° to 22.54°, and from 0 to 8 km for longitude, latitude, and altitude, respectively; that is, a voxel of 8 × 7 in the horizontal direction and 10 layers in the vertical direction. As shown in Figure 1, thirteen GNSS sites (T430, HKKT, HKLT, HKSL, HKNP, HKMW, HKPC, HKLM, HKOH, HKSC, HKST, HKSS, HKWS) in this region were used to provide SWV in the tomography modeling, and one GNSS (HKQT) and one radiosonde (45004) site were selected to validate the results of the water vapor tomography.

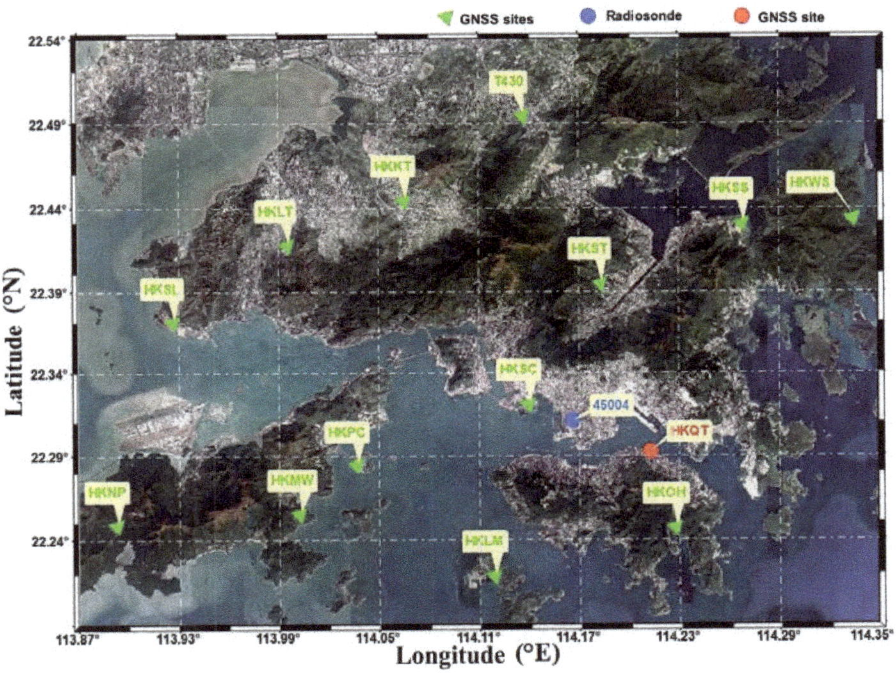

Figure 1. Distribution of GNSS, radiosonde sites, and the horizontal structure of the voxels.

We utilized the GAMIT 10.71 software to estimate the tropospheric parameters including ZTD and gradient parameters using the four GNSS systems. In this process, the elevation cutoff angle was set to 15°, the IGS precise ephemeris was adopted. Three MEGX stations (JFNG, URUM, and LHAZ) were incorporated into the solution model to reduce the strong correlation of tropospheric parameters caused by the short baseline. The processing strategies were set to LC_AUTCLN and BASELINE modes, meaning that the ionosphere-free linear combination was selected and the orbital parameters were fixed, respectively. The tropospheric parameters, including troposphere delay gradients and ZTD at 4 and 2 h intervals, are estimated and interpolated to a 30 s sampling rate in the GAMIT software. After calculating the ZHD using the measured pressure recorded by an automatic meteorological device, the SWV values of each satellite system were obtained by using Equations (1) and (2).

In this experiment, the GNSS observation data of one month from DOY 121 to 151, 2021 were selected to conduct the modeling and solution of water vapor tomography. For each tomographic solution, the period covered is 0.5 h. To assess the performance of water vapor tomography based on different satellite systems and different combinations of satellite system, each tomographic solution has 15 results, including those of a signal satellite system, the combination of two satellite systems, the combination of three satellite systems, and the combination of four satellite systems, namely GPS (G), BDS (C), GLONASS (R), Galileo (E), GPS+BDS (GC), GPS+GLONASS (GR),

GPS+Galileo (GE), BDS+GLONASS (CR), BDS+Galileo (CE), GLONASS+Galileo (RE), GPS+BDS+GLONASS (GCR), GPS+BDS+Galileo (GCE), GPS+GLONASS+Galileo (GRE), BDS+GLONASS+Galileo (CRE), and GPS+BDS+GLONASS+Galileo (GCRE). Note that both BDS-2 and BDS-3 were included in the experiment.

3.2. Experimental Analysis

The number of satellite signal rays available for the four satellite systems in each tomographic solution is counted and their averages during the 31 days from DOY 121 to 151, 2021 are shown in Figure 2. It can be seen that the BDS has the largest number (704) of available signal rays, followed by GPS, Galileo, and GLONASS with the average values of 507, 329, and 351, respectively. The percentages of available signal rays in BDS that exceed GLONAA and Galileo are more than 100%, achieving 114% and 101%, respectively. Compared with GPS, the value also reaches 39%. The number of signal rays used in GPS is the most stable during the experimental period, the difference between the maximum and minimum value is less than 100 with the standard deviation (STD) being only 23. While the other three satellite systems have obvious fluctuations in the number of available signal rays, with the differences between the maximum and minimum value far greater than 100 and the STDs reach 51, 48, and 34 for BDS, GLONASS, and Galileo, respectively. Note that the average number of signal rays used in Galileo is greater than that of GLONASS, but there are still days when GLONASS has more available signal rays than Galileo. When the satellite systems are combined, only the available signal rays of RE have just reached the level of BDS and the other combinations are all obviously improved compared with these single systems, especially since the average number of signal rays used in the combination of four systems could be close to 2000.

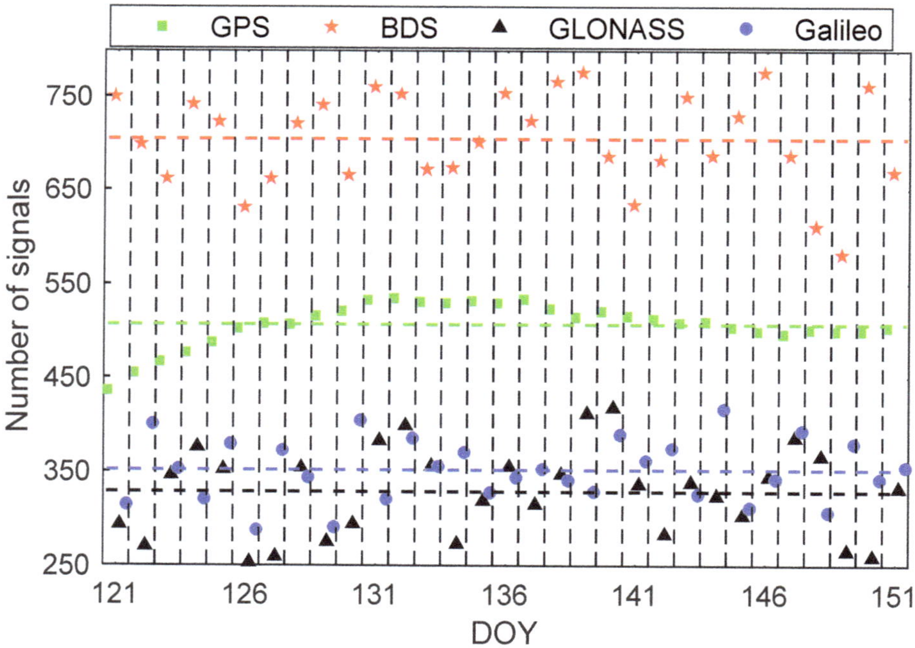

Figure 2. Average number of signal rays used in each solution based on different satellite systems during the experimental period.

The number of voxels passed through by signal rays for the four satellite systems in each tomographic solution is also counted, and their average values are shown in Figure 3. It was observed that the GPS has the largest number of voxels crossed by signal rays, followed by BDS, GLONASS, and Galileo with average values of 425, 424, 392, and 377, respectively. Corresponding to the 560 voxels in the entire tomographic region, the coverage rate of the four satellite systems reaches 75.9%, 75.4%, 70%, and 67.3%, respectively. Note that GPS and GLONASS with fewer available signal rays have more penetrated voxels than BDS and Galileo, respectively, and in fact, their differences are relatively small. In addition, the number of voxels crossed by signal rays for the four satellite systems all show a certain fluctuation during the experimental period.

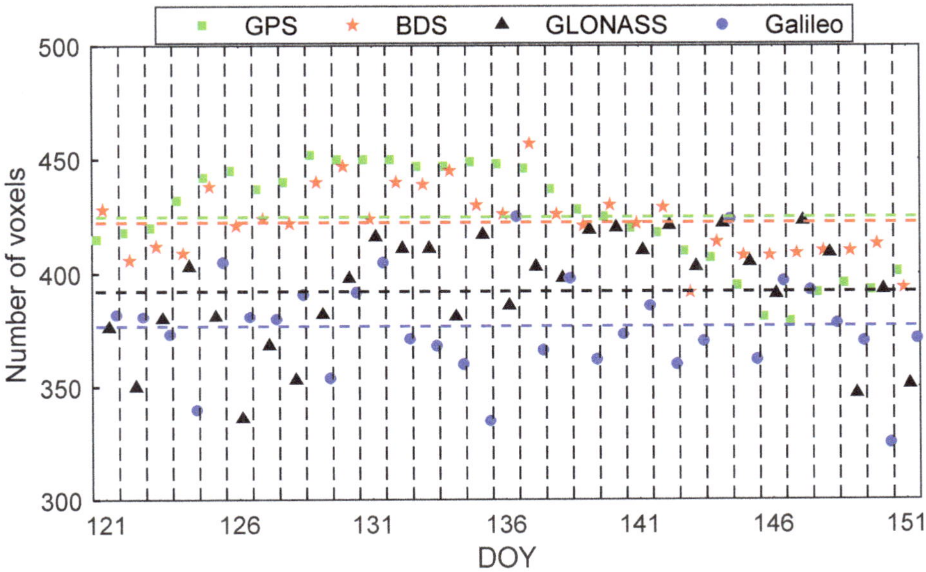

Figure 3. Average number of voxels penetrated by signal rays in each solution based on different satellite systems during the experimental period.

When combining the satellite systems, the number of crossed voxels and their coverage rate is counted and shown in the form of a histogram in Figure 4. It can be seen that the number and coverage rate of voxels are increased after the combinations compared with single satellite system. In addition, the performances of the three-satellite systems combination are better than those of the two-satellite systems combination, and four satellite systems combination outperforms the three-satellite systems combination. Specifically, combination of GCRE achieved the best performance with the number and coverage rate of voxels of 468 and 83.6%, respectively.

In the tomographic experiment, we found the existence of voxels that were only penetrated by a few signal rays, thus the concept of voxels crossed by sufficient signal rays was introduced from the relevant literature [7]. Based on the fact that a ray crossed a minimum number of voxels when the signal ray passed vertically through the tomographic region, the minimum probability that a voxel will be penetrated by a ray could be calculated. In this experiment, the value is 10/560, namely 1.79%. Then, the value of minimum probability multiplied by total SWV used is regarded as the criteria to determine whether a voxel is crossed by sufficient signal rays. Therefore, the number of voxels passed through by sufficient signal rays for the four single satellite systems and their combinations are counted and listed in Table 1 during the experimental period. It was observed that GPS had the largest number of voxels penetrated by sufficient signal rays among the four single systems, and only 7 voxels more than Galileo with the least effective voxels. After the combinations, the number of voxels increased but very little and the value of the combination of four system was only 278. Regarding the coverage rate, the difference of those 15 values in Table 1 is even smaller.

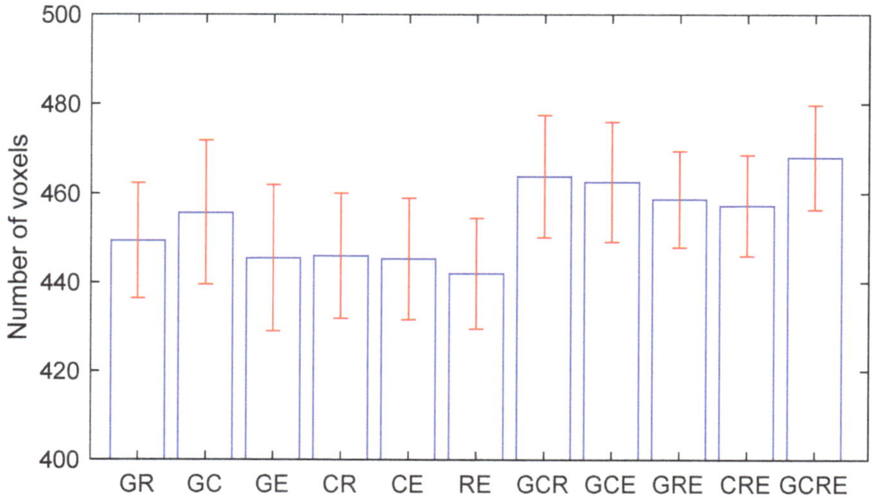

Figure 4. Average number of voxels penetrated by signal rays in each solution based on different combination during the experimental period.

Table 1. Average number of voxels penetrated by sufficient signal rays based on different satellite systems and different combinations during the experimental period.

Combinations	Number	Coverage Rate
G	271	48.4%
C	268	47.9%
R	265	47.3%
E	264	47.1%
GR	278	49.6%
GC	273	48.8%
GE	274	48.9%
CR	272	48.6%
CE	270	48.2%
RE	271	48.4%
GCR	277	49.5%
GEC	276	49.3%
GRE	276	49.3%
CRE	275	49.1%
GCRE	278	49.6%

Further, the situation that each voxel passed through by signal rays in a certain tomographic solution (UTC 11:45–12:15, DOY 137, 2021) is shown in detail in Figure 5, in which the black and white rectangles represent the voxels crossed by sufficient and in sufficient signal rays, respectively. Note that only the four single satellite systems and the combination of the four systems are illustrated in this figure. It is observed in the figure that the distribution of the black and white rectangles for different systems is very similar, especially in the lower and middle layers. From this point, for water vapor tomography in Hong Kong, the selection of a single satellite system or multi-GNSS combination has little effect on the structure of the tomographic model.

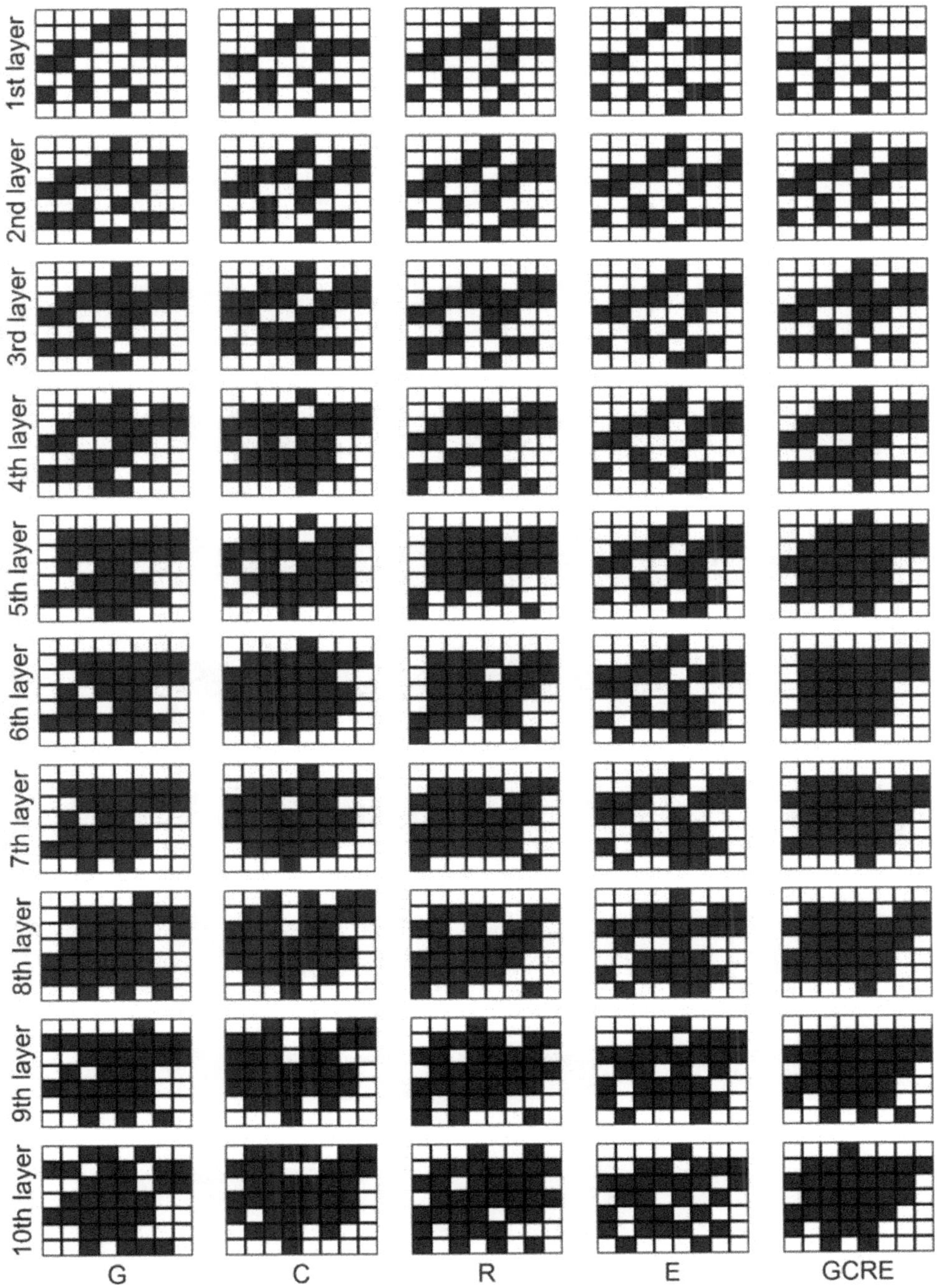

Figure 5. Distribution of voxel with sufficient signal rays at each layer for GPS, BDS, GLONASS, Galileo, and a combination of GCRE.

4. Discussion

To assess the performance of water vapor tomography using different satellite systems, SWV of the GNSS sites for validation were computed using these 15 tomographic results and the distances of signal ray in each voxel based on the observation equation established in Equation (5). The 15 tomography-computed SWV were then compared with the GAMIT-estimated SWV (as a reference). Figure 6 shows the change of tomography-computed vs. GAMIT-estimated SWV residuals with elevation angle during the experimental period for the four single systems. The change of the SWV residuals has the same trend in the four satellite systems, and they decrease as the elevation angle increases. It is observed that the residuals of four systems all ranged from −10 to 10 mm, and most of them concentrated between −2.0 and 2.0 mm. The percentage of absolute residuals smaller than 2.0 mm are 86.9%, 88.1%, 85.7%, and 85.3% for GPS, BDS, GLONASS, and Galileo, respectively. The largest absolute residual of the four satellite systems is 8.46, 9.63, 9.37, and 9.91 mm, respectively. We obtained the SWV residuals for various combinations of satellite systems, which also follow a decreasing trend with increasing elevation. These ranges and concentrated areas of the SWV residuals are unchanged compared with the four single satellite systems.

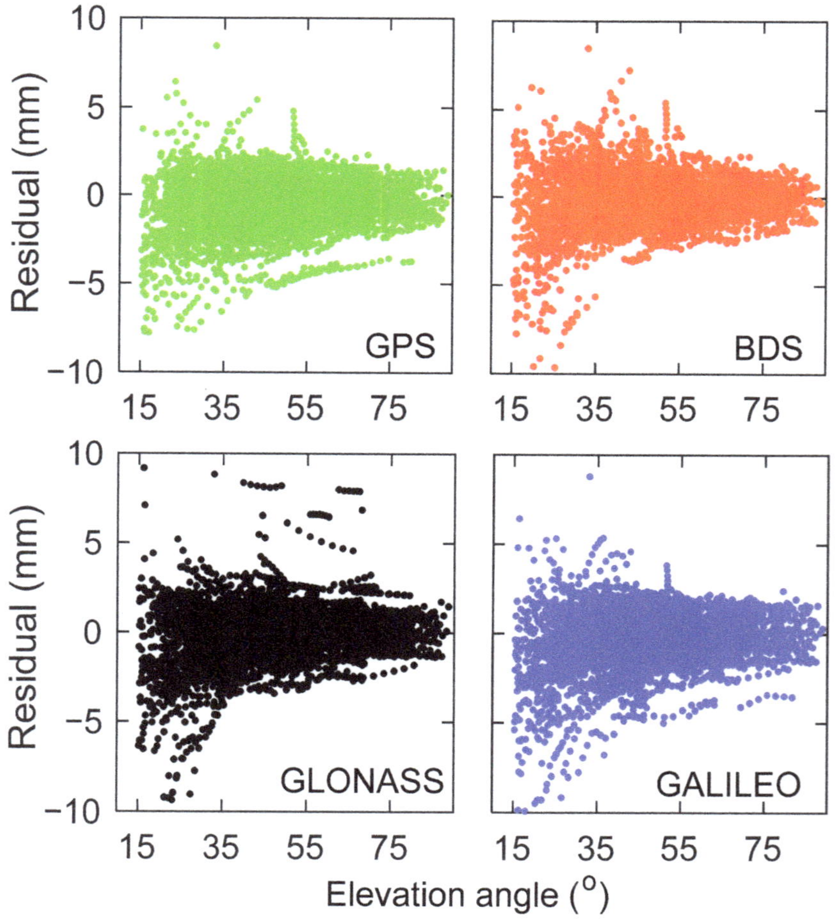

Figure 6. Scatter diagram of the change of SWV residuals with elevation angle for the four satellite systems.

To further assess their performance, SWV values were grouped into individual elevation bins of 5°, i.e., all SWVs with an elevation angle between 15 and 20° were evaluated as a single unit. Thus, the RMSE of each elevation bin for these 15 tomographic results was calculated and is shown in Figure 7. It can be seen from the left panel that the GLONASS and Galileo performance is not as good as the BDS and GPS at low elevation angles. As the elevation angle increases, their differences become very small. BDS achieved the best RMSE with a value of 1.59 mm, followed by GPS, Galileo, and GLONASS. In fact, the differences between these RMSEs are small and the values do not exceed 0.2 mm. Considering the magnitude range of SWV, these differences can be negligible. After the combinations, the RMSE of SWV residuals in each elevation bin were shown in the middle and right panels, which are the combination of two systems and multi systems, respectively. The RMSE difference of the SWV residuals for various combinations is relatively small in each elevation bin. Specifically, the RMSEs of whole SWV residuals are 1.66, 1.59, 1.75, 1.74, 1.68, 1.64, 1.67, 1.62, 1.63, 1.60, 1.59, 1.64, 1.65, 1.65, and 1.63 mm for G, C, R, E, GC, GR, GE, CR, CE, RE, GCR, GCE, GRE, CRE, and GCRE, respectively. Considering the magnitude range of SWV values, the differences of RMSE mentioned above not more than 0.2 mm could be negligible. Therefore, it is concluded that the tomographic results of different satellite systems and different combinations have little difference in SWV validation.

Radiosonde data are well suited as a reference to validate the accuracy of the water vapor tomography results, since they can provide a water vapor density profile with high precision based on the atmospheric parameters obtained at different altitudes. Figure 8 illustrates the water vapor density comparisons between radiosonde data and these 15 tomographic results for different altitudes on UTC 11:45–12:15, DOY 137, 2021, which is consistent with the time of tomographic solution shown in Figure 5. It is clear from the profiles that the water vapor density decreased with increasing height. The water vapor density profiles reconstructed by these 15 tomographic results conform with those derived from radiosonde data. From Figure 8, it is difficult to observe the difference in the water vapor density reconstructed by different satellite combinations. Therefore, the radiosonde comparison of 31 days from DOY 121 to 151, 2021 was conducted and the statistical results were listed in Table 2 to further illustrate their performances. From the mean value of RMSE, the difference between the WVD results reconstructed by single system tomography is 0.05 gm^{-3}, and BDS and GPS outperforms GLONASS and Galileo slightly. Compared with the single system, improvement can be observed from the WVD results reconstructed after the satellite system combination. The largest improvement appears from the Galileo with a RMSE of 1.46 gm^{-3} to the combination of GCR with a RMSE of 1.30 gm^{-3}. The number of satellite systems in the combination (two, three, or four satellite systems) did not present an obvious impact on the WVD results reconstructed by water vapor tomography.

Table 2. Statistical results of the water vapor density composition between radiosonde and tomographic results of different combinations during the experimental period.

Combinations	RMSE (gm^{-3}) Coverage Rate		
	Max	Min	Average
G	1.98	0.52	1.42
C	2.05	0.57	1.41
R	2.08	0.51	1.45
E	2.13	0.58	1.46
GR	1.92	0.53	1.33
GC	1.96	0.48	1.34
GE	2.04	0.61	1.41
CR	2.09	0.56	1.34
CE	1.95	0.57	1.36
RE	2.01	0.61	1.40
GCR	1.95	0.44	1.30
GEC	1.97	0.50	1.34
GRE	2.01	0.49	1.36
CRE	2.02	0.59	1.37
GCRE	1.97	0.53	1.32

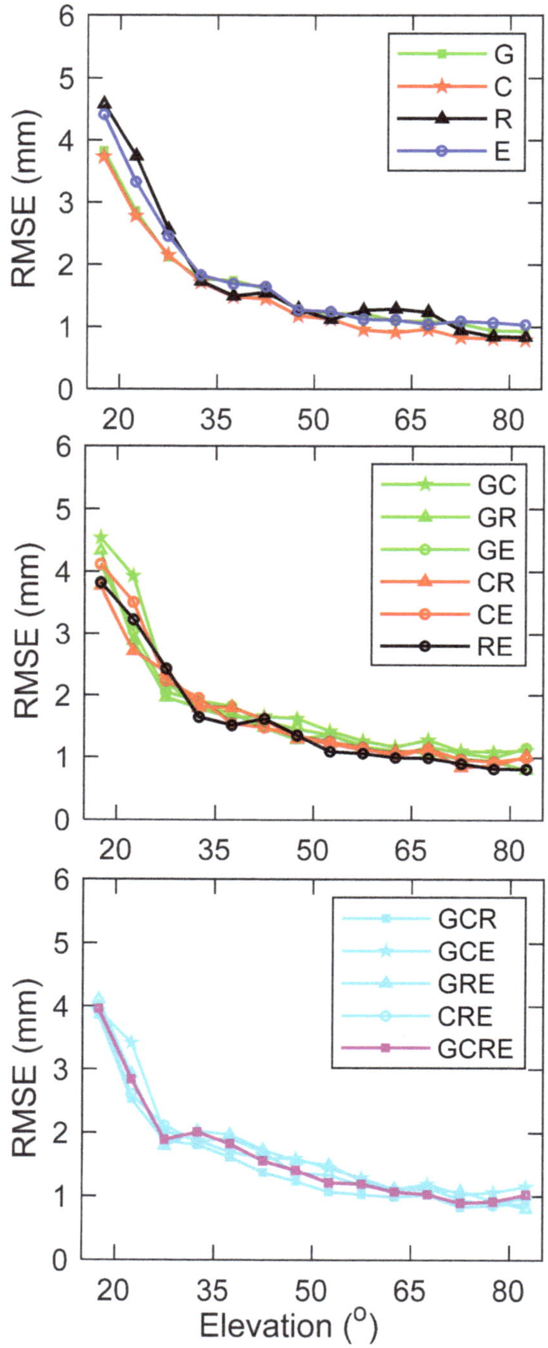

Figure 7. Comparison of SWV residuals in each elevation bin for various combinations.

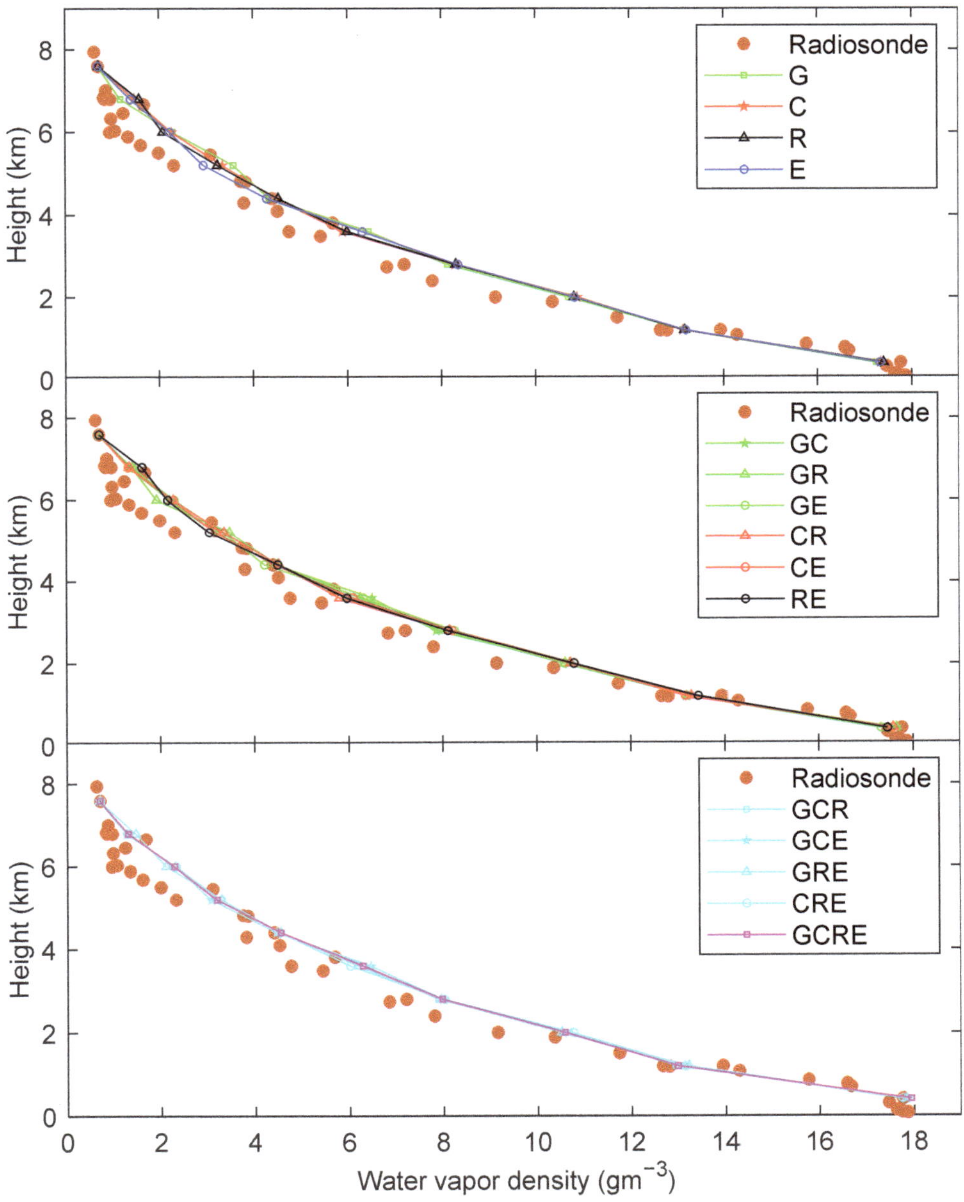

Figure 8. Water vapor density comparisons between radiosonde and 15 tomographic results.

5. Conclusions

In this paper, the performances of the four navigation satellite systems and their combinations in the water vapor tomography were analyzed and assessed using the GNSS data of SatRef in Hong Kong. In the tomographic modeling, the signal rays that can be used, the voxels crossed by signal rays, and the number and distribution of the effective voxels were computed and counted for these different combinations. For the tomographic results, the GAMIT-estimated SWV of HKQT and the water vapor density derived from radiosonde were selected as references to assess these 15 tomographic solutions.

In the experimental period, the average number of available signal rays was 507, 704, 329, and 351 for GPS, BDS, GLONASS, and Galileo, respectively. Combining satellite systems in water vapor tomography can increase the number of available signal rays, especially as the value of four-system combination reaches close to 2000. The average number of voxels crossed by signal rays are 425, 424, 392, and 377 for GPS, BDS, GLONASS, and Galileo, respectively, showing that the number of penetrated voxels is not entirely determined by the number of available signal rays. The combinations improved the number of voxels crossed by signal rays; for example, the number and coverage rate of penetrated voxels achieved by GCRE are 468 and 83.6%, respectively. When the voxels with sufficient signal rays are concerned, these 15 tomographic solutions differ very little in both number and coverage rate. The distribution diagram of effective voxels based on black and white rectangle also indicated the small differences in the 15 solutions. The numerical statistics showed that the RMSE in SWV comparison are 1.66, 1.59, 1.75, 1.74, 1.68, 1.64, 1.67, 1.62, 1.63, 1.60, 1.59, 1.64, 1.65, 1.65, and 1.63 mm for G, C, R, E, GC, GR, GE, CR, CE, RE, GCR, GCE, GRE, CRE, and GCRE, respectively. In the comparison with radiosonde data, the average RMSE are 1.42, 1.41, 1.45, 1.46, 1.33, 1.34, 1.41, 1.34, 1.36, 1.40, 1.30, 1.34, 1.36, 1.37, and 1.32 gm^{-3} for these 15 tomographic results. The above comparisons indicated that the differences in the tomographic results of a single satellite system are small, and the combinations of satellite systems have limited improvement in the water vapor tomography results.

In the follow-up research, the impact of different satellite systems and their combination on water vapor tomography need to be explored in more representative regions. In addition, the number, distribution, and density of GNSS stations in the research region is another important factor determining the structure of the tomographic model. Thus, it is necessary to pay more attention to the influence of the GNSS sites on water vapor tomographic results in the case of a determined satellite system.

Author Contributions: Conceptualization, F.Y.; data curation, J.W., H.W., X.G. and L.W.; funding acquisition, F.Y.; investigation, J.W.; methodology, F.Y.; resources, J.W., H.W., X.G. and L.W.; software, B.H.; validation, F.Y.; writing—original draft, F.Y.; writing—review and editing, F.Y., J.W., H.W., X.G., L.W. and B.H. All authors have read and agreed to the published version of the manuscript.

Funding: This research was funded by Beijing Natural Science Foundation (No. 8224093), China Postdoctoral Science Foundation (No. 2021M703510), Fundamental Research Funds for the Central Universities (No. 2021XJDC01), Key Laboratory of South China Sea Meteorological Disaster Prevention and Mitigation of Hainan Province (No. SCSF202109), Open Fund of State Key Laboratory of Geodesy and Earth's Dynamics Innovation Academy for Precision Measurement Science and Technology (No. SKLGED2022-3-1), and Beijing Key Laboratory of Urban Spatial Information Engineering (No. 2022017). We thank all anonymous reviewers for their valuable, constructive, and prompt comments.

Data Availability Statement: The datasets generated and analyzed during the current study are available from the corresponding author on reasonable request.

Acknowledgments: We thank the Hong Kong Geodetic Survey Services for providing the GNSS data of the Hong Kong Satellite Positioning Reference Station Network (SatRef). We also thank all the anonymous reviewers for their valuable, constructive, and prompt comments.

Conflicts of Interest: The authors declare no conflict of interest.

References

1. Bevis, M.; Businger, S.; Herring, T.; Rocken, C.; Anthes, R.; Ware, R. GPS meteorology: Remote sensing of atmospheric water vapor using the Global Positioning System. *J. Geophys. Res. Atmos.* **1992**, *97*, 15787–15801. [CrossRef]
2. Rocken, C.; Ware, R.; Hove, T.V.; Solheim, F.; Alber, C.; Johnson, J.; Bevis, M.; Businger, S. Sensing atmospheric water vapor with the global positioning system. *Geophys. Res. Lett.* **1993**, *20*, 2631–2634. [CrossRef]

3. Duan, J.; Bevis, M.; Fang, P.; Bock, Y.; Herring, T. GPS meteorology: Direct estimation of the absolute value of precipitable water. *J. Appl. Meteorol.* **1996**, *35*, 830–838. [CrossRef]
4. Braun, J.; Rocken, C.; Meetrens, C.; Ware, R. Development of a Water Vapor Tomography System Using Low Cost L1 GPS Receivers. In Proceedings of the 9th ARM Science Team Meeting, US Dep of Energy, San Antonio, TX, USA, 22–26 March 1999.
5. Flores, A.; Ruffini, G.; Rius, A. 4D tropospheric tomography using GPS slant wet delays. *Ann. Geophy.* **2000**, *18*, 223–234. [CrossRef]
6. Zhang, W.; Zhang, S.; Chang, G.; Ding, N.; Wang, X. A new hybrid observation GNSS tomography method combining the real and virtual inverted signals. *J. Geod.* **2021**, *95*, 128. [CrossRef]
7. Yang, F.; Guo, J.; Shi, J.; Meng, X.; Zhao, Y.; Zhou, L.; Zhang, D. A GPS water vapor tomography method based on a genetic algorithm. *Atmos. Meas. Tech.* **2020**, *13*, 355–371. [CrossRef]
8. Chen, B.; Liu, Z. Voxel-optimized regional water vapor tomography and comparison with radiosonde and numerical weather model. *J. Geod.* **2014**, *88*, 691–703. [CrossRef]
9. Yao, Y.; Zhao, Q. A novel, optimized approach of voxel division for water vapor tomography. *Meteorol. Atmos. Phys.* **2017**, *129*, 57–70. [CrossRef]
10. Trzcina, E.; Rohm, W. Estimation of 3D wet refractivity by tomography, combining GNSS and NWP data: First results from assimilation of wet refractivity into NWP. *Q. J. R. Meteorol. Soc.* **2019**, *145*, 1034–1051. [CrossRef]
11. Sa, A.; Rohm, W.; Fernandes, R.; Trzcina, E.; Bos, M.; Bento, F. Approach to leveraging real-time GNSS tomography usage. *J. Geod.* **2021**, *95*, 8. [CrossRef]
12. Adavi, Z.; Mashhadi-Hossainali, M. 4D tomographic reconstruction of the tropospheric wet refractivity using the concept of virtual reference station, case study: Northwest of Iran. *Meteorol. Atmos. Phys.* **2014**, *126*, 193–205. [CrossRef]
13. Yang, F.; Guo, J.; Shi, J.; Zhou, L.; Xu, Y.; Chen, M. A method to improve the distribution of observations in GNSS water vapor tomography. *Sensors* **2018**, *18*, 2526. [CrossRef] [PubMed]
14. Zhao, Q.; Yao, Y. An improved troposphere tomographic approach considering the signals coming from the side face of the tomographic area. *Ann. Geophy.* **2017**, *35*, 87–95. [CrossRef]
15. Zhang, W.; Zhang, S.; Ding, N.; Zhao, Q. A Tropospheric Tomography Method with a Novel Height Factor Model Including Two Parts: Isotropic and Anisotropic Height Factors. *Remote Sens.* **2020**, *12*, 1848. [CrossRef]
16. Hu, P.; Huang, G.; Zhang, Q.; Yan, X.; Li, Z. A multi-GNSS water vapor tomography method considering boundary incident signals. *Acta Geod. Et Cartogr. Sin.* **2020**, *49*, 557–568.
17. Zhao, Q.; Yao, Y.; Yao, W. A method to establish the tomography model considering the data of GNSS stations outside the research are. *Acta Geod. Et Cartogr. Sin.* **2021**, *50*, 285–294.
18. Zhao, Q.; Zhang, K.; Yao, Y.; Li, X. A new troposphere tomography algorithm with a truncation factor model (TFM) for GNSS networks. *GPS Solut.* **2019**, *23*, 64. [CrossRef]
19. Ye, S.; Xia, P.; Cai, C. Optimization of GPS water vapor tomography technique with radiosonde and cosmic historical data. *Ann. Geophy.* **2016**, *34*, 789–799. [CrossRef]
20. Shafei, M.; Hossainali, M. Application of the GPS reflected signals in tomographic reconstruction of the wet refractivity in Italy. *J. Atmos. Sol.-Terr. Phys.* **2020**, *207*, 105348. [CrossRef]
21. Benevides, P.; Nico, G.; Catalao, J.; Miranda, P. Bridging InSAR and GPS tomography: A new differential geometrical constraint. *IEEE Trans. Geosci. Remote Sens.* **2016**, *54*, 697–702. [CrossRef]
22. Heublein, M.; Alshawaf, F.; Erdnub, B.; Zhu, X.; Hinz, S. Compressive sensing reconstruction of 3D wet refractivity based on GNSS and InSAR observations. *J. Geod.* **2019**, *93*, 197–217. [CrossRef]
23. Haji-Aghajany, S.; Amerian, Y.; Verhagen, S.; Rohm, W.; Ma, H. An optimal troposphere tomography technique using the WRF model outputs and topography of the area. *Remote Sens.* **2020**, *12*, 1442. [CrossRef]
24. Xiong, S.; Ma, F.; Ren, X.; Chen, J.; Zhang, X. LEO Constellation-Augmented Multi-GNSS for 3D Water Vapor Tomography. *Remote Sens.* **2021**, *13*, 3056. [CrossRef]
25. Zhang, W.; Zheng, N.; Zhang, S.; Ding, N.; Qi, M.; Wang, H. GNSS water vapor tomography algorithm constrained with high horizontal resolution PWV data. *Geomat. Inf. Sci. Wuhan Univ.* **2021**, *46*, 1627–1635.
26. Zhang, W.; Zhang, S.; Zheng, N. A new integrated method of GNSS and MODIS measurements for tropospheric water vapor tomography. *GPS Solut.* **2021**, *25*, 79. [CrossRef]
27. Bender, M.; Dick, G.; Ge, M.; Deng, Z.; Wickert, J.; Kahle, H.; Raabe, A.; Tetzlaff, G. Development of a GNSS water vapour tomography system using algebraic reconstruction techniques. *Adv. Space Res.* **2011**, *47*, 1704–1720. [CrossRef]
28. Wang, X.; Wang, X.; Dai, Z.; Ke, F.; Cao, Y. Tropospheric wet refractivity tomography based on the BeiDou satellite system. *Adv. Atmos. Sci.* **2014**, *31*, 355–362. [CrossRef]
29. Xia, P.; Ye, S.; Jiang, P. Research on three-dimensional water vapor tomography using GPS/GLONASS PPP method. *J. Geod. Geody.* **2015**, *35*, 72–76.
30. Benevides, P.; Nico, G.; Catalao, J.; Miranda, P. Analysis of Galileo and GPS integration for GNSS tomography. *IEEE Trans. Geosci. Remote Sens.* **2017**, *55*, 1936–1943. [CrossRef]
31. Dong, Z.; Jin, S. 3D water vapor tomography in Wuhan from GPS, BDS and GLONASS observations. *Remote Sens.* **2018**, *10*, 62. [CrossRef]

32. Zhao, Q.; Yao, Y.; Cao, X. W Accuracy and reliability of tropospheric wet refractivity tomography with GPS, BDS, and GLONASS observations. *Adv. Space Res.* **2018**, *63*, 2836–2847. [CrossRef]
33. Adavi, Z.; Mashhadi-Hossainali, M. 4D-tomographic reconstruction of water vapor using the hybrid regularization technique with application to the North West of Iran. *Adv. Space Res.* **2015**, *55*, 1845–1854. [CrossRef]
34. Yang, F.; Guo, J.; Meng, X.; Shi, J.; Zhang, D.; Zhao, Y. An improved weighted mean temperature (Tm) model based on GPT2w with Tm lapse rate. *GPS Solut.* **2020**, *24*, 46. [CrossRef]
35. Huang, L.; Jiang, W.; Liu, L.; Chen, H.; Ye, S. A new global grid model for the determination of atmospheric weighted mean temperature in GPS precipitable water vapor. *J. Geod.* **2019**, *93*, 159–176. [CrossRef]
36. Bevis, M.; Businger, S.; Chiswell, S. GPS meteorology: Mapping zenith wet delays onto precipitable water. *J. Appl. Meteorol.* **1994**, *33*, 379–386. [CrossRef]
37. Zhang, B.; Fan, Q.; Yao, Y.; Xu, C.; Li, X. An Improved Tomography Approach Based on Adaptive Smoothing and Ground Meteorological Observations. *Remote Sens.* **2017**, *9*, 886. [CrossRef]
38. Yang, F.; Meng, X.; Guo, J.; Yuan, D.; Chen, M. Development and evaluation of the refined zenith tropospheric delay (ZTD) models. *Satell. Navig.* **2021**, *2*, 21. [CrossRef]
39. Saastamoinen, J. Atmospheric correction for the troposphere and stratosphere in radio ranging satellites. *Use Artif. Satell. Geod.* **1972**, *15*, 247–251.
40. Elosegui, P.; Rius, A.; Davis, J.; Ruffini, G.; Keihm, S.; Burki, B.; Kruse, L. An experiment for estimation of the spatial and temporal variations of water vapor using GPS data. *Phy. Chem. Earth Parts A/B/C* **1998**, *23*, 125–130. [CrossRef]
41. Guo, J.; Yang, F.; Shi, J.; Xu, C. An optimal weighting method of Global Position System (GPS) Troposphere Tomography. *IEEE J. Sel. Top Appl. Earth Obs. Remote Sens.* **2016**, *9*, 5880–5887. [CrossRef]

remote sensing

MDPI

Communication

Phase Centre Corrections of GNSS Antennas and Their Consistency with ATX Catalogues

Lukasz Borowski [1], Jacek Kudrys [2,3], Bartosz Kubicki [1], Martina Slámová [4] and Kamil Maciuk [2,3,*]

1 Department of Construction Materials Engineering and Geoengineering, Faculty of Civil Engineering and Architecture, Lublin University of Technology, 20-618 Lublin, Poland; l.borowski@pollub.pl (L.B.); b.kubicki@pollub.pl (B.K.)
2 Department of Integrated Geodesy and Cartography, AGH University of Science and Technology, 30-059 Krakow, Poland; jkudrys@agh.edu.pl
3 Joint International Tourism College, Hainan University—Arizona State University, Haikou 570228, China
4 UNESCO Department for Ecological Awareness and Sustainable Development, Faculty of Ecology and Environmental Sciences, Technical University in Zvolen, 960 01 Zvolen, Slovakia; martina.slamova@tuzvo.sk
* Correspondence: maciuk@agh.edu.pl

Abstract: Changes of the antenna models on permanent global navigation satellite system (GNSS) stations can lead to jumps and discontinuities in the coordinate time series. In this paper, the results of research on the adequacy of the antenna phase centre corrections (PCC) variations are presented by analysing its component—the antennas' phase centre offset (PCO). For this purpose, height differences were determined using different and independent methods: EUREF Permanent Network (EPN) combined solutions, Precise Point Positioning (PPP), and the single baseline solution. The results of GNSS processing were referenced to direct geometric levelling outputs. The research was conducted only within the global positioning system (GPS) system due to the compatibility of one of the receivers, and the experiment was based on a comparison of the height differences between four GNSS antennas located on the roof of a building: two permanent station antennas and two auxiliary points. The antennas were located at similar heights; precise height differences were determined by geometric levelling, both at the beginning and the end of the session. Post-processing was conducted with the use of the GPS system, precise ephemeris, the adopted antenna correction model, and a zero-elevation mask. For one of the antennas, a change of the antenna characteristic model from IGS08 to IGS14 leads to an 8-mm difference in height. Older antennas used in the national (or transnational) permanent network need individual PCC.

Keywords: PCC; GNSS; GPS; antenna; calibration; EPN

Citation: Borowski, L.; Kudrys, J.; Kubicki, B.; Slámová, M.; Maciuk, K. Phase Centre Corrections of GNSS Antennas and Their Consistency with ATX Catalogues. *Remote Sens.* **2022**, *14*, 3226. https://doi.org/10.3390/rs14133226

Academic Editors: José Fernández, Juan F. Prieto and Serdjo Kos

Received: 6 June 2022
Accepted: 2 July 2022
Published: 5 July 2022

1. Introduction

The correct calibration of global navigation satellite system (GNSS) antennas is the basis for precise satellite positioning and plays a key role in many aspects of positioning and navigation [1,2]. In general, GNSS antennas have a well-defined antenna phase centre (APC, Figure 1) and should be described by the azimuth (α) and elevation (β) functions [3]. The APC determined by phase centre variation (PCV) is a shift in position depending on the observed elevation angle and azimuth with regards to the satellite [4]. The antenna reference point (ARP) is a physical point where the antenna height above the physical point is measured [5]. The antenna phase centre offset (PCO) is the difference between the ARP and the mean electrical APC defined by the intersection of the vertical antenna axis of symmetry with the bottom of the antenna [6]. The combination of PCO and PCV is called phase centre corrections (PCC).

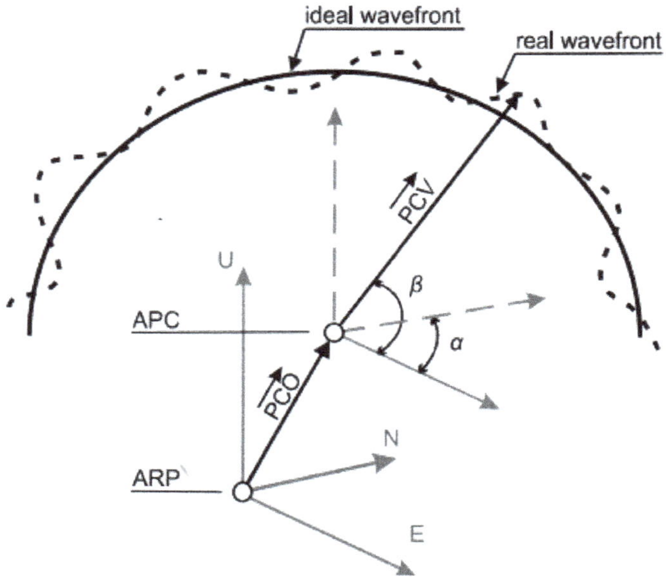

Figure 1. A phase shift due to antenna model change.

Depending on the accuracy and quality of solutions needed to be achieved, absolute and relative antenna calibration models are used. For many years, relative phase centre corrections were made based on measurements on a short baseline referred to as a reference antenna (The Dorne Margolin T of Allen Osborne Associates antenna—AOAD/M_T), with the arbitrary assumption that the PCV of the reference antenna is zero [7,8]. Absolute correction for an antenna can be obtained either by measurement in an anechoic chamber [9] or by field measurements on a short baseline using a robot mount [10].

As the research shows, the problem of the antenna calibrations is not new, but it still experiences issues with proper assumptions regarding calibrations and the correctness of calculations. Different antenna calibration models impact, among other things, estimated station positions. Some empirical research tests show a relative vertical offset exists with the same model of antennas of up to 1 mm, while horizontal offsets are negligible [11]. The research conducted by Baire et al. showed agreement to the level of 2 mm in the case of horizontal offsets and 5 mm in the case of the height component based on daily static Precise Point Positioning (PPP) sessions [3]. A variety of studies have been conducted for the establishment of new calibration methods, e.g., [11–14]. The differences in model calibrations between the robot and anechoic chamber have an impact on the coordinates: up to +/− 2 mm for the horizontal and +/− 10 mm for the height component [15]. Over 10 mm of the height component difference occurs when individual antenna corrections by GEO++ are comparable to igs08.atx [16]. Moreover, in specific conditions, it is worth considering the impact of other environmental phenomena, such as multipath interference [17,18], antenna mounting [19,20], the near-field effect, or seasonal snow cover [21]. Analyses performed for EUREF permanent network (EPN) stations show that the change to the new catalogue of absolute calibration models (from IGS08.ATX to IGS14.ATX) might have an impact on stations' coordinates. Phase centre corrections (PCC) are an integral part of high accuracy GNSS applications [22]. The research showed on more than 40 PCC models that the magnitudes of the PCO shifts among different versions of the antenna files (Antex) were above 1 mm [23]. The differences between the antenna model (mean) and the individual PCC may cause a discrepancy of 10 mm for the horizontal and vertical components, but this does not usually exceed 2 mm for the horizontal and 4 mm for the vertical ones [24].

Based on this knowledge, an analysis of the selected antenna models, TRM57971.00 NONE and ASH70195C_M SNOW, was performed. These antennas are used on EPN's KRA1 (TRM57971.00 NONE) and KRAW (ASH70195C_M SNOW) stations (Kraków, Poland). The stations are located close to each other (3.5 m), which is a convenient situation for the verification of their individual characteristics by comparison to the catalogue ones. The study consisted of several analyses. The first was focused on differences of the studied models in the igs08.atx and igs14.atx catalogues. The second was the difference in height between the antenna ARP, as calculated by different satellite methods. Additionally, height differences were referenced to the auxiliary stations (KR01 and KR02) and to precise spirit levelling. The third analysis gathered archival material based on several experiments that have been conducted since 2011.

2. Methods

The paper is focused on the changes of the ellipsoid height, which are measured to the ARP. The height and the horizontal component are revealed in the current international terrestrial reference frame (ITRF2014). Along with the reference frame change, GNSS receiver antenna calibration models are also changed to be consistent with the current realisation of the frame. In the epoch under analysis, the International GNSS Service (IGS) provided data on 316 GNSS permanent stations. After the introduction of the ITRF2014, 113 of these changed their reported heights due to the antenna calibration parameter changes. The values of the estimated height offset are shown in Figure 2, which are based on IGS data.

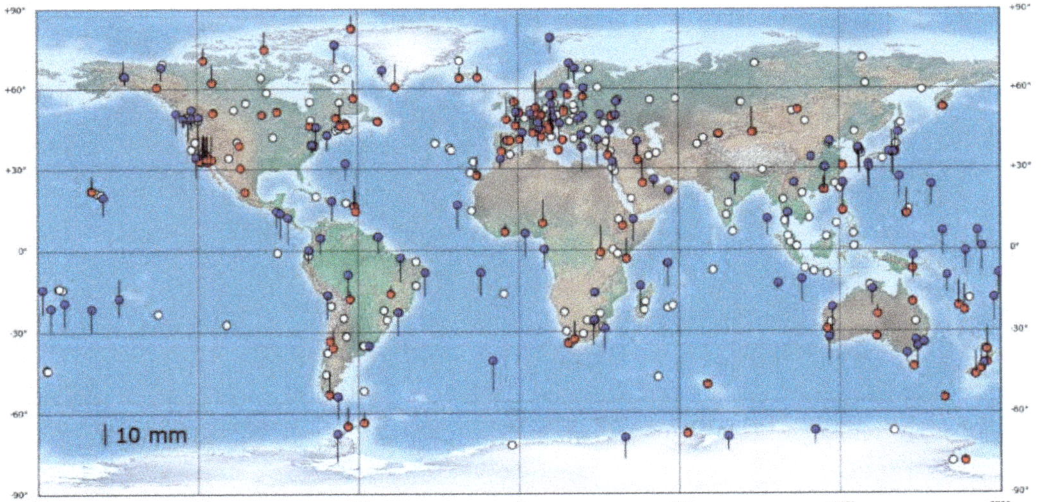

Figure 2. Height offset due to antenna model changes; red marks indicate positive height changes, blue marks indicate negative changes, and white indicates that no changes occurred. The 'whiskers' show the amount of height offset.

As an example, the difference between the I08.ATX and I14.ATX calibration models for the TRM57971.00 NONE antenna is shown in Figure 3. The −1.8-mm PCO difference for the G01 (GPS L1) frequency and high PCV elevation dependent differences for the R01 (GLONASS L1) frequency are clearly seen.

Figure 4 shows PCV variations of the TRM57971.00 NONE antenna between the ATX08 and ATX14 antenna models for two (G01 and G02) GPS signal frequencies. Azimuth elevation differences vary from −1.5 mm to 1.5 mm. The largest differences are in the northwest vs. southeast directions, which are up to 1.5 mm, while, in the perpendicular direction (northeast vs. southwest), the situation is the opposite, with a difference of up to −1.5 mm. Such maximum differences appear only for the nearest-to-horizon satellites, which have the smallest impact on the GNSS processing due to weighting. Differences for satellites with the highest elevation (located in the centre of the circles) are close to 0.

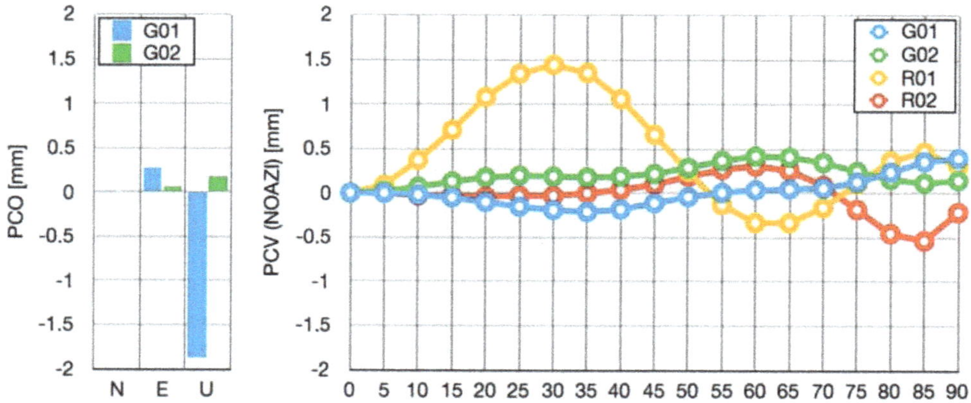

Figure 3. PCO and PCV difference between I08.ATX and I14.ATX calibration data for the TRM57971.00 NONE antenna.

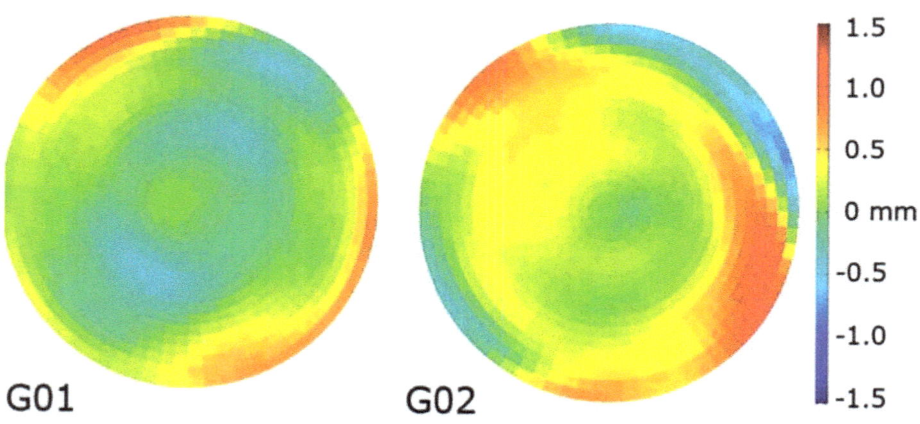

Figure 4. PCV variations between ATX08 and ATX14 for two GPS signal frequencies for the TRM57971.00 NONE antenna.

In general, changes in the station heights mean that the antenna model changes or suggests an inaccurate calibration model of the antenna's PCO or/and PCV. Detailed research has been performed for the ASH701945_C antenna. This type of the antenna, with various domes, is still used at stations including IENG, KELY (NONE), BOGI, KRAW (SNOW), and SKEO (OSOD). The ASH701945C_M SNOW antenna was calibrated by Geo++ GmbH on 25 March 2011 with the use of a single antenna [4]. Six others had individual calibrations, not KRAW or BOGI. It is possible that the antenna used at the KRAW station has distinctive characteristics other than those determined by Geo++. To check on this, an experiment on the KRAW antenna was conducted. In contrast to robot or anechoic chamber experiments, the authors used an analysis of the short baselines between the different permanent (KRAW and KRA1) and auxiliary stations (KR01 and KR02).

3. KRAW and KRA1 Stations

The pair of antennas KRAW and KRA1 are located on a five-storey university building about 3.5 m away from each other. This is an unusual circumstance, which is beneficial for GNSS networks [25], but very few permanent reference stations are located in this way. The oldest station KRAW has continuously collected data since the end of 2002, while the younger KRA1 has been collecting data since 2010 [26]. In 2005, the characteristics of the gravity field (acceleration and gradient) were measured for the KRAW station, and a station normal height was determined [27]. Up to 2018 to 2019, the station's equipment did not change (brand and model). The KRAW used an Ashtech UZ-12 receiver with an ASH70195C_M SNOW antenna. In 2019, a new receiver was introduced, a LEICA RX1200 GGPRO (without changing the antenna). The KRA1 uses a Trimble NETR5 receiver with a TRM57971.00 NONE antenna. In 2018, the KRA1 antenna was replaced by another TRM57971.00 NONE model with individual PCC. For the KRAW station, the ASH70195C_M SNOW antenna was not calibrated individually.

During the experiments on quasi-geoid modelling, it was found that the antenna's calibrations might be inaccurate [28]. It was suspected that the ellipsoidal height difference between the stations is of the opposite sign to their coordinates in the EPN network [29]. To verify this situation, a test survey was organised. It was based on a comparison of the height differences between the stations (KRAW and KRA1) and temporary stations KR01 and KR02 (with TRM57971.00 NONE antennas), as well as an analysis of the archival data. The Trimble antennas at KR01 and KR02 are commonly used in GNSS permanent networks. Exactly the same type was tested at Wuhan University by the absolute method (the field robot); the test showed the high repeatability of the obtained calibrations: $\sigma_N = 0.17$ mm, $\sigma_E = 0.12$ mm, and $\sigma_U = 0.30$ mm [12]. The same model of antennas on KRA1, KR01, and KR02 guaranteed that the impact of the differences between the antennas would not be significant. The only difference was the power supply system (different model). The antennas were located at similar height levels, with a distance of about 3.5 m from each other (Figure 5).

Figure 5. Location of the KRAW and KRA1 permanent stations on the roof of the AGH-UST buildings with the temporary KR01 and KR02 stations.

The observation sessions took place over 12 days in December 2016. The precise differences in height were defined by GPS post-processing. As the height difference reference value, the results of geometric levelling (Zeiss KoNi 007) were used. The levelling was conducted both at the beginning and the end of the GPS session.

4. Results

The difference in heights between KRAW, KRA1, KR01, and KR02 was calculated by post-processing of the observations. Due to the KRAW Ashtech UZ-12 receiver capability, only the GPS system was used. Post-processing was based on the final IGS orbits, the igs08.atx antenna corrections model, and a zero-elevation mask. Height differences obtained from precise levelling were compared with those derived from EPN-combined solutions, PPP calculated with Bernese GNSS Software (BGS) [30], Nevada Geodetic Laboratory (NGL, [31]), a network solution, a single baseline solution (GNSS Solutions 3.80.8, Trimble Business Centre 3.50), and other archive materials. In GNSS Solutions, only the elevation-dependent PCV version of the antenna characteristics was used.

Height differences obtained from precise levelling and GNSS solutions showed several discrepancies. Height differences between the KRA1 antenna reference point and the three other measured stations (KRAW, KR01, and KR02) from GNSS post-processing are affected by the inaccuracy of the antenna PCO and PCV. The analysis shows that the height of the KRAW station antenna (ASH70195C_M SNOW) from GNSS processing (EPN combined and PPP solutions) is 16 mm too high in comparison to the levelling results (Figure 6). This may indicate the inadequacy of the ASH70195C_M SNOW calibration data. A lower disagreement was obtained in the case of the TRM57971.00 NONE antenna (KR01 and KR02 stations). This discrepancy was about 5 mm for PPP solutions and 2 mm for network solutions. The differences obtained by single baseline solutions using the Trimble Business Centre (TBC) and GNSS Solution engineering software show far fewer discrepancies, not exceeding 0.4 mm. This is due to the very short baseline between the measured points.

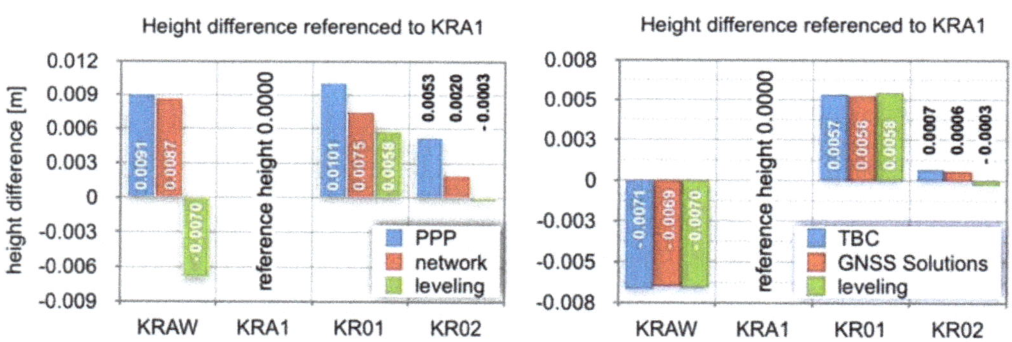

Figure 6. Differences between the heights obtained with PPP and a network solution (**left**) and single baseline solutions (**right**) [32].

The height difference of KRAW–KRA1 was calculated based on the EPN weekly combined SINEX (Software INdependent Exchange format) solutions from 2010 to 2016 (1564 to 1877 GPS weeks, Figure 7). During weeks when igs08 and igb08 were in use, there were some inaccurate solutions, which may be treated as gross errors. Despite that, clearly visible is the change in ellipsoidal height difference after the introduction of the IGS14 system (after GPS week 1933 [33]). The offset is up to about 8 mm (Figure 7). It is surprising that, during the short time between the introduction of the IGS14 system and individual phase corrections for the KRA1 antenna (week 1986), the height difference was closest to levelling (−0.5 mm) from any network calculation.

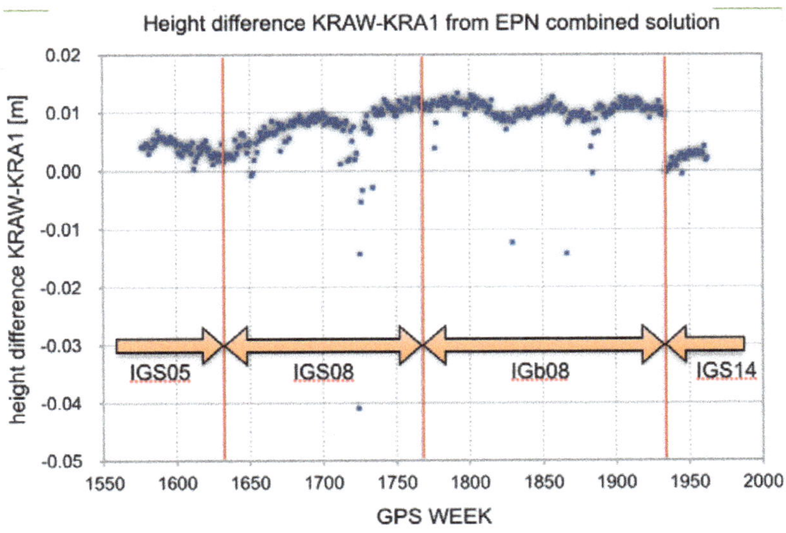

Figure 7. The time series for the height difference between KRAW–KRA1 calculated by EPN on the basis of several antenna correction catalogues (IGS ATX files) [32].

Alternatively, an independent PPP calculation was used—the Nevada Geodetic Laboratory (NGL). The facility provides permanent, fully, and freely available PPP results based on globally distributed permanent GNSS station data. Based on these data, the ellipsoidal height differences between station KRAW and KRA1 for the period 2009.7–2022.0 GPS weeks 1569–2190) were analysed (Figure 8). The results showed a mean difference +1.83 mm (green line) and were compared to precise geometric levelling −6.2 mm (red line). Both data provided inaccurate results, which was not a surprise, as they were based on the different antenna characteristics (IGS Antex files) of the KRAW antenna.

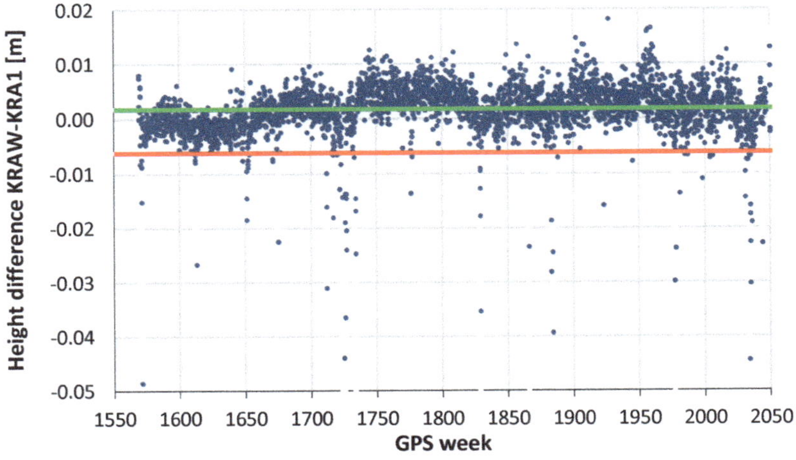

Figure 8. Height difference between KRA1–KRAW from NGL–PPP (navy dots), systematic difference from levelling −6.2 mm (red line), and mean difference 1.83 mm (green line) [31].

The archival materials regarding the height differences between the KRAW–KRA1 stations consist of: two independent sets of post-processing results from 2008 to 2011 calculated during the Polish integration of the base geodetic network with reference permanent stations ASG-EUPOS (BGS), EPN combined solutions 2013–2014, PPP (BGS) 2012–2014, and the post-processing of various short GPS observations (Trimble Business Centre 3.50—TBC) 2011–2016 (Table 1). The short GPS observations consist of various 4-h, 8-h, and 12-h sessions, the calculations of which were also checked in GNSS Solutions 3.80.3 and Geonet 2006 software [29]. The differences between them are negligible, which allows us to show a single (mean) value achieved by the most often-used software. The compatibility of the short observation post-processing (TBC, GNSS Solutions, or Geonet 2006) and spirit levelling (−6.2 mm) results is noticeable when the more sophisticated solutions (e.g., PPP network solutions) give opposite values. The most contrary are values from PPP (BGS) 2012–14; the differences in the levelling result were as much as 20 mm. All of them used IGS08 antenna characteristics or only the elevation-dependent PCV model (GNSS Solutions). Surprisingly, such popular "engineering" software (e.g., TBC) achieved better results, comparable to spirit levelling, than the BGS (both short baseline calculations and PPP). The origin of the advantage of this type of software is not clear. It is probably that, for very close stations, the single baseline solutions, based only on the L1 frequency, give better results.

Table 1. Archive values of differences between the KRA1–KRAW stations.

Archive Data	Δh (mm)	σ (mm)
Warsaw University of Technology 2008–2011 (BGS) [34]	5.5	-
PAN Space Research Centre 2008–2011 (BGS) [35]	7.3	-
EPN Combined Solution 2013–2014 [29]	11.0	1.5
PPP (BGS) 2012–2014 [29]	14.7	-
Trimble Business Centre 2011–2016 [29]	−6.9	0.5
MUT Combined Solutions igs14.atx 2017 [36]	−0.5	2.2
MUT * Combined Solutions igs14.atx 2018–2022 (KRA1—individual PCC) [36]	1.1	2.7

* Military University of Technology Analysis Coordination Centre.

Finally, the results of the height difference between the KRAW–KRA1 station were estimated by geometrical levelling, using a series of observations from 2011 to 2022 (Figure 9). Its most probable value is −6.2 mm ± 0.2 mm (Figure 10).

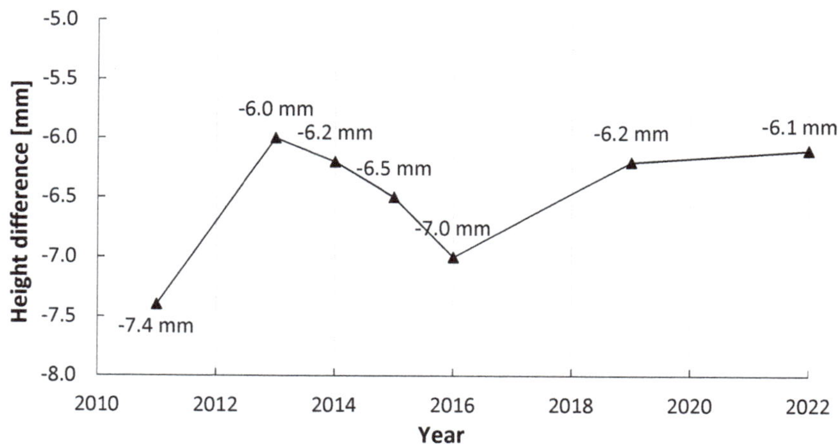

Figure 9. Height difference between KRA1 and KRAW from the levelling from 2011 to 2022.

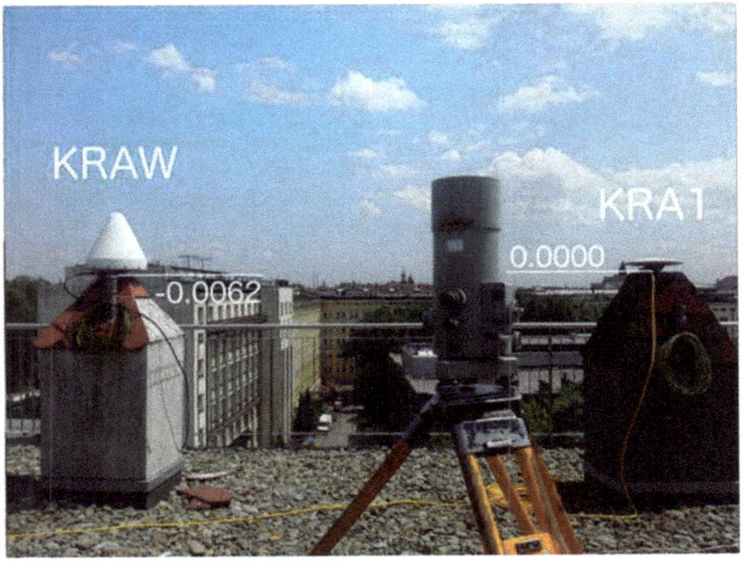

Figure 10. Height difference (m) between the KRAW and KRA1 antenna reference points from precise levelling.

5. Discussion

Modern manufactured antennas are of better repeatability than older models, i.e., exhibiting less deviation from the pattern (average) in antex files. This is the case for, e.g., the Trimble Zephyr Geodetic 2 (TRM57971.00 NONE) antenna used in the experiment. In selected stations (e.g., KRAW and BOGI), antennas are still being used that date back to the 1990s. Their characteristics may differ from the accepted standards. Therefore, it can be postulated that individual characteristics should be developed for them, so that they do not introduce errors to the network in which they work (IGS, EPN, etc.). However, their individual calibrations are a matter of debate due to costs.

We checked the KRAW station based on a KRA1–KRAW overpass survey and the additional temporary stations KR01 and KR02. Archival materials from various authors were also collected. As a result of the analysis of these materials, it can be stated that the antenna (Ashtech ASH70195C_M SNOW) of the KRAW station has different characteristics from those revealed in the IGS files of the subsequent versions. Azimuth elevation is mostly dependent on the lowest-over-horizon satellites [4,37], which have the smallest weights in processing. Thus, this aspect has the smallest impact. PPP processing shows a systematic shift in the KRAW–KRA1 height difference at the level of +1.8 mm, while measurements directly from levelling show a −6.2-mm difference. This may result in incorrect altitude results and have an adverse effect on, e.g., fitting the global geopotential model into the national height frame. Therefore, old stations that are included in the national (or transnational) permanent network need individual PCC. The tested network solutions, both PPP and baseline solutions [34,35], were not resistant to the impact of real KRAW antenna characteristics, different from any analysed ATX catalogue. As a result, the permanent stations' ellipsoid heights in network solutions may be subject to systematic errors. Even using advanced and renowned software (Bernese GNSS Solutions), such errors cannot be detected. The analysis showed that a calculation strategy based on a short baseline L1-only solution may detect the error, even if the popular "engineering" software is used (Trimble Business Centres, GNSS Solutions, Geonet).

The question arises as to what the optimal solution for old antennas is—the determination of individual characteristics (PCC) or replacement with a new antenna (also with individual PCC). In other words, should the old stations be partially or fully modernized? For the analysed station KRAW (antenna ASH70195C_M SNOW), a full upgrade had the advantage of introducing four GNSS systems compared to the existing GPS-only. Especially, introduction of the Galileo system may benefit the station solution [38,39], e.g., decreasing the positioning error statistics for GPS + Galileo combinations [40] despite Galileo still not providing the same satellite availability as GPS [41]. Concluding the results and achievements in the investigation of PCC by other authors [16,19,21,24], we are inclined to say that the time of antennas from the 1990s should be completed and full modernization carried out, despite the higher costs.

Author Contributions: Conceptualization, L.B.; methodology, L.B. and J.K.; software, L.B. and J.K.; validation, L.B., J.K. and K.M.; formal analysis, L.B., J.K. and K.M.; investigation, L.B., J.K. and K.M.; resources, L.B. and J.K.; data curation, L.B. and J.K.; writing—original draft preparation, L.B., J.K. and K.M.; writing—review and editing, L.B., J.K., K.M., M.S. and B.K.; visualization, L.B., J.K., B.K. and K.M.; supervision, L.B., J.K., K.M. and M.S.; project administration, L.B.; and funding acquisition, L.B., J.K., M.S., B.K. and K.M. Percentage contributions: L.B., 60%; J.K., 15%; B.K., 5%; M.S., 5%; and K.M., 15%. All authors have read and agreed to the published version of the manuscript.

Funding: This work was funded by the Initiative for Excellence Research University grant from the AGH University of Science and Technology, by research subvention no. 16.16.150.545, and by the National Science Centre as part of MINIATURA 5, application no. 2021/05/X/ST10/00058. This work was supported by the grant VEGA 1/0736/21 of the Scientific Grant Agency of the Ministry of Education Science Research and Sport of the Slovak Republic and the Slovak Academy of Sciences.

Data Availability Statement: Not applicable.

Conflicts of Interest: The authors declare no conflict of interest.

References

1. Specht, M.; Specht, C.; Dąbrowski, P.; Czaplewski, K.; Smolarek, L.; Lewicka, O. Road tests of the positioning accuracy of INS/GNSS systems based on MEMS technology for navigating railway vehicles. *Energies* **2020**, *13*, 4463. [CrossRef]
2. Specht, C.; Specht, M.; Dąbrowski, P. Comparative analysis of active geodetic networks in poland. In Proceedings of the 17th International Multidisciplinary Scientific GeoConference SGEM 2017, Vienna, Austria, 27–29 November 2017.
3. Baire, Q.; Bruyninx, C.; Legrand, J.; Pottiaux, E.; Aerts, W.; Defraigne, P.; Bergeot, N.; Chevalier, J.M. Influence of different GPS receiver antenna calibration models on geodetic positioning. *GPS Solut.* **2014**, *18*, 529–539. [CrossRef]
4. Dawidowicz, K. Comparison of Using Relative and Absolute PCV Corrections in Short Baseline GNSS Observation Processing. *Artif. Satell.* **2011**, *46*, 19–31. [CrossRef]
5. Hofmann-Wellenhof, B.; Lichtenegger, H.; Wasle, E. *GNSS—Global Navigation Satellite Systems*; Springer: Vienna, Austria, 2008; ISBN 978-3-211-73012-6.
6. Luo, X. Mathematical Models for GPS Positioning. In *GPS Stochastic Modelling*; Springer Theses; Springer: Berlin/Heidelberg, Germany, 2013; pp. 55–116; ISBN 978-3-642-34835-8.
7. Breuer, B.; Campbell, J.; Görres, B.; Hawig, R.; Wohlleben, R. Kalibrierung von GPS-Antennen für hochgenaue geodätische Anwendungen. *Z. Satell. Position. Navig. Kommun.* **1995**, *2*, 49–59.
8. Schmid, R.; Dach, R.; Collilieux, X.; Jäggi, A.; Schmitz, M.; Dilssner, F. Absolute IGS antenna phase center model igs08.atx: Status and potential improvements. *J. Geod.* **2016**, *90*, 343–364. [CrossRef]
9. Schupler, B.R.; Clark, T.A. How different antennas affect the GPS observable. *GPS World* **1991**, *2*, 32–36.
10. Görres, B.; Campbell, J.; Becker, M.; Siemes, M. Absolute calibration of GPS antennas: Laboratory results and comparison with field and robot techniques. *GPS Solut.* **2006**, *10*, 136–145. [CrossRef]
11. Daneshmand, S.; Sokhandan, N.; Zaeri-Amirani, M.; Lachapelle, G. Precise calibration of a GNSS antenna array for adaptive beamforming applications. *Sensors* **2014**, *14*, 9669–9691. [CrossRef]
12. Hu, Z.; Zhao, Q.; Chen, G.; Wang, G.; Dai, Z.; Li, T. First Results of Field Absolute Calibration of the GPS Receiver Antenna at Wuhan University. *Sensors* **2015**, *15*, 28717–28731. [CrossRef]
13. Filler, V.; Kostelecký, J. GPS antenna calibration at the Geodetic Observatory Pecny, Czech Republic. *Rep. Geod. Geoinform.* **2001**, *1*, 233–244.
14. Bergstrand, S.; Jarlemark, P.; Herbertsson, M. Quantifying errors in GNSS antenna calibrations: Towards in situ phase center corrections. *J. Geod.* **2020**, *94*, 1–15. [CrossRef]
15. Krzan, G.; Dawidowicz, K.; Wielgosz, P. Antenna phase center correction differences from robot and chamber calibrations: The case study LEIAR25. *GPS Solut.* **2020**, *24*, 44. [CrossRef]

16. Araszkiewicz, A.; Kiliszek, D.; Podkowa, A. Height variation depending on the source of antenna phase centre corrections: LEIAR25.R3 case study. *Sensors* **2019**, *19*, 4010. [CrossRef] [PubMed]
17. Li, L.; Li, B.; Chen, H.; Wang, F. Multipath Effect on Phase Center Calibration of GNSS Antenna. In *China Satellite Navigation Conference (CSNC) 2015 Proceedings: Volume III*; Sun, J., Liu, J., Fan, S., Lu, X., Eds.; Springer: Berlin/Heidelberg, Germany, 2015; pp. 259–268.
18. Rovera, G.D.; Abgrall, M.; Uhrich, P.; Defraigne, P.; Bertrand, B. GNSS antenna multipath effects. In Proceedings of the 2018 European Frequency and Time Forum (EFTF), Turin, Italy, 10–12 April 2018; pp. 208–212.
19. Dawidowicz, K.; Krzan, G.; Baryła, R.; Swiatek, K. The Impact of GNSS Antenna Mounting during Absolute Field Calibration on Phase Center Correction—JAV_GRANT-G3T Antenna Case Study. In Proceedings of the 10th International Conference "Environmental Engineering, Vilnius, Lithuania, 27–28 April 2017; pp. 27–28.
20. Echeverri, J.M.Q.; Pareja, T.F.; De Vicente y Oliva, J. Effectiveness of centering devices of geomatics instruments. *Procedia Manuf.* **2019**, *41*, 437–444. [CrossRef]
21. Kostelecký, J. On determination of GPS antenna phase centres at the geodetic observatory Pecny. *Reports Geod.* **2002**, *1*, 145–152.
22. Breva, J.; Kröger, J.; Kersten, T.; Schön, S. Validation of phase center corrections for new GNSS signals obtained with absolute antenna calibration in the field. In Proceedings of the EGU General Assembly 2019, Vienna, Austria, 7–12 April 2019; p. 14143.
23. Zhou, R.; Hu, Z.; Zhao, Q.; Cai, H.; Liu, X.; Liu, C.; Wang, G.; Kan, H.; Chen, L. Consistency Analysis of the GNSS Antenna Phase Center Correction Models. *Remote Sens.* **2022**, *14*, 540. [CrossRef]
24. Araszkiewicz, A.; Völksen, C. The impact of the antenna phase center models on the coordinates in the EUREF Permanent Network. *GPS Solut.* **2017**, *21*, 747–757. [CrossRef]
25. Brockmann, E.; Ineichen, D.; Schaer, S. Benefits of double stations in permanent GNSS networks. In Proceedings of the EGU General Assembly Conference Abstracts, Vienna, Austria, 7–12 April 2013; EUREF: Budapest, Hungary, 2013.
26. Kudrys, J. Permanent GNSS Observations at AGH-UST Satellite Observatory. *Rep. Geod. Geoinform.* **2016**, *100*, 101–107. [CrossRef]
27. Banasik, P. Wyznaczenie wysokości normalnej i charakterystyk pola cieżkościowego dla stacji permanentnej KRAW. *Geomat. Environ. Eng.* **2007**, *1*, 53–60.
28. Borowski, Ł. Determination of normal heights of ASG-EUPOS stations illustrated with an example od KRAW station. In *4th Doctoral Seminar on Geodesy and Cartography*; Biryło, M., Ed.; Wydawnictwo UWM: Olsztyn, Poland, 2012; pp. 19–25. ISBN 978-83-7299-766-1.
29. Borowski, Ł. Zastosowanie Sieci ASG-EUPOS Do Modelowania Lokalnej Quasi-Geoidy. Ph.D. Thesis, AGH University of Science and Technology, Krakow, Poland, 2015.
30. Dach, R.; Lutz, S.; Walser, P.; Fridez, P. *Bernese GNSS Software Version 5.2*; University of Bern, Bern Open Publishing: Berne, Switzerland, 2015; Volume 47, ISBN 1879621142.
31. Blewitt, G.; Hammond, W.; Kreemer, C. Harnessing the GPS Data Explosion for Interdisciplinary Science. *Eos* **2018**, *99*, 485. [CrossRef]
32. Borowski, Ł.; Kudrys, J. Adequacy of GNSS antenna calibration model—Case study Abstract. In Proceedings of the EUREF 2017 Symposium, Wrocław, Poland, 17–19 May 2017.
33. Dawidowicz, K. IGS08.ATX to IGS14.ATX change dependent differences in a gnss-derived position time series. *Acta Geodyn. Geomater.* **2018**, *15*, 363–378. [CrossRef]
34. Liwosz, T.; Rogowski, J.; Kruczyk, M.; Rajner, M.; Kurka, W. *Wyrównanie Kontrolne Obserwacji Satelitarnych Gnss Wykonanych na Punktach Asg-Eupos, Euref-Pol, Euvn, Polref i Osnowy i Klasy Wraz z Oceną Wyników*; Warsaw Univeristy of Technology: Warsaw, Poland, 2011.
35. Jaworski, L. *Sprawozdanie Techniczne IV Etapu Kampani Integracyjnej—Opracowanie i Wyrównanie Obserwacji GNSS*; Centrum Badań Kosmicznych PAN: Warszawa, Poland, 2011.
36. Military University of Technology EPN Analysis Coordination Center. Available online: http://www.gnss.wat.edu.pl/ (accessed on 15 June 2022).
37. Bassa, C.G.; Hainaut, O.R.; Galadí-Enríquez, D. Analytical simulations of the effect of satellite constellations on optical and near-infrared observations. *Astron. Astrophys.* **2022**, *657*, A75. [CrossRef]
38. Mikoś, M.; Nowak, A.; Lackowski, M.; Sośnica, K. Wkład systemu satelitarnego Galileo w rozwój geodezji. *Przegląd Geod.* **2022**, *1*, 23–26. [CrossRef]
39. Maciuk, K. Advantages of Combined GNSS Processing Involving a Limited Number of Visible Satellites. *Sci. J. Sil. Univ. Technol. Ser. Transp.* **2018**, *98*, 89–99. [CrossRef]
40. Prochniewicz, D.; Wezka, K.; Kozuchowska, J. Empirical stochastic model of multi-GNSS measurements. *Sensors* **2021**, *21*, 4566. [CrossRef] [PubMed]
41. Prochniewicz, D.; Grzymala, M. Analysis of the impact of multipath on Galileo system measurements. *Remote Sens.* **2021**, *13*, 2295. [CrossRef]

remote sensing

MDPI

Communication

Comparative Study of Predominantly Daytime and Nighttime Lightning Occurrences and Their Impact on Ionospheric Disturbances

Louis Osei-Poku [1,2], Long Tang [3], Wu Chen [1,2,*], Mingli Chen [4] and Akwasi Afrifa Acheampong [5]

1 Department of Land Surveying and Geo-Informatics, The Hong Kong Polytechnic University, Kowloon, Hong Kong; louis.oseipoku@connect.polyu.hk
2 Research Institute of Land and Space, The Hong Kong Polytechnic University, Kowloon, Hong Kong
3 School of Civil and Transportation Engineering, Guangdong University of Technology, Guangzhou 510006, China; ltang@gdut.edu.cn
4 Department of Building Environment and Energy Engineering, The Hong Kong Polytechnic University, Kowloon, Hong Kong; mingli.chen@polyu.edu.hk
5 Department of Geomatic Engineering, Kwame Nkrumah University of Science and Technology, Kumasi AK-448-4746, Ghana; aaacheampong.coe@knust.edu.gh
* Correspondence: lswuchen@polyu.edu.hk

Abstract: Space weather events adversely impact the operations of Global Navigation Satellite Systems (GNSS). Understanding space weather mechanisms, interactions in the atmosphere, and the extent of their impact are useful in developing prediction and mitigation models. In this study, the hourly lightning occurrence and its impact on ionospheric disturbances, quantified using the Rate of Total electron content Index (ROTI), were assessed. The linear correlation between diurnal lightning activity and ROTI in the coastal region of southern China where lightning predominates in the daytime was initially negative contrary to a positive correlation in southern Africa where lighting predominates in the evening. After appreciating and applying the physical processes of gravity waves, electromagnetic waves and the Trimpi effect arising from lightning activity, and the time delay impact they have on the ionosphere, the negative correlation was overturned to a positive one using cross-correlation. GNSS has demonstrated its capability of revealing the impact lightning has on the ionosphere at various times of the day.

Keywords: GNSS; lightning; ROTI; gravity wave; daytime; nighttime

Citation: Osei-Poku, L.; Tang, L.; Chen, W.; Chen, M.; Acheampong, A.A. Comparative Study of Predominantly Daytime and Nighttime Lightning Occurrences and Their Impact on Ionospheric Disturbances. *Remote Sens.* **2022**, *14*, 3209. https://doi.org/10.3390/rs14133209

Academic Editors: José Fernández, Juan F. Prieto, Serdjo Kos and Xiaogong Hu

Received: 20 May 2022
Accepted: 29 June 2022
Published: 4 July 2022

1. Introduction

The wide spatio-temporal advantages of Global Navigation Satellite Systems (GNSS) have made it an effective tool not only in positioning and navigation but also in studying and monitoring weather events. Geomagnetic and solar radiation, which are the key drivers of ionospheric plasma variations have been extensively observed using GNSS [1–4]. Tropospheric weather events such as cyclones [5–8], earthquakes [9–13], and/or man-made events such as rocket launches [14] and explosive bursts [15] have also been observed to cause plasma variations through the use of GNSS.

Thunderstorms/lightning, another troposphere weather event, has also garnered interest in the GNSS scientific community for adopting the advantages of GNSS to gain more insight into its activities. In recent GNSS lightning studies, Osei-Poku et al. [16] evaluated common total electron content (TEC) detrending techniques during lightning events. Rahmani et al. [17] probed the vertical coupling effect of a thunderstorm on the lower ionosphere. The ionospheric response to thunderstorms in West Africa has been reported by Ogunsua et al. [18], whereas Tang et al. [19] and Liu et al. [20] presented gravity waves resulting from thunderstorms. Blanc et al. [21] also showed gravity wave measurements in the ionosphere following thunderstorms in Europe. Lay [22] reported acoustic wave activity above thunderstorms. Amin [23] also studied the hourly lightning activity and its effect on the ionosphere using GNSS in southern Africa.

The quest to harness GNSS for more insightful studies about lightning continues. As a recommendation in their work, Amin [23] suggested the use of more long-term data to study the correlation between hourly lightning activity and irregular ionospheric events as their study had a limited dataset with few case studies. The use of long-term data would provide more consistent, reliable, and insightful information In this regard, this current study uses four-year long-term data to study the relationship between the hourly lightning activity and the rate of the TEC index (ROTI) in the low-latitude coastal sea region of southern China. Similar techniques used in Amin [23] are deployed for an effective comparison. The results are compared to Amin [23] and those in other geographical regions. The initial results in this present study showed a negative linear correlation between hourly lightning activity and ROTI as opposed to the positive linear correlation in Amin [23]. One reason could be attributed to lightning interactions with the ionosphere in the different geographical regions. In the subsequent sections, the data and methods used are described. Further investigations are made to find and discuss the physical mechanisms underlying the differences in the results. The derived conclusions are then presented.

2. Data and Methods

2.1. Lightning Data

The lightning data is from a local lightning network in the coastal region of southern China. The network consists of about 17 Vaisala LS8000 sensors that provide the geolocation and source peak currents of lightning in the area [24]. The number of times a current is recorded is positively correlated with lightning activity [25]; a day with a lightning count greater than 10,000 is deemed a "lightning day" [16]. The data span is from 2014 to 2017.

2.2. GNSS Data

The local GNSS data were obtained from the Hong Kong Satellite Reference (HK SatRef). The GNSS receivers have a sampling rate of 30 s. Information on the network is given in the works of Ji et al. [26] and Kumar et al. [27]. More information on HK SatRef is available at the website of the Hong Kong Survey Department (https://www.geodetic.gov.hk/en/rinex/downv.aspx, (accessed on 14 June 2019)).

2.3. Method

2.3.1. ROTI

ROTI defined by Pi et al. [28] as the root mean square of the TEC rate is used to characterize ionospheric irregularity. ROTI could be used as a proxy for the scintillation [29,30]. ROTI is computed from the GNSS data as follows. First, a 15° elevation cut-off angle was used to eliminate the multipath effect [6]. A geometry-free linear combination of pseudo- and carrier-phase signals was then used to compute slant TEC (STEC) at 30 s sampling intervals. STEC was converted to vertical TEC (VTEC) by applying a mapping function using Equation (1) below, where R_e is the earth's radius, θ is the elevation angle at the ionospheric pierce point (IPP) of the signal–receiver path, and h_i is the ionospheric single layer, approximated at 350 km.

$$\mathrm{VTEC} = \sqrt{1 - \left(\frac{R_e \cos\theta}{R_e + h_i}\right)^2} * \mathrm{STEC} \tag{1}$$

Finally, at five-minute intervals of the rate of TEC (ROT), ROTI was computed according to Equation (2), where ROT and ROTI are in TEC units (TECu: 1 TECu = 10^{16} e/m^2) and the notation $\langle\cdot\rangle$ is the averaging operation [31].

$$\mathrm{ROTI} = \sqrt{< \mathrm{ROT}^2 > - < \mathrm{ROT} >^2} \tag{2}$$

ROTI as shown in Equation (2) is usually for a single satellite-receiver pair. ROTI values exceeding 0.2 TECu are used to indicate that ionospheric scintillation has happened [32]. However, ROTI average ($ROTI_{avg}$) is the average value of ROTI over 30 min for all satellites received by a single station; following Oladipo et al. [33], a scintillation is deemed to occur for a $ROTI_{avg}$ value exceeding a threshold of 0.8 TECu. To effectively compare this study to that of Amin [23], $ROTI_{avg}$ is adopted.

2.3.2. Hourly Occurrence

For the diurnal hourly occurrence, the number of times lightning occurred and $ROTI_{avg}$ was greater than the threshold were accumulated within hourly intervals for all days in each year. A correlation is then looked for between the lightning and $ROTI_{avg}$ hourly occurrences.

2.3.3. Selection Criteria

To avoid the geomagnetic storm and solar radiation effects, only days with disturbance storm time (Dst) greater than −30 nT [34] and solar flux index (F10.7 index) less than 150 sfu (solar flux units, 1 sfu = 10^{-22} watt per square meter-hertz) [35] were selected. Table 1 shows the total number of days for the years 2014–2017 that had lightning activity only.

Table 1. Number of lightning days in each year void of geomagnetic storm and solar radiation effects.

Year	Number of Lightning Days
2014	85
2015	81
2016	108
2017	102

3. Results

According to Tang et al. [19], the stations of HK Satref are quite close, hence their observations are similar. Only the observations from one station (HKOH) are presented.

Figure 1 shows the $ROTI_{avg}$ from 2014 to 2017. In Figure 1, the highest values are mostly in the nighttime and between 0.07–0.2 TECu; this indicates that a scintillation is often a nighttime event. This is coherent with Tang et al. [31] and Ji et al. [26] who have shown that a greater percentage of ROTI lies between 0.02 and 0.05 TECu in the Hong Kong region. In this present study, the $ROTI_{avg}$ threshold is set at 0.075 TECu instead of 0.8 TECu, unlike Oladipo et al. [33] but the same as Nishioka et al. [36], who suggested a scintillation threshold of 0.075 TECu in the Asian region. More so, the years 2014–2017 are at the declining phase of the 24th solar cycle [37] where scintillation values are low compared to the high solar active years of 2002 and 2012 in Oladipo et al. [33] and Amin [23], respectively. Also, Jacobsen [38] and Liu et al. [39] have demonstrated that GNSS receiver types, configurations, and sampling rates influence ROTI values. These differences in ROTI values (thresholds) arising from the technique used, GNSS receiver configurations, and geographic location should be important factors to consider when developing regional and global models.

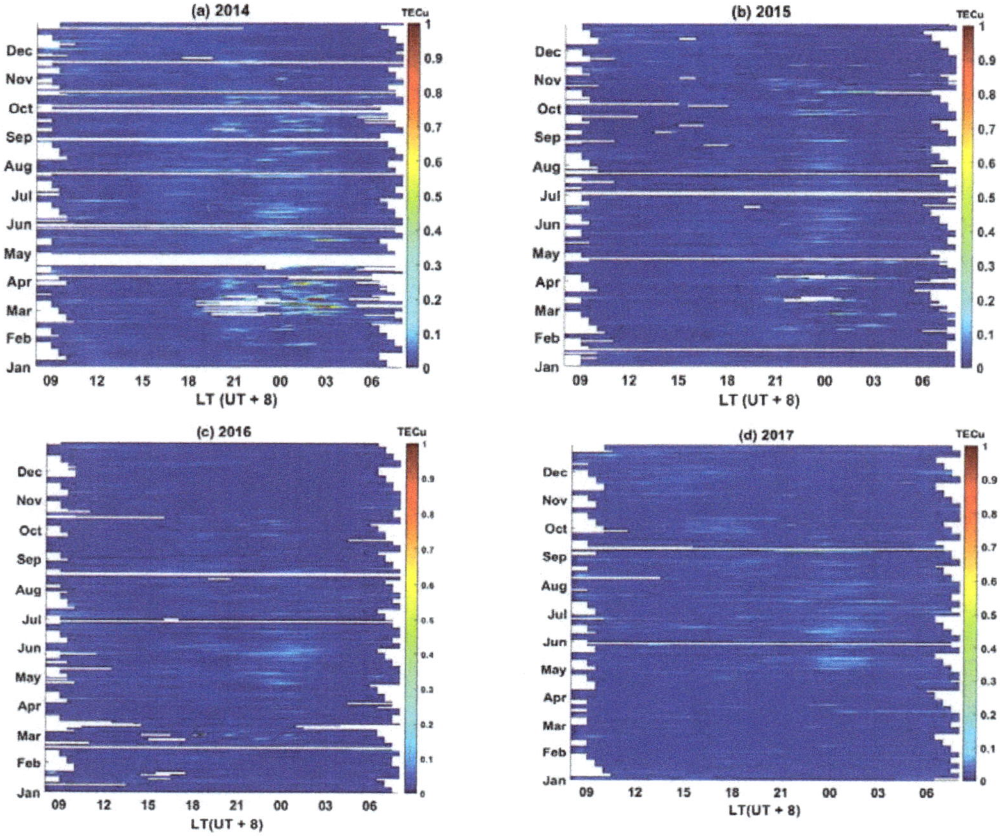

Figure 1. $ROTI_{avg}$ for the years 2014–2017. The abscissa axis shows the local time (LT: UT + 8) in hours. The ordinate axis shows the months of the year. The color bar shows the $ROTI_{avg}$ values in TECu.

On the diurnal hourly occurrence, the number of times lightning occurred and the $ROTI_{avg}$ was greater than 0.075 TECu were accumulated within hourly intervals for all days in each year. Panels a to d in Figure 2 show the annual hourly diurnal $ROTI_{avg}$ occurrence against that of the lightning occurrence. All years showed a similar trend. The trend reveals that the lighting occurrence peaked before that of $ROTI_{avg}$. This resulted in negative linear correlation values of −0.364, −0.41, −0.371, and −0.421 with significant values (p-value) of 0.05, 0.04, 0.05, and 0.04 at a confidence interval of 95% (α = 0.05) in chronological order from 2014 to 2017 as seen in panels e to h in the right column in Figure 2. The p-values indicate that despite being negative, the linear correlation is statistically significant. This implies that as lightning increases, ROTI decreases and vice versa. These are contrary to observations in Amin [23] where both lightning and $ROTI_{avg}$ peaked at the same time in the evening resulting in positive linear correlation values. As the objective of this study is to find the relationship between lightning and ROTI, reasons are discussed to explain the differences between this observation and that of [23].

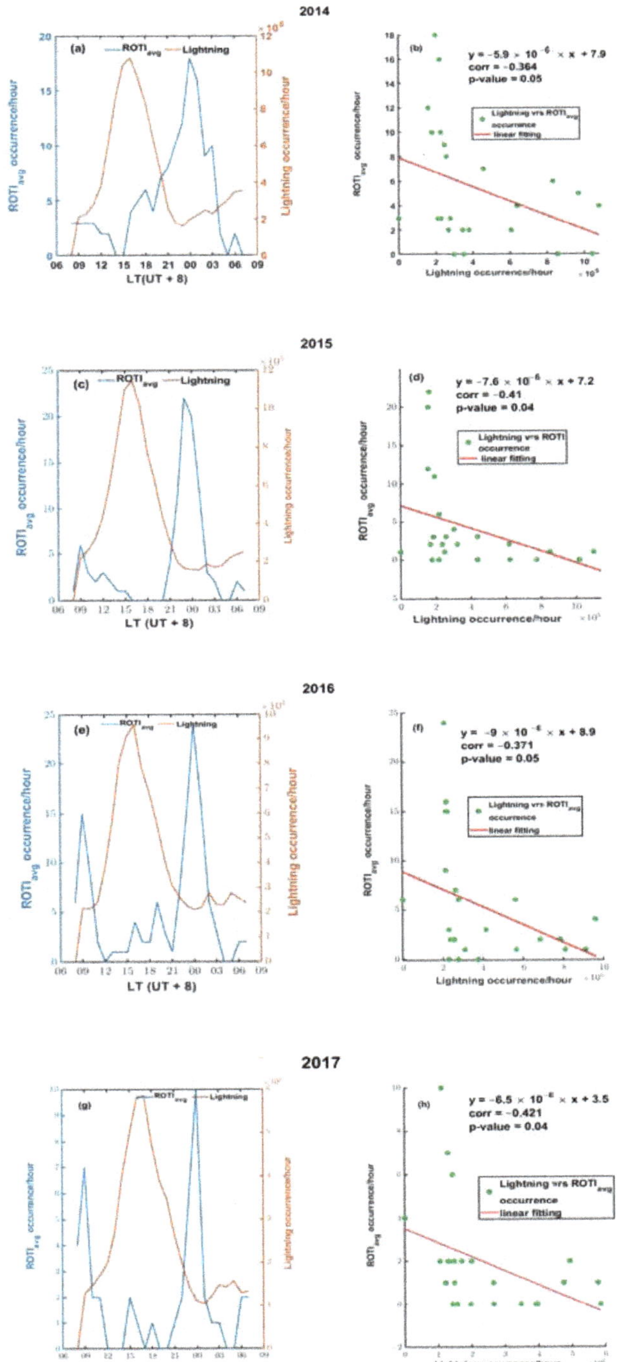

Figure 2. Panels in the left column (**a–d**) show the diurnal hourly occurrence of $ROTI_{avg}$ and lightning. Panels in the right column (**e–h**) are scatter diagrams showing the linear correlation between the diurnal hourly occurrence of lightning and $ROTI_{avg}$ for the respective years.

4. Discussions

The objective of this work is to appreciate and understand the relationship between hourly lightning activity and ionospheric irregularities (ROTI) through the use of long-term data. The initial results seen in Figure 2 show a negative linear correlation contrary to the positive linear correlation reported in Amin [23]. One reason for the observed differences in linear correlation could be attributed to the interaction between local variations in mesoscale and the topographic nature of the geographic region [40,41] resulting in different diurnal trends of lightning. That is, different regions tend to respond to lightning differently. From the observed data and results in other studies, there seems to be a common trend in the diurnal pattern of lightning. Places in the northern hemisphere seem to have a common pattern. Likewise, places in the southern hemisphere. For instance, Figure 5 in Williams et al. [42] shows a similar pattern of diurnal lightning as observed in this current study, and their Figure 6 is similar to that of Figure 5.13 of Amin [23]. In Figure 5 of Williams et al. [42], Australia, which is in the southern hemisphere, shows a broad peak of lightning from noon, which lasts until about 23:00 local time (LT). This pattern is similar to that of South Africa, also in the southern hemisphere. Brazil also shows a similar trend as seen in Pinto et al. [40]. In Figure 6 of Williams et al. [42], French Guyana, which is in the northern hemisphere and in South America, exhibits a lightning peak of around 15 LT, which lasts for about 3 h (till 18 LT). This pattern is similar to that of Hong Kong. Other similar trends are observed in the Indian region [43] and the United States [41,44], all in the northern hemisphere. The broad width of the lightning peak observed in the southern hemisphere regions should cover their peak time of $ROTI_{avg}$, which happens in the evening (mostly between 20 and 23 h). Statistically, lightning and $ROTI_{avg}$ having similar trends and with their peaks coinciding at the same time stands a high chance of giving a positive linear correlation value, whereas peaks that occur far apart at different times would give a negative linear correlation value.

Correlation may not necessarily mean causation. Possible mechanisms influencing the diurnal correlations in this work and that of Amin [23] are the lightning activity interactions with the ionosphere through lightning-related current and electrical discharges [45,46] and processes of gravity and electromagnetic wave [24,47–49].

A thunderstorm with lightning activity may affect the electron density in the lower ionosphere (particularly the D-layer) through two basic physical processes: the gravity wave (GW) due to the convection and thunder [47] and the electromagnetic wave (EM) due to the lightning [24]. The GW will generate a disturbance of the mesosphere (60–90 km) and the lower thermosphere (above 90 km), thus of the neutral atmosphere, which drags electrons from the ionosphere. Although the GW is a transversal wave, any changes in its amplitude generated from a lightning discharge will perform as a longitudinal wave that propagates upwards at the speed of the order of sound in air. The higher the air density, the stronger and faster the GW. Therefore, the GW due to the lightning and thunder and the thunderstorm convection disturbance would be stronger and faster as it propagates in the lower (higher air density) D-layer during the daytime than it does in the higher (lower air density) D-layer during the nighttime. The GW due to the lightning and thunder and the thunderstorm convection disturbance on ROTI lags the lightning activity due to its very low propagation speed at a high altitude. Theoretical studies have shown that the GW needs about one to several hours to propagate to higher altitudes depending on speed and period [50]. Using the wave model of Row [51] and Francis [52], which assumes that the GW propagates upwards through the atmosphere, Taylor et al. [53] found that the GW from a thunderstorm took about six hours before reaching the airglow layers (80 km above ground surface), which implies a vertical velocity of about 5 ms^{-1}. In Figure 2, the time delay between lightning and $ROTI_{avg}$ is about 5–7 h, similar to that observed by [53]. At an altitude of 350 km and a time delay of 5–7 h, the vertical velocity of the GW would be approximately 10–15 ms^{-1}. These observations show that the average vertical velocity of the GW is about 5–15 ms^{-1} confirming the low propagation speed of the GW. Also, the GW is predominant in the horizontal component which takes a relatively longer time to reach higher atmosphere heights [19,20,54]. This could best account for the difference in peak times of lightning and the ROTI pattern in this study, where lightning is predominantly in the afternoon. Again, Lay [22] made an observation where lightning occurring predominantly in the daytime (14–16 LT) had its related ionospheric disturbances around 00–02 LT similar to the observations in the left panels in Figure 2. Thus, lightning that occurs predominantly in the afternoon sees the ionospheric disturbance some hours later leading to an initial negative linear correlation.

Lightning-generated electromagnetic waves (EM) on the other hand propagate at the speed of light and reach the lower ionosphere almost immediately after the lightning activity has taken place. The strength of the impact of the EM on ROTI depends on the density of the electrons present. The higher the electron density, the stronger the impact. Theoretical simulations have shown that the EM could lead to a reduction in electron density at lower altitudes and an increase in electron density at higher altitudes in an ionospheric D-layer due to Joule heating effects [55–57], which have also been proven by lightning sferics observations [24,58,59]. Lightning sferics also produce very-low-frequency (VLF) discharges that contribute to the Trimpi effect. The Trimpi effect refers to transient perturbations caused by electron precipitation on sub ionospheric propagating waves [60]. The perturbations associated with sferics take only about 0.6 s to begin. Also, the Trimpi effect is only observed in nighttime ionospheric conditions and not daytime [61,62]. Therefore, the lightning-generated EM and the Trimpi effect would have a stronger impact on ROTI during the nighttime than during the daytime. The appearance of the impact of the EM and Trimpi effect on ROTI is almost on par with the time the lightning activity happened. Thus, lightning that occurs predominantly in the evening sees an ionospheric disturbance almost instantly leading to a positive linear correlation.

To further illustrate the time of lightning dominance and the time lags of its associated ionospheric disturbance, two diurnal zone divisions are made. That is, 07–17 and 18–06 LT to represent the daytime and evening time zones, respectively [30]. In Figure 2, 2014–2017 have similar lightning and $ROTI_{avg}$ trends. Some days are selected to illustrate this further as most of the individual days have similar observations. The assessment is done by finding the cross-correlation between lightning and ROTI instead of the linear correlation. Cross-correlation studies different variables to identify their similarity and draws characteristics relative to each other based on time to derive new information. Pseudorandom noise codes (PRN) from the individual satellite-receiver pairs available at the time ROTI lags the lightning are presented to provide extra evidence of ionospheric disturbances. The TEC of the PRN is detrended using the Savitzky–Golay filter of order 6 and window length of 120 min [16]. Detrended TEC (DTEC) is filtered with a bandpass of frequencies between 1 and 2.8 MHz and 4.2 and 8.2 MHz to derive ionospheric gravity (IGW) and ionospheric acoustic (IAW) waves, respectively, from lightning [17,63]. A filtered DTEC amplitude above 0.08 TECu and 0.025 TECu [63] and ROTI greater than 0.2 TECu signify the presence of IGW, IAW, and ionospheric disturbances, respectively. The selected days for the daytime and evening are presented in Figures 3 and 4, respectively. The left panels of the top rows for each day in Figures 3 and 4 show the $ROTI_{avg}$ and lightning counts. The right panels of the top rows show the cross-correlation between the $ROTI_{avg}$ and lightning counts from the left panel. The bottom rows show the DTEC, IGW, IAW, and ROTI of the PRN, which was available at the time ROTI lags the lightning. The magenta lines are the thresholds of IGW, IAW, and ROTI.

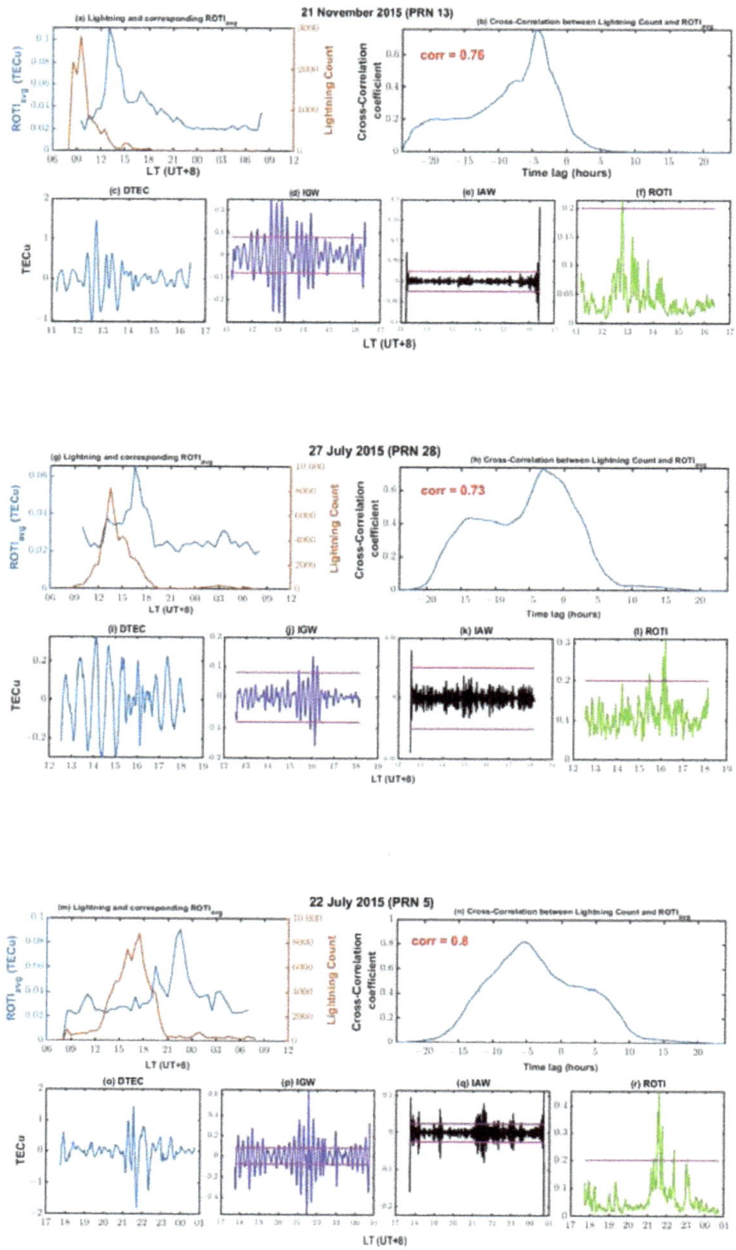

Figure 3. Some selected days (21 November 2015, 27 July 2015, 22 July 2015) where lightning was predominantly in the daytime (07–17 local time) and its cross-correlation with ROTI$_{avg}$ (**top rows of each day**). The associated ionospheric disturbances are also indicated by DTEC, IGW, IAW, and ROTI (**bottom rows of each day**) by satellites that passed some hours after the lightning activity. The magenta lines show the threshold of ±0.08 TECu, ±0.025 TECu and 0.2 TECu for IGW, IAW and ROTI respectively.

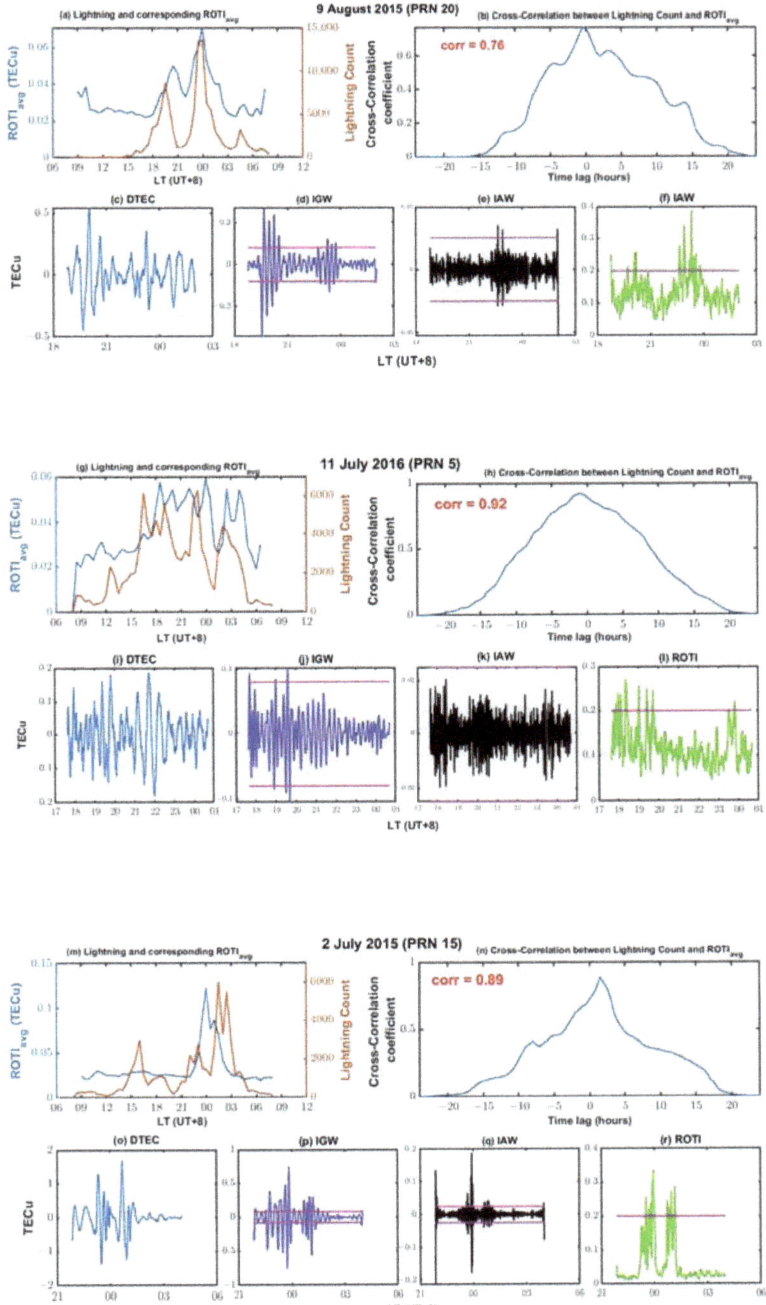

Figure 4. Some selected days (9 August 2015, 11 July 2016, 2 July 2015) where lightning was predominantly in the evening (18–06 local time) and its cross-correlation with $ROTI_{avg}$ (**top rows of each day**). The associated ionospheric disturbances indicated by DTEC, IGW, IAW, and ROTI (**bottom rows of each day**) by satellites that passed at the time of lightning activity. The magenta lines show the threshold of ±0.08 TECu, ±0.025 TECu and 0.2 TECu for IGW, IAW and ROTI respectively.

Figure 3 shows the days with lightning predominantly in the daytime. In panels b, h, and n in Figure 3, $ROTI_{avg}$ is seen to have a negative time lag of about 4 h to the lightning activity but is strongly positively correlated. The time of the disturbances as seen in the PRNs in the bottom rows confirms the delay. The negative time lags; however, positive coefficients indicate that the disturbances are associated with lightning, which can be attributed to the delayed impact of the gravity wave mechanism from the predominantly daytime lightning on the ionosphere.

Figure 4 shows the days with lightning predominantly in the evening. From panels b, h, and n in Figure 4, $ROTI_{avg}$ is seen to have a time lag of zero to a few minutes to the lightning activity and with strong positive correlation coefficients. The time of the disturbances as seen in the PRNs in the bottom rows is on par with the lightning activity. The time lag of zero to a few minutes and the positive coefficients indicate that the disturbances are associated with lightning, which can be explained by the almost immediate impact of the electromagnetic wave mechanism and the Trimpi effect resulting from the predominantly nighttime lightning on the ionosphere.

Following the mechanisms of the gravity wave, electromagnetic wave, and the Trimpi effect as discussed above, cross-correlation of the lightning and $ROTI_{avg}$ counts from the left panels in Figure 2 yielded positive correlation values of 0.89, 0.82, 0.72, and 0.61 for 2014 to 2017, respectively as shown in Figure 5. The values are similar to 0.86 reported by Amin [23]. This shows that the diurnal hourly lightning activity and associated $ROTI_{avg}$ have a positive correlation.

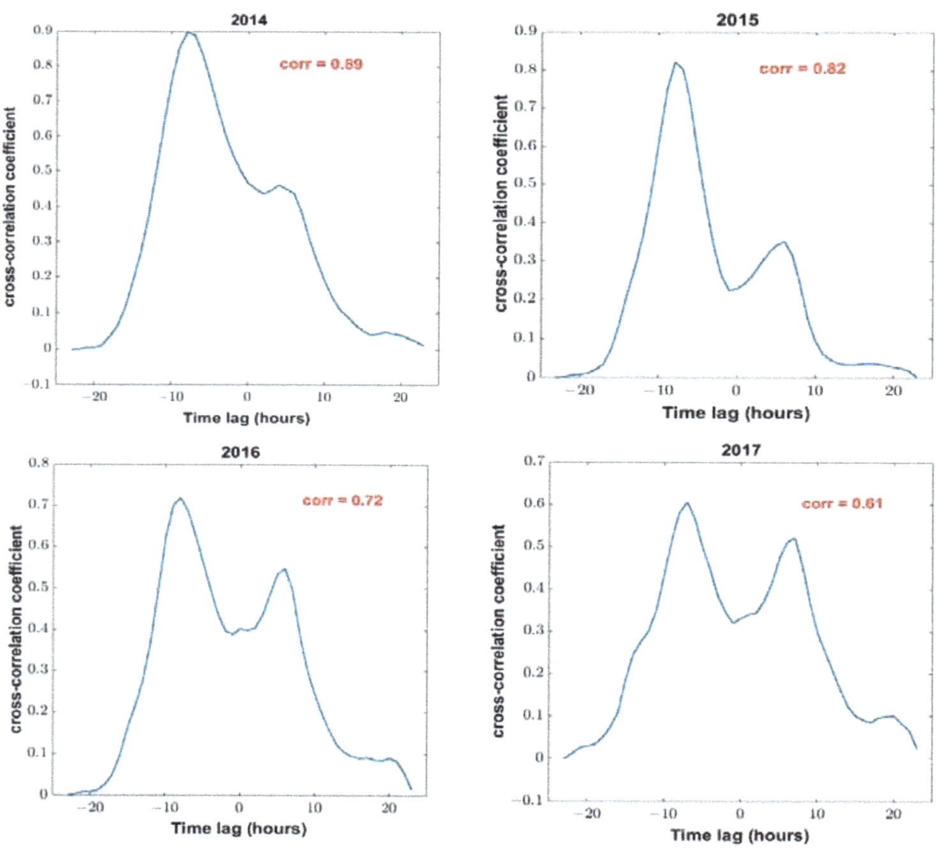

Figure 5. Cross-correlation of lightning and $ROTI_{avg}$ counts from left panels in Figure 2.

The observations in Figures 3–5 confirm that ionospheric disturbances from lightning occurring in the afternoon take a relatively longer time to manifest, whereas those due to lightning occurring in the evening are almost immediate. Using cross-correlation provides a better means of accessing the relationship between lightning activity and its associated ROTI at various times of the day. The cross-correlation coefficient is statistically significant when its absolute value is greater than the expected cross-correlation coefficient (ECCF), which is computed as [64].

$$\mathrm{ECCF} = \frac{2}{\sqrt{n - |k|}} \tag{3}$$

where n is the total number of observations and k is the number of observations between zero and the point of the maximum cross-correlation value on the lag axis. Negative and positive k are to the left and right of zero on the lag axis, respectively.

Table 2 is a summary of the cross-correlation coefficients of days and years in Figures 3–5 and their ECCF. The observations of the days in Figures 3 and 4 are at 30 min intervals, whereas those of the years in Figure 5 are at hourly intervals. The cross-correlation coefficients being greater than the ECCFs show that lightning activity is significantly positively correlated with its associated ROTI regardless of the time of day the lightning was predominant.

Table 2. Cross-correlation coefficient of days and years in Figures 3–5 and their ECCF.

		Cross-Correlation Coefficient	n	k	ECCF
Days in Figure 3	21 November 2015	0.76	48	−9	0.32
	27 July 2015	0.76	48	−6	0.31
	22 July 2015	0.8	48	−11	0.33
Days in Figure 4	9 August 2015	0.76	48	0	0.3
	2 July 2015	0.89	48	−1	0.29
	11 July 2016	0.92	48	−1	0.29
Years in Figure 5	2014	0.89	24	−8	0.5
	2015	0.82	24	−8	0.5
	2016	0.71	24	−8	0.5
	2017	0.6	24	−7	0.4

5. Conclusions

In this work, a correlational study between hourly lightning activity and ROTI was carried out using long-term data. The data were from 2014 to 2017 with the study area being Hong Kong. The linear correlation between the hourly lightning activity and ROTI was negative compared to a similar study in southern Africa, where the correlation was positive. The initial differences in linear correlation values between this present study and that in southern Africa are attributed to the time differences at which lightning and ROTI peaked. Although lightning peaked in the daytime in Hong Kong, its associated ROTI peaked in the evening. For southern Africa, both lightning and ROTI peaked at the same time in the evening. Probing further, this current study found that the ROTI resulting from lightning predominantly in the daytime lagged due to slow propagation of gravity wave mechanisms but did not lag when lightning was predominantly in the evening due to electromagnetic waves and the Trimpi mechanisms. The temporal discrepancy could be explained by either or both mechanisms and it would require simulations to confirm their likelihood. Using datasets obtained from radio occultation missions such as COSMIC would be effective for quantifying these mechanisms and their perturbations at the various layers of the ionosphere. Unlike the linear correlation that showed that predominantly daytime lightning is negatively correlated with its ROTI, cross-correlation offered a better means to access the lightning activity–ROTI relationship. Cross-correlation revealed the time delays and showed that the lightning activity and its associated ROTI are positively correlated, which can be explained by the physical mechanisms. The results and observations have shown that GNSS can reveal the impact lightning activity has on the ionosphere at various times of the day. The models for ionospheric scintillation simulations, prediction, and forecasting purposes based on lightning activities should consider and incorporate this observation when being developed.

Author Contributions: Conceptualization and research design were done by L.O.-P. and W.C. L.O.-P., L.T., M.C. and A.A.A. contributed to data processing and analysis. W.C. and M.C. supervised the work. Writing and editing of the manuscript were done by all authors. All authors have read and agreed to the published version of the manuscript.

Funding: This research was funded by the University Grants Committee of Hong Kong under the scheme Research Impact Fund on the project R5009-21, and the Research Institute of Land and Space, Hong Kong Polytechnic University, and the National Natural Science Foundation of China (Grant number 41804021).

Data Availability Statement: The Dst-index data (IAGA 2002-like format) can be obtained from the Data Analysis Center for Geomagnetism and Space Magnetism, Kyoto University, operating WDC for Geomagnetism, Kyoto (http://wdc.kugi.kyoto-u.ac.jp/dstae/index.html). The Bz component of the interplanetary magnetic field was obtained from the GSFC/SPDF OMNIWeb interface at http://omniweb.gsfc.nasa.gov. The GNSS data can be obtained from Hong Kong SatRef of the Lands Department of the Hong Kong Government (https://www.geodetic.gov.hk/en/rinex/downv.aspx). All data center websites were accessed on 14 June 2019.

Acknowledgments: The research institutes and government departments where data was obtained for this study are duly acknowledged for their continuous support of scientific studies. The authors also thank the anonymous reviewers for their useful and insightful comments for improving the manuscript.

Conflicts of Interest: The authors declare no conflict of interest.

References

1. Cherniak, I.; Zakharenkova, I. Large-Scale Traveling Ionospheric Disturbances Origin and Propagation: Case Study of the December 2015 Geomagnetic Storm. *Space Weather* **2018**, *16*, 1377–1395. [CrossRef]
2. Lu, Y.; Wang, Z.; Ji, S.; Chen, W. Assessing the Positioning Performance under the Effects of Strong Ionospheric Anomalies with Multi-GNSS in Hong Kong. *Radio Sci.* **2020**, *55*, 1–18. [CrossRef]
3. Afraimovich, E.L.; Astafyeva, E.I.; Demyanov, V.V.; Edemskiy, I.K.; Gavrilyuk, N.S.; Ishin, A.B.; Kosogorov, E.A.; Leonovich, L.A.; Lesyuta, O.S.; Palamartchouk, K.S.; et al. A review of GPS/GLONASS studies of the ionospheric response to natural and anthropogenic processes and phenomena. *J. Space Weather Space Clim.* **2013**, *3*, A27. [CrossRef]
4. Lay, E.H.; Shao, X.M.; Carrano, C.S. Variation in total electron content above large thunderstorms. *Geophys. Res. Lett.* **2013**, *40*, 1945–1949. [CrossRef]
5. Chen, G.; Zhou, C.; Liu, Y.; Zhao, J.; Tang, Q.; Wang, X.; Zhao, Z. A statistical analysis of medium-scale traveling ionospheric disturbances during 2014–2017 using the Hong Kong CORS network. *Earth Planets Space* **2019**, *71*, 52. [CrossRef]
6. Yang, Z.; Liu, Z. Observational study of ionospheric irregularities and GPS scintillations associated with the 2012 tropical cyclone Tembin passing Hong Kong. *J. Geophys. Res. Space Phys.* **2016**, *121*, 4705–4717. [CrossRef]
7. Polyakova, A.S.; Perevalova, N.P. Investigation into impact of tropical cyclones on the ionosphere using GPS sounding and NCEP/NCAR Reanalysis data. *Adv. Space Res.* **2011**, *48*, 1196–1210. [CrossRef]
8. Mao, T.; Wang, J.; Yang, G.; Yu, T.; Ping, J.; Suo, Y. Effects of typhoon Matsa on ionospheric TEC. *Chin. Sci. Bull.* **2010**, *55*, 712–717. [CrossRef]
9. Kong, J.; Yao, Y.; Zhou, C.; Liu, Y.; Zhai, C.; Wang, Z.; Liu, L. Tridimensional reconstruction of the Co-Seismic Ionospheric Disturbance around the time of 2015 Nepal earthquake. *J. Geod.* **2018**, *92*, 1255–1266. [CrossRef]
10. Miyazaki, S.I.; Larson, K.M. Coseismic and early postseismic slip for the 2003 Tokachi-oki earthquake sequence inferred from GPS data. *Geophys. Res. Lett.* **2008**, *35*, L04302. [CrossRef]
11. Craig, T.J.; Calais, E. Strain accumulation in the New Madrid and Wabash Valley seismic zones from 14 years of continuous GPS observation. *J. Geophys. Res. Solid Earth* **2014**, *119*, 9110–9129. [CrossRef]
12. Calais, E.; Minster, J.B. GPS, earthquakes, the ionosphere, and the Space Shuttle. *Phys. Earth Planet. Inter.* **1998**, *105*, 167–181. [CrossRef]
13. Afraimovich, E.L.; Ding, F.; Kiryushkin, V.V.; Astafyeva, E.I.; Jin, S.; Sankov, V.A. TEC response to the 2008 Wenchuan Earthquake in comparison with other strong earthquakes. *Int. J. Remote Sens.* **2010**, *31*, 3601–3613. [CrossRef]
14. Mendillo, M. The effect of rocket launches on the ionosphere. *Adv. Space Res.* **1981**, *1*, 275–290. [CrossRef]
15. Kundu, B.; Senapati, B.; Matsushita, A.; Heki, K. Atmospheric wave energy of the 2020 August 4 explosion in Beirut, Lebanon, from ionospheric disturbances. *Sci. Rep.* **2021**, *11*, 2793. [CrossRef]

16. Osei-Poku, L.; Tang, L.; Chen, W.; Mingli, C. Evaluating Total Electron Content (TEC) Detrending Techniques in Determining Ionospheric Disturbances during Lightning Events in A Low Latitude Region. *Remote Sens.* **2021**, *13*, 4753. [CrossRef]
17. Rahmani, Y.; Alizadeh, M.M.; Schuh, H.; Wickert, J.; Tsai, L.-C. Probing vertical coupling effects of thunderstorms on lower ionosphere using GNSS data. *Adv. Space Res.* **2020**, *66*, 1967–1976. [CrossRef]
18. Ogunsua, B.O.; Srivastava, A.; Bian, J.; Qie, X.; Wang, D.; Jiang, R.; Yang, J. Significant Day-time Ionospheric Perturbation by Thunderstorms along the West African and Congo Sector of Equatorial Region. *Sci. Rep.* **2020**, *10*, 8466. [CrossRef]
19. Tang, L.; Chen, W.; Chen, M.; Osei-Poku, L. Statistical observation of thunderstorm-induced ionospheric Gravity waves above low-latitude areas in the northern hemisphere. *Remote Sens.* **2019**, *11*, 2732. [CrossRef]
20. Liu, T.; Yu, Z.; Ding, Z.; Nie, W.; Xu, G. Observation of Ionospheric Gravity Waves Introduced by Thunderstorms in Low Latitudes China by GNSS. *Remote Sens.* **2021**, *13*, 4131. [CrossRef]
21. Blanc, E.; Ceranna, L.; Hauchecorne, A.; Charlton-Perez, A.; Marchetti, E.; Evers, L.G.; Kvaerna, T.; Lastovicka, J.; Eliasson, L.; Crosby, N.B.; et al. Toward an Improved Representation of Middle Atmospheric Dynamics Thanks to the ARISE Project. *Surv. Geophys.* **2018**, *39*, 171–225. [CrossRef]
22. Lay, E.H. Ionospheric Irregularities and Acoustic/Gravity Wave Activity Above Low-Latitude Thunderstorms. *Geophys. Res. Lett.* **2018**, *45*, 90–97. [CrossRef]
23. Amin, M.M. Influence of Lightning on Electron Density Variation in the Ionosphere Using WWLLN Lightning Data and GPS Data. Master's Thesis, University of Cape Town, Cape Town, South Africa, 2015.
24. Qin, Z.; Chen, M.; Lyu, F.; Cummer, S.A.; Gao, Y.; Liu, F.; Zhu, B.; Du, Y.P.; Usmani, A.; Qiu, Z. Prima Facie Evidence of the Fast Impact of a Lightning Stroke on the Lower Ionosphere. *Geophys. Res. Lett.* **2020**, *47*, e2020GL090274. [CrossRef]
25. Salem, M.A.; Liu, N.; Rassoul, H.K. Modification of the lower ionospheric conductivity by thunderstorm electrostatic fields. *Geophys. Res. Lett.* **2016**, *43*, 5–12. [CrossRef]
26. Ji, S.; Chen, W.; Wang, Z.; Xu, Y.; Weng, D.; Wan, J.; Fan, Y.; Huang, B.; Fan, S.; Sun, G.; et al. A study of occurrence characteristics of plasma bubbles over Hong Kong area. *Adv. Space Res.* **2013**, *52*, 1949–1958. [CrossRef]
27. Kumar, S.; Chen, W.; Chen, M.; Liu, Z.; Singh, R.P. Thunderstorm-/lightning-induced ionospheric perturbation: An observation from equatorial and low-latitude stations around Hong Kong. *J. Geophys. Res. Space Phys.* **2017**, *122*, 9032–9044. [CrossRef]
28. Pi, X.; Mannucci, A.J.; Lindqwister, U.J.; Ho, C.M. Monitoring of global ionospheric irregularities using the worldwide GPS network. *Geophys. Res. Lett.* **1997**, *24*, 2283–2286. [CrossRef]
29. Yang, Z.; Liu, Z. Correlation between ROTI and Ionospheric Scintillation Indices using Hong Kong low-latitude GPS data. *GPS Solut.* **2016**, *20*, 815–824. [CrossRef]
30. Yang, Z.; Liu, Z. Low-Latitude Ionospheric Density Irregularities and Associated Scintillations Investigated by Combining COSMIC RO and Ground-Based Global Positioning System Observations Over a Solar Active Period. *J. Geophys. Res. Space Phys.* **2018**, *123*, 3998–4014. [CrossRef]
31. Tang, L.; Louis, O.-P.; Chen, W.; Chen, M. A ROTI-Aided Equatorial Plasma Bubbles Detection Method. *Remote Sens.* **2021**, *13*, 4356. [CrossRef]
32. Wu, Q.; Chou, M.-Y.; Schreiner, W.; Braun, J.; Pedatella, N.; Cherniak, I. COSMIC observation of stratospheric gravity wave and ionospheric scintillation correlation. *J. Atmos. Sol. Terr. Phys.* **2021**, *217*, 105598. [CrossRef]
33. Oladipo, O.A.; Schüler, T. Equatorial ionospheric irregularities using GPS TEC derived index. *J. Atmos. Sol. Terr. Phys.* **2013**, *92*, 78–82. [CrossRef]
34. Loewe, C.A. Classification and mean behavior of magnetic storms. *J. Geophys. Res. A Space Phys.* **1997**, *102*, 14209–14213. [CrossRef]
35. Yu, S.; Liu, Z. The ionospheric condition and GPS positioning performance during the 2013 tropical cyclone Usagi event in the Hong Kong region. *Earth Planets Space* **2021**, *73*, 66. [CrossRef]
36. Nishioka, M.; Saito, A.; Tsugawa, T. Occurrence characteristics of plasma bubble derived from global ground-based GPS receiver networks. *J. Geophys. Res. Space Phys.* **2008**, *113*. [CrossRef]
37. Patel, K.; Singh, A.; Singh, S.B.; Singh, A.K. Causes responsible for intense and severe storms during the declining phase of Solar Cycle 24. *J. Astrophys. Astron.* **2019**, *40*, 4. [CrossRef]
38. Jacobsen, K.S. The impact of different sampling rates and calculation time intervals on ROTI values. *J. Space Weather Space Clim.* **2014**, *4*, A33. [CrossRef]
39. Liu, Z.; Yang, Z.; Xu, D.; Morton, Y.J. On Inconsistent ROTI Derived From Multiconstellation GNSS Measurements of Globally Distributed GNSS Receivers for Ionospheric Irregularities Characterization. *Radio Sci.* **2019**, *54*, 215–232. [CrossRef]
40. Pinto, I.R.C.A.; Pinto, O.; Rocha, R.M.L.; Diniz, J.H.; Carvalho, A.M.; Filho, A.C. Cloud-to-ground lightning in southeastern Brazil in 1993: 2. Time variations and flash characteristics. *J. Geophys. Res. Atmos.* **1999**, *104*, 31381–31387. [CrossRef]
41. Lopez, R.E.; Holle, R.L. Diurnal and Spatial Variability of Lightning Activity in Northeastern Colorado and Central Florida during the Summer. *Mon. Weather Rev.* **1986**, *114*, 1288–1312. [CrossRef]
42. Williams, E.R.; Heckman, S.J. The local diurnal variation of cloud electrification and the global diurnal variation of negative charge on the Earth. *J. Geophys. Res. Atmos.* **1993**, *98*, 5221–5234. [CrossRef]
43. Kandalgaonkar, S.S.; Tinmaker, M.I.R.; Kulkarni, J.R.; Nath, A. Diurnal variation of lightning activity over the Indian region. *Geophys. Res. Lett.* **2003**, *30*. [CrossRef]

44. Easterling, D.R.; Robinson, P.J. The Diurnal Variation of Thunderstorm Activity in the United States. *J. Appl. Meteorol. Climatol.* **1985**, *24*, 1048–1058. [CrossRef]
45. Kuo, C.L.; Lee, L.C. Ionospheric plasma dynamics and instability caused by upward currents above thunderstorms. *J. Geophys. Res. Space Phys.* **2015**, *120*, 3240–3253. [CrossRef]
46. Collier, A.B.; Hughes, A.R.W.; Lichtenberger, J.; Steinbach, P. Seasonal and diurnal variation of lightning activity over southern Africa and correlation with European whistler observations. *Ann. Geophys.* **2006**, *24*, 529–542. [CrossRef]
47. Alexander, M.J.; Holton, J.R.; Durran, D.R. The Gravity Wave Response above Deep Convection in a Squall Line Simulation. *J. Atmos. Sci.* **1995**, *52*, 2212–2226. [CrossRef]
48. Vadas, S.L.; Liu, H.-I. Generation of large-scale gravity waves and neutral winds in the thermosphere from the dissipation of convectively generated gravity waves. *J. Geophys. Res. Space Phys.* **2009**, *114*, A10310. [CrossRef]
49. Pasko, V.P.; Inan, U.S.; Bell, T.F.; Reising, S.C. Mechanism of ELF radiation from sprites. *Geophys. Res. Lett.* **1998**, *25*, 3493–3496. [CrossRef]
50. Occhipinti, G.; Rolland, L.; Lognonné, P.; Watada, S. From Sumatra 2004 to Tohoku-Oki 2011: The systematic GPS detection of the ionospheric signature induced by tsunamigenic earthquakes. *J. Geophys. Res. Space Phys.* **2013**, *118*, 3626–3636. [CrossRef]
51. Row, R.V. Acoustic-gravity waves in the upper atmosphere due to a nuclear detonation and an earthquake. *J. Geophys. Res.* **1967**, *72*, 1599–1610. [CrossRef]
52. Francis, S.H. Global propagation of atmospheric gravity waves: A review. *J. Atmos. Terr. Phys.* **1975**, *37*, 1011–1054. [CrossRef]
53. Taylor, M.J.; Hapgood, M.A. Identification of a thunderstorm as a source of short period gravity waves in the upper atmospheric nightglow emissions. *Planet. Space Sci.* **1988**, *36*, 975–985. [CrossRef]
54. Chimonas, G. Enhancement of sporadicEby horizontal transport within the layer. *J. Geophys. Res.* **1971**, *76*, 4578–4586. [CrossRef]
55. Taranenko, Y.N.; Inan, U.S.; Bell, T.F. Interaction with the lower ionosphere of electromagnetic pulses from lightning: Heating, attachment, and ionization. *Geophys. Res. Lett.* **1993**, *20*, 1539–1542. [CrossRef]
56. Marshall, R.A.; Inan, U.S.; Chevalier, T.W. Early VLF perturbations caused by lightning EMP-driven dissociative attachment. *Geophys. Res. Lett.* **2008**, *35*. [CrossRef]
57. Marshall, R.A. Effect of self-absorption on attenuation of lightning and transmitter signals in the lower ionosphere. *J. Geophys. Res. Space Phys.* **2014**, *119*, 4062–4076. [CrossRef]
58. Cheng, Z.; Cummer, S.A.; Baker, D.N.; Kanekal, S.G. Nighttime D region electron density profiles and variabilities inferred from broadband measurements using VLF radio emissions from lightning. *J. Geophys. Res. Space Phys.* **2006**, *111*. [CrossRef]
59. Shao, X.M.; Lay, E.H.; Jacobson, A.R. Reduction of electron density in the night-time lower ionosphere in response to a thunderstorm. *Nat. Geosci.* **2012**, *6*, 29–33. [CrossRef]
60. Armstrong, W.C. Recent advance from studies of the Trimpi effect. *Antarct. Jour.* **1983**, *18*, 281–290.
61. Leyser, T.B.; Inan, U.S.; Carpenter, D.L.; Trimpi, M.L. Diurnal variation of burst precipitation effects on subionospheric VLF/LF signal propagation near L = 2. *J. Geophys. Res. Space Phys.* **1984**, *89*, 9139–9143. [CrossRef]
62. Rodger, C.J. Subionospheric VLF perturbations associated with lightning discharges. *J. Atmos. Sol. Terr. Phys.* **2003**, *65*, 591–606. [CrossRef]
63. Lay, E.H.; Shao, X.M.; Kendrick, A.K.; Carrano, C.S. Ionospheric acoustic and gravity waves associated with midlatitude thunderstorms. *J. Geophys. Res. A Space Phys.* **2015**, *120*, 6010–6020. [CrossRef]
64. Interpret all Statistics and Graphs for Cross Correlation. Available online: https://support.minitab.com/en-us/minitab/18/help-and-how-to/modeling-statistics/time-series/how-to/cross-correlation/interpret-the-results/all-statistics-and-graphs/ (accessed on 10 June 2022).

remote sensing

MDPI

Article

Retrieval of Soil Moisture Content Based on Multisatellite Dual-Frequency Combination Multipath Errors

Shihai Nie, Yanxia Wang *, Jinsheng Tu, Peng Li, Jianhui Xu, Nan Li, Mengke Wang, Danni Huang and Jia Song

College of Geographic Information and Tourism, Chuzhou University, Chuzhou 239000, China; nsh1017@chzu.edu.cn (S.N.); tujinsheng@chzu.edu.cn (J.T.); lipeng@chzu.edu.cn (P.L.); xujh@chzu.edu.cn (J.X.); linan@chzu.edu.cn (N.L.); mengkewang@sdust.edu.cn (M.W.); hdn325@chzu.edu.cn (D.H.); songjia@chzu.edu.cn (J.S.)

* Correspondence: lidarcenter@chzu.edu.cn; Tel.: +86-159-5501-2916

Abstract: Global navigation satellite system interferometric reflectometry (GNSS-IR) is a new type of microwave remote sensing technology that can measure soil moisture content (SMC). GNSS-IR soil moisture retrieval methods based on the satellite signal-to-noise ratio (SNR) and triple-frequency signal combination have the following shortcomings: SNR does not always exist in the original GNSS file, and the number of triple-frequency signal observation satellites is small, resulting in GNSS-IR soil moisture observation time resolution being low. Based on the above problems, in this study, we constructed a soil moisture inversion method based on multisatellite dual-frequency combined multipath error is proposed: the multipath error calculation model of dual-frequency carrier phase (L4 Ionosphere Free, L4_IF) and dual-frequency pseudorange (DFP) without ionospheric effect is constructed. We selected the data of the five epochs before and after the time point of the effective satellite period to construct the multipath error model and error equation, and we solved the delay phase for soil moisture retrieval. We verified the method using Plate Boundary Observatory (PBO) P041 site data. The results showed that the Pearson correlation coefficients (R) of L4_IF and DFP methods at P041 station are 0.97 and 0.91, respectively. To better verify the results' reliability and the proposed method's effectiveness, the soil moisture data of the MFLE station about 210 m away from P041 station are used as the verification data in this paper. The results showed that the delay phase solved by multipath error and soil moisture strongly correlate. Pearson correlation coefficients (R) of L4_IF and DFP methods at MFLE station are 0.93 and 0.86, respectively. In order to improve the inversion accuracy of GNSS-IR soil moisture, this paper constructs the prediction model of soil moisture by using the linear regression (ULR), back propagation neural network (BPNN) and radial basis function neural network (RBFNN), and evaluates the accuracy of each model. The results showed that the soil moisture retrieval method based on multisatellite dual-frequency combined multipath error can replace the traditional retrieval method and effectively improve the time resolution of GNSS-IR soil moisture estimation. To perform highly dynamic monitoring of soil moisture, higher retrieval accuracy can only be obtained with a small epoch multipath error.

Keywords: GNSS-IR; soil moisture content; multisatellite combination; dual-frequency pseudorange; dual-frequency carrier phase combination; multipath error; phase delay

Citation: Nie, S.; Wang, Y.; Tu, J.; Li, P.; Xu, J.; Li, N.; Wang, M.; Huang, D.; Song, J. Retrieval of Soil Moisture Content Based on Multisatellite Dual-Frequency Combination Multipath Errors. *Remote Sens.* **2022**, *14*, 3193. https://doi.org/10.3390/rs14133193

Academic Editors: Serdjo Kos, José Fernández and Juan F. Prieto

Received: 28 April 2022
Accepted: 1 July 2022
Published: 3 July 2022

Publisher's Note: MDPI stays neutral with regard to jurisdictional claims in published maps and institutional affiliations.

1. Introduction

Soil moisture content (SMC) is the physical quantity that characterizes the degree of soil wetting and drying. As an indispensable environmental factor on the surface, SMC plays an active role in weather forecast, climate research, slope stability prediction, and accurate prediction of flood disasters [1–4]. Global Navigation Satellite System interferometric reflection (GNSS-IR) is a new microwave remote sensing technology that mainly uses the interference effect generated by the direct and surface reflection signals obtained at the GNSS receiver to invert the surface parameters according to the characteristics of the interference signals [5–9].

In recent years, many scholars have made remarkable achievements in soil moisture retrieval using GNSS-IR technology. Larson et al. observed that the ground-reflected signal captured by a geodetic-quality Global Positioning System (GPS) antenna is sensitive to soil moisture, so the GPS signals can be used to sense soil moisture. They first proposed using GPS SNR retrieval of soil moisture, confirming the feasibility of SNR data retrieval [10–12]. Zhang et al. used Bei Dou Navigation Satellite System (BDS) and GPS SNR to invert volumetric soil moisture (VSM) changes and wheat growth, and compared the results with the original observation values. The experimental results showed that GPS L1/L2 and BDS B1/B2/B3 frequencies in VSM retrieval are consistent with in situ VSM [13]. Liang et al. estimated the near-surface soil moisture using SNR data, corrected the original phase by obtaining the amplitude and phase of the SNR interferogram, weakened the influence on vegetation change, and established a genetic algorithm back propagation neural network (BPNN) model for soil moisture retrieval. Their experiments showed that the correlation between retrieval results and soil moisture was substantially improved [14]. Han et al. proposed a semiempirical signal-to-noise ratio model as a curve-fitting model to reconstruct direct and reflected signals from SNR data and extract frequency and phase information. The results showed that the soil moisture retrieval effect of the reconstructed signal, with a height angle of 5–15°, was more accurate, and the fitting quality increased by about 45% [15].

Jin et al. used SNR data to solve the different frequency phases of the SNR sequence by spectrum analysis and the least-squares method and fused the dual-frequency phase observation values with the entropy method. Finally, the fusion results were combined with the measured soil moisture to establish an empirical model to retrieve soil moisture. The results showed that the dual-frequency fusion method can effectively improve retrieval accuracy [16]. Ran et al. used the detrended signal-to-noise ratio (DSNR) sequence and proposed an arc-editing method to edit the DSNR sequence. Only the DSNR data with the typical interference mode chord waveform were retained, and the arc editing method of SMC retrieval was compared with the conventional method. The experimental results showed that the proposed method had higher SMC retrieval accuracy than the traditional method and could improve the retrieval accuracy of SMC in undulating terrain [17]. Han et al. proposed a method to reduce the impact of direct signal components by signal reconstruction and normalization according to the variation law of SNR. The results showed that under the high rough surface conditions, the normalized amplitude strongly correlated with in situ soil moisture. The quadratic model was used to invert soil moisture from the normalized amplitude, and the retrieval error was less than 0.085 $cm^3\ cm^{-3}$ [18]. Li et al. proposed a new soil moisture estimation method based on SNR data. A solution to SNR AAF was constructed based on the relationship between the amplitude attenuation factor (AAF) of the signal-to-noise ratio in in situ observations and soil moisture. The results showed that the measured soil moisture value was in good agreement with the estimated soil moisture range of 0.35–0.45 $cm^3\ cm^{-3}$, and the RMSE was less than 0.012 $cm^3\ cm^{-3}$ [19].

Roussel et al. obtained the amplitude and phase from SNR data and proposed a Global Navigation Satellite System reflectometer interference pattern technique to estimate the temporal variation in soil moisture content around a single earth antenna. Satellite signal-to-noise ratio (SNR) observation data with two satellite altitude angles of 2–30° and 30–70° were used. The experimental results showed that the method could effectively invert soil moisture [20]. Yu et al. proposed a combined linear method for snow depth retrieval using the phase of GPS triple-frequency signals. This method is independent of geometric freedom and is not affected by the ionospheric delay. The results showed that the accuracy of snow depth retrieval based on triple-frequency multipath error and SNR was high [21]. Zhang et al. proposed two new SMC estimation methods: triple-frequency carrier phase (TRFCP) and triple-frequency pseudorange (TRFP). The experimental results showed that the phase delay estimated by the two methods strongly correlates with Plate boundary observation (PBO) SMC [22]. Shen et al. proposed a BDS Medium Earth Orbit

(MEO) and Inclined Geosynchronous Satellite Orbit (IGSO) satellite multisatellite soil moisture retrieval method based on SNR observations. This method weakens the influence of environmental differences in different directions by considering the satellite repetition period. The experimental results showed that the estimation results of BDS IGSO and MEO soil moisture agreed with the in situ soil moisture fluctuation. They verified that the BDS MEO satellite could effectively capture sudden rainfall [23].

According to the above research results, most scholars have focused on analysing SNR data for GNSS-IR soil moisture retrieval. However, GNSS-IR soil moisture retrieval based on SNR observations has the following problems: SNR is useless for most GNSS users, as SNR does not always exist in the original GNSS file [9]; the performance of GNSS-IR with SNR as the system input depends, to a large extent, on the observation quality of SNR and whether the direct component (trend term) of SNR is successfully removed [21]. However, the actual SNR is often impure by abnormal noise, so the multipath SNR was obtained using a low-order polynomial to draw the trend term to characterize multipath information. Owing to the above two reasons, GNSS-IR performance based on SNR time series may be seriously inaccurate. At present, some scholars have used the triple-frequency signal combination for GNSS-IR snow depth detection and soil moisture retrieval, but the number of triple-frequency signal observation satellites is small, resulting in the time resolution of GNSS-IR retrieval of surface physical parameters being low. The duration of the signal-to-noise ratio for the effective satellite elevation angle is shorter, which is basically maintained at about 0.5–2 h, which is not conducive to enabling highly dynamic monitoring of soil moisture.

Based on the above problems, to compensate for the shortcomings of SNR observation, including too few triple-frequency signal satellites and low time resolution, by combining the current multimode and multifrequency development pattern of GNSS, GNSS can play a positive role in environmental monitoring. As such, GNSS-IR soil moisture retrieval based on multifrequency linear combination observation has not yet been studied. In this study, we constructed a soil moisture retrieval method based on multisatellite dual-frequency combined multipath error by constructing the dual-frequency pseudorange (DFP) and the dual-frequency carrier phase (L4 Ionosphere Free, L4_IF) multipath error calculation model affected by deionosphericity. We selected the five epochs before and after the time point of the effective satellite period and used a total of eleven epoch data to construct the multipath error model and error equation. We solved the delay phase for soil moisture retrieval. We used Plate Boundary Observatory (PBO) P041 (Boulder, CO, USA) site data to verify the proposed method and further verified the method by using the soil moisture data of MFLE station close to P041 station, which effectively improves the time resolution of GNSS-IR soil moisture estimation. To realize the high dynamic monitoring of soil moisture, this paper uses the multipath error of fewer epochs to calculate the delay phase for SMC inversion. Because soil moisture is often affected by vegetation cover, soil temperature, air humidity and other factors, to better improve the inversion accuracy of GNSS-IR soil moisture, we use ULR, BPNN and RBFNN to construct soil moisture prediction models and evaluate the accuracy of each model.

2. Materials and Methods

2.1. GNSS-IR SMC Retrieval Principle

2.1.1. GNSS Multipath Error Principle

In the actual measurement, the signal received by the GNSS receiver antenna is not only the signal directly from the satellite but also the signal reflected by the surface. The direct and the reflected signals entering the receiver antenna interfere with each other, which causes the observation value to deviate from the actual value and produce the multipath error. According to the roughness of the reflective surface of the ground, the two main types of reflection are diffuse reflection and specular reflection [22,24]. For convenience, we assume that multipath error is only caused by specular reflections. Figure 1 shows that direct signal and reflected signal generate corresponding interference effects at the receiver to form a composite interference signal.

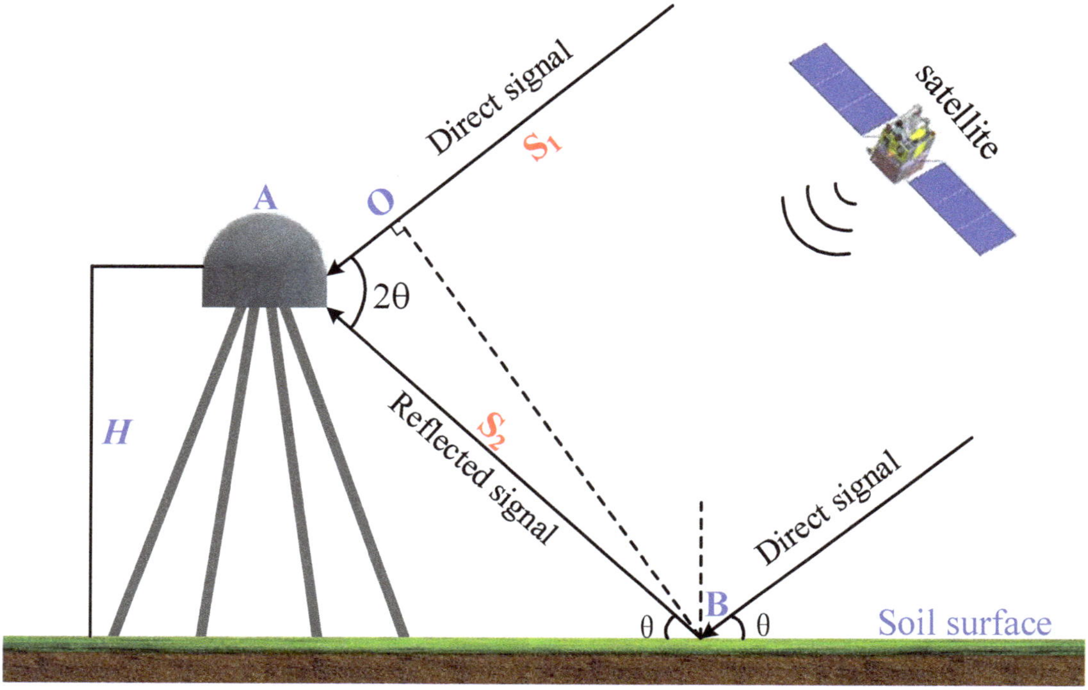

Figure 1. Schematic diagram of GNSS-IR SMC retrieval. After the satellite sends the signal, the right-handed circular polarized (RHCP) antenna receives the direct signal and the surface reflected signal, producing an interference effect at the receiver. A is the GNSS receiver antenna; B is the reflection point position of GNSS satellite signals passing through the ground; O is the footing point of the vertical line between the over-reflection point and the GNSS satellite direct signal; θ is the elevation angle of the satellite; S_1 is the direct satellite signal received by the receiver antenna; S_2 is the reflected signal reflected by the object's surface around the receiver antenna and enters the antenna; H is the vertical height from the antenna phase center to the ground.

The elevation of the GPS satellite (θ) determines the path delay ΔS, which is the extra distance travelled by the reflected signal compared to that of the direct signal.

$$\Delta S = 2H\sin\theta, \tag{1}$$

By substituting carrier wavelength λ into Equation (1), the phase delay $\delta\varphi(t)$ corresponding to the path delay can be expressed as:

$$\delta\varphi(t) = 2\pi\frac{\Delta S}{\lambda} = \frac{4\pi H}{\lambda}\sin\theta(t), \tag{2}$$

where λ is the carrier wavelength; t is the observation epoch. The phase delay is related to the antenna height, carrier wavelength and satellite elevation angle; that is, for a given antenna height and carrier signal, the phase delay is a function of the elevation angle. Equations (1) and (2) only consider the geometric delay and ignore the phase contribution of the Fresnel reflection and antenna radiation.

The amplitude A_C and phase difference $\beta(t)$ of the reflected signal compared to the direct signal can be expressed as:

$$\begin{cases} A_C = \sqrt{A_d^2 + A_m^2 + 2A_d \cdot A_m \cdot \cos\delta\varphi(t)} \\ \beta(t) = \tan^{-1}\left[\frac{\kappa\cdot\sin\delta\varphi(t)}{1+\kappa\cdot\cos\delta\varphi(t)}\right] \end{cases}, \tag{3}$$

where $A_m = \kappa \cdot A_d$, κ is the amplitude attenuation factor (AAF); S_C is the composite signal; A_C is the amplitude of S_C. The factor $\beta(t)$ is the composite excess phase with respect to the direct phase, which can be approximately expressed as [21]:

$$\beta(t) = \left(\frac{A_m}{A_d}\right) \cdot \sin\delta\varphi(t) = \kappa \cdot \sin\delta\varphi(t), \tag{4}$$

When only specular reflection is considered, the composite signal formed by the superposition of the direct signal and the reflected signal S_C can be expressed as [21]:

$$S_C = S_L + S_M = A_d\cos(\omega_0 t) + A_m\cos[\omega_0 t + \delta\varphi(t)] = A_C\cos[\omega_0 t + \beta t], \tag{5}$$

where S_L and S_M are the direct and reflected signals, respectively; A_d and A_m are the amplitudes of the direct and reflected signals, reflectively; ω_0 is the angular frequency of the signal.

2.1.2. Calculation of Multipath Error of Linear Combination of Observations

If only the influence of atmospheric delay error is considered, the GNSS code measurement pseudorange and carrier phase observation equation can be approximately described as [25]:

$$P_i = \rho + c(\delta t_R - \delta t_S) + T + I_i + E + MP_i + \varepsilon_{pi}, \tag{6}$$

$$\lambda_i\varphi_i = \rho + c(\delta t_R - \delta t_S) + T - I_i + E + M\varphi_i + \varepsilon_{\varphi_i} - \lambda_i N_i, \tag{7}$$

where i is the carrier number; φ is the carrier phase observation value; ρ is the geometric distance between the receiver and the satellite; c is the propagation velocity of electromagnetic waves in a vacuum; δt_R and δt_S are the clock difference between the receivers and satellite clock, respectively; N is the ambiguity of the whole week; λ is the carrier wavelength; M_P and M_φ are the pseudorange and carrier multipath error, respectively; T and I are the tropospheric and ionospheric delay errors, respectively; ε represents other unmodeled errors.

2.1.3. Error Calculation of Dual-Frequency Pseudorange Multipath

As shown in Equations (6) and (7), when i is 1 and 2, Equations (6) and (7) can be differentiated, respectively. Considering $|MP_i| \gg |M\varphi_i|$, $|\varepsilon_{pi}| \gg |\varepsilon_{\varphi i}|$ is as shown in Equations (8) and (9).

$$P_1 - \lambda_1\varphi_1 = 2I_1 + MP_1 + \varepsilon_{\varphi 1} + \lambda_1 N_1, \tag{8}$$

$$P_2 - \lambda_2\varphi_2 = 2I_2 + MP_2 + \varepsilon_{\varphi 2} + \lambda_2 N_2, \tag{9}$$

The relationship between carrier frequency f and ionospheric delay I is:

$$I_2 = \frac{f_1^2}{f_2^2} I_1, \tag{10}$$

By subtracting the L_1 and L_2 carrier phase observation equations and taking into account Equation (10), we obtain:

$$\lambda_1\varphi_1 - \lambda_2\varphi_2 = I_2 - I_1 + \lambda_1 N_1 - \lambda_2 N_2 = \left(\frac{f_1^2}{f_2^2} - 1\right) I_1 + \lambda_2 N_2 - \lambda_1 N_1, \tag{11}$$

Therefore, as shown in Equation (12):

$$I_1 = \frac{(\lambda_1\varphi_1 - \lambda_2\varphi_2 + \lambda_1 N_1 - \lambda_2 N_2) \times f_2^2}{f_1^2 - f_2^2}, \tag{12}$$

Therefore, the pseudorange multipath error equation at L_1 band is:

$$\begin{aligned} MP_1 &= P_1 - \lambda_1\varphi_1 - 2I_1 - \varepsilon_{P1} - \lambda_1 N_1 \\ &= P_1 - \tfrac{f_1^2+f_2^2}{f_1^2-f_2^2}\lambda_1\varphi_1 + \tfrac{2f_2^2}{f_1^2-f_2^2}\lambda_2\varphi_2 + K(N_1, N_2, \varepsilon_{P1}), \end{aligned} \tag{13}$$

Similarly, the pseudorange multipath error equation in L_2 band is:

$$MP_2 = P_2 - \frac{2f_1^2}{f_1^2 - f_2^2}\lambda_1\varphi_1 + \frac{f_1^2 + f_2^2}{f_1^2 - f_2^2}\lambda_2\varphi_2 + K(N_1, N_2, \varepsilon_{P2}), \tag{14}$$

where MP_1 and MP_2 are the multipath errors of the L_1 and L_2 carriers, respectively; P_1 and P_2 are the pseudorange observations of the L_1 and L_2 carriers, respectively; f_1 and f_2 are the frequencies of the L_1 and L_2 carriers, respectively; λ_1 and λ_2 are the L_1 and L_2 carrier wavelengths, respectively; φ_1 and φ_2 are the observed values of L_1 and L_2 carrier phases, respectively; N_1 and N_2 are the ambiguities of L_1 and L_2 carriers, respectively; ε_{P1} and ε_{P2} are other unmodeled errors.

In Equations (13) and (14), $K(N_1, N_2)$ is the integer ambiguity combination, which is usually a constant and does not affect the overall trend in the pseudorange multipath error. If no cycle slip occurs, it can be omitted. Therefore, the main factors affecting the quality of pseudorange multipath error are the noise in pseudorange observations and the accuracy of carrier phase observations and code pseudorange observations. If cycle slip occurs in carrier phase observations, it affects the value of pseudorange multipath error. Therefore, for the pseudo-range multipath error in each period, cycle slip detection and repair are required. We selected the multipath error MP_2 of the L_2 carrier as the research object.

The dual-frequency pseudorange multipath can be calculated by the linear combination of the pseudorange observations and the carrier phase observations, which can be expressed as [26]:

$$MP_1 = P_1 - \frac{f_1^2 + f_2^2}{f_1^2 - f_2^2}\lambda_1\varphi_1 + \frac{2f_2^2}{f_1^2 - f_2^2}\lambda_2\varphi_2, \tag{15}$$

$$MP_2 = P_2 - \frac{2f_1^2}{f_1^2 - f_2^2}\lambda_1\varphi_1 + \frac{f_1^2 + f_2^2}{f_1^2 - f_2^2}\lambda_2\varphi_2, \tag{16}$$

2.1.4. Multipath Error Calculation of Dual-Frequency Carrier Phase Linear Combination

For carrier phase observables, isolating the multipath from several other effects, such as the satellite antenna pseudorange and ionospheric and tropospheric delay, is difficult. However, when considering the multipath, the carrier phase contains two quantities: the phase of the direct signal and the composite excess phase with respect to the direct phase. Thus, the GNSS phase observation for carriers L_1 and L_2 can be expressed as [27]:

$$\begin{cases} L_1 = \lambda_1\psi_1 + \lambda_1\beta_1(t) \\ L_2 = \lambda_2\psi_2 + \lambda_2\beta_2(t) \end{cases} \tag{17}$$

where ψ_1 and ψ_2 represent the phase of the direct signal; $\beta_1(t)$ and $\beta_2(t)$ are the composite excess phase with respect to the direct phase.

In GNSS positioning, the carrier phase of the direct signal can be expressed as:

$$\lambda_i \psi_i = \rho + I_i + T + \Delta, \tag{18}$$

where i = 1 or 2, which is the two signals with different frequencies. The geometric distance between the GNSS satellite and antenna is represented by ρ, I_i represents ionospheric delays, T represents tropospheric delays, and Δ accounts for all other carrier-independent effects. To isolate the multipath, the dual-frequency carrier-phase combination L_4 is the $L_1 - L_2$ combination of carrier phase measurement, and its mathematical expression is [9,27]:

$$L_4 = L_1 - L_2 = I_1 - I_2 + \lambda_1 \beta_1(t) - \lambda_2 \beta_2(t), \tag{19}$$

Equation (19) eliminates the satellite clock error, receiver clock error, and tropospheric delay, as well as the geometric distance between the satellite and the receiver. That is the observed value of L_4 is affected by both ionospheric delay and phase error. Therefore, when analyzing the multipath error, a suitable method must be adopted to weaken the influence of the ionospheric delay on the multipath error. The high-order polynomial is used to fit the L_4 multipath error, and the dual-frequency linear combination multipath error that is not affected by ionospheric delay is obtained, which is the L4-free (L4_IF) observation value.

2.2. Establishment of Model Error Equation

According to Equations (1) and (2), the phase delay and path delay are functions of the elevation angle and reflector height. The reflector height changes with the SMC, which contributes to the change in the phase delay and path delay [12]. demonstrated that the agreement between H and the SMC was not as strong as that between $\delta\varphi(t)$ and the SMC. Therefore, the phase delay can be used to characterize the change in the SMC. To accomplish this, Equation (5) must be linearized first. Thus, the error equations can be expressed as:

$$\beta(t) + V_{\beta(t)} = \kappa_0 \sin \delta\varphi_0 + \sin \delta\varphi_0 V_\kappa + \kappa_0 \cos \delta\varphi_0 V_{\delta\varphi}, \tag{20}$$

where $\beta(t)$ can be easily calculated through Equations (16), (17) and (19). $V_{\beta(t)}$ is the residuals corresponding to the DFP multipath error and L4_IF multipath error. κ_0 denotes the initial value of the AAF. The initial value of the phase delay, i.e., $\delta\varphi_0$, can be calculated by incorporating the wavelength (λ), antenna height (H), and elevation angle (θ) into Equations (1) and (2), respectively. V_κ and $V_{\delta\varphi}$ are two unknown correction parameters corresponding to κ_0 and $\delta\varphi_0$ that need to be solved. To ensure the reliability of the solution and to avoid the influence of the excessive difference in multipath errors on the correction parameter solving, we assumed that the SMC remained the same in the short term (e.g., within 5 min) and employed a total of 11 multipath errors to solve the correction parameters through unweighted least squares adjustment. It is worth mentioning that the number of multipath errors is not limited to 11, but should be greater than or equal to the number of the correction parameters to be solved.

2.3. Solving the Phase Delay

The least-squares adjustment method is used to solve the delay phase. The L4_IF method is taken as an example to illustrate the solving process of the phase delay. For the 11 multipath errors of one observation session for a single GPS satellite, 11 error equations, such as Equation (20), can be obtained. Then, these error equations can be expressed by the following matrix:

$$\underset{11\times 1}{V_{\beta(t)}} = \underset{11\times 2}{A}\underset{2\times 1}{X} - \underset{11\times 1}{l}, \tag{21}$$

where

$$\underset{11\times 1}{V_{\beta(t)}} = \left[V_{\beta 1}\ V_{\beta 2}\ V_{\beta 3}\ \cdots\ V_{\beta 11}\right]^T, \tag{22}$$

$$\underset{2\times 1}{X} = \left[V_k\ V_{\delta\varphi}\right]^T, \tag{23}$$

$$\underset{11\times 2}{A} = \begin{bmatrix} \sin\delta\varphi_0 & \sin\delta\varphi_0 & \sin\delta\varphi_0 & \cdots & \sin\delta\varphi_0 \\ \kappa_0\cos\delta\varphi_0 & \kappa_0\cos\delta\varphi_0 & \kappa_0\cos\delta\varphi_0 & \cdots & \kappa_0\cos\delta\varphi_0 \end{bmatrix}^T, \tag{24}$$

$$\underset{11\times 1}{l} = \left[\beta_1 - \kappa_0\sin\delta\varphi_0\ \beta_2 - \kappa_0\sin\delta\varphi_0\ \beta_3 - \kappa_0\sin\delta\varphi_0\ \cdots\ \beta_{11} - \kappa_0\sin\delta\varphi_0\ \right]^T, \tag{25}$$

where $\beta(t)$ represents the L4_IF multipath error, $V_{\beta(t)}$ is the residual for $\beta(t)$. The subscripts $1, 2, 3, \ldots, 11$ denote the serial number of the selected 11 multipath errors. Thus, based on unweighted least-squares adjustment ($V^TV = \min$). X can be obtained by the following formula.

$$X = \begin{bmatrix} V_\kappa \\ V_{\delta\varphi} \end{bmatrix} = \left(A^TA\right)^{-1} \times \left(A^Tl\right) = \underset{2\times 2}{N^{-1}}\underset{2\times 1}{W}, \tag{26}$$

where $N = A^TA; W = A^Tl$. Because its rank equals one, N is rank deficient, which leads to a non-unique solution of X. Therefore, to obtain a unique solution, pseudo-inverse adjustment was adopted.

$$X = \begin{bmatrix} V_\kappa \\ V_{\delta\varphi} \end{bmatrix} = \underset{2\times 2}{N^{-}}\underset{2\times 1}{W}, \tag{27}$$

where N^{-1} and N^{-} is the Kelley inverse and the pseudo-inverse of N, respectively. Thus, the adjusted phase delay and AAF can be obtained:

$$\begin{bmatrix} \hat{\kappa} \\ \widehat{\delta\varphi} \end{bmatrix} = \begin{bmatrix} \kappa_0 + V_\kappa \\ \delta\varphi_0 + V_{\delta\varphi} \end{bmatrix}, \tag{28}$$

The diurnal phase delay representing the trend of the soil moisture change, namely the mean of the adjusted phase delay of twelve observation sessions, can be obtained by the following formula:

$$\overline{\delta\varphi_i} = \frac{\widehat{\delta\varphi}_{i,1} + \widehat{\delta\varphi}_{i,2} + \widehat{\delta\varphi}_{i,3} + \cdots + \widehat{\delta\varphi}_{i,12}}{12}, \tag{29}$$

where the subscript i equals $1, 2, 3, \ldots, 78$, and represents the ith day; the numbers 1, 2, 3, ..., 12 represent the serial number of observation sessions on a single day.

2.4. Data Sources

The experimental data are from the USA. Plate Boundary Observatory (PBO). We selected the GPS observation data of the P041 station (Figure 2) and the SMC data collected by the P041 and MFLE stations (https://cires1.colorado.edu/portal/) (accessed on 10 September 2021). The days of the year (DOYs) of selected observation data are between 45 and 131 of 2014 (Note: 57, 67–69, 71, 82, 94, 104, 105 days of soil moisture data are missing and have been eliminated).

The PBO SMC consisted of the site averaged SMC on a daily timescale, and the median SMC value of all satellite tracks for each day was used [22]. (https://data.unavco.org/archive/gnss/products/) (accessed on 15 September 2021). Located in Colorado, USA, and positioned at an altitude of 1728.8 m, the P041 station (39.9495°N, 105.1943°W). The MFLE station is about 210 m away from the P041 station, its altitude is 1727.3 m, and its position is longitude and latitude (39.9476°N, 105.1944°W). The soil moisture data of the MFLE station is used to further verify this method's reliability. In view of this, this paper takes

P041 station as the research station, MFLE station as the verification station, and MFLE station is very close to P041 station, as shown in Figure 2. The surrounding environment of P041 and MFLE stations are shown in Figure 3 (https://www.unavco.org) (accessed on 25 September 2021). The GNSS receiver and related parameters of the P041 station are shown in Table 1.

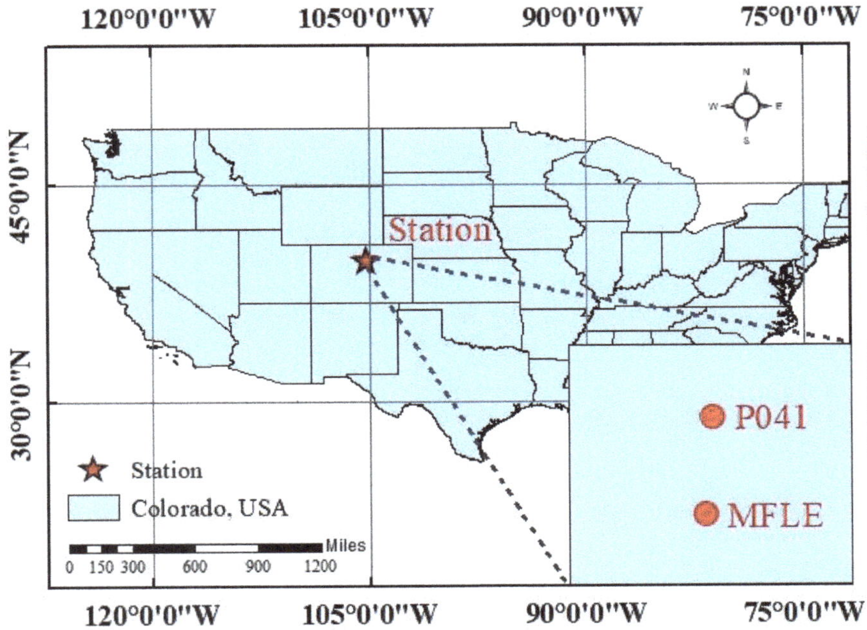

Figure 2. GNSS test site locations. P041 and MFLE stations, Boulder, CO, USA.

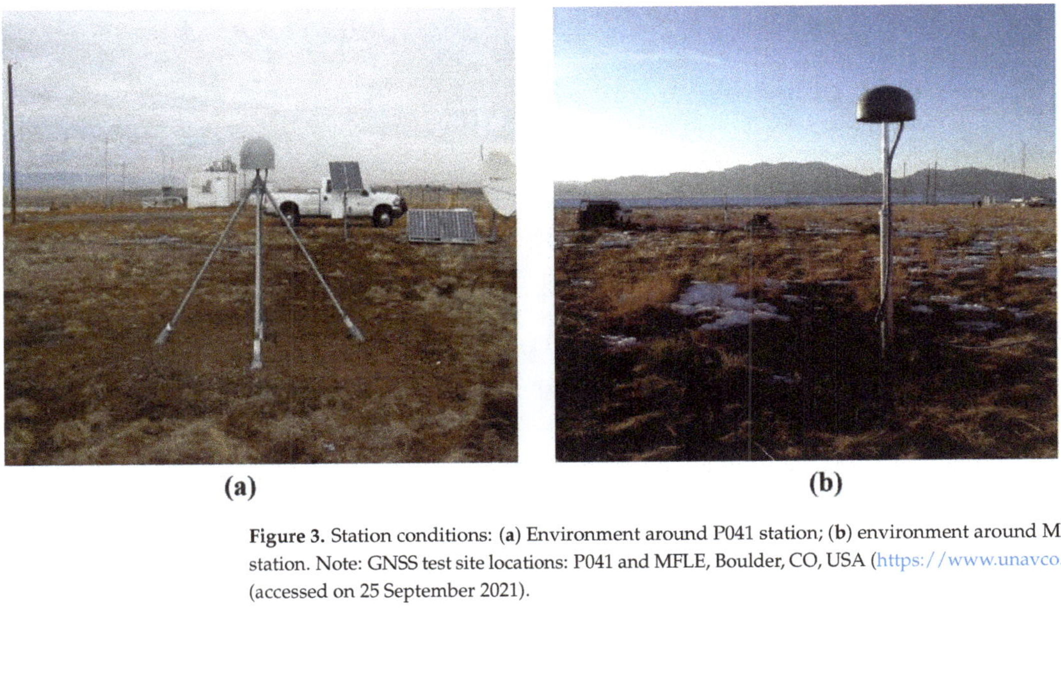

Figure 3. Station conditions: (**a**) Environment around P041 station; (**b**) environment around MFLE station. Note: GNSS test site locations: P041 and MFLE, Boulder, CO, USA (https://www.unavco.org) (accessed on 25 September 2021).

Table 1. Receiver and its related parameters at the P041 site.

Project	Parameter
Type of receiver	POLARX5
Sampling interval	15 s
Type of antenna	TRM59800.80
Antenna height	1.90 m

The P041 and MFLE stations can be seen in Figure 3a,b, respectively. They are located in a flat and open area, and the surrounding vegetation is scarce, and the land cover consists of exposed soils and short grasses. This paper selects the experimental data for the late winter and early spring. Therefore, the surface reflection signal is less affected by vegetation attenuation.

Figure 4 shows the observed SMC and precipitation data series corresponding to the time (DOYs: 45-131,2014) of selecting GNSS observation data from P041 and MFLE stations, which are presented in a line graph and a histogram, respectively. As demonstrated in Figure 4, six significant indigenous precipitation events occurred during the experiment, with a maximum precipitation of 24.8 mm, mainly during DOYs 63–66, 92–93, 102–103, 109–110, 127–128 and 130–131. Continuous precipitation led to a significant nonlinear increase in the SMC. As precipitation decreased or stopped, there was a decrease in the SMC. Evidently, precipitation was the primary factor that caused sudden changes in the SMC. The precipitation at the P041 and MFLE stations during the experimental period was appropriate and suitable for SMC retrieval.

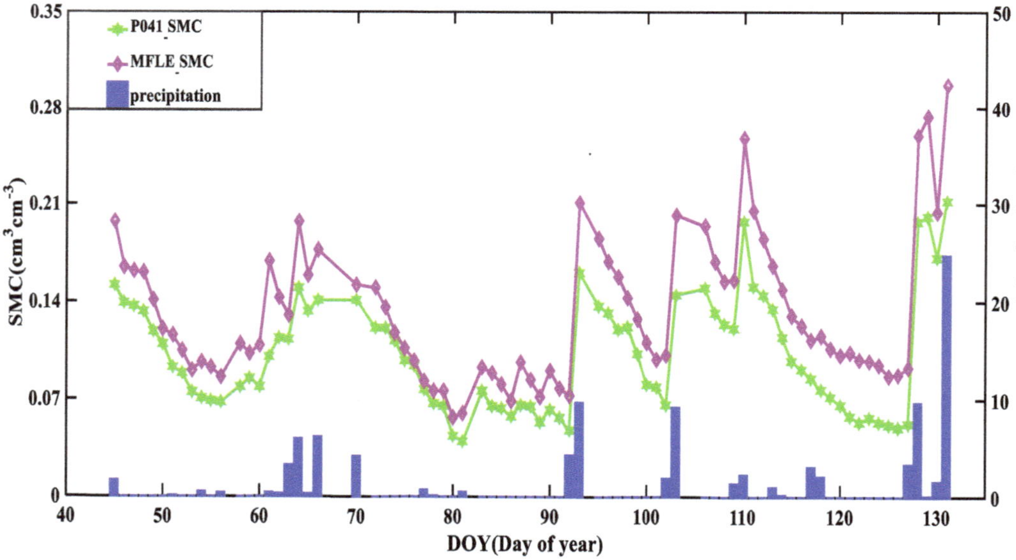

Figure 4. The soil moisture rainfall diagram during the experimental period.

3. Experiment and Results

3.1. Experimental Technical Scheme

Figure 5 shows the flow chart of the SMC retrieval technique used in this study. The technical route we followed can be divided into three parts: (1) We first pre-processed the GNSS-IR data: the carrier phase, pseudorange observation data, azimuth, elevation angle, and epoch extraction. We extracted other data parameters from the observation (OBS) file and navigation (NAV) file collected by GNSS receivers. (2) We determined the initial values of phase delay and amplitude attenuation factor, then constructed the dual-frequency carrier phase linear combination multipath error model and dual-frequency pseudorange multipath error model, and then calculated the multipath errors of the two models separately. We fit the ionospheric linear combination (L4), namely the simple difference between L_1 and L_2 (L4 $= L_1 - L_2$), with a high order to obtain L4_IF. We used a 10-degree polynomial fitting. (3) By constructing the error equation, we solved the delay phase. Then, we performed the delayed phase combination representing the variation trend in SMC and the retrieval SMC analysis. (4) Taking the multipath error of the five epochs before and after as the observation value, we calculated the dual-frequency pseudorange multipath error and the dual-frequency carrier-phase combination error and determined the initial phase and the amplitude attenuation factor (AAF). According to Equation (5), we established the model error equation. We used the Lomb–Scargle periodogram (LSP) and least-squares adjustment method to solve the delay phase. In solving the delay phase, the Pearson correlation coefficient (R) between the phase delay and SMC was used to characterize the changing trend between the delay phase and soil moisture.

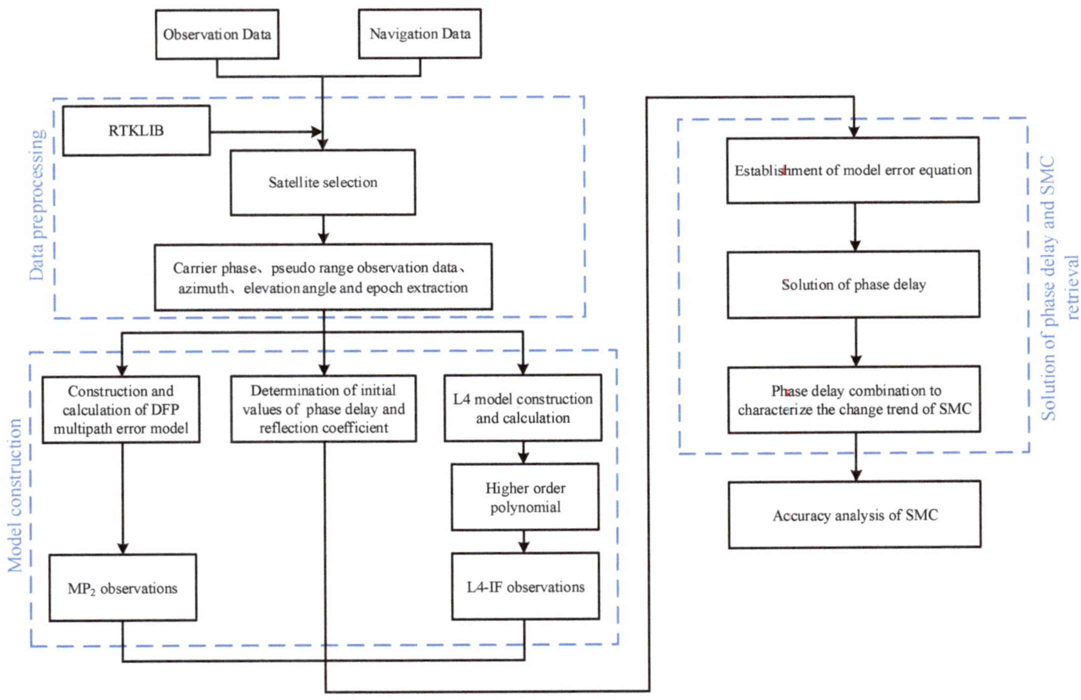

Figure 5. The technical process we followed for multisatellite dual-frequency combined multipath error SMC retrieval. Note: RTKLIB is an open-source program package for standard and precise positioning with GNSS (global navigation satellite system). RTKLIB consists of a portable program library and several APs (application programs) utilizing the library (http://www.rtklib.com/) (accessed on 10 February 2022).

3.2. *SMC Retrieval*

3.2.1. Choice of Elevation Angle

Given the lack of measured soil moisture data, we selected the soil moisture value from the P041 station estimated with the new GPS L2C carrier SNR as the reference value. The soil moisture value of the PBO site is based on a value per day and is the median value estimated by all satellites [22]. As soil moisture changes more during the day than that at night and more GPS satellites are visible during the day than at night, to simplify the calculation, generally, only four and eight observation periods are selected at night and day, respectively. As such, we used twelve observation periods every day. To ensure the solution's reliability and avoid excessive differences in multipath errors from affecting the parameter solution, we assumed that the soil moisture remained constant for a short time (e.g., within 5 min). We used the multipath error of the element used as the observation value for 11 multipath errors in total.

To provide a more intuitive view of the number of GPS satellites visible during the day compared to at night, Figure 6 shows the change in elevation of almost all GPS satellites at site P041 on 19 February 2014 (DOY: 2014–050). Figure 6 shows that more GPS satellites are visible during the day than at night, and the duration of a single GPS satellite being visible during 1 day is generally about 2.5–8 h, and the period at a low satellite elevation angle (<30°) is shorter, being maintained at about 0.5–2 h.

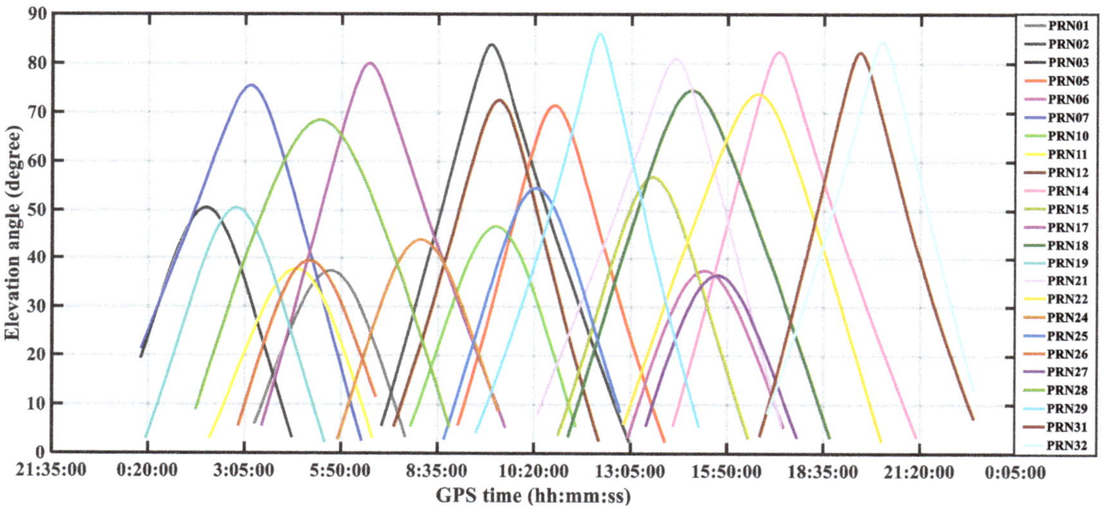

Figure 6. Changes in the elevation angles of almost all GPS satellites at P041.

Figure 7 shows the linear combination of GPS satellite G22 (DOY: 2014-050) observations at the test site. The multipath error varied with the elevation angle of the satellite. The black curve represents the change trend in MP_2 multipath error and L4_IF multipath error over time, and the red parabola represents the change trend in satellite elevation angle with time. Figure 8 shows that the satellite elevation angle occurs between 13:30 and 21:30. The time gradually increases to the maximum value and then slowly decreases until the satellite disappears. The changes in the MP_2 multipath error and L4_IF multipath error are closely related to the satellite elevation angle. When the satellite elevation angle is low, the MP_2 value oscillation amplitude can reach several meters, and the L4_IF value oscillation amplitude is relatively large. When it is high, the oscillation amplitudes of the MP_2 value and L4_IF value decrease accordingly. For increased accuracy, we selected the data at the lower elevation angle as the experimental data, as shown in Figure 8.

Figure 8 shows the multipath error of the linear combination observations of the GPS satellite G22 (DOY: 2014-050) at a satellite elevation of 5–25° for the test site GPS satellite. Figure 8 shows that as the satellite elevation angle increased, the multipath error showed a downward trend, which indirectly indicated that the carrier pseudorange and carrier phase observations of low-elevation satellites are more susceptible to the influence of multipath errors; it provides a theoretical basis for the selection of GNSS-IR satellite altitude angle. The satellite elevation angles in this experiment were all limited to 5–25°.

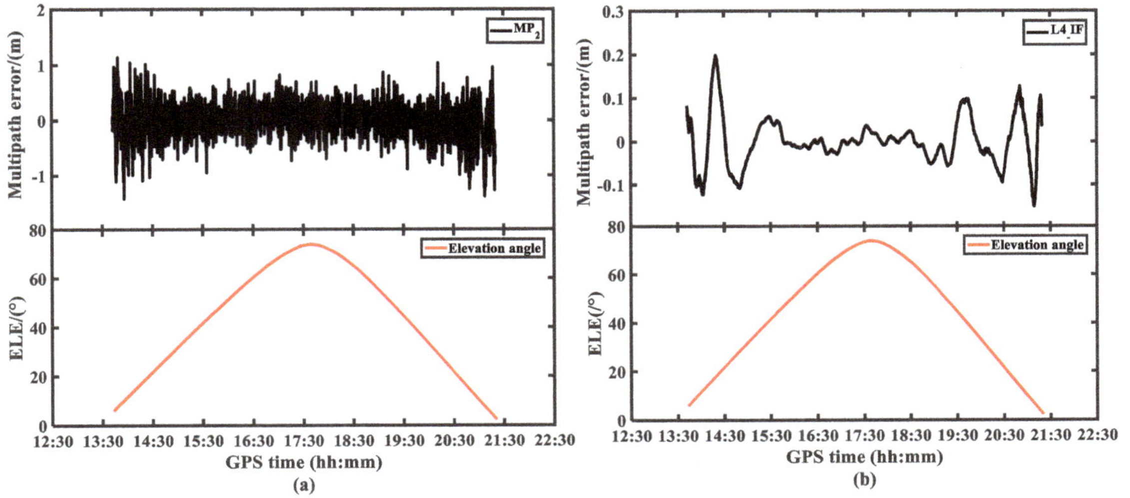

Figure 7. The multipath error varies with satellite elevation angle: (**a**) dual-frequency pseudorange multipath error; (**b**) dual-frequency carrier phase linear combination of multipath errors.

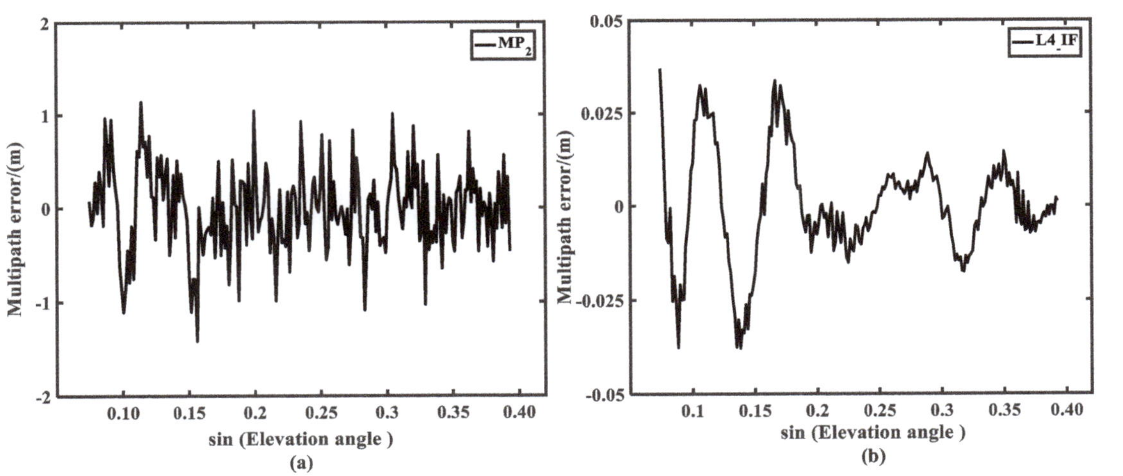

Figure 8. The multipath error of low elevation angle varies with the sine value of satellite elevation angle: (**a**) the dual-frequency pseudo multipath error MP_2; (**b**) the dual-frequency carrier phase linear combination multipath error L4_IF for removing ionospheric delay.

3.2.2. Choice of Azimuth

Satellite signal reflection point trajectory is a function of satellite elevation, azimuth, and antenna height. The reflection point trajectory reflects the position of the satellite relative to the receiver at a certain moment, from which the spatial information of the reflected soil can be obtained. Therefore, the advantages of the satellite reflection point trajectory map can be used, and the satellite azimuth angle can be considered to ensure as much consistency of the multiperiod multipath environment as possible. We selected the area with more visible satellites as the azimuth angle range at the experimental site.

Given the spatial differences in soil moisture and surface environment in the study area, the soil moisture at different locations was not entirely consistent, which led to the difference between the retrieval results based on the observation values at different reflection locations. In addition, not all DFP observations and L4_IF observations of the satellite elevation state contain the physical information of the surface reflector. Therefore, we needed to select observation values based on the study area environment, satellite elevation angle, and satellite azimuth angle to improve the accuracy of GNSS-IR soil moisture retrieval. As shown in Figure 9, as the reflection area of the soil was roughly the same, and on the premise of ensuring the number of satellites in each period, we selected an azimuth angle range of 30–330°. The satellites all had continuous observation values when their elevation angle was 5–25°, providing relatively sufficient observation data.

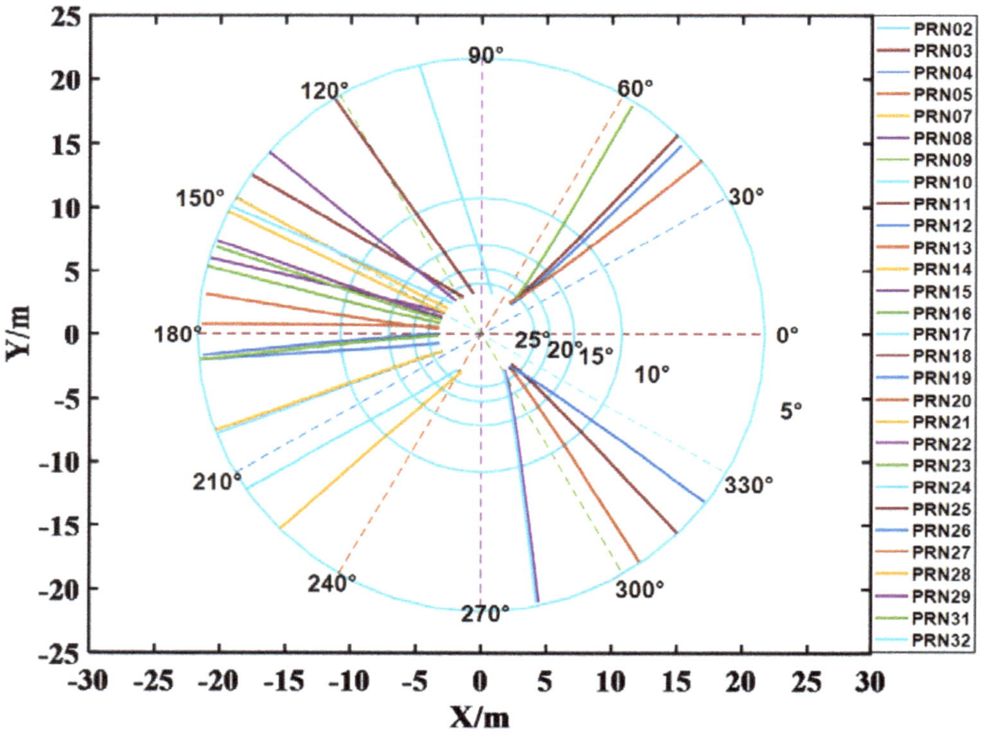

Figure 9. The projected trajectory of the P041 (DOY: 2014-065) satellite at an elevation of 5–25°.

3.2.3. Selection of Effective Satellites

The Fresnel reflection region of a GNSS signal is a set of ellipses related to the elevation, azimuth, and antenna height of a satellite. The first Fresnel reflection region contributes the most to the reflected signal, and the reflection medium change in the reflection zone strongly affects the relevant physical characteristics of the reflected signal.

The semimajor and semiminor axes of the ellipse of the first Fresnel reflection region of the ground-based GNSS-R can be determined by [28]:

$$S_x = h \cdot \cot\theta, \tag{30}$$

$$S_y = 0, \tag{31}$$

$$a = \frac{\sqrt{\lambda h \sin\,\theta}}{\sin^2\theta}, \tag{32}$$

$$b = \frac{\sqrt{\lambda h \sin\theta}}{\sin\theta}, \tag{33}$$

where S_x is the projection position of the reflection point on the ground connecting the satellite and the receiver; S_y is the position of the receiver projection point; a and b are semimajor and semiminor axes of the first Fresnel reflection region; h is the distance from the antenna phase center to the reflection surface; λ is the carrier wavelength; θ is the elevation angle of the GNSS satellite.

Figure 10 shows the first Fresnel reflection region ellipse group of the GPS L_2 carrier when the satellite azimuth was 0°, the satellite elevation was 5–25°, and the receiver antenna height was 1.9 m. The horizontal axis represents the distance from the receiver antenna to the direction of the satellite, and the vertical axis represents the distance perpendicular to the direction of the receiver antenna and satellite. When the satellite elevation angle increases, the Fresnel reflection region shrinks, and the reflection center is closer to the GNSS receiver. The area of the ellipse can be obtained from the semimajor and semiminor axes of the ellipse, that is, the area of the first Fresnel reflection region. Using the first Fresnel reflection region can avoid the influence of other multipath sources, such as tall buildings, vegetation, and water; maximize the unity of the reflection medium in the reflection zone; and be used for the design of the experimental plan. Therefore, the Fresnel reflection region provides a certain basis for determining the location of soil moisture sample collection points and selecting effective satellites. The change in the reflection medium in the Fresnel reflection region considerably affects the signal received by the receiver.

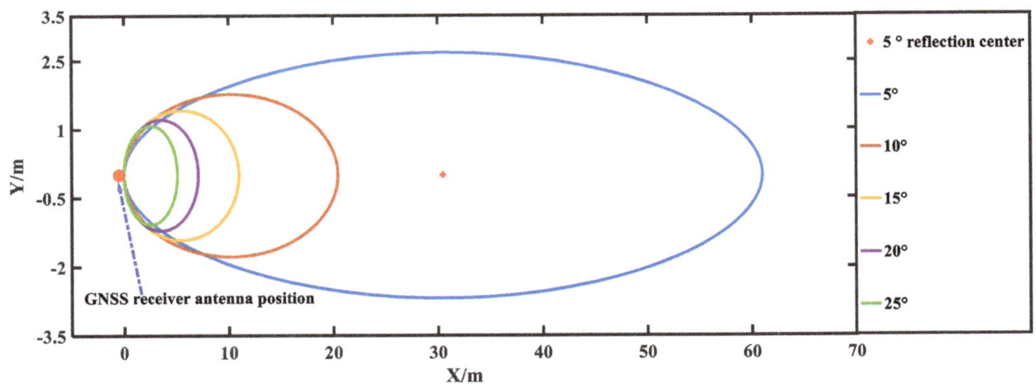

Figure 10. First Fresnel reflection region for GPS L_2 frequency at different elevation angles.

Figure 11 shows the first Fresnel reflection region map of station P041 (DOY: 2014-065). When the satellite elevation angle was 10°, the reflection point was up to 21 m away from the receiver. If slopes or other interference sources are near the measurement area, the first Fresnel reflection region should be introduced to reduce the excessive soil moisture retrieval error caused by different soil reflection media. The lower the satellite elevation angle, the larger the region of the first Fresnel reflection. When the elevation angle range is limited, the azimuth angle of each GPS satellite does not change much. Therefore, the calculation of the first Fresnel reflection region can ensure the unity of the medium in the reflection region, meaning that the homogeneity of the soil can be more accurately assessed and can provide a certain basis for other factors such as experimental planning and satellite selection.

The maximum power spectral density can be used to characterize the quality of a multipath error signal to a certain extent. Spectrum analysis using the Lomb–Scargle method can be used to judge satellite signal quality [22,29]. A satellite track should have a relatively stable singular dominant frequency for the same reflecting surface. In general, the dominant frequency's power spectral density (PSD) should be at least twice as high as the power of the noise or the second most powerful frequency in the periodogram [22,30].

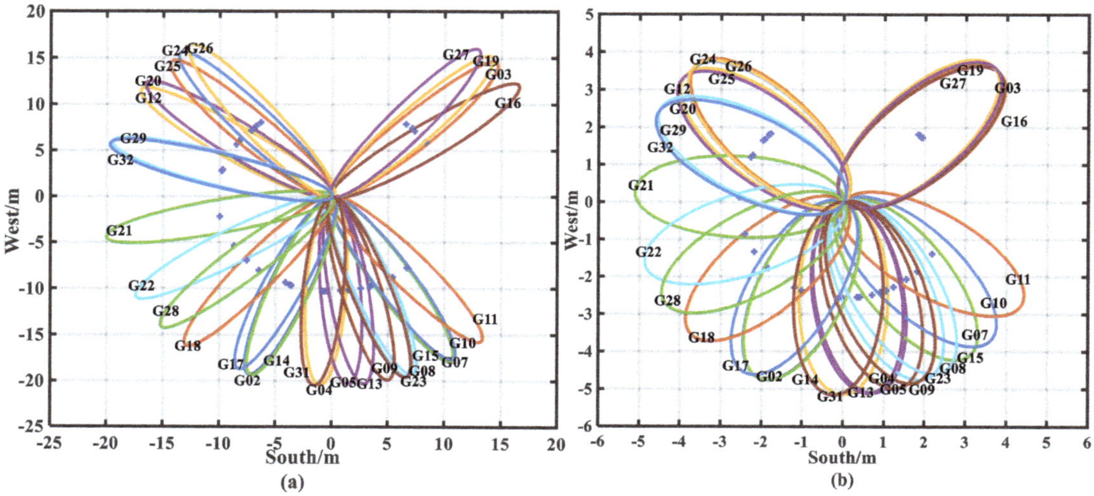

Figure 11. First Fresnel reflection region map of station P041 (DOY: 2014-065): GPS satellite distribution in the first Fresnel reflection region of station P041 when satellite elevation angle was (**a**) 10°; (**b**) 25°.

Figure 12a,b show that the power that meets the dominant frequency should be at least twice the power of the secondary main peak frequency, and that the quality of satellite data is higher. We used the PSD of dual-frequency pseudorange and dual-frequency carrier phase combined multipath error as the main auxiliary tool for satellite selection. Figure 12c,d lack any dominant frequency, which could have introduced substantial errors in the subsequent parameter estimation. Therefore, we also discarded such satellites when selecting satellites.

Based on the above analysis, after limiting the satellite elevation angle to about 5°–25° and after further considering the dual-frequency multipath error for satellite selection, we used twelve time periods every day: eight evenly distributed in the daytime and four and evenly distributed in the nighttime. Although the multipath error duration of a single GPS satellite in a day is generally about 2–8 h, the corresponding duration of the low satellite elevation angle multipath error is shorter, basically maintained at about 1 h. As a result, even fewer GPS satellites are available at low elevations. Therefore, we selected only one GPS satellite for each observation period. Combining satellite elevation angle, satellite reflection trajectory, satellite azimuth, first Fresnel reflection area, station environment, spectrum analysis, etc., Table 2 shows the results of the twelve time periods on 6 March 2014.

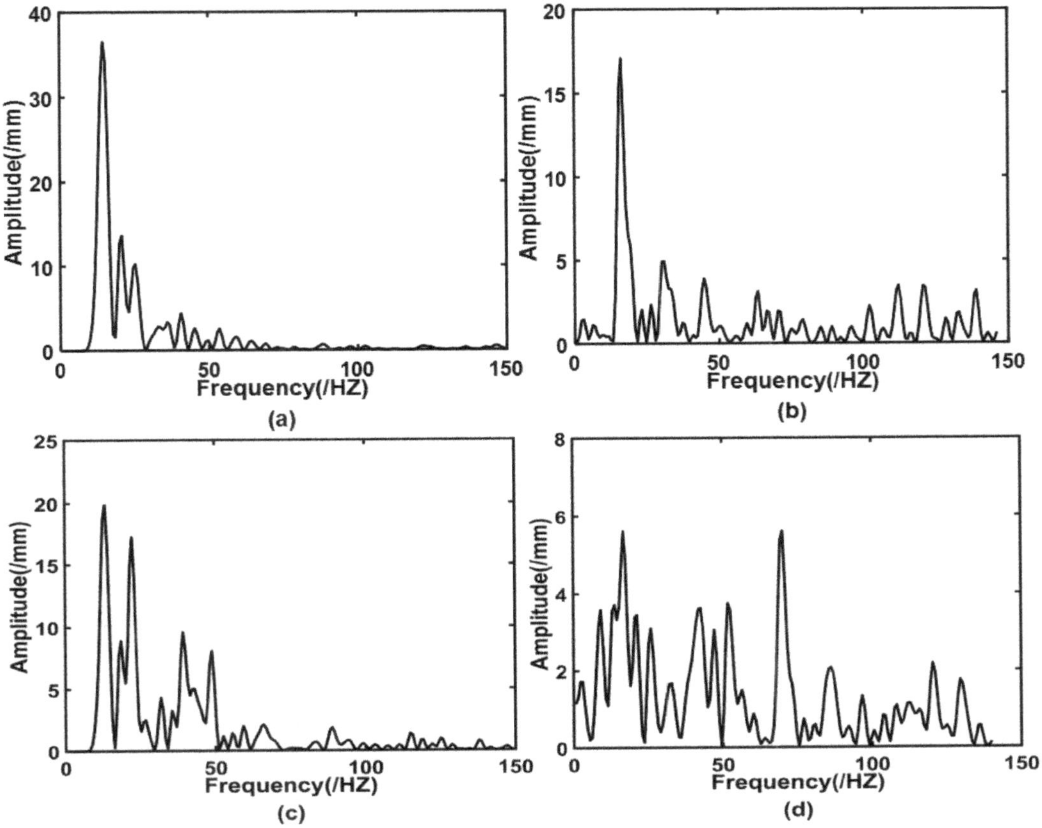

Figure 12. Multisatellite dual-frequency combined multipath error Lomb–Scargle periodogram. Spectrum analysis diagrams of the (**a**,**c**) dual-frequency carrier phase combined multipath error (L4_IF;) (**b**,**d**) dual-frequency pseudorange multipath error (MP_2).

Table 2. Satellite selection results at site P041 on 6 March 2014 (DOY: 2014-065).

Observation Period Number	GPS Satellite Number (PRN)	Azimuth (/°)	Height Angle (/°)	GPS Time (hh:mm:ss)
1	PRN16	170.75–171.16	6.08–7.05	01:28:45–01:31:15
2	PRN03	44.55–44.68	14.72–15.69	03:28:45–03:31:15
3	PRN26	239.89–240.82	24.45–25.25	05:58:45–06:01:15
4	PRN20	50.47–51.46	12.70–12.94	07:58:45–08:01:15
5	PRN24	221.92–222.65	13.37–14.22	09:58:45–10:01:15
6	PRN04	67.64–68.26	7.66–8.46	11:28:45–11:31:15
7	PRN15	156.52–157.09	14.87–15.82	12:28:45–12:31:15
8	PRN05	57.21–57.69	11.33–12.21	14:28:45–14:31:15
9	PRN16	275.01–275.94	6.19–6.73	15:28:45–15:31:15
10	PRN31	184.73–184.91	8.41–9.47	17:58:45–18:01:15
11	PRN25	73.92–75.11	22.27–22.47	19:58:45–20:01:15
12	PRN16	170.75–171.17	6.09–7.05	20:58:45–21:01:15

3.3. GNSS-IR SMC Retrieval Results

To obtain the parameters that characterize the change trend in soil moisture, we used the multipath error as the system input, and the amplitude attenuation factor and phase delay were the parameters to be estimated. We ignored the impact of the satellite elevation, soil moisture changes in the short term, and the amplitude attenuation factor and phase. The delay was a constant in the short term, and we obtained the error equation by linearizing Equation (5). The initial value of the phase delay was determined using Equation (2), where the satellite elevation angle is the satellite elevation angle corresponding to the first soil moisture sample collection time, considering that the soil surface reflectivity is between 0.3 and 0.8 [21]. In this study, we focused on the amplitude and phase of multipath error, so the actual value of κ is not required. Taking an κ initial value $\kappa_0 = 0.3$, we ignored the complex effect of antenna gain mode.

Considering the small changes in the sine values of antenna height and satellite height and our focus on the changing trend in the phase delay, we used the same initial phase delay value for all periods of data processing. For the solution of the parameters, we adopted the indirect adjustment method based on least squares, and the data processing followed these principles:

(1) Cycle slip detection and repair were conducted on the observed carrier phase value;
(2) We assumed that the amplitude attenuation factor and phase delay did not change in a short time;
(3) To avoid the large difference between the delay phase and the acquisition time of soil moisture, considering the necessary observation number, taking the acquisition time of soil moisture as a reference, we took the multipath error of five epochs before and after as the observation value. That is, we ignored the change in the delay phase with satellite elevation angle in the short term. We regarded the multipath error selected in each period as the repeated observation of the same parameter. According to Equations (15) and (16), the dual-frequency pseudorange multipath error MP could be calculated; According to Equation (19), the dual-frequency carrier-phase combination multipath error L4 could be calculated. We performed high-order fitting (we used 10-degree polynomial fitting) of L4 to remove the influence of ionospheric delay and obtain L4_IF. According to Equation (2), the initial value of the delay phase and the initial value of AAF (κ_0) are determined. According to Equation (5), we constructed the error equation of the corresponding method, and performed the Lomb–Scargle periodogram (LSP) and least square adjustment method. We solved each period to obtain a delay phase representing the change trend in soil moisture.

Combining the results of daily satellite selection, for the selected single satellite, we calculated the multipath error according to the dual-frequency signal of 11 epochs before and after the corresponding time. Combining the above principles and methods, we obtained the delay phase corresponding to the time of soil moisture acquisition through adjustment. Therefore, one phase delay could be calculated for each observation period, and 12 phase delays could be calculated for 12 observation periods in a day. By averaging the 12 results, we obtained the average phase delay, which is the daily phase delay. On a daily time scale, the L4_IF method and the DFP method were calculated separately to obtain the corresponding daily phase delay, and Time trends of soil moisture and delayed phase at P041 and MFLE stations were plotted (Figures 13a,b and 14a,b).

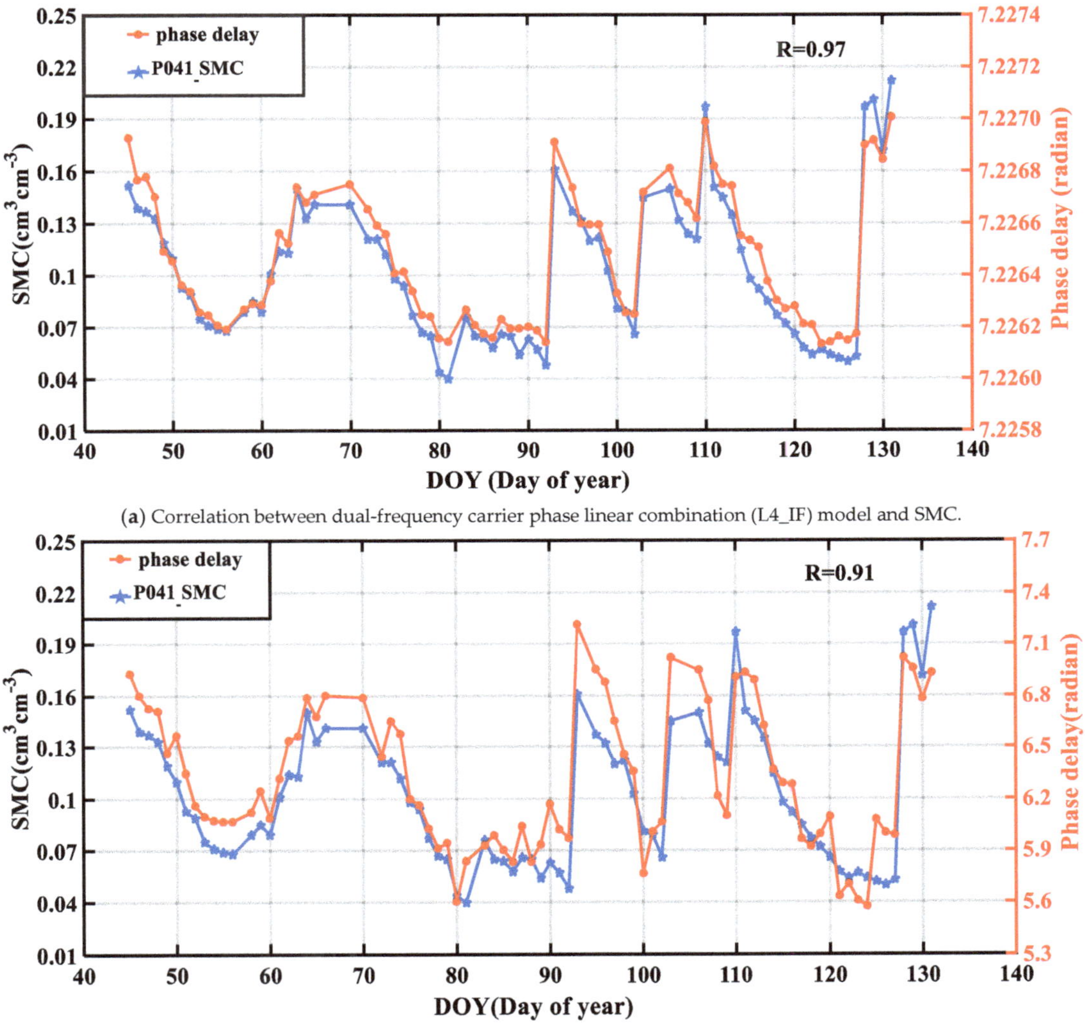

(**a**) Correlation between dual-frequency carrier phase linear combination (L4_IF) model and SMC.

(**b**) Correlation between DFP multipath model and SMC.

Figure 13. P041 station; correlation between Delay Phase and SMC.

To further test the L4_IF method and the DFP method, 50 d of the P041 and MFLE SMC, respectively, and the estimated phase delay were employed to build an empirical model, and the other 28 d of data were used to verify the model. The modelling method employed was unary linear. In the relationship between soil moisture and phase delay, in this study, we used the phase delay as the independent variable (x) and the soil moisture as the dependent variable (y) to perform univariate linear regression modeling. We obtained the scatter plot and regression shown in Figure 15.

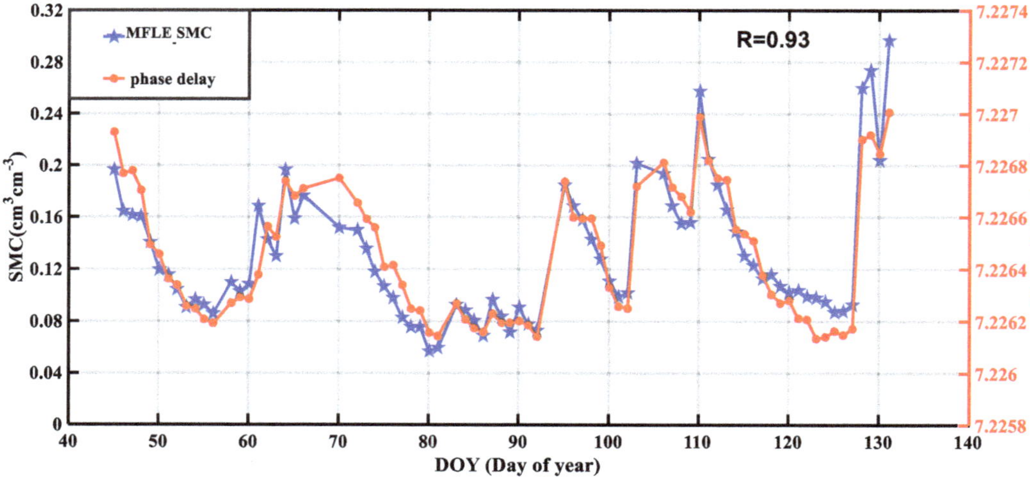

(**a**) Correlation between dual-frequency carrier phase linear combination (L4_IF) model and SMC.

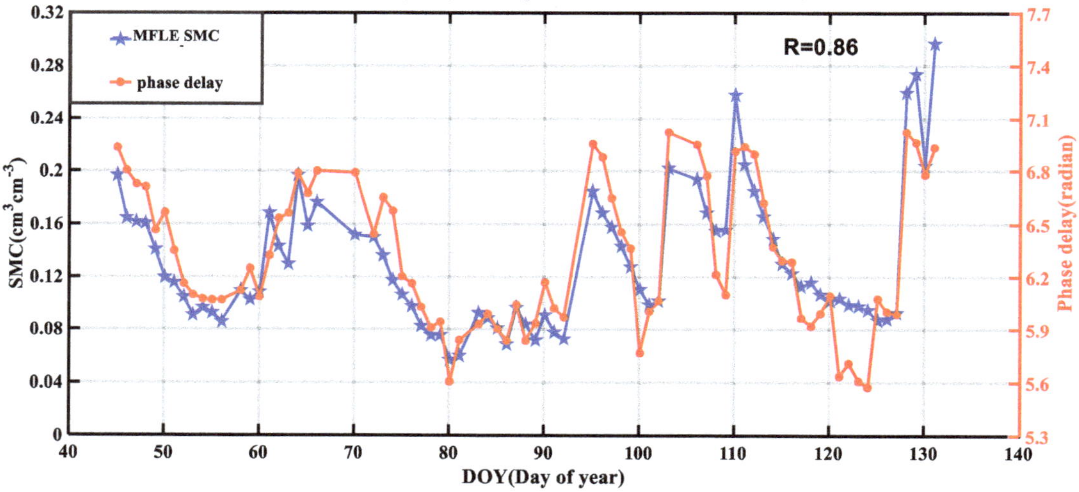

(**b**) Correlation between DFP multipath model and SMC.

Figure 14. MFLE station; correlation between Delay Phase and SMC.

To better reflect the soil moisture retrieval performance of the two methods of P041 and MFLE stations, we performed precision statistical analysis on each method separately, and the statistical results are shown in Table 3.

Soil moisture is often affected by vegetation cover, soil temperature, air humidity and other factors. In addition, due to the difference between the actual reflection conditions of GNSS satellite signals and the hypothetical ideal reflection conditions, the relationship between multi-path induced phase delay and soil moisture becomes complicated, which makes the ordinary linear model not necessarily able to describe the trend of soil moisture. Compared with other analytical models, the neural network model is less sensitive to the influence of soil flatness, surface vegetation and soil surface roughness. Therefore, in order to improve the inversion accuracy of GNSS-IR soil moisture, BPNN and RBFNN were used to construct soil moisture prediction models, and the prediction results were compared with those of ULR model, and the accuracy of each model was evaluated.

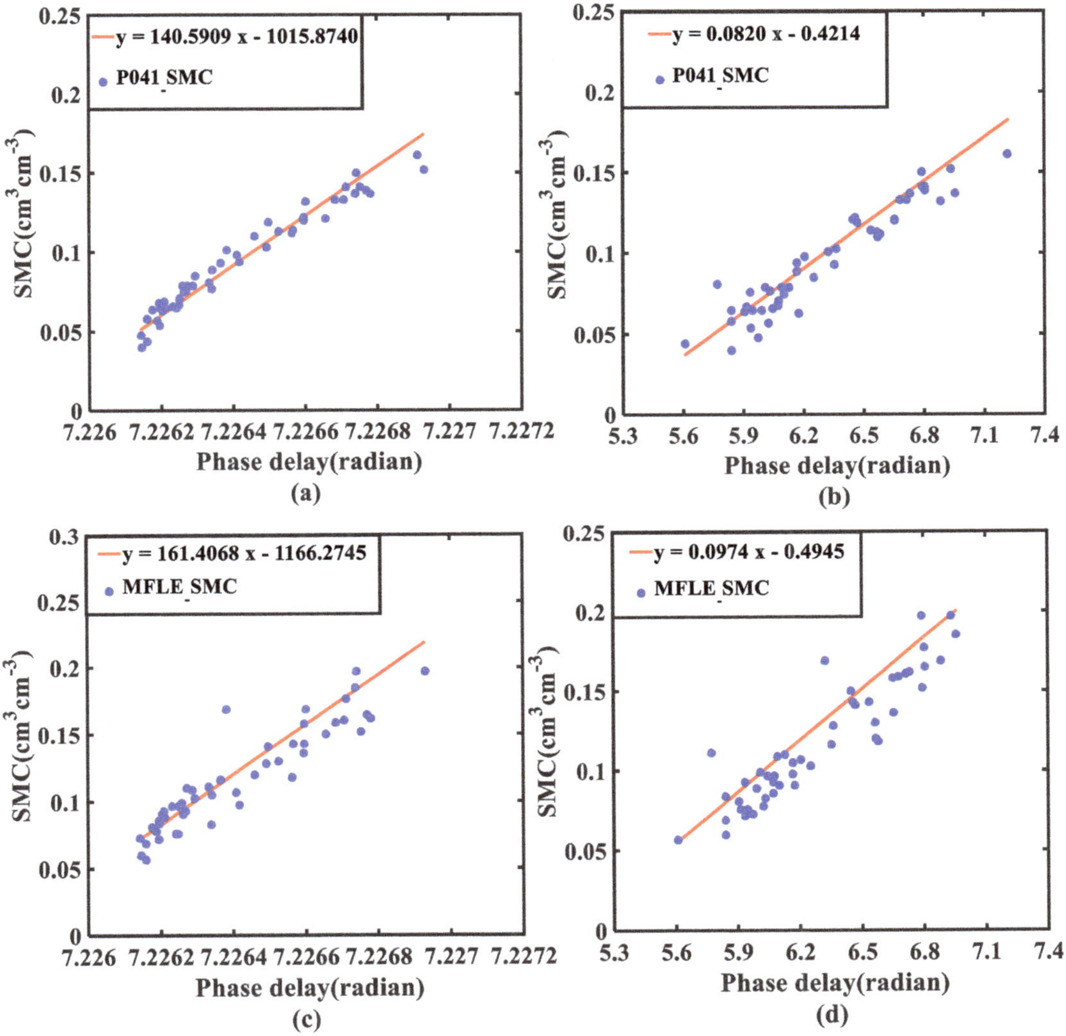

Figure 15. For the P041 station; scatter diagram of the SMC and regression equation based on (**a**) the L4_IF method; (**b**) the DFP multipath model. For the MFLE station; scatter diagram of the SMC and regression equation based on (**c**) the L4_IF method; (**d**) the DFP multipath model.

Table 3. Statistical table for accuracy comparison between L4_IF and DFP methods.

Station ID	Method	Correlation Coefficient (*R*)	STD (cm^3 cm^{-3})	MAE (cm^3 cm^{-3})	RMSE (cm^3 cm^{-3})
P041	L4_IF	0.97	0.040	0.037	0.014
	DFP	0.91	0.040	0.038	0.026
MFLE	L4_IF	0.93	0.047	0.043	0.029
	DFP	0.86	0.047	0.049	0.042

Note: The STD in Table 3 is the error standard deviation; the MAE in Table 3 is mean absolute error; the RMSE in Table 3 is root-mean squared error.

The phase delay and measured soil moisture of P041 and MFLE stations calculated based on L4_IF method and DFP method were used as data sources, and the training set and test set were divided according to the ratio of 7:3. That is, 50 d data were used as training set data, and 28 d data were used as test set data. BPNN and RBFNN are used to model. Figures 16 and 17 show the comparison between the predicted values of the three models of the two methods corresponding to the P041 station and the MFLE station and the measured values of soil moisture.

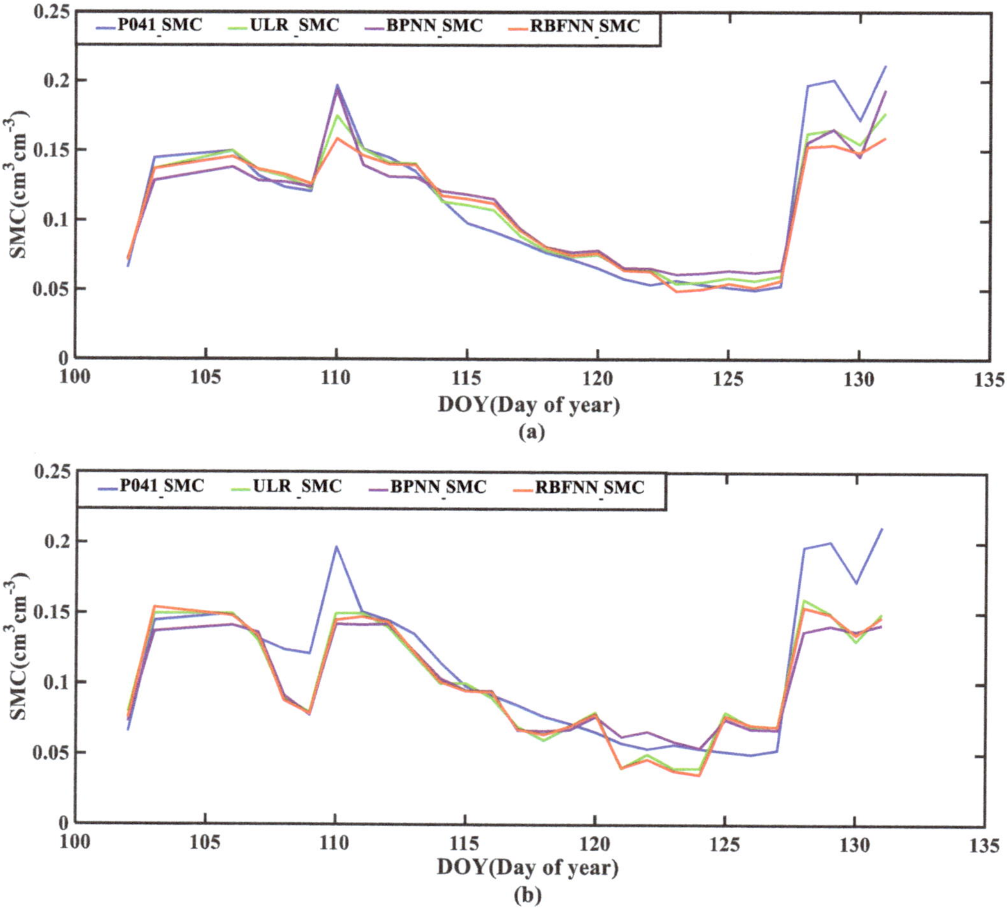

Figure 16. For P041 station; comparison between the predicted values of three models and measured values of soil moisture (**a**) the L4_IF method; (**b**) the DFP multipath model.

In order to better reflect the prediction results of three prediction models for soil moisture and the comparison results between different models, the accuracy of L4_IF and DFP methods corresponding to the three models of P041 and MFLE stations were statistically analyzed. The statistical results are shown in Table 4.

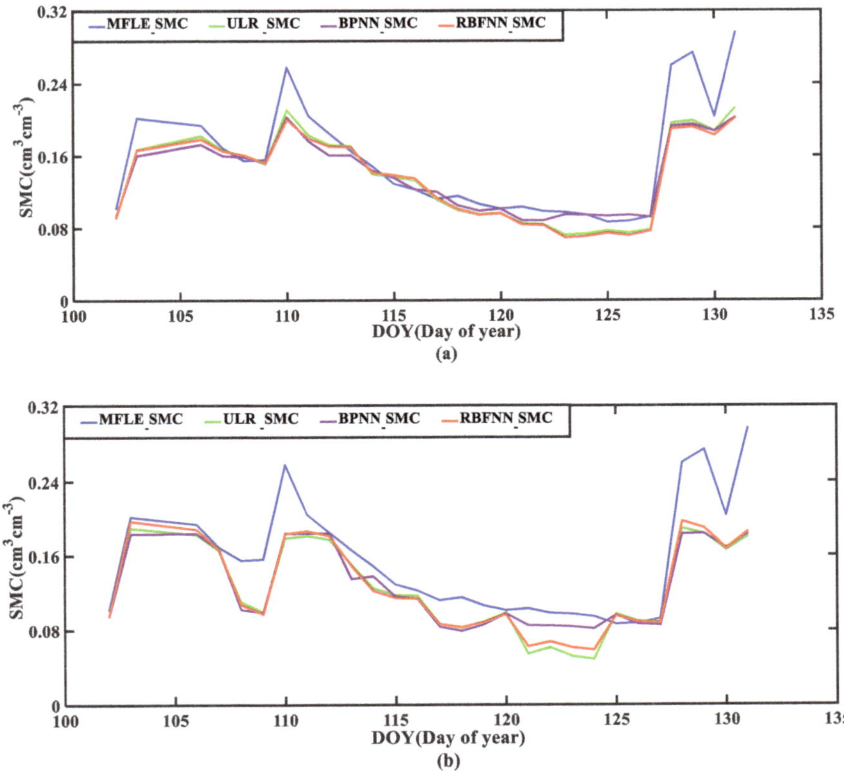

Figure 17. For the MFLE station, comparison between predicted values of three models and measured values of soil moisture (**a**) the L4_IF method; (**b**) the DFP multipath model. Note: The P041_SMC in Figure 16a,b are SMC value of the P041 station; the MFLE_SMC in Figure 17a,b are SMC values of the MFLE station; the ULR_SMC in Figures 16 and 17 are unitary linearity regression model predictive value; the BPNN_SMC in Figures 16 and 17 are back propagation neural network model predictive value; the RBFNN_SMC in Figures 16 and 17 are radial basis function neural network model predictive value.

Table 4. Statistical table for precision comparison of three models of L4_IF and DFP methods.

Station ID	Method	Model	STD (cm^3 cm^{-3})	MAE (cm^3 cm^{-3})
P041	L4_IF	ULR	0.040	0.037
		BPNN	0.039	0.034
		RBFNN	0.039	0.035
	DFP	ULR	0.040	0.038
		BPNN	0.033	0.033
		RBFNN	0.039	0.038
MFLE	L4_IF	ULR	0.047	0.043
		BPNN	0.039	0.037
		RBFNN	0.045	0.042
	DFP	ULR	0.047	0.049
		BPNN	0.042	0.046
		RBFNN	0.047	0.049

Note: The STD in Table 4 is error standard deviation; the MAE in Table 4 is mean absolute error; the ULR in Table 4 is unitary linearity regression; the BPNN in Table 4 is back propagation neural network; the RBFNN in Table 4 is radial basis function neural network.

4. Discussion

From the above experimental results can be drawn, Figures 13a,b and 14a,b show multiple apparent peaks in the phase delay and soil moisture, and the overall change trend has a strong consistency. For the P041 station, the correlation coefficients between the soil moisture and phase delay calculated based on the L4_IF multipath error and the DFP multipath error are 0.97 and 0.91, respectively. For the MFLE station, the correlation coefficients between the soil moisture and phase delay calculated based on the L4_IF multipath error and the DFP multipath error are 0.93 and 0.86, respectively, which are statistically significant. Figure 15a–d shows that the soil moisture values obtained using the L4_IF and DFP methods both fluctuate around the linear regression equation built by each, and the deviation is slight. The L4_IF and DFP methods have a statistical significance level of 0.01 and 0.05, respectively, regarding soil moisture retrieval. The probability values of the two methods are close to zero ($P \approx 0$), indicating that the correlation between the estimated phase delay and soil moisture at the P041 and MFLE stations is significant. Figures 16 and 17 show that the fitting degree between the predicted and measured values of the BPNN model and RBFNN model is better than that of the ULR model, and the error is smaller.

Table 3 shows that the standard deviation and absolute average error of the L4_IF and DFP methods are the same, but the root mean square error of the L4_IF method is smaller than that of the DFP method. Therefore, the L4_IF method is generally more accurate than the DFP method. The observation precision of the carrier phase is much higher than that of the pseudorange. To better compare the L4_IF and DFP methods, we used the error propagation law to compare the error level of L4_IF and DFP. Assuming that the observation error of the dual-frequency carrier phase has the same standard deviation of 1 mm ($\sigma_0 = \sigma_1 = \sigma_2 = 1\text{mm}$), the standard deviation of the GPS C/A code measurement pseudorange observation value is 2.93 m. Therefore, the error standard deviation of the L4_IF method is $\sigma_{L4} = \sqrt{2}\sigma_0 = 1.41$ mm, and that of the DFP method is $\sigma_{MP_2} = 2.93$ m $\sigma_{L4} < \sigma_{MP_2}$, also means that the quality of the multipath error of the L4_IF method is higher than that of the DFP method, so the L4_IF method obtained a higher correlation coefficient and root mean square error.

Table 4 shows that for the L4_IF and DFP methods of P041 and MFLE stations, the prediction results of the BPNN model are better than the RBFNN model and ULR model, and the RBFNN model is slightly better than the ULR model. The reasons may be as follows. Firstly, soil moisture is affected by vegetation cover, soil temperature, air humidity and other factors, and multipath reflection is not an imaginary mirror reflection, which makes it difficult to characterize other nonlinear characteristics between delay phase and soil moisture by using simple linear or exponential models. Secondly, the BPNN algorithm can correct the dielectric constant difference caused by weakening soil roughness, and has a certain inhibitory effect on surface fluctuation, soil roughness and vegetation.

5. Conclusions

Based on the analysis of the multipath error generation mechanism and the calculation model of GNSS measurement, we constructed a soil moisture retrieval method based on multisatellite dual-frequency combined multipath error. We proposed a method of estimating soil moisture using two combined dual-frequency signals. We used the L4_IF and DFP methods, which are independent of geometric distance, to estimate soil moisture. We verified the proposed method using the archived data set and SMC data of the P041 and MFLE stations. We drew the following conclusions through experimental analysis:

(1) The delay phases obtained by the multipath error solution and the soil moisture are strongly correlated. For the P041 station, the R values of the L4_IF and DFP methods are 0.97 and 0.91, respectively. For the MFLE station, the R values of the L4_IF and DFP methods are 0.93 and 0.86, respectively. Because the observation error of the L4 linear combination is low and the change in the ionospheric delay in the short term is small, we used the high-order fitting to further weaken the influence of the ionospheric

delay. The L4_IF method has higher R and soil moisture estimation accuracy than the DFP method. When the BPNN model, RBFNN model and ULR model are used to predict soil moisture, the results show that the prediction results of the BPNN model are better than the RBFNN model and ULR model, and the RBFNN model is slightly better than the ULR model. The results show that BPNN can improve the inversion accuracy of GNSS-IR soil moisture.

(2) Since the calculation of the phase delay only requires a small amount of multipath error compared to the soil moisture retrieval based on the SNR, the proposed method does not require the diagnostic signal frequency, and only a tiny number of epoch multipath errors needs to be used to calculate the delay phase. So, achieving high-time-resolution GNSS-IR SMC retrieval is easier. Therefore, this method can be used to easily obtain high-time-resolution and accurate soil moisture estimations.

(3) Given the changes in soil moisture, the reflectivity of the surface changes, which in turn will lead to changes in the amplification, attenuation factor κ, and phase delay. A further new finding is that the phase delay and the amplification attenuation factor κ based on the L4_IF method show the same change trend, and the Pearson correlation coefficient between them is 1. Conversely, the phase delay based on the MP_2 method and the amplification attenuation factor κ show the opposite trend, and the Pearson correlation coefficient between them is -1. These results show that the phase delay is closely related to the amplification attenuation factor κ. In other words, the amplification and attenuation factor κ can also be used for soil moisture estimation, which can obtain the same result as the phase delay. For the sake of brevity, this article does not list the research results of the amplification and attenuation factor κ.

If SNR does not exist in the original file of GNSS and the number of GNSS satellites with triple-frequency signals is still limited, the L4_IF method and DFP method proposed in this paper can be used as alternative methods for monitoring through GNSS-IR so as to enrich the data source of GNSS-IR soil moisture inversion and improve the ability of GNSS to serve environmental monitoring. However, it should be noted that not all satellite observations are suitable for estimating soil moisture, and different satellite selections will lead to different results. Therefore, the effective selection of satellites is still a challenge. Because this method does not need to diagnose the signal frequency and only needs less epoch multipath error to calculate the delay phase, it is easier to achieve higher time resolution of GNSS-IR soil moisture inversion. Soil moisture inversion based on a multipath error enriches GNSS-IR data sources and enhances the reliability of GNSS-IR.

Author Contributions: Conceptualization, Y.W., J.T. and P.L.; methodology, S.N.; formal analysis, J.X., N.L., M.W. and P.L.; data curation, J.X., D.H. and J.S.; writing—original draft preparation, S.N.; writing—review and editing, Y.W. and J.T.; supervision, Y.W. and J.T. All authors have read and agreed to the published version of the manuscript.

Funding: This research was funded by the Anhui Educational Commission Key Project, grant numbers: KJ2020A00706, KJ2021A1077, KJ2021A1078, KJ2021A1084; the Chuzhou Science and Technology Guidance Program Project, grant number: 2021ZD009; the Anhui Educational Commission General Project, grant number: KJ2021B08, KJ2021B07.

Data Availability Statement: The GNSS site data were provided under the PBO Observation Program of the United States, and the measured soil moisture data were obtained from https://cires1.colorado.edu/portal (accessed on 10 September 2021) and https://data.unavco.org/archive/gnss/products (accessed on 10 September 2021).

Acknowledgments: We thank the PBO Observation Program of the United States for providing the GNSS data, and the University of Colorado for providing data for soil moisture comparison analysis.

Conflicts of Interest: The authors declare no conflict of interest.

References

1. Seneviratne, S.I.; Corti, T.; Davin, E.L.; Hirschi, M.; Jaeger, E.B.; Lehner, I.; Orlowsky, B.; Teuling, A.J. Investigating soil moisture–climate interactions in a changing climate: A review. *Earth-Sci. Rev.* **2010**, *99*, 125–161. [CrossRef]
2. Ray, R.L.; Jacobs, J.M. Relationships among remotely sensed soil moisture, precipitation and landslide events. *Nat. Hazards.* **2007**, *43*, 211–222. [CrossRef]
3. Wanders, N.; Karssenberg, D.; de Roo, A.; de Jong, S.M.; Bierkens, M.F.P. The suitability of remotely sensed soil moisture for improving operational flood forecasting. *Hydrol. Earth Syst. Sci.* **2014**, *18*, 2343–2357. [CrossRef]
4. Schaufler, G.; Kitzler, B.; Schindlbacher, A.; Skiba, U.; Sutton, M.A.; Zechmeister-Boltenstern, S. Greenhouse gas emissions from European soils under different land use: Effects of soil moisture and temperature. *Eur. J. Soil Sci.* **2010**, *61*, 683–696. [CrossRef]
5. Jin, S.; Zhang, Q.; Zhang, X. New Progress and Application Prospects of Global Navigation Satellite System Reflectometry (GNSS + R). *Acta Geod. Cartogr. Sin.* **2017**, *46*, 1389–1398. [CrossRef]
6. Wan, W.; Chen, X.; Peng, X.; Huang, B.; Ming, X.; Hong, L.; Min, Z.; Pang, X.; Ting, Y.; Chang, Y.; et al. Overview and outlook of GNSS remote sensing technology and applications. *J. Remote Sens.* **2016**, *20*, 858–874. (In Chinese) [CrossRef]
7. Ge, L.; Han, S.; Rizos, C. Multipath Mitigation of Continuous GPS Measurements Using an Adaptive Filter. *GPS Solut.* **2000**, *4*, 19–30. [CrossRef]
8. Li, Y.; Chang, X.; Yu, K.; Wang, S.; Li, J. Estimation of snow depth using pseudorange and carrier phase observations of GNSS single-frequency signal. *GPS Solut.* **2019**, *23*, 118. [CrossRef]
9. Ozeki, M.; Heki, K. GPS snow depth meter with geometry-free linear combinations of carrier phases. *J. Geod.* **2012**, *86*, 209–219. [CrossRef]
10. Larson, K.M.; Small, E.E.; Gutmann, E.; Bilich, A.; Axelrad, P.; Braun, J. Using GPS multipath to measure soil moisture fluctuations: Initial results. *GPS Solut.* **2008**, *12*, 173–177. [CrossRef]
11. Larson, K.M.; Small, E.E.; Gutmann, E.; Bilich, A. Use of GPS receivers as a soil moisture network for water cycle studies. *Geophys. Res. Lett.* **2008**, *35*, 851–854. [CrossRef]
12. Larson, K.M.; Braun, J.J.; Small, E.E.; Zavorotny, V.; Gutmann, E.; Bilich, A. GPS multipath and its relation to near-surface soil moisture content. *IEEE J. Sel. Top. Appl. Earth Obs. Remote Sens.* **2010**, *3*, 91–99. [CrossRef]
13. Zhang, S.; Wang, T.; Wang, L.; Zhang, J.; Peng, J.; Liu, Q. Evaluation of GNSS-IR for Retrieving Soil Moisture and Vegetation Growth Characteristics in Wheat Farmland. *J. Surv. Eng.* **2021**, *147*, 04021009. [CrossRef]
14. Liang, Y.; Ren, C.; Wang, H.; Huang, Y.; Zheng, Z. Research on soil moisture inversion method based on GA-BP neural network model. *Int. J. Remote Sens.* **2019**, *40*, 2087–2103. [CrossRef]
15. Han, M.; Zhu, Y.; Yang, D.; Hong, X.; Song, S. A Semi-Empirical SNR Model for Soil Moisture Retrieval Using GNSS SNR Data. *Remote Sens.* **2018**, *10*, 280. [CrossRef]
16. Jin, L.; Yang, L.; Han, M.; Hong, X.; Sun, B.; Liang, Y. Soil moisture inversion method based on GNSS-IR dual frequency data fusion. *J. Beijing Univ. Aeronaut. Astronaut.* **2019**, *45*, 1248. [CrossRef]
17. Ran, Q.; Zhang, B.; Yao, Y.; Li, J. Editing arcs to improve the capacity of GNSS-IR for soil moisture retrieval in undulating terrains. *GPS Solut.* **2022**, *26*, 1–11. [CrossRef]
18. Han, M.; Zhu, Y.; Yang, D.; Chang, Q.; Hong, X.; Song, S. Soil moisture monitoring using GNSS interference signal: Proposing a signal reconstruction method. *Remote Sens. Lett.* **2020**, *11*, 373–382. [CrossRef]
19. Li, Y.; Yu, K.; Jin, T.; Chang, X.; Zhang, Q.; Xu, C.; Li, J. Soil moisture estimation using amplitude attenuation factor of low-cost GNSS receiver based SNR observations. In Proceedings of the 2021 IEEE International Geoscience and Remote Sensing Symposium IGARSS, Brussels, Belgium, 11–16 July 2021; pp. 7654–7657. [CrossRef]
20. Roussel, N.; Frappart, F.; Ramillien, G.; Darrozes, J.; Baup, F.; Lestarquit, L.; Ha, M.C. Detection of soil moisture variations using GPS and GLONASS SNR data for elevation angles ranging from 2° to 70°. *IEEE J. Sel. Top. Appl. Earth Obs. Remote Sens.* **2016**, *9*, 4781–4794. [CrossRef]
21. Yu, K.; Ban, W.; Zhang, X.; Yu, X. Snow Depth Estimation Based on Multipath Phase Combination of GPS Triple-Frequency Signals. *IEEE Trans. Geosci. Remote Sens.* **2015**, *53*, 5100–5109. [CrossRef]
22. Zhang, X.; Nie, S.; Zhang, C.; Zhang, J.; Cai, H. Soil moisture estimation based on triple-frequency multipath error. *Int. J. Remote Sens.* **2021**, *42*, 5953–5968. [CrossRef]
23. Shen, F.; Sui, M.; Zhu, Y.; Cao, X.; Ge, Y.; Wei, H. Using BDS MEO and IGSO Satellite SNR Observations to Measure Soil Moisture Fluctuations Based on the Satellite Repeat Period. *Remote Sens.* **2021**, *13*, 3967. [CrossRef]
24. Bilich, A.; Larson, K.M. Mapping the GPS multipath environment using the signal-to-noise ratio (SNR). *Radio Sci.* **2007**, *42*, 1–16. [CrossRef]
25. Zhang, Z.; Guo, F.; Zhang, X. Triple-frequency multi-GNSS reflectometry snow depth retrieval by using clustering and normalization algorithm to compensate terrain variation. *GPS Solut.* **2020**, *24*, 52. [CrossRef]
26. Lv, J.; Zhang, R.; Yu, B.; Pang, J.; Liao, M.; Liu, G. A GPS-IR Method for Retrieving NDVI from Integrated Dual-Frequency Observations. *IEEE Geosci. Remote Sens. Lett.* **2021**, *19*, 1–5. [CrossRef]
27. Wang, N.; Xue, T.; Gao, F.; Xu, G. Sea Level Estimation Based on GNSS Dual-Frequency Carrier Phase Linear Combinations and SNR. *Remote Sens.* **2018**, *10*, 470. [CrossRef]
28. Katzberg, S.J.; Torres, O.; Grant, M.S.; Masters, D. Utilizing calibrated GPS reflected signals to estimate soil reflectivity and dielectric constant: Results from SMEX02. *Remote Sens. Environ.* **2006**, *100*, 17–28. [CrossRef]

29. Yu, K.; Li, Y.; Chang, X. Snow Depth Estimation Based on Combination of Pseudorange and Carrier Phase of GNSS Dual-Frequency Signals. *IEEE Trans. Geosci. Remote Sens* **2019**, *57*, 1817–1828. [CrossRef]
30. Chew, C.; Small, E.E.; Larson, K.M. An algorithm for soil moisture estimation using GPS-interferometric reflectometry for bare and vegetated soil. *GPS Solut.* **2015**, *20*, 525–537. [CrossRef]

remote sensing

MDPI

Article

Regional Ionospheric Corrections for High Accuracy GNSS Positioning

Tam Dao [1,*], Ken Harima [2,3], Brett Carter [1], Julie Currie [1], Simon McClusky [2], Rupert Brown [3], Eldar Rubinov [3] and Suelynn Choy [1]

1 SPACE Research Centre, School of Science, Royal Melbourne Institute of Technology (RMIT) University, Melbourne 3000, Australia; brett.carter@rmit.edu.au (B.C.); julie.currie@rmit.edu.au (J.C.); suelynn.choy@rmit.edu.au (S.C.)

2 Geoscience Australia, Canberra 2609, Australia; ken.harima@ga.gov.au (K.H.); simon.mcclusky@ga.gov.au (S.M.)

3 FrontierSI, Melbourne 3000, Australia; rbrown@frontiersi.com.au (R.B.); erubinov@frontiersi.com.au (E.R.)

* Correspondence: tam.dao@rmit.edu.au

Abstract: Centimetre-level accurate ionospheric corrections are required for a high accuracy and rapid convergence of Precise Point Positioning (PPP) GNSS positioning solutions. This research aims to evaluate the accuracy of a local/regional ionospheric delay model using a linear interpolation method across Australia. The accuracy of the ionospheric corrections is assessed as a function of both different latitudinal regions and the number and spatial density of GNSS Continuously Operating Reference Stations (CORSs). Our research shows that, for a local region of 5° latitude ×10° longitude in mid-latitude regions of Australia (~30° to 40°S) with approximately 15 CORS stations, ionospheric corrections with an accuracy of 5 cm can be obtained. In Victoria and New South Wales, where dense CORS networks exist (nominal spacing of ~100 km), the average ionospheric corrections accuracy can reach 2 cm. For sparse networks (nominal spacing of >200 km) at lower latitudes, the average accuracy of the ionospheric corrections is within the range of 8 to 15 cm; significant variations in the ionospheric errors of some specific satellite observations during certain periods were also found. In some regions such as Central Australia, where there are a limited number of CORSs, this model was impossible to use. On average, centimetre-level accurate ionospheric corrections can be achieved if there are sufficiently dense (i.e., nominal spacing of approximately 200 km) GNSS CORS networks in the region of interest. Based on the current availability of GNSS stations across Australia, we propose a set of 15 regions of different ionospheric delay accuracies with extents of 5° latitude ×10° longitude covering continental Australia.

Keywords: Global Navigation Satellite Systems (GNSS); Continuously Operating Reference Station (CORS); Precise Point Positioning (PPP); ionospheric corrections; ionospheric model

Citation: Dao, T.; Harima, K.; Carter, B.; Currie, J.; McClusky, S.; Brown, R.; Rubinov, E.; Choy, S. Regional Ionospheric Corrections for High Accuracy GNSS Positioning. *Remote Sens.* **2022**, *14*, 2463. https://doi.org/10.3390/rs14102463

Academic Editors: José Fernández, Juan F. Prieto and Serdjo Kos

Received: 21 April 2022
Accepted: 18 May 2022
Published: 20 May 2022

1. Introduction

Applications of the Global Navigation Satellite Systems (GNSS) Precise Point Positioning (PPP) technique are rapidly increasing due to the methods computational efficiency, low communications bandwidth requirement and relative accuracy. PPP users can obtain position solutions with centimetre- to decimetre-level accuracy in static and kinematic modes [1–3]. The main advantage of PPP compared to differential based GNSS positioning techniques (e.g., Real Time Kinematics, RTK) is that PPP is based on the State-Space Representation (SSR) of corrections approach [4,5]. The SSR approach allows for GNSS-PPP-related errors such as satellite orbits and clocks to be modelled using a global sparse network of GNSS Continuously Operating Reference Stations (CORS) infrastructure. This removes the need for users to operate nearby base stations (i.e., base line < 100 km) or within a local GNSS CORS network. However, a drawback of PPP is its relatively long solution convergence time, which limits its adoption in real-time applications requiring almost instantaneous centimetre-level positioning accuracy [4]. Tens of minutes to a few hours is required for PPP solutions to provide coverage to centimetre-level accuracy [6,7]. A significant amount of research effort has been invested in the past decade to reduce the solution convergence time [8,9]. One known solution is to strengthen the PPP underlining measurement model by including externally computed, accurate ionospheric delay estimates into the end-user PPP algorithms [10–14].

For single-frequency PPP, ionospheric models such as the Klobuchar and the Global Ionospheric Maps (GIMs) estimated from a sparse global CORS network are applied. The accuracy of the ionospheric corrections derived from these models is approximately 15–20 TECu and 2–4 TECu, respectively [15], where one Total Electron Content unit (1 TECu) = 16 cm of ionospheric correction on L1 frequency. While these ionospheric correction models are useful in aiding precise single-frequency positioning, ionospheric corrections at 2–4 TECu are not sufficiently accurate to accelerate PPP convergence times from a few hours to minutes. PPP that uses two or more frequencies eliminates the need for an ionospheric model, as dual-frequency GNSS receivers can remove the first order (~99%) ionospheric delays by forming the ionosphere-linear combination. However, this linear combination of two signals, whilst removing the effects of the ionospheric delays, increases the noise amplification factor by three. This weakens the measurement model in dual-frequency PPP and results in a long solution convergence time of typically a couple of hours [12,16].

To facilitate a fast solution convergence time of a few minutes for PPP, the ionospheric corrections need to be accurate to less than one decimetre, that is, <~0.6 TECu. This requires the use of high-density local/regional CORS infrastructure networks to accurately estimate the ionosphere delay corrections, as in the case of network-based RTK. Hence, the terminology of "PPP-RTK" being a hybrid between the global PPP solutions and high-density local-network-based RTK solutions. The convergence time of a PPP-RTK solution depends on the accuracy of the ionosphere corrections (typically < 10 cm), which is dependent on the combination of the density of the local CORS networks and its latitudinal extent [17].

Geoscience Australia, through its Positioning Australia program, aims to accelerate the adoption and development of GNSS positioning technology and applications in Australia. The aim will be achieved by integrating and upgrading the country's GNSS space-based infrastructure via the deployment of a Satellite-based Augmentation System (SBAS), and ground-based CORS infrastructure networks to provide high accuracy, high reliability fit-for-purpose GNSS positioning services throughout its territory [18]. The Analysis Centre Software (ACS), called 'Ginan', is a GNSS processing toolkit developed by Geoscience Australia for processing GNSS observations for geodetic applications, as well as computations of GNSS corrections products to enable high-accuracy GNSS applications. Ginan is a free-to-use open source and is made up of two distinct software entities: The Precise Orbit Determination (POD) and the Parameter Estimation Algorithm (PEA). PEA uses a Kalman filter to estimate precise satellite orbits from a global CORS network, as well as satellite clocks, vertical TEC and zenith troposphere delays. A country like Australia presents challenges when designing and developing a high-accuracy national GNSS positioning service. Australia is a vast landmass, most of which is sparsely populated and cannot economically justify hosting a high-density CORS network. Furthermore, Australia extends from 10° to almost 44° south, providing ionospheric disturbances at different scales.

This research assesses the achievable accuracy of a local ionospheric delay model for Australia based on the linear interpolation method. The assessment will be conducted by taking into consideration the number and spatial density of the GNSS CORS networks. The model is populated with points where the ionospheric effects are known—because these have been calculated based on readings taken at CORS. The effect of the ionosphere at a given point is calculated using a linear interpolation to that point from the known (CORS) points. As Australia has a large and sparsely populated landmass (its population and infrastructure are mainly concentrated along the coast), the density of the GNSS CORS networks across the continent is not uniformly distributed. Thus, it is also of interest to identify the optimum number of GNSS CORS stations, and the corresponding spatial density required to achieve centimetre-level ionospheric corrections, thus realising the ambition of the Australian Government Positioning program. The goal of this research is to contribute towards our understanding of the regions in Australia that may require the additional investment and densification of CORS infrastructure for enabling high-accuracy positioning. The accuracy of the derived ionospheric corrections will be evaluated as functions of different latitudinal regions and the spatial density of the GNSS CORS networks across Australia. The Ginan toolkit will be used to generate between-satellite single-difference ionosphere delay estimates. These estimates will then be used to derive a series of regional ionospheric grid maps that are 5° latitude × ~10° longitude in size.

2. Methodology

2.1. Ionosphere Delay Estimation Using PPP Technique

PPP calls for the explicit modelling and estimation of biases on the GNSS measurements. For this purpose, the undifferenced uncombined PPP measurement model is characterised as follows [19]:

$$L_{r,f}^{s} = R_r^s + c(dt_r - dt^s) + \tau_r^s - \frac{\lambda_f^2}{\lambda_1^2} I_r^s + \left(b_{r,f} + \lambda_f N_{r,f}^s - b_f^s\right) + \varepsilon_{L,f}^{s} \tag{1}$$

$$P_{r,f}^{s} = R_r^s + c(dt_r - dt^s) + \tau_r^s + \frac{\lambda_f^2}{\lambda_1^2} I_r^s + B_{r,f} - B_f^s + \varepsilon_{P,f}^{s} \tag{2}$$

where $L_{r,f}^{s}$ and $P_{r,f}^{s}$ are the carrier phase and pseudorange measurements corresponding to satellite s, station r and frequency f; R_r^s is the geometric distance; c is the speed of light; dt_r and dt^s are the receiver and satellite clock errors, respectively; τ_r^s is the slant troposphere delay between satellite and receiver; I_r^s is the ionosphere delay along the line-of-sight from a satellite to a receiver; $N_{r,f}^{s}$ is the carrier phase ambiguity; λ_f is the wavelength for the frequency f; $b_{r,f}$ and $B_{r,f}$ are the satellite hardware delays of the carrier phase and pseudorange measurements, respectively; b_f^s and B_f^s are the satellite hardware delays of the carrier phase and pseudorange measurements, respectively; $\varepsilon_{L,f}^{s}$ and $\varepsilon_{P,f}^{s}$ are the carrier phase and pseudorange measurement noises, respectively.

It is worth noting that the effects of the antenna phase centre offset and phase centre variation, solid earth tides, phase windup and relativistic effects have been corrected in the measurements. The Ginan toolkit is now capable of generating rapid and ultra-rapid orbit parameters, but, for the purposes of this research, the International GNSS Service (IGS) final products were used to correct for the satellite orbital errors.

The ionosphere-free combination of pseudoranges $\left(P_{r,IF}^{s}\right)$ and carrier phase measurements ($L_{r,IF}^{s}$) is taken as:

$$P_{r,IF}^{s} = \frac{\lambda_2^2}{\lambda_2^2 - \lambda_1^2} P_{r,1}^{s} - \frac{\lambda_1^2}{\lambda_2^2 - \lambda_1^2} P_{r,2}^{s} \tag{3}$$

$$L_{r,IF}^{s} = \frac{\lambda_2^2}{\lambda_2^2 - \lambda_1^2} L_{r,1}^{s} - \frac{\lambda_1^2}{\lambda_2^2 - \lambda_1^2} L_{r,2}^{s} \tag{4}$$

and replacing the satellite and receiver clock with:

$$dt_{IF,r} = dt_r + \frac{\lambda_2^2}{\lambda_2^2 - \lambda_1^2} B_{r,1} - \frac{\lambda_1^2}{\lambda_2^2 - \lambda_1^2} B_{r,2} \tag{5}$$

$$dt_{IF}^{s} = dt^s + \frac{\lambda_2^2}{\lambda_2^2 - \lambda_1^2} B_1^s - \frac{\lambda_1^2}{\lambda_2^2 - \lambda_1^2} B_2^s \tag{6}$$

Therefore, the $\left(L^s_{r,IF}\right)$ and $\left(P^s_{r,IF}\right)$ can be written as follows:

$$L^s_{r,IF} = R^s_r + c(dt_{IF,\,r} - dt^s_{IF}) + \tau^s_r + A^s_{r,IF} \tag{7}$$

$$P^s_{r,IF} = R^s_r + c(dt_{IF,r} - dt^s_{IF}) + \tau^s_r \tag{8}$$

Equations (7) and (8) can be used to estimate the station positions, the satellite clock offset $cdt^s_{IF} = cdt^s + B^s_{IF}$, station clock offset $cdt_{IF,r} = cdt_r + B_{r,IF}$ and the tropospheric delays τ^s_r, as well as the real valued ambiguity $A^s_{r,IF}$ consisting of ambiguity plus the satellite and receiver hardware biases as shown in (9).

$$A^s_{r,IF} = b_{r,IF} - B_{r,IF} - b^s_{IF} + B^s_{IF} + \frac{\lambda_1\lambda_2}{\lambda_1+\lambda_2}N^s_{r,1} + \frac{\lambda_2\lambda_1^2}{\lambda_2^2-\lambda_1^2}(N^s_{r,1} - N^s_{r,2}) \tag{9}$$

The use of a wide-lane combination for ambiguity resolution is helpful due to their longer wavelength. Here, the Melbourne–Wübbena combination [20–22]

$$A^s_{r,MW} = \frac{L'^s_{r,1}}{\lambda_1} - \frac{L'^s_{r,2}}{\lambda_2} + \frac{\lambda_1-\lambda_2}{\lambda_1+\lambda_2}\left(\frac{P'^s_{r,1}}{\lambda_1} + \frac{P'^s_{r,2}}{\lambda_2}\right) \tag{10}$$

is used to isolate and resolve the wide-lane ambiguities $N^s_{r,WL} = N^s_{r,1} - N^s_{r,2}$. Equation (10) can be rewritten as

$$A^s_{r,MW} = \beta_{r,MW} - \beta^s_{MW} + N^s_{r,WL} \tag{11}$$

where $\beta_{r,MW}$ and β^s_{MW} are Melbourne–Wübbena combination receiver and satellite biases.

Once the wide-lane ambiguities have been solved, the ambiguity on the GPS L1 carrier phase can be isolated and resolved as well, as shown in (12).

$$\frac{\lambda_1+\lambda_2}{\lambda_1\lambda_2}A^s_{r,IF} + \frac{\lambda_1}{\lambda_1-\lambda_2}N'^s_{r,WL} = \beta_{r,IF} - \beta^s_{IF} + \frac{\lambda_1\lambda_2}{\lambda_1+\lambda_2}N'^s_{r,1} \tag{12}$$

For enabling global PPP solutions, the satellite positions or orbits, the satellite clock offset dt^s_{IF} and the phase bias estimates for GPS L1 and GPS L2 (b'^s_1, b'^s_2) can be used to deliver ambiguity-resolved PPP solutions to the end user.

$$b'^s_1 = \lambda_1\beta^s_{IF} - \frac{\lambda_1^2}{\lambda_2-\lambda_1}\beta^s_{MW} \tag{13}$$

$$b'^s_2 = \lambda_2\beta^s_{IF} - \frac{\lambda_2^2}{\lambda_2-\lambda_1}\beta^s_{MW} \tag{14}$$

To obtain the ionosphere delay estimates used in this research, the carrier phase measurements for L1 and L2 were corrected using (15) and (16).

$$L''^s_{r,1} = L'^s_{r,1} - \left(b'_{r,1} + \lambda_1 N^s_{r,1} - b^s_1\right) \tag{15}$$

$$L''^s_{r,2} = L'^s_{r,2} - \left(b'_{r,2} + \lambda_2 N'^s_{r,1} - \lambda_2 N^s_{r,WL} - b'^s_2\right) \tag{16}$$

The phase biases can be precisely estimated with the use of a global CORS network, and there are no errors associated with ambiguities once they are solved. Thus, the noise level of the corrected measurements is expected to be dominated by the un-modelled errors in the carrier phase measurement.

The geometry-free combination of these measurements can be used to obtain a precise estimate of the ionospheric delay (I^s_r) on the L1 = 1575.45 MHz carrier.

$$I^s_r = \frac{\lambda_1^2}{\lambda_2^2-\lambda_1^2}\left(L''^s_{r,1} - L''^s_{r,2}\right) + \frac{\lambda_1^2}{\lambda_2^2-\lambda_1^2}(B^s_2 - B^s_1) - \frac{\lambda_1^2}{\lambda_2^2-\lambda_1^2}(B_{r,2} - B_{r,1}) \tag{17}$$

It is worth noting that the ionosphere delay measurements obtained this way are slant ionospheric delay measurements between the line of sight from a GPS satellite to a receiver. These measurements are subject to the receiver and satellite hardware. For a carrier phase measurement, the delay estimation of 1.0 cm on L1 equates to 0.16 TECu.

2.2. GINAN Parameter Estimation Algorithm Processing

The Parameter Estimation Algorithm (PEA) is one of two components in the Ginan toolkit to process GNSS raw observations. The PEA reads observations in both RINEX and RTCM3 formats and is capable of estimating satellite clocks and zenith troposphere delays. The PEA application can be run on a Linux server that has sufficient memory (>8 GB) to cope with the data from a network of 80–250 stations [23].

In this research, the slant ionospheric delay corrections were derived from the geometry-free combination of GPS measurements (Equation (17)). To remove the receiver hardware bias, between-satellite single-differences (SD) of the ionospheric delay corrections at each CORS station in the test region were used, except one station, which is used for model evaluation purposes. The satellite hardware bias was still included in the ionospheric delay measurement in comparisons.

To compute the SD ionospheric corrections at every epoch, the measurements were subtracted from each visible satellite with respect to one of the reference satellites, which was the most common satellite in view in a local region.

$$I_{SD} = I_{si} - I_{so} \tag{18}$$

where I_{si} and I_{so} are the ionospheric delays for each visible satellite (si) and reference satellite (so), respectively.

The SD ionospheric corrections (I_{SD}) for each satellite include outliers (1% of data) occurring in the processing. To remove the outliers, we discarded the values larger than three standard deviations from the mean of all I delays of that satellite observed by all stations in the local network.

Once the SD ionospheric corrections for each epoch were computed, linear interpolation of the SD ionospheric corrections was carried out for all stations in a test region, except for one station (treated as the evaluation station). The interpolation forms a surface of ionospheric delays that fits with all ionospheric pierce points for at least 5 CORS stations in the test region to map the corrections for each single GPS satellites. From this mapping, the interpolated ionospheric delay corrections at the evaluation station were calculated and subsequently compared with the actual measured values. The differences between the interpolated ionospheric corrections and measured values at the evaluation CORS station were computed to evaluate the achievable accuracy of the ionospheric linear interpolation model. In this study, the accuracy presents the reliability of the ionospheric delays estimated from the regional mapping and the actual measurement. The mean accuracy was then taken as an average of all the accuracy retrieved on a whole day.

3. Results

3.1. Evaluation of the Ionospheric Corrections in Different Parts of Australia

Figure 1 shows the current distribution of the Australian GNSS CORS networks along with its states and territories. As Australia has a large and sparsely populated landmass, the density of the GNSS CORS networks is not uniformly distributed. Hence, it is of interest to evaluate the achievable accuracy of a linearly interpolated ionospheric corrections model based on the existing configuration of GNSS CORS networks. Furthermore, as the ionospheric effects vary as a function of latitude, it is therefore of importance to evaluate the ionospheric corrections in different regions. A series of analyses were undertaken based on selected test regions defined by latitudinal changes from low to mid-latitudes. The width (change in latitude) of test regions was defined as 5° latitude starting from 10°S to 45°S, and the length (change in longitude) varied according to the respective test regions. In general, it varied between 8° to 12° longitude, starting from 110°E to 155°E.

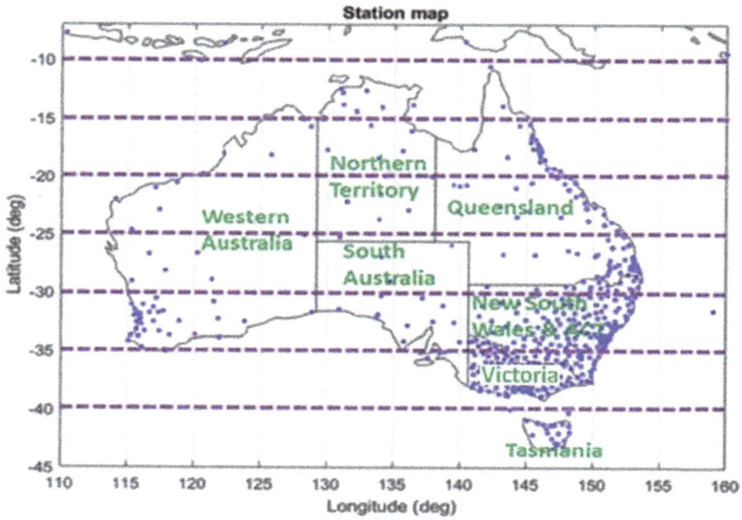

Figure 1. A map showing the geographical distribution of the Australian GNSS CORS networks (blue point). The test regions have a latitudinal width of 5° north south, whereas the longitudinal length varies between 8 to 12° east west.

More than 100 Australian CORS stations with emphasis on state-wide infrastructure networks were used to assess the latitudinal variation in the interpolated ionospheric delay corrections over Australia. An additional 70 global CORS stations were also selected to supplement and strengthen the processing model. Only GPS measurements were processed and used for estimating ionospheric corrections. An example of the geographic distribution of the selected CORS stations used is presented in Figure 2.

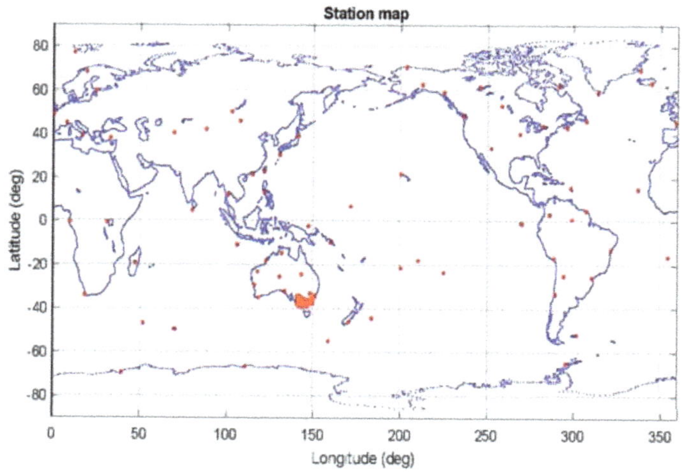

Figure 2. A map showing the geographical distribution of the GNSS stations (red aasterisks) used to compute the ionospheric corrections for the Victoria region. Respective statewide CORS stations in New South Wales (NSW), Australian Capital Territory (ACT), Western Australia (WA), North Territory (NT), Queensland (QLD) and South Australia (SA) were used to derive ionospheric corrections for those regions.

3.1.1. Victoria (VIC), New South Wales (NSW) and Canberra (ACT)

The State of Victoria is in the mid-latitude region between 34°S to 39°S. Victoria has the densest GNSS CORS network in Australia, with an average distance between stations of 80 km. One hundred and twelve state-wide Victorian stations were selected to assess the variability of the ionospheric corrections. The time series plots of the left of Figure 3 show the precision of the SD ionospheric corrections on 6 May 2021 at two Victorian CORS stations, ALBU and ANGS. The bar charts on the right of Figure 3 show the accuracy (difference) of the ionospheric model at these two stations. The accuracy was computed based on the differences between the interpolated ionospheric corrections and measured values at the evaluation CORS station. It is important to note that the measured slant ionospheric corrections at the evaluation stations were not used in generating the interpolation model. In addition, only values from 3 to 24 UT (LT = UT + 10) were considered in the evaluation, as the PPP network processing required a couple of hours for the solution to converge and stabilise. For a dense network in Victoria, a linear interpolation method of SD slant ionospheric corrections can reach an average accuracy of less than 5 cm, with a mean accuracy of 1.2 cm for ALBU station and 0.8 cm for ANGS station. There were some occasions where large variations can be found up to 10 cm, but 99% of the delay accuracy was distributed within 5 cm.

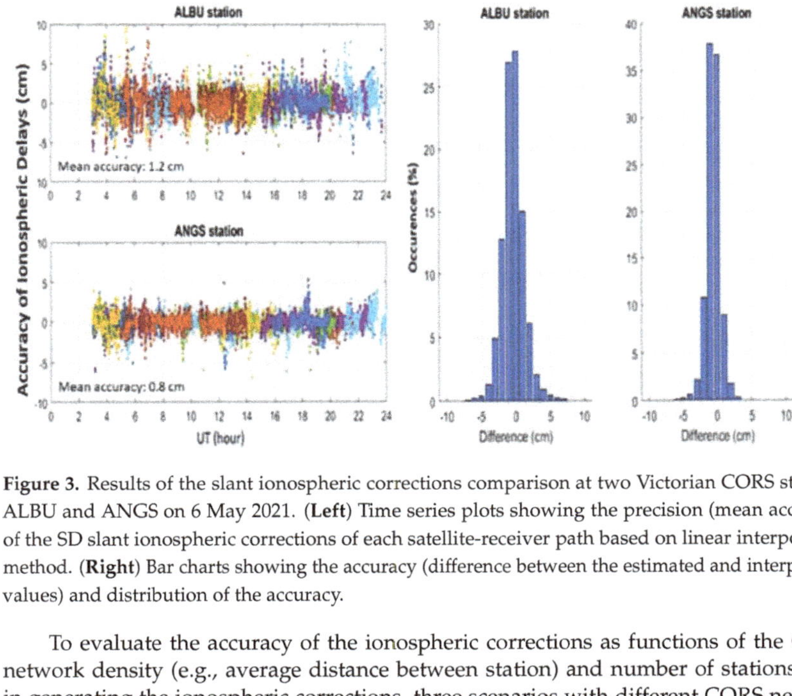

Figure 3. Results of the slant ionospheric corrections comparison at two Victorian CORS stations, ALBU and ANGS on 6 May 2021. (**Left**) Time series plots showing the precision (mean accuracy) of the SD slant ionospheric corrections of each satellite-receiver path based on linear interpolation method. (**Right**) Bar charts showing the accuracy (difference between the estimated and interpolated values) and distribution of the accuracy.

To evaluate the accuracy of the ionospheric corrections as functions of the CORS network density (e.g., average distance between station) and number of stations used in generating the ionospheric corrections, three scenarios with different CORS network configurations were simulated as shown in Figure 4a–c. The three scenarios were (a) 112 stations with an average between-station distance of 80 km; (b) 40 stations with an average between-station distance of 120 km; (c) 15 stations with an average between-station distance of 200 km. Figure 4d–f shows the mean accuracy for each station in the corresponding network. From the analysis, large differences were mainly observed at stations located at the region border and away from the network. In scenario (c), two stations (i.e., LIPO and CA10), located at the latitude of 34°S and far away from the network, showed larger differences than in scenarios (a) and (b).

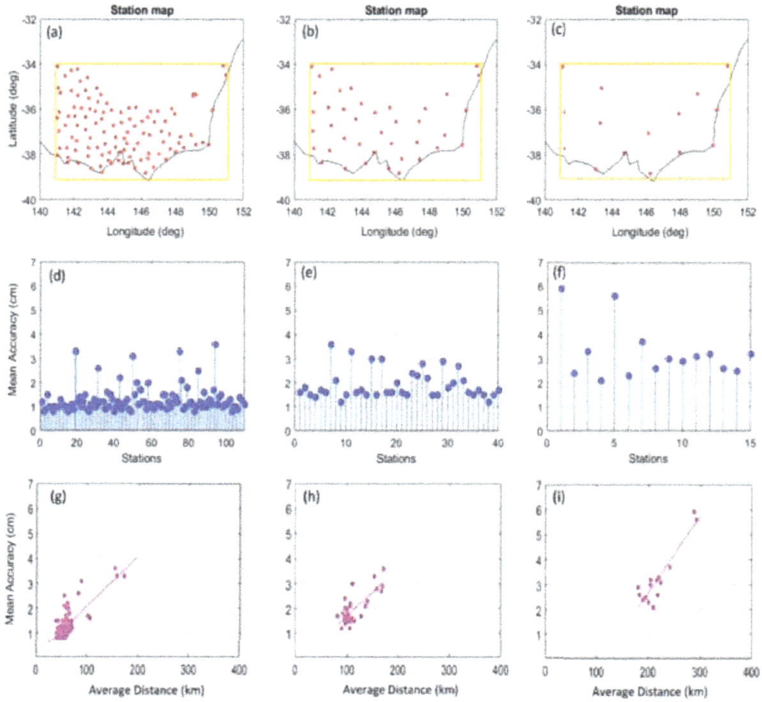

Figure 4. (**Top**: **a**–**c**) Three test scenarios in Victoria to assess the relation between the accuracy of the interpolated ionospheric corrections as function of the density of the CORS network configurations. The orange rectangles show the test area of 5° latitude ×10° longitude, and the red asterisks present for the CORS stations. (**Middle**: **d**–**f**) Plots showing the mean accuracy for each station in the corresponding network. (**Bottom**: **g**–**i**) Plots showing the accuracy of the interpolated ionospheric corrections as a function of the average separation distance between the testing station and its three nearest stations in the local network.

To examine the relation between the mean accuracy and the position of a testing station in a regional network, we calculated an average distance from the testing station to three nearest CORS stations, herein known as the average distance. Figure 4g–i presents the accuracy as a function of the average distance of the testing station in the network. For a region of 5° latitude ×10° longitude with 112 stations (g), the ionospheric corrections accuracy of 1 to 2.5 cm can be obtained with the average distance below 80 km, whereas approximately 3 cm and 4 cm mean accuracies can be reached with average distances of approximately 100 km and up to 200 km, respectively. These values are also the same as for the network of 40 stations (h). For a network of 15 stations (i), 13 stations with an average distance around 200 to 250 km have a mean accuracy from 2 to 4 cm, whereas two CORS stations situated at the border of the test region between VIC and NSW, approximately 300 km from the three nearest stations, have a mean accuracy of up to 6 cm. Therefore, it can be concluded that the number of stations and spatial distribution play major roles in influencing the achievable accuracy of the ionospheric corrections model.

To further validate the scenario with 15 stations evenly distributed in the 5° × 10° area, it was of interest to investigate if the inclusion of additional stations around the border would strengthen the ionospheric modelling around the border region, thus improving the overall accuracy of the corrections. Figure 5 (left) is a map of 15 CORSs with 10 VIC stations plus 5 neighbouring NSW stations. In this set up, the CORSs were well distributed across the test region compared to the previous test scenarios in Figure 4c. The average distance between the testing station to the three nearest stations in this test region is from 200 to 300 km. With this network configuration, an average of 5 cm accuracy can be achieved for all stations in the local area. The NSW and ACT areas also have a dense GNSS CORS network with an average distance of 120 km and an even distribution. Using this simple interpolation, we can achieve an accuracy below 5 cm.

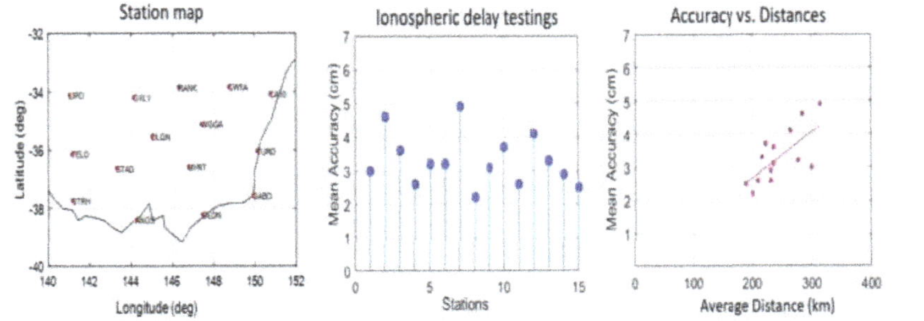

Figure 5. (**Left**) Maps of 15 CORS stations including names of ten stations in VIC and five stations (CWRA, RANK, WGGA, DLQN and OXLY) from the neighbouring NSW. The red asterisks present coordinates of those stations. (**Middle**) The average accuracy of the interpolated ionospheric corrections for each station on 6 May 2021. (**Right**) The accuracy as a function of the average separation distance of the testing station to the three nearest stations in the network, same as Figure 4.

3.1.2. North Territory (NT) and East Coast of Queensland (QLD)

Northern Territory (NT) and the northern part of Queensland (QLD) are in the low-latitude region (~0° to 30°S), where the electron density is expected to be higher than the mid-latitude region. In the NT and west QLD, the GNSS networks are sparse. In fact, the majority of the GNSS stations in QLD are located along the east coast, where the population concentrates. In NT, there are approximately nine stations covering an area of 5° latitude ×8° longitude. The yellow box in Figure 6 shows the test region defined for NT.

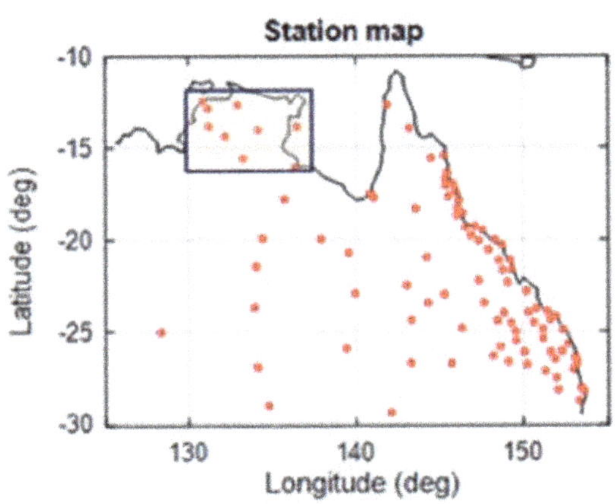

Figure 6. Map showing the GNSS CORS stations (red stars) in NT and QLD. The yellow box indicated the nominated test region for NT.

Figure 7 shows the differences between the interpolated ionospheric corrections and the measured values for each of the nine NT CORS stations on 6 May 2021. Based on the observations on this day, all stations show a high variation of up to 5 m, particularly between 3:30 to 8 UT, corresponding to 12:00 to 17:30 LT. These high variations occurred during the afternoon, when the Equatorial Ionospheric Anomaly (EIA) is at maximum density and were observed from some GPS satellites located on the north of the testing station. This caused a large deviation since the signals of those satellites pass through the EAI, where the high electron density content is.

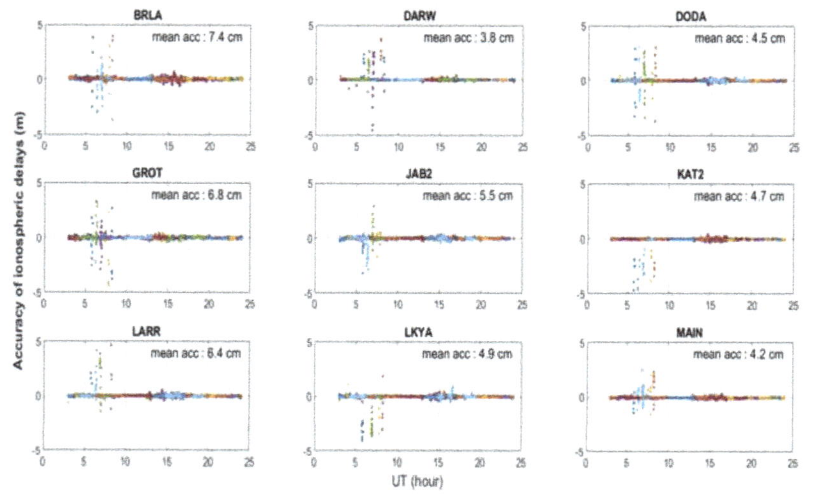

Figure 7. Plots showing the differences between the interpolated and measured ionospheric corrections as function of time at the nine NT CORS GNSS stations on 6 May 2021. The colours represent for different satellite observations.

Upon closer inspection of the results, as shown in Figure 8 (note that the y-axis scale is in centimetres), differences of up to 50 cm are also found between 13 to 17 UT, corresponding to 22:30 to 2:30 LT. These high variations occurred around midnight LT and were observed from most GPS satellites of the testing station. This potentially indicates the effects of mid-night ionospheric density disturbance occurring in the low-latitude region, which were not present in the mid-latitude region.

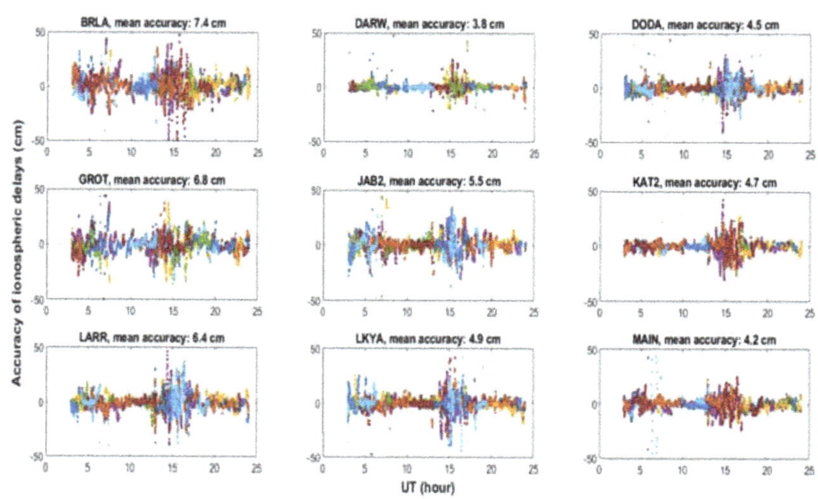

Figure 8. Plots showing the differences between the interpolated and measured ionospheric corrections as function of time at the nine NT CORS GNSS stations on 6 May 2021. This figure is the same as Figure 7, but the y-axis scale is decreased from 5 m to 50 cm.

To further investigate of this phenomenon, GNSS station data along the east coast of QLD were processed to evaluate and determine the latitudinal extent of these ionospheric disturbances. The testing region included a dense CORS network from 15° to 30°S within 142° to 152°E. Three test areas of 5° latitude ×5° longitude and 5° latitude ×8° longitude were defined for the QLD east coast (Figure 9), each consisting of 14 CORS stations. The red stations with IDs in Figure 9 are the testing stations used in the ionospheric corrections interpolation. A high variation in ionospheric delays can be seen at stations with latitudes of up to 20°S during these time periods; and the variation diminished as the latitude increased.

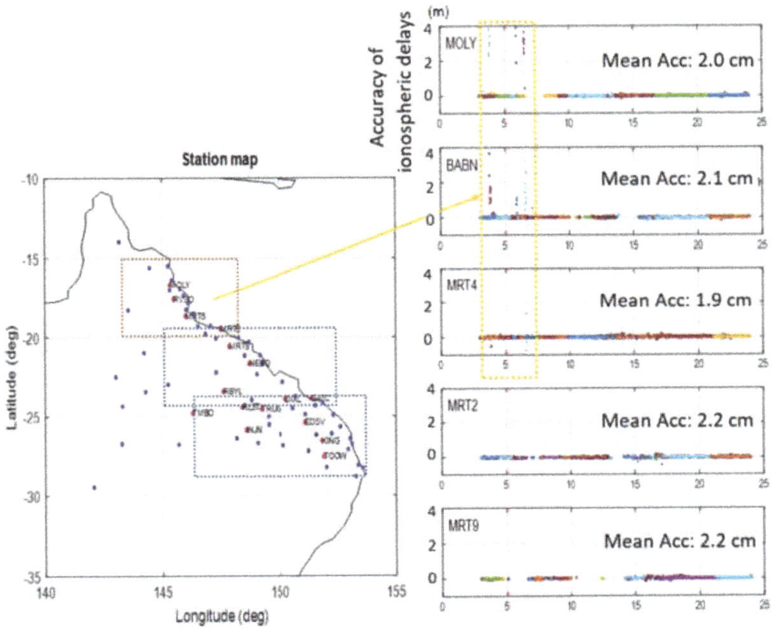

Figure 9. Plots showing the differences between the interpolated and measured ionospheric corrections as a function of time at some testing stations on 6 May 2021. The dash boxed in the station map show three testing areas. The red stations with names on the station map are the testing stations. The colours in the right boxes represent for different satellite observations. The dash-orange rectangle indicates some high variation in ionospheric delays.

3.1.3. South Australia (SA), Western Australia (WA) and Central Australia

Additional test regions in South Australia (SA), Western Australia (WA) and Central Australia as shown in Figure 10 were selected for evaluation. Each of these test regions cover 5° latitude with a varying longitude of 8° to 10° depending on the availability of the stations' data. Based on the results presented in this figure, the blue regions provide a higher average accuracy of less than 8 cm. The orange regions, on the other hand, give an average accuracy between 5 and 15 cm. Those accuracies correspond to the latitudes (mid/low) and the distribution of GPS stations (high/less) in the test regions.

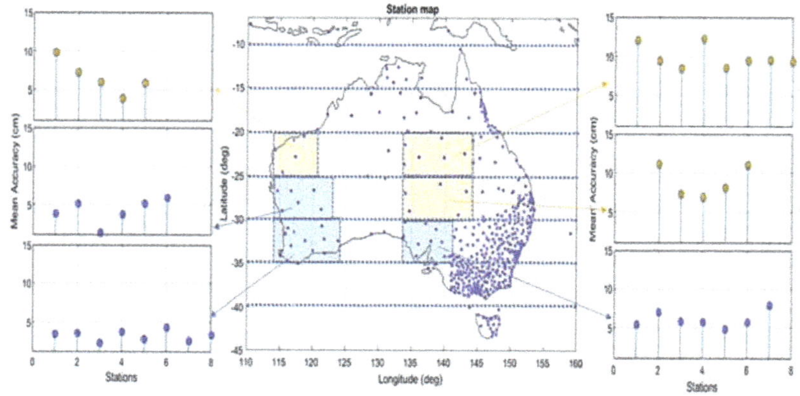

Figure 10. Evaluation of the accuracy of the ionospheric corrections in the test regions of South Australia (SA), Western Australia (WA) and Central Australia. The blue and orange boxes show the testing regions where the mean accuracy achieved is below 8 cm and up to 15 cm, respectively. The arrows link from the testing stations to the corresponding figures of mean accuracy of each station in the test region.

3.2. *Day-to-Day Accuracy of the Ionospheric Corrections during Ionosphere Quiet and Disturbed Days*

To assess the accuracy of the ionospheric corrections for a day-to-day period, seven days of GNSS measurements from 6 to 9 May (DOY 126 to 129) and 11 to 13 May (DOY 131 to 133) 2021 were processed for two test regions in WA and QLD as shown in Figure 11. DOY 130 was excluded due to the unavailability of station data for the testing. These two test regions have a similar number and density of GNSS CORS stations and are in different latitudes.

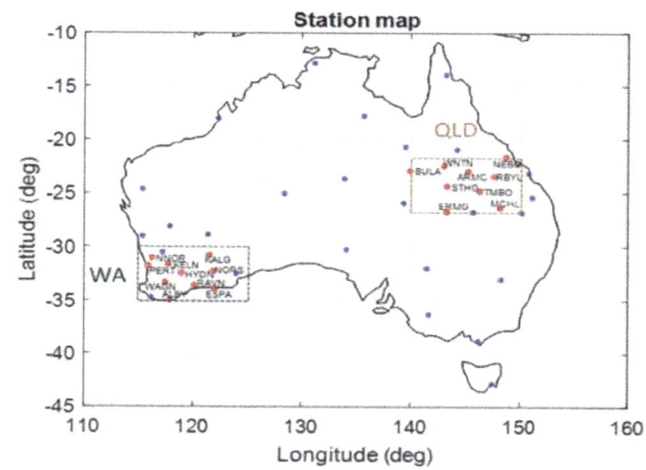

Figure 11. Two long-term testing areas in WA and QLD (dash boxes). The red stars are testing stations in each region. The blue stars are stations around Australia processed with PEA. Note that 15 stations around and inside the boxes were used for conducting interpolation at each testing region.

Figure 12 shows the mean accuracy of the interpolated ionospheric corrections for QLD (top) and WA (bottom) for the seven-day study period. In the WA region, the day-to-day mean accuracy seems stable and varies within 5 cm. In QLD, a higher variation is observed in the day-to-day mean accuracy, from 2 cm to 10 cm. On DOY 133, the mean accuracy at the NEBO station reaches 12 cm and is likely caused by its location situated at the border. The reason for the differences observed in these two regions could be based on the latitude as well as the distribution of the CORS stations in that region. The QLD test region is situated in the low latitude, whereas the WA test region is in the mid-latitude. Even though the number of CORS stations used in the ionospheric corrections interpolation in two regions is similar, the WA CORS stations are located nearer to each other (approximately 120 km) compared to those of QLD (more than 200 km).

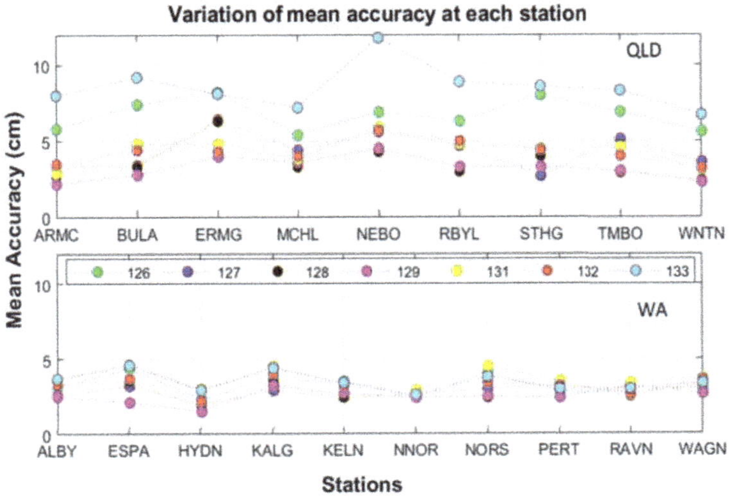

Figure 12. The mean accuracy of the interpolated ionospheric corrections for the testing stations at QLD (**top**) and WA (**bottom**) during 6 to 9 May (DOY 126 to 129) and 11 to 13 May (DOY 131 to 133) 2021.

The one-week study period included a geomagnetic storm with Kp = 7 on 12 May 2021 (DOY 132). Day 132 is the most disturbed day and day 126 (6 May) is in the list of quiet days in May 2021 [24,25]. However, based on the results presented in Figure 12, it appeared that there were little differences observed in the mean accuracy. In fact, the mean accuracy on some quiet days can be higher than those on the disturbed day.

For further investigation, data from two days: one ionospheric quiet day on 6 May 2021, and one ionospheric disturbed day on 12 May 2021, and from different testing regions across Australia, were selected for comparison. The results presented in Figure 13 are for NT, WA, NSW, and VIC. From this study, no large differences in accuracy were observed between the ionospheric quiet and disturbed day. However, a larger standard deviation was observed during the disturbed day than the quiet day and, in some instances, reached 70 cm in NT, which indicates that the ionospheric delays were more varied during the disturbed period.

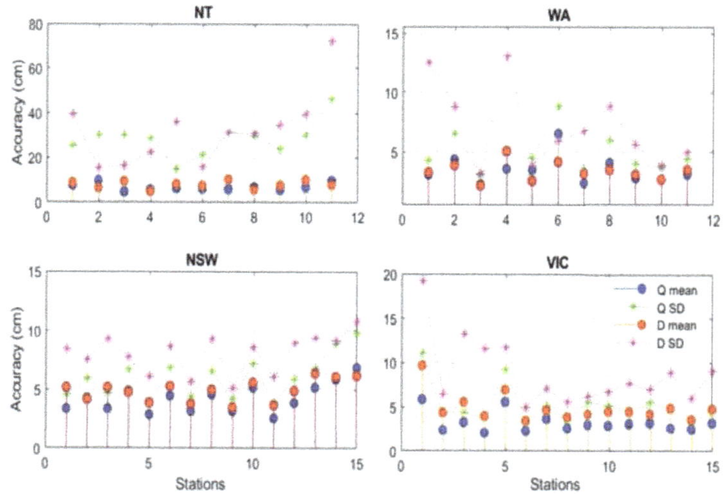

Figure 13. Comparison of average accuracy in quiet day (Q on 6 May 2021) and disturbance day (D on 12 May 2021). The blue and red markers show the mean accuracy of the testing stations, whereas the green and pink stars present the standard deviation of the accuracy on the quiet and disturbance day, respectively.

4. Discussion

The achievable accuracy of a local ionospheric delay model using the linear interpolation method in Australia has been evaluated to understand the impact of the ionosphere on GNSS positioning in different latitudinal regions, as well as the number and spatial density of GNSS stations in Australia. With dense CORS networks (nominal spacing of 100 km or less) in VIC, NSW and ACT, 1 to 5 cm accurate ionospheric corrections can be obtained with a simple linear interpolation ionospheric model. In fact, approximately a one-centimetre-level accuracy can be obtained most of the time if all stations in the current networks in this region are processed and used in the ionosphere modelling. Referring to Section 3.1, it was found that, with at least 15 GNSS stations (nominal spacing of 200 km and well geographically distributed) for a region of 5° latitude ×10° longitude, a mean accurate ionospheric correction within 5 cm can be obtained when a user stands on this region.

For other regions in mid latitudes, the mean accuracy obtained varied depending on the distributions of CORS stations and their availability. The mean accuracy obtained from the valuation can be around 5 cm or up to 15 cm (Figure 10). For those mid regions, the mean accuracy is not varied much for day-to-day variation. It largely depends on the spatial distribution of the station network. The linear interpolation in our method is feasible from five CORS stations. Therefore, for the testing stations with the numbers of CORS stations around 10, to maximise the achievements, some surrounding stations nearby the border of the network can be used to interpolate. Therefore, the area can be extended, not only using stations in the size of testing.

For low-latitude regions, overall, a mean accuracy of 8 cm can be obtained. However, high variations in ionospheric corrections were found during the afternoon (or daytime LT) or midnight. As presented in Section 3.2, Figure 7 shows the discrepancy of ionospheric delays during 5 to 8 UT on 6 May 2021 for some specified satellites observed at all stations. The data from those stations were further examined during the week and it was found that these variations often occurred in a low latitude from 3 to 8 UT (daytime in LT). To better understand this phenomenon, we selected the location of high variations in ionospheric delays of each satellite during daytime, which has an accuracy larger than 50 cm. It was found that those high variations occurred at the north of the testing station, where the

GNSS signal paths were of a low elevation. The north side of stations at NT is the equatorial regions and NT is located in the equatorial ionospheric anomaly, so those variations may be caused by electron density irregularities that affected some receiver-satellite paths. The magnitude of those values varied day by day. Figure 14 presents the findings of two example stations, BRLA and DODA. The subplots on the top show the differences between the interpolated and measured ionospheric corrections as a function of time at BRLA and DODA on 6 May 2021, which are the same as those presented in Figure 7. The subplots on the bottom of Figure 14 are the sky paths of all GPS satellites received during 5 to 10 UT observed at the two stations. More than a 50 cm accuracy is marked by red crosses, which are in low elevations at the north of the testing station (the black asterisk at each figure).

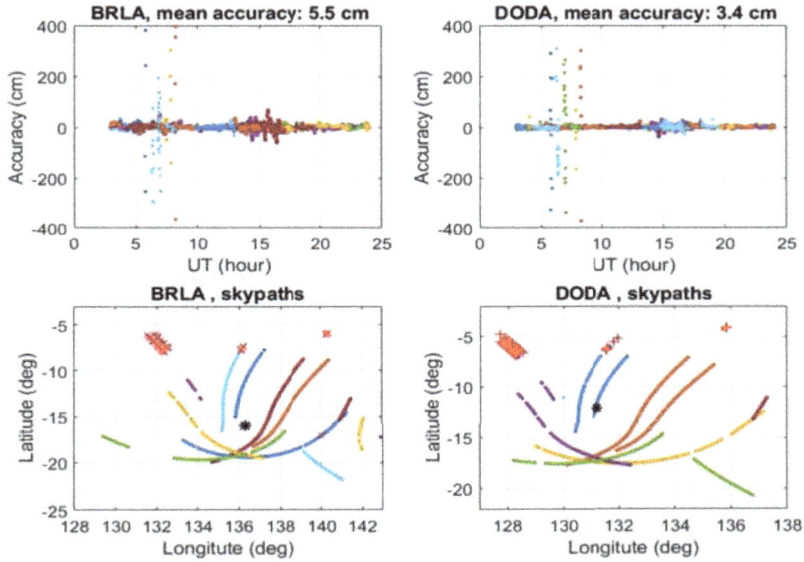

Figure 14. (**Top**) The differences between the interpolated and measured ionospheric corrections as a function of time of stations BRLA and DODA on 6 May 2021. (**Bottom**) The sky paths of all satellites received during 5 to 10 UT observed by BRLA and DODA stations. The line colours represent each GPS satellite. The red crosses show an accuracy higher than 50 cm. The black asterisk at each figure shows the location of testing station BRLA or DODA.

High variations were also observed for most satellites during midnight LT as presented in Figure 8. The 50 cm deviations that occurred around midnight on 6 May could be caused by a substorm-like activity that regularly took place during quiet days [26]. Based on the SME index [27], the substorm-like activity coincided with these high variations. Wave activity, which can occur from substorm activity, such as travelling ionospheric disturbances, are often larger in amplitude in the equatorial/low-latitude region compared with the midlatitude region [28]. Therefore, these may explain the high variations observed.

The mean accuracy of the interpolated ionospheric corrections compared with the measurement was found to be comparable between the ionospheric quiet of 6 May and disturbed days in the minor storm of 12 May. However, a larger standard deviation was observed during the disturbed day than the quiet day, indicating a variability of ionospheric delays during the disturbed period in the Australian region. The impact of the storm is different globally. In this minor storm, the impact was noticed in the South American sector with 60%, whereas less impact was found in Australia [29].

From this research, the desirable size of the region/grid for the mapping of the SD ionospheric corrections using the linear interpolation of ionospheric corrections is 5° latitude ×10° longitude. This recommendation is based on the current configuration and availability of GNSS CORS networks across Australia. Continental Australia can be divided into 15 regional/local maps, labelled as 1 to 15 and marked by different colours in Figure 15.

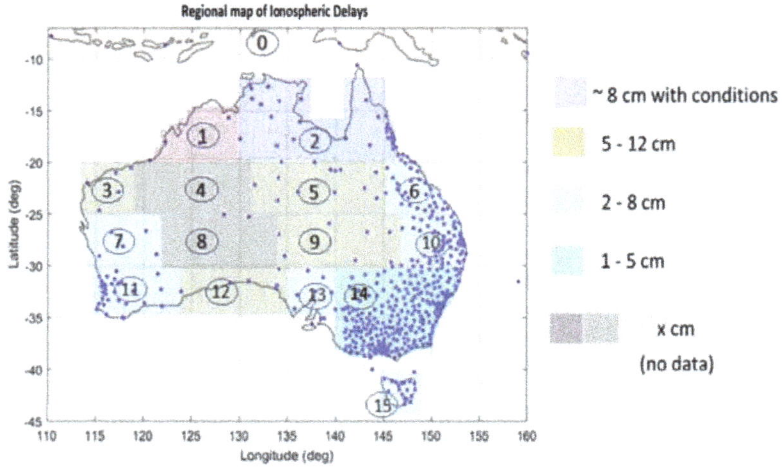

Figure 15. A map showing the recommended regions for mapping of the SD ionospheric corrections and the achievable accuracy using a simple linear interpolation method. The blue dots represent CORS stations in our testing. The labels of 1 to 15 describe 15 regional maps that are divided based on the available testing stations.

The regions in this map can be explained as:

(1) Region 0 covers multiple islands of Indonesia and there were no available GNSS stations; therefore, it is not included for analysis.

(2) Regions 1 and 2 are in the low latitudes of 10° to 20°S, where the high electron density and the variation in the equatorial ionospheric anomaly are found. The ionospheric corrections around midnight or low-elevation GNSS satellites below 20 deg during daytime can be high. With the current available CORS networks in NT and northern QLD, 8 cm level ionospheric corrections can be obtained in Region 2, while additional GNSS stations will be required to be installed in Region 1 (northern WA) if centimetre-level accurate ionospheric corrections will be needed in this area.

(3) Regions 3, 5, 9 and 12 are in the mid latitudes from approximately 20° to 35°S. The mid-latitude regions were less impacted by equatorial ionospheric disturbances. In these regions, the current available GNSS CORS networks are sparse. Nevertheless, it was found that 5 to 12 cm level accurate ionospheric corrections can be obtained in these regions. In the eastern corner of WA, where the borders of SA and NT meet, i.e., Regions 4 and 8, no evaluation was undertaken due to an insufficient number of GNSS stations.

(4) Regions 6, 7, 10, 11, 13, 14 and 15 are also in the mid latitudes covering 20° to 45°S. For these regions, a high number and dense CORS networks exist. Therefore, the average obtainable accuracy of the ionospheric corrections is within 2 to 8 cm.

This map is a reference based on the temporal valuable data in different regions in Australia. The precise accuracy may change based on the number of CORS station networks and the availability of data in each region. However, we can basically estimate how good the ionospheric delays are using a simple interpolation of the local network compared to the real measurement.

5. Conclusions

The aim of this research was to assess the achievable accuracy of a local ionospheric delay model for Australia using the linear interpolation method. The assessment was conducted by taking into consideration the impact of the ionosphere in different latitudinal regions, i.e., low and mid latitudes, as well as the number and spatial density of the GNSS CORS networks that exist in Australia. To summarise, local/regional ionospheric modelling using the linear interpolation method can produce the centimetre-level accurate ionospheric correction required for high-accuracy GNSS positioning. Based on our observations, centimetre-level accurate ionospheric corrections can be achieved if there are sufficiently dense (i.e., nominal spacing of approximately 200 km) GNSS CORS networks in the region. The achievable accuracy could be dependent on the latitudinal region (i.e., low or mid latitude) and the time of day, as well as the number and spatial density of the GNSS CORS network. For a large geographic country like Australia where CORS networks are not uniformly distributed, consideration for CORS network investment will mainly depend on the socioeconomic benefits and return of investment. To obtain centimetre-level ionospheric corrections across Australia using a simple linear interpolation, we propose a framework of 15 regional ionospheric maps of 5° latitude ×10° longitude with a minimum of 15 CORS stations in each map region to cover continental Australia.

Author Contributions: Conceptualisation, S.C. and K.H.; methodology, T.D. and K.H.; software, T.D., K.H. and S.M.; validation, T.D. and K.H.; formal analysis, T.D.; investigation, T.D. and J.C.; resources, T.D., S.C. and B.C.; data curation, T.D.; writing—original draft preparation, T.D.; writing—review and editing, S.C.; visualisation, T.D.; supervision, S.C. and B.C.; project administration, E.R. and R.B.; funding acquisition, S.C. All authors have read and agreed to the published version of the manuscript.

Funding: This research is funded by FrontierSI and Geoscience Australia, Project No. 1002A, "Ionospheric modelling for the ACS and NPI".

Data Availability Statement: The RINEX files were obtained from International GNSS services (IGS) and the Geoscience Australia. The supported data for processing can be downloaded from the Crustal Dynamics Data Information System (CDDIS) of the National Aeronautics and Space Administration (NASA); The PEA in Ginan toolkit can be downloaded at https://github.com/GeoscienceAustralia/ginan (accessed on 1 April 2022).

Acknowledgments: The Ginan toolkit development team in Geoscience Australia are greatly acknowledged.

Conflicts of Interest: The authors declare no conflict of interest.

References

1. Choy, S.; Zhang, K.; Silcock, D. An Evaluation of Various Ionospheric Error Mitigation Methods used in Single Frequency PPP. *Positioning* **2008**, *1*, 62–71. [CrossRef]
2. Choy, S.; Bisnath, S.; Rizos, C. Uncovering common misconceptions in GNSS Precise Point Positioning and its future prospect. *GPS Solut.* **2016**, *21*, 13–22. [CrossRef]
3. Choy, S. High accuracy precise point positioning using a single frequency GPS receiver. *J. Appl. Geod.* **2011**, *5*, 59–69. [CrossRef]
4. Zumberge, J.F.; Heflin, M.B.; Jefferson, D.C.; Watkins, M.M.; Webb, F.H. Precise point positioning for the efficient and robust analysis of GPS data from large networks. *J. Geophys. Res.* **1997**, *102*, 5005–5017. [CrossRef]
5. Kouba, J.; Héroux, P. Precise point positioning using IGS orbit and clock products. *GPS Solut.* **2001**, *5*, 12–28. [CrossRef]
6. Samper, M.D.; Merino, M.M. Advantages and Drawbacks of the Precise Point Positioning (PPP) Technique for Earthquake, Tsunami Prediction and Monitoring. In Proceedings of the ION 2013 Pacific PNT Meeting, Honolulu, HI, USA, 25 April 2013; pp. 9–26.
7. Wabbena, G.; Schmitz, M.; Bagge, A. PPP-RTK: Precise Point Positioning Using State-Space Representation in RTK Networks. In Proceedings of the 18th International Technical Meeting of the Satellite Division of The Institute of Navigation (ION GNSS 2005), Long Beach, CA, USA, 16 September 2005; pp. 2584–2594.
8. Leick, A.; Rapoport, L.; Tatarnikov, D. *GPS Satellite Surveying*; John Wiley & Sons: Hoboken, NJ, USA, 2015.
9. Alkan, R.M.; Erol, S.; Ozulu, I.M.; Ilci, V. Accuracy comparison of post-processed PPP and real-time absolute positioning techniques. *Geomat. Nat. Hazards Risk* **2020**, *11*, 178–190. [CrossRef]
10. Rovira-Garcia, A.; Juan, J.M.; Sanz, J.; González-Casado, G.; Bertran, E. Fast Precise Point Positioning: A System to Provide Corrections for Single and Multi-frequency Navigation. *J. Inst. Navig.* **2016**, *63*, 231–247. [CrossRef]
11. Juan, J.M.; Hernández-Pajares, M.; Sanz, J.; Ramos-Bosch, P.; Aragon-Angel, A.; Orus, R.; Ochieng, W.; Feng, S.; Jofre, M.; Coutinho, P.; et al. Enhanced Precise Point Positioning for GNSS Users. *IEEE Trans. Geosci. Remote Sens.* **2012**, *50*, 4213–4222. [CrossRef]

12. Psychas, D.; Verhagen, S.; Liu, X.; Memarzadeh, Y.; Visser, H. Assessment of ionospheric corrections for PPP-RTK using regional ionosphere modelling. *Meas. Sci. Technol.* **2019**, *30*, 014001. [CrossRef]
13. Li, X.; Zhang, X.; Ge, M. Regional reference network augmented precise point positioning for instantaneous ambiguity resolution. *J. Geod.* **2011**, *85*, 151–158. [CrossRef]
14. Hernández-Pajares, M.; Juan, J.M.; Sanz, J.; Orus, R.; Garcia-Rigo, A.; Feltens, J.; Komjathy, A.; Schaer, S.C.; Krankowski, A. The IGS VTEC maps: A reliable source of ionospheric information since 1998. *J. Geod.* **2011**, *83*, 263–275. [CrossRef]
15. Rovira-Garcia, A.; Juan, J.M.; Sanz, J.; González-Casado, G.; Ibáñez, D. Accuracy of ionospheric models used in GNSS and SBAS: Methodology and analysis. *J. Geod.* **2016**, *90*, 229–240. [CrossRef]
16. Banville, S.; Collins, P.; Zhang, W.; Langley, R.B. Global and regional ionospheric corrections for faster PPP convergence. *J. Inst. Navig.* **2014**, *61*, 115–124. [CrossRef]
17. Choy, S.; Harima, K. Satellite delivery of high-accuracy GNSS precise point positioning service: An overview for Australia. *J. Spat. Sci.* **2019**, *64*, 197–208. [CrossRef]
18. Brown, N.; Dawson, J.; Ruddick, R. Positioning Australia for the Future. *Engineering* **2020**, *6*, 857–859. [CrossRef]
19. Harima, K.; Choy, S.; Elneser, L.; Kogure, S. Local augmentation to wide area PPP systems: A case study in Victoria, Australia. In Proceedings of the IGNSS Conference, Sydney, Australia, 6–8 December 2016.
20. Melbourne, W. The case for ranging in GPS based geodetic systems. In Proceedings of the 1st International Symposium on Precise Positioning with the Global Positioning System, Rockville, MD, USA, 15–19 April 1985; pp. 373–386.
21. Wübbena, G. Software developments for geodetic positioning with GPS using TI 4100 code and carrier measurements. In Proceedings of the 1st International Symposium on Precise Positioning with the Global Positioning System, Rockville, MD, USA, 15–19 April 1985; pp. 403–412.
22. Subirana, J.S.; Zornoza, J.J.; Hernández-Pajares, M. Detector Based in Code and Carrier Phase Data: The Melburne–Wübbena Combination. ESA Navipedia. Available online: https://gssc.esa.int/navipedia/index.php/Detector_based_in_code_and_carrier_phase_data:_The_Melbourne-W%C3%BCbbena_combination (accessed on 12 April 2022).
23. The PEA Packet. Available online: https://bitbucket.org/geoscienceaustralia/pea/src/master/ (accessed on 15 June 2021).
24. Johnston, H.F. Mean, K-Indices from Twenty-One Magnetic Observatories and Five Quiet Days and Five Disturbed Days for 1942. *Terr. Magn. Atmos. Electr.* **1943**, *47*, 219. [CrossRef]
25. The International 5 and 10 Quietest and 5 Most Disturbed Days in Each Month. Available online: https://wdc.kugi.kyoto-u.ac.jp/qddays/ (accessed on 15 April 2022).
26. Newell, T.P.; Gjerloev, J.W. Evaluation of SuperMAG auroral electrojet indices as indicators of substorms and auroral power. *J. Geophys. Res.* **2011**, *116*, A12211. [CrossRef]
27. Gjerloev, J.W. The SuperMAG data processing technique. *J. Geophys. Res.* **2012**, *117*, A09213. [CrossRef]
28. Pradipta, R.; Valladares, C.E.; Carter, B.A.; Doherty, P.H. Interhemispheric propagation and in-teractions of auroral traveling ionospheric disturbances near the equator. *J. Geophys. Res. Space Phys.* **2016**, *121*, 2462–2474. [CrossRef]
29. Valdés-Abreu, J.C.; Diaz, M.A.; Báez, J.C.; Stable-Sánchez, Y. Effects of the 12 May 2021 Geomgagnetic Storm on Georeferencing Precision. *Remote Sens.* **2022**, *14*, 38. [CrossRef]

remote sensing

Article

A Graph Convolutional Incorporating GRU Network for Landslide Displacement Forecasting Based on Spatiotemporal Analysis of GNSS Observations

Yanan Jiang [1,2], Huiyuan Luo [2], Qiang Xu [2,*], Zhong Lu [3], Lu Liao [4], Huajin Li [5] and Lina Hao [1]

1 College of Earth Science, Chengdu University of Technology, Chengdu 610059, China; jiangyanan@cdut.edu.cn (Y.J.); haolina13@cdut.edu.cn (L.H.)
2 State Key Laboratory of Geohazard Prevention and Geoenvironment Protection, Chengdu University of Technology, Chengdu 610059, China; luohuiyuan1@stu.cdut.edu.cn
3 Huffington Department of Earth Sciences, Southern Methodist University, Dallas, TX 75275, USA; zhonglu@smu.edu
4 Technology Service Center of Surveying and Mapping, Sichuan Bureau of Surveying, Mapping and Geoinformation, Chengdu 610081, China; liaolusc@uestc.edu.cnm
5 School of Architecture and Civil Engineering, Chengdu University, Chengdu 610106, China; lihuajin@cdu.edu.cn
* Correspondence: xq@cdut.edu.cn

Abstract: Landslide displacement prediction is crucial for the early warning of slope failure but remains a challenging task due to its spatiotemporal complexity. Although temporal dependency has been well studied and discussed, spatial dependence is relatively less explored due to its significant variations of the spatial structure of landslides. In this study, a novel graph convolutional incorporating GRU network (GC-GRU-N) is proposed and applied to landslide displacement forecasts. The model conducts attribute-augmented graph convolution (GC) operations on GNSS displacement data with weighted adjacency matrices and an attribute-augmented unit to combine features, including the displacements, the distance, and other external influence factors to capture spatial dependence. The output of multi-weight graph convolution is then applied to the gated recurrent unit (GRU) network to learn temporal dependencies. The related optimal hyper-parameters are determined by comparison experiments. When applied to two typical landslide sites in the Three Gorge Reservoir (TGR), China, GC-GRU-N outperformed the comparative models in both cases. The ablation experiment results also show that the attribute augmentation, which considers external factors of landslide displacement, can further improve the model's prediction performance. We conclude that the GC-GRU-N model can provide robust landslide displacement forecasting with high efficiency.

Keywords: spatiotemporal analysis; landslide displacement prediction; attribute-augmented; deep learning

Citation: Jiang, Y.; Luo, H.; Xu, Q.; Lu, Z.; Liao, L.; Li, H.; Hao, L. A Graph Convolutional Incorporating GRU Network for Landslide Displacement Forecasting Based on Spatiotemporal Analysis of GNSS Observations. *Remote Sens.* **2022**, *14*, 1016. https://doi.org/10.3390/rs14041016

Academic Editors: Serdjo Kos, José Fernández and Juan F. Prie

Received: 12 January 2022
Accepted: 14 February 2022
Published: 19 February 2022

1. Introduction

Landslides are a harmful environmental and geological phenomenon, occurring frequently worldwide [1,2]. They are gradually formed by the long-term interactions of both natural and human factors under specific geologic and geographic conditions. The occurrence of landslides is irreversible, and a severe landslide may induce a series of geological environmental disasters and form a disaster chain, posing severe threats to human life and built infrastructures [3]. Thus, analyzing and predicting geological hazards using monitoring data collected from various sources is essential to mitigate these severe devastations.

Time series landslide displacement, directly reflecting the deformation and stability of a slope, is the most critical dataset to understand landslide characteristics and infer its future development [4]. For instance, the Global Navigation Satellite System (GNSS), measuring surface motion at a very high frequency and accuracy, is a powerful tool to help diagnose the progression of landslide movement. For this reason, these datasets enable deep-learning models to be trained for predicting the future state of surface deformation.

Deep-learning approaches, as suggested, are superior to traditional statistical methods in many applications, especially in time-series prediction [5,6]. Representative statistical models, such as multivariate regression models (MLR) [7], auto-regressive integrated moving average (ARIMA) [8], and others, are widely used for a single time series forecasting [8–12], while they neglect the potential relationship among multiple time series in the monitoring network under similar geological conditions. By comparison, deep-learning methods integrate various processing layers and produce abstract learning features and nonlinear dependencies from multidimensional datasets [6,11,13], making them sound alternatives in landslide displacement forecasting.

Given the outstanding performance in time-series forecasts, recurrent neural networks (RNN) and their variants [14], such as long short-term memory (LSTM) and gated recurrent unit (GRU), have achieved impressive results in the prediction of land displacements based on GNSS time-series data [9,10]. These models can solve the problems of nonlinear dynamic characteristics in complex time series, thus being particularly suitable to predict time-series landslide displacement. The workflows of these methods are approximately the same: select representative GNSS monitoring stations and plot the curve of displacement–time to analyze the deformation; next, the monitoring data of specific stations will be adopted for modelling one station by one station; as each displacement–time curve only shows the evolution characteristics of a single monitoring point, the model predictions only reflect the displacement behaviour of a single location.

However, landslide-displacement prediction is a spatiotemporal task because the evolution of the landslide process often exhibits spatial and temporal characteristics. The existing time-series forecast model only explores temporal features, ignoring the underlying spatial correlations. Thus, it is difficult to comprehensively assess the displacement changes of the entire monitoring system and reveal the future state of the landslide as a whole. Several studies have utilized convolutional neural networks (CNNs) to explore the spatial dependencies and build prediction models in traffic forecasting problems [6,15–17]. However, CNN-based models only consider the absolute distance relationship among stations in Euclidean space. Compared with CNN, graph convolutional networks (GCNs) can handle neighbourhood information in non-Euclidean spaces, providing a more feasible way to model spatial dependencies within a monitoring network [13,17,18].

Based on the problems mentioned above and inspired by current encouraging results in traffic forecasting problems, there is a need to combine GCN and RNN models to build a collaborative prediction model to capture spatial and temporal features for spatial–temporal forecast problems. However, displacement prediction of a landslide relies not only on historical GNSS measurements and the spatial correlations of the monitoring network but also on internal geological conditions and various external factors, such as hydrologic conditions [19–22], anthropogenic factors, etc. For example, in China's Three Gorges Reservoir area, many landslides are triggered and accelerated by seasonal precipitation and the fluctuation of reservoir water level [17,19]; thus, the impact factors in predicting landslide deformation are indispensable during modelling.

To the best of the authors' knowledge, there is currently no related work focusing on addressing the prediction of rainfall reservoir-induced landslide displacement from a holistic perspective combining the external incentive factors. Inspired by current encouraging results in traffic forecasting problems [13,18], we propose a novel deep-learning method named graph convolutional incorporating gated recurrent unit network (GC-GRU-N). In the GC-GRU-N, the monitored GNSS time-series displacements, the distance, and other external triggering factors are integrated to construct the GCN module handling the spatial dependency; the GRU module model's temporal dependence captures long-term dependencies by considering landslide displacement time series. This architecture is expected to inherit the merits from both GCN in extracting spatial dependencies and GRU in capturing temporal correlation features. The main contributions are twofold. First, we have extended GCN for spatial data imputation in the GNSS network deployed on the landslides. Second, we introduce a graph deep-learning framework to predict landslide displacement in time and space.

2. Methods

Landslide displacement forecasting is a spatiotemporal prediction task because the evolution of landslide movement often exhibits spatial and temporal characteristics. This paper proposes a deep-learning framework to predict the landslide displacement based on the spatiotemporal analysis of the time series monitoring data. This framework is expected to inherit the merits from both GCN in extracting spatial dependencies and GRU in capturing temporal correlation features.

The workflow is shown in Figure 1. According to the GNSS monitoring network structure and the obtained time-series datasets, pro-processing is conducted to obtain spatial and temporal attributes as the model inputs. Then, the GCN module is employed to handle spatial dependencies, while the GRU module is used to capture temporal dependencies. This paper uses tensorFlow2.1, Python3.6, and Matlab2020 to conduct the experiments and analysis.

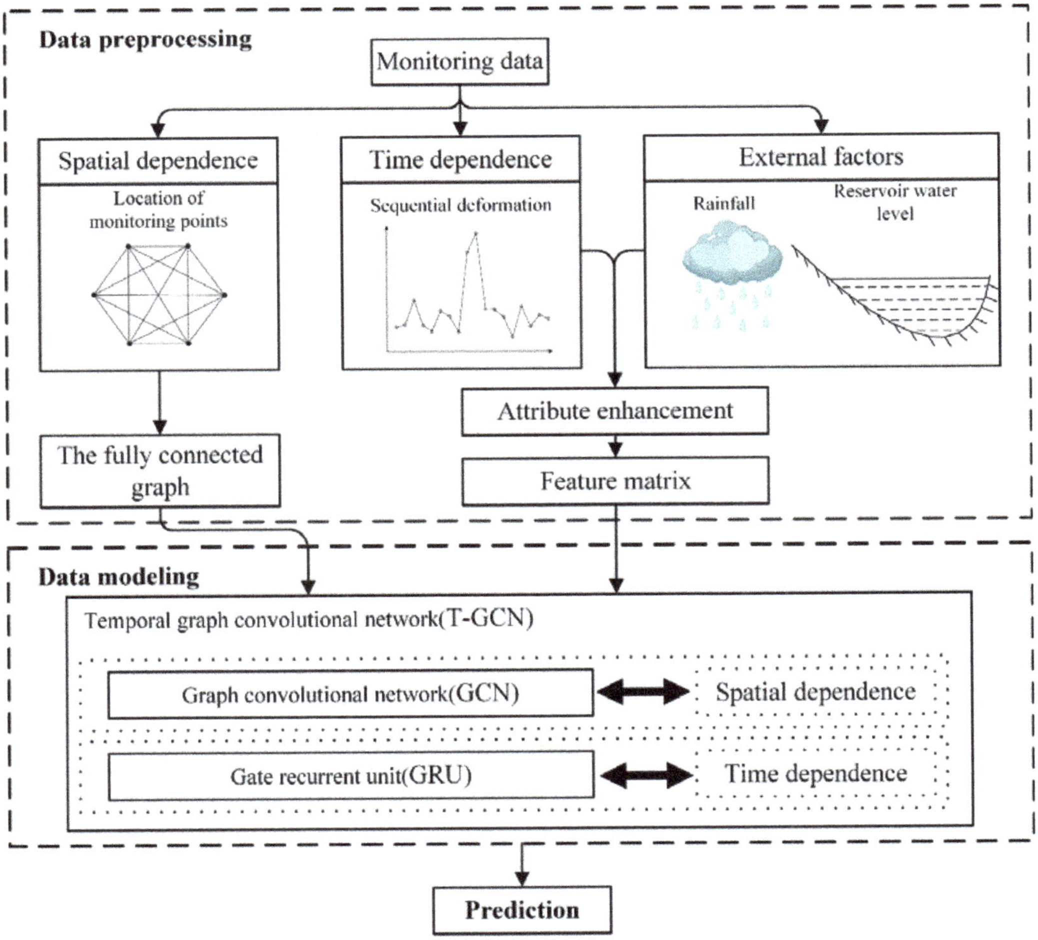

Figure 1. The workflow of the proposed model.

2.1. Study Area and Dataset

Since the Three Gorges Reservoir (TGR) was used in 2003, the fluctuated water level has changed the rock and soil physical and mechanical properties around the reservoir [19]. Over 4200 landslides are distributed in this region, and the majority of these landslides show characteristics of multiple triggers and reactivations [4]. The Baishuihe landslide and the Shuping landslide (Figure 2) are two typical recurrence reservoir landslides that have attracted the concern of researchers for a long time. As shown in Figure 2, both landslides are located on the south bank of the Yangtze River and spread into the Yangtze River.

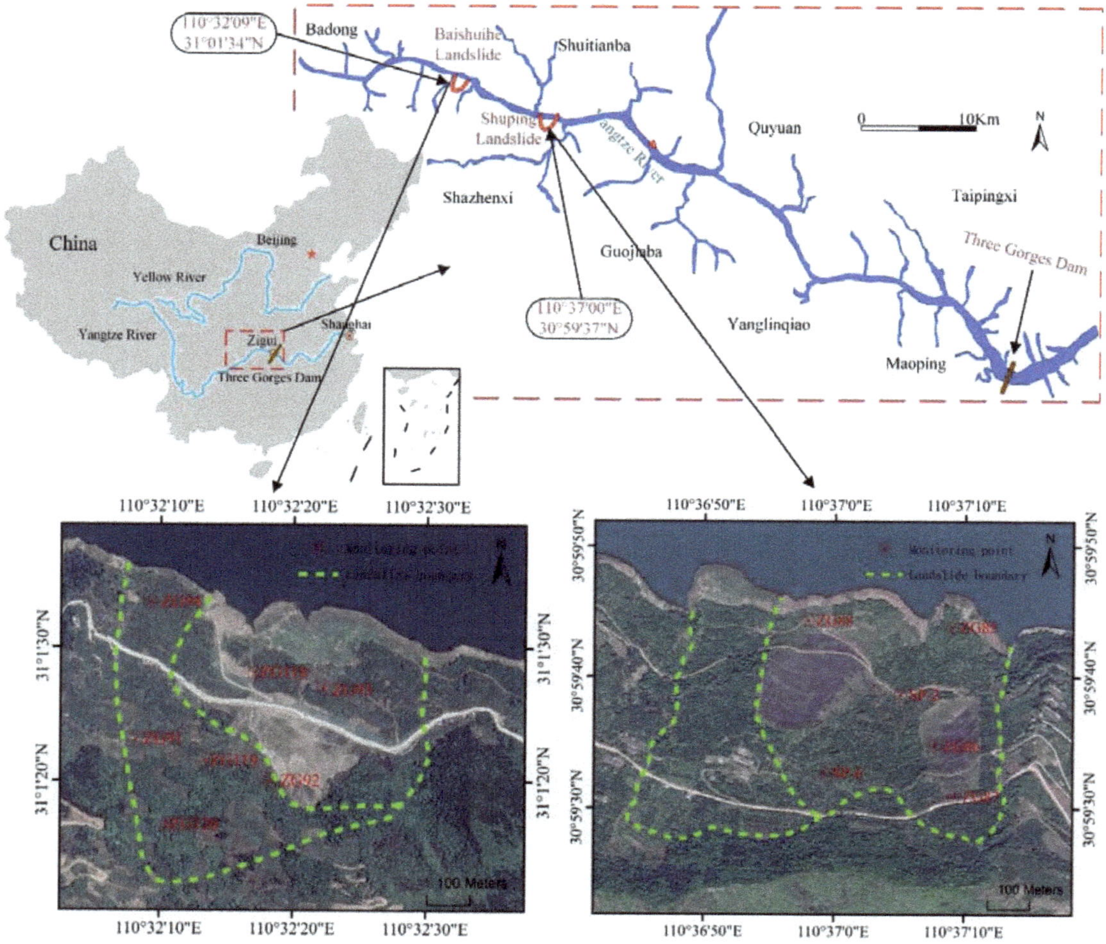

Figure 2. Location of the study area and the overall view of the two landslides.

The Shuping landslide has an elevation of between 65 m and 400 m and is about 650 m wide. It is a south–north-oriented slope with a gradient varying from 22° in the upper part to 35° in the lower part. The overall sliding mass is about 27 million m3 with a thickness of approximately 40–70 m [19]. According to the field investigation, this landslide is divided by a valley into two blocks (Figure 2).

While the maximal dimensions of the Baishui landslide in the north–south and east–west are 780 m and 700 m, respectively, it has a volume of about 12.6 million m^3 with an average thickness of approximately 30 m [20]. The field investigation and monitoring data have confirmed that the landslide has a relatively flat central part with more significant gradients in the upper and lower parts of the landslide. It can also be categorized as two blocks (Figure 2).

The two landslides were re-activated by the first impoundment of the TGR, and since then, visible cracks have gradually formed [19,20]. Two GNSS networks were deployed to study the displacement characteristics during landslide evolution (Figure 2). The displacement dataset was collected monthly by the Trimble GPS receiver with a plane accuracy of 5 ± 1 ppm. The measurements of the reservoir water level were collected daily by the water level indicator provided by the China Three Gorges Project Development Corporation. The precipitation observations were collected daily by the rain gauge provided by Zigui County Meteorological Bureau. These collected datasets (from July 2003 to March 2013 for the Baishuihe landslide; from September 2007 to May 2015 for the Shuping landslide) are presented in Figures 3 and 4. It can be inferred from Figures 3 and 4 that:

(1) The TGR became fully operational in November 2008 when its highest water level reached 175 m. Since then, the reservoir water level has fluctuated between 145 m and 175 m in a year, exhibiting seasonal changes due to artificial flood control.
(2) The rainy season of the study area lasts from June to October each year. The rainfall data also display seasonal variations due to monsoon influences. In contrast, the reservoir began impounding at the end of the wet season in October and quickly reached the maximum water level and maintained this from November to February, with a cycle opposite to the precipitation conditions.
(3) The historical GNSS measurements of both landslides also show evident seasonal patterns. The displacements increase from April to September per year and remain relatively stable from October to April in the next subsequent year. The displacements rise with the drawdown of the reservoir water level and during the period of increased rainfall in the wet season.

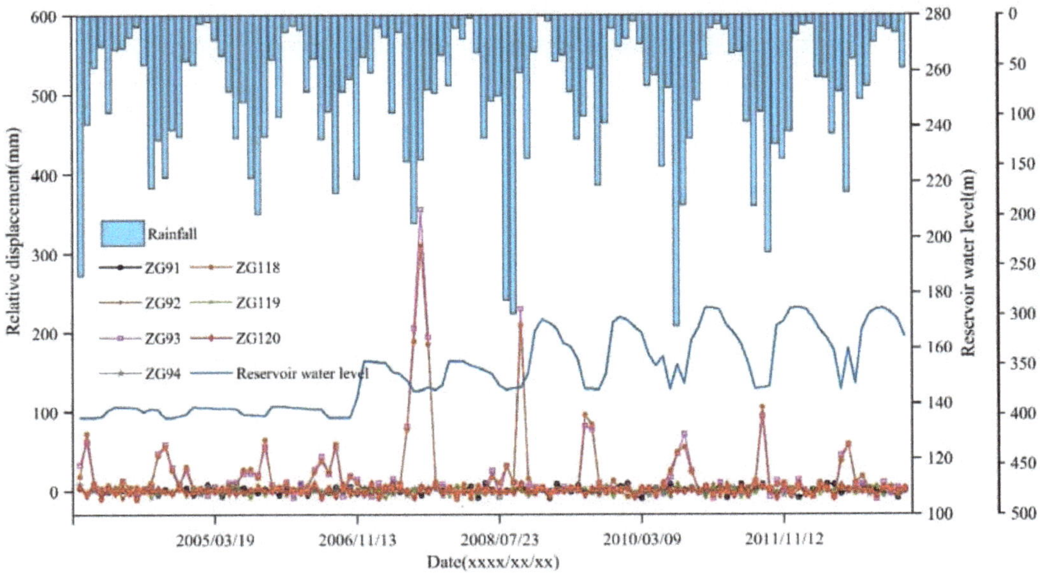

Figure 3. Monitoring data in time series of the Baishuihe landslide.

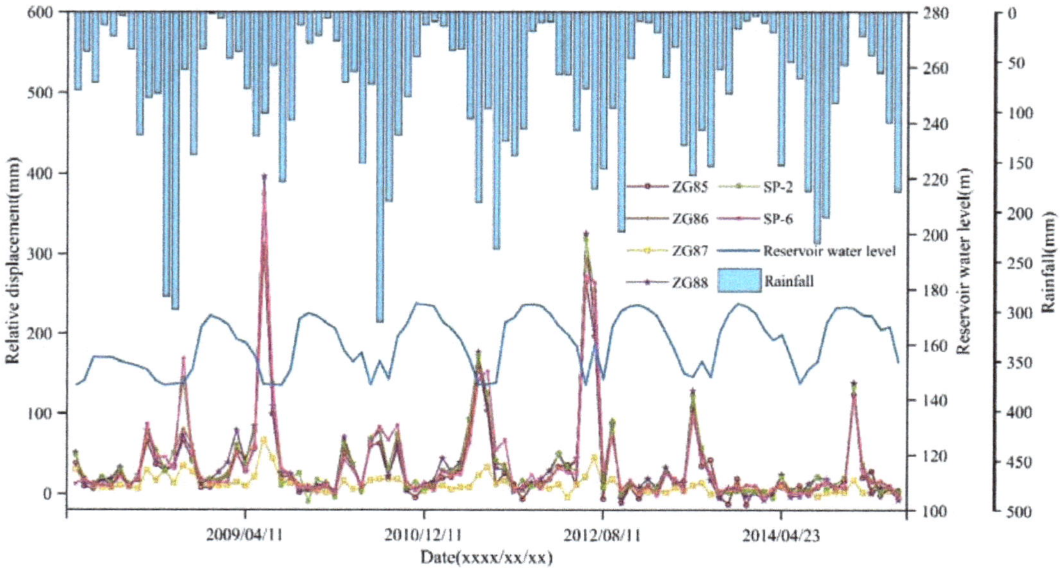

Figure 4. Monitoring data in time series of the Shuping landslide.

Thus, the seasonal characteristics of the evolution of the landslides are a joint effort of the precipitation and the fluctuation of reservoir water levels, with a period of about a year.

2.2. Data Processing

2.2.1. Representation of the Spatial Correlation

Our framework defines the GNSS monitoring network structure as a weighted graph G = (V, E, W). The monitoring sites are regarded as nodes, symbolized by V, and E is a finite set of edges representing the connection between the nodes. The numbers of the edges are $N(N-1)/2$, where N is the number of monitoring stations of the network, $W \in R^{N \times N}$ is a weighted adjacency matrix representing the correlation between the nodes (Figure 5).

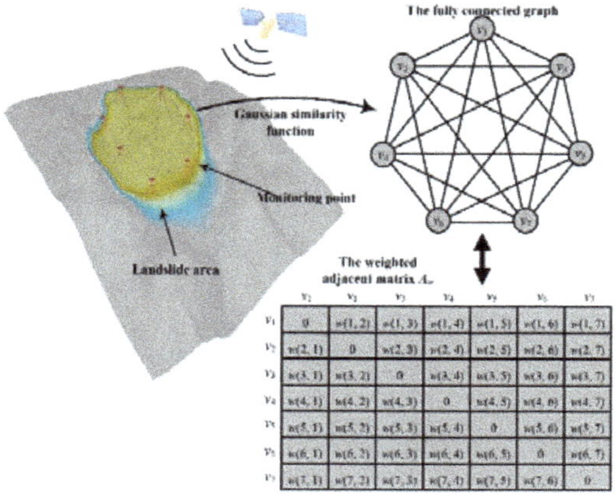

Figure 5. The diagram of a weighted graph and adjacency matrix.

Generally, the deformation characteristics of a landslide at different parts vary with the monitoring site's location. The spatial correlation of monitoring sites in the GNSS network graph shows a strong place dependence. Thus, the weighted adjacency matrix is calculated using the Gaussian similarity functions based on spatial proximity. As given in Equation (1), weights w_{ij} of edges e_{ij} representing the spatial correlation between nodes (v_i, v_j) can be calculated.

$$w(i,j) = \exp(-\|v_i - v_j\|^2 / 2\sigma^2) \tag{1}$$

where $\|v_i - v_j\|^2$ denotes the spatial dependence between nodes (v_i, v_j), and σ is the standard deviation controlling the width of the neighbourhoods.

The weighted adjacency matrix can be expressed as Equation (2), where a more significant weight indicates a higher correlation between the two neighbourhood nodes.

$$A_w = \begin{pmatrix} 0 & w(1,2) & \cdots & w(1,N) \\ w(2,1) & 0 & \cdots & w(2,N) \\ \vdots & \vdots & \ddots & \vdots \\ w(N,1) & w(N,2) & \cdots & 0 \end{pmatrix} \tag{2}$$

2.2.2. Representation of the Temporal Correlation

The spatial and temporal attributes are two critical elements of landslide displacement prediction. This section will explore the node features that can represent the temporal correlation. Once the displacement data are collected through the GNSS monitoring system, preprocessing is needed before analysis. Outlier removal and missing value imputation are first carried out, followed by denoising and normalization. This study applies a wavelet-based denoising method to remove the random noise and improve the data quality. Then, the monitoring date is normalized into the range from 0 to 1 by max-min normalization to eliminate dimensional effects.

A feature matrix $X \in R^{N \times P}$ is defined, which contains the time-series information of the monitoring stations (nodes). Where N is the number of monitoring stations in the network, P denotes the number of node time-series features, such as the length of the historical time series. $X \in R^{N \times t}$ represents the displacement at each monitoring station at time t. Thus, the input $[X_{t\text{-}n}, \ldots, X_{t-1}, X_t]$ is a sequence of n historical displacement dataset, and $[X_{t+1}, \ldots, X_{t+T}]$ is the predicted displacement in the following T time series.

2.2.3. Attribute Augmentation by Incorporating External Factors

Generally, the dynamic movement of a landslide is subject to internal geological conditions and external triggering factors [4,21]. As for landslides on the reservoir bank of TGR, the fluctuation of the reservoir water level and varying precipitation are two main external factors influencing landslide behaviours [22,23]. However, the studies using GCN to learn spatial dependencies often adhere to a single measure (e.g., distance) to represent the weights in the adjacency matrix [24,25] without considering the effects of the external trigging factors, which inevitably hampers the model performance given the complexity of landslide deformation patterns.

In this study, we apply attribute-augmented graph convolution operations on GNSS observations. The attribute-augmented unit integrates features of the displacements time series, the synchronous precipitation and the water level fluctuation to represent the contribution of the external dynamic triggering factors. The augmented matrix with weighted adjacency matrices is incorporated into the forecast model to enhance the extraction of realistic spatiotemporal dependency.

An attribute matrix $D \in R^{N \times (k*t)}$ stands for k external factors at time t. It considers that the effects of the triggering factors on the landslide displacements show significant time lags. We use an extended time window $m + 1$ to express the attribute information instead of the original one at time t; that is, the attribute matrix D^k is denoted by

$D^k_{t-m,t} = [D^k_{t-m}, D^k_{t-m-1}, \ldots, D^k_t]$. Then, the attribute-augmented matrix S can be inferred by combining the feature matrix X and the attribute matrix D:

$$S_t = [X_t, D^1_{t-m,t}, D^2_{t-m,t}, \ldots, D^k_{t-m,t}] \quad (3)$$

Thus, the displacement prediction task can be regarded as learning the function f to predict the displacements, as shown in Equation (4):

$$f(G, X|D) = [X_{t+1}, \ldots, X_{t+T}] \quad (4)$$

2.3. Data Modelling

2.3.1. Spatial Dependence Modeling by GCN

Acquiring complex spatial dependence is one of the critical problems in spatiotemporal predicting. Traditional CNN-based models can capture local spatial features but are only usable in Euclidean space, such as a regular grid [15]. The GNSS monitoring network deployed on a landslide is a graph structure rather than a two-dimensional grid, which means the traditional CNN cannot capture the spatial dependence correctly.

Recently, the GCN model has received widespread attention, extending convolutional operations to non-Euclidean domains. It has been gradually applied in image classification and traffic road networks and has demonstrated that the spatial structure captured by GCNs improves the forecasting accuracy [26].

Given an adjacency matrix A and a feature matrix X, the GCN model builds a filter to handle specific spectral information in the Fourier domain. The filter, working on the nodes (e.g., monitoring sites) of a graph (e.g., GNSS network structure), focuses on the nodes' spatial features and measures their closeness by its first-order neighbourhood.

The GCN model learns the topological relationship between the nodes and their surrounding nodes for each node. Then, the encoder generates the latent representations for all geographical units of the monitoring network and the attributes of the nodes with graph convolution to obtain spatial dependence. As a result, similar units obtain similar representations, which are then used by the decoder for predicting.

As shown in Figure 6a, the GCN model can be constructed by stacking multiple convolutional layers to learn higher-order similarities between the nodes in the graph. The propagation of the GCN can be formulated as:

$$y_{l+1} = \sigma(\widetilde{D}^{-\frac{1}{2}} \widetilde{A} \widetilde{D}^{-\frac{1}{2}} y_l W_l) \quad (5)$$

where $\sigma(.)$ represents the nonlinear sigmoid function, such as the ReLU. $\widetilde{A} = A + I$ is a self-connection structure matrix with an identity matrix I. $\widetilde{D}$ is the diagonal node degree matrix of $\widetilde{A}$, represents as $\widetilde{D} = \sum_j \widetilde{A}_{ij}$. W_l is a weight matrix for the *l-th* neural network layer. y_l is a node-level output of l layer with $y_0 = X$.

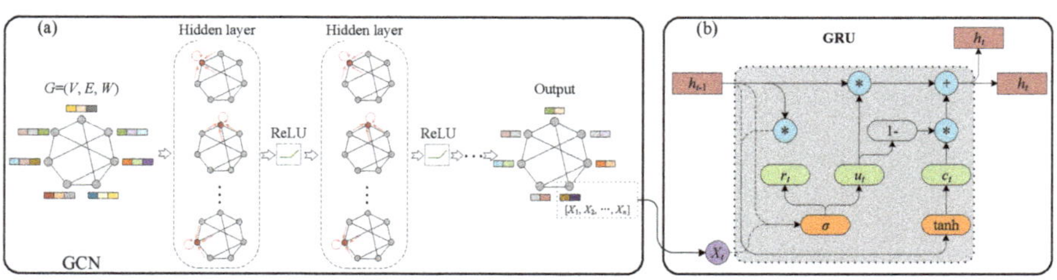

Figure 6. The architectures of two deep-learning models: (**a**) the graph convolution networks; (**b**) the gated recurrent units.

2.3.2. Temporal Dependence Model by GRU

Acquiring temporal dependence is another crucial problem in landslide displacement spatiotemporal predicting. The recurrent neural network (RNN) has achieved impressive results given the outstanding time series forecasts [14]. However, traditional RNN has limitations for long-term forecasts due to deficiencies, such as gradient disappearance and explosion [27].

The LSTM model [28] and the GRU model [29] are proposed to address these problems. The basic principles of LSTM and GRU are almost identical. They all incorporate gated mechanisms and can handle longer sequences of tasks. Compared to LSTM, GRU combines the forget gate and the input gate into an update gate, decreasing the data flow; thus, it has a more straightforward structure, fewer parameters, and faster convergence speed [30]. Therefore, we have chosen the GRU model to learn temporal dependence from the displacement time series data.

The data flow of the GRU is illustrated in Figure 6b. x_t denotes the model inputs at time t, h_{t-1} is the hidden state at time $t-1$, all the subscript letter indicates time, r_t stands for the reset gate, u_t represents the update gate, c_t is the memory content stored, and h_t denotes the output state. The update gate u_t is used to control the degree of the status information transmission from time $t-1$ to time t; the larger the value of u_t, the more critical the previous state. The reset gate tensor r_t controls the influence of time $t-1$ on-time t; the smaller the value of r_t, the weaker the effect.

The GRU obtains the monitoring information at time t, by taking the hidden status at time $t-1$ and the current monitoring information while capturing the monitoring information at the present moment; the model still retains the changing trend of historical monitoring information and can capture temporal dependence.

2.3.3. Spatiotemporal Model Using GC-GRU-N

To capture both spatial and temporal features from landslide monitoring data, we propose a new deep learning model (GC-GRU-N) based on graph convolutional network (GRU) and gated recurrent unit (GRU). In the model, an attribute-augmented graph convolution operation with weighted adjacency matrices and an attribute-augmented unit is employed to represent the spatiotemporal correlations of the monitoring network, and the obtained results will be used as the model inputs.

The architecture of the model is shown in Figure 7. The upper part shows the process of spatiotemporal displacement prediction. The GCN module is used to model spatial dependencies, while the GRU module is used to model temporal dependencies. A fusion layer is implemented to incorporate extracted features from both space- and time-domain. The model predicted displacement could be represented as $f(A,S)$ with A the weighted adjacency matrix and S the attribute-augmented matrix.

The lower part (marked by a dashed box) gives a specific structure of a T-GCN cell. In each model cell, h_{t-1} denotes the output at time $t-1$, $gc(.)$ is the graph convolution process, u_t, r_t are the update gate and reset gate at time t, and h_t denotes the output at time t. The calculation process of spatiotemporal displacement prediction is shown below:

$$u_t = \sigma(W_u \cdot [gc(S_t, A), h_{t-1}] + b_u) \tag{6}$$

$$r_t = \sigma(W_r \cdot [gc(S_t, A), h_{t-1}] + b_r) \tag{7}$$

$$c_t = \tanh(W_c \cdot [gc(S_t, A), (r_t * h_{t-1})] + b_c) \tag{8}$$

$$h_t = u_t * h_{t-1} + (1 - u_t) * c_t \tag{9}$$

where $\sigma(.)$ and $\tanh(.)$ represents the sigmoid function, W and b stand for the weights and biases in the training process, respectively. * denotes the matrix multiplication.

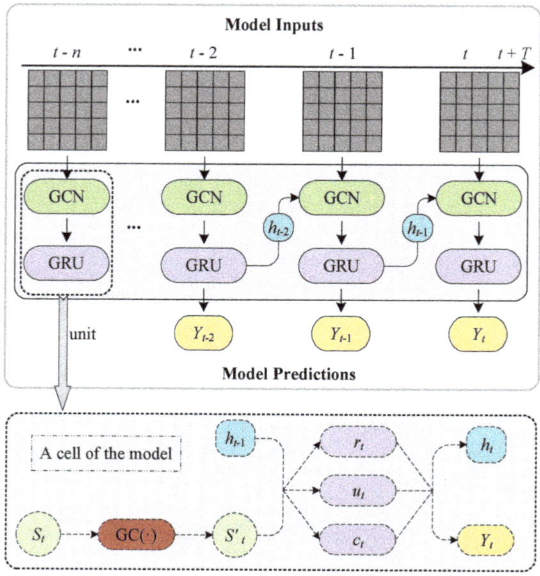

Figure 7. The overall process of spatiotemporal prediction.

2.3.4. Evaluation Metrics of the Prediction

To evaluate the model performance, we introduce three evaluation indicators, namely the mean absolute error (*MAE*), the mean absolute scaled error (*MASE*), and the root mean square error (*RMSE*) [31]. These metrics are widely used in the regression tasks, defined as follows:

$$MAE = \frac{1}{n}\sum_{t=1}^{n}|Y_t - \hat{Y}_t| \tag{10}$$

$$MASE = mean\left(\left|\frac{e_j}{\frac{1}{n-1}\sum_{t=2}^{n}|Y_t - Y_{t-1}|}\right|\right) \tag{11}$$

$$RMSE = \sqrt{\frac{1}{n}\sum_{t=1}^{n}(Y_t - \hat{Y}_t)^2} \tag{12}$$

where n is the length of time series, Y_t represents the actual measurement, $\hat{Y}_t$ denotes the predicted value, and $e_j = \sum_{t=1}^{n}|Y_t - \hat{Y}_t|$ indicates the forecast error for a given period j (the number of forecasts). *MAE* can reflect the absolute error of the prediction result. *MASE* is a favourable property to calculate the time-series forecast errors, which can be used to compare forecasts across data sets with different scales. *RMSE* can more accurately reflect the similarity between the predicted and the observed sequence.

3. Experiments and Results

3.1. Analysis of the Spatiotemporal Correlation

Generally, the deformation of a landslide varies spatially at different parts of the landslide body (e.g., [19,20]). The spatial–temporal correlation of each monitoring site in the GNSS network shows a strong location dependence. This section analyzes the measurements on each monitoring site to determine their spatiotemporal relations. To do so, we introduce grey relational analysis (GRA) to help estimate the correlation of monitoring points [19]; it is believed that the value of grey relational degree (GRD) greater than 0.6 denotes a close correlation.

For the Shuping landslide, all monitoring stations (Figure 2) are deployed on the active block. As shown in Table 1, for any neighbouring connection, their GRD values are greater than 0.6, implying a strong spatial–temporal correlation among them. We select two pairs of adjacent monitoring points (ZG85 with SP-2 and ZG85 with ZG87) and plot the observed deformation time series. The results are illustrated in Figure 8a,b, showing the strong consistency of each pair. The calculated GRD value is 0.8 and 0.87, respectively, suggesting a strong spatiotemporal correlation between them. From these GNSS observations on the east block (Figures 2 and 4), the landslide deforms locally obviously, indicting an unstable state. Still, it does not mean that the whole landslide is moving, or that the landslide is likely to occur, unless it is already entering an accelerated deformation stage with an increasingly accelerated velocity.

Table 1. Grey relation analysis results of the Shuping landslide.

Point	ZG85	ZG86	ZG87	ZG88	SP-2	SP-6
ZG85	1	0.78	0.87	0.82	0.80	0.84
ZG86	0.78	1	0.88	0.75	0.86	0.85
ZG87	0.87	0.88	1	0.88	0.87	0.88
ZG88	0.82	0.75	0.88	1	0.83	0.82
SP-2	0.80	0.86	0.87	0.83	1	0.75
SP-6	0.84	0.85	0.88	0.82	0.75	1

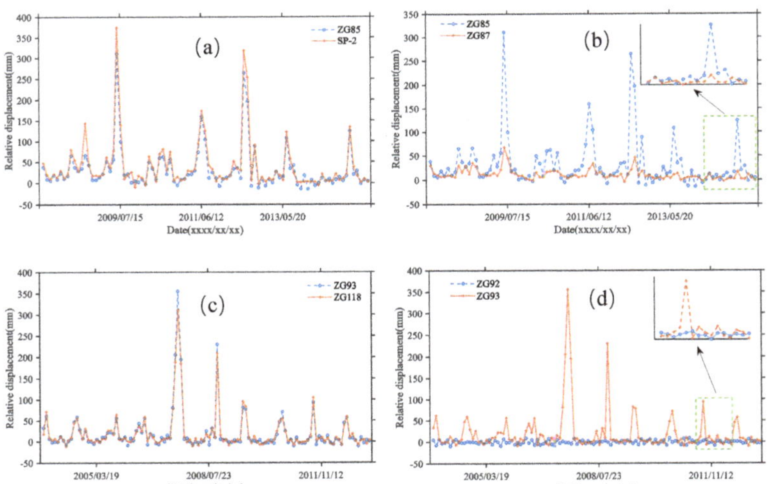

Figure 8. (**a**–**d**) Deformation over time of adjacent monitoring points of different pairs.

For the Baishuihe landslide, two stations, namely ZG93 and ZG118, are installed in the more active block, with the rest monitoring stations at the other block (Figure 2). It can be seen from Figures 2 and 8c that monitoring sites located at the more active block exhibit almost identical displacement trends (GRD = 0.74). At the other monitoring sites in another block, the measured displacements are relatively small, fluctuating to within 20 mm per month. The GRD between these sites is larger than 0.6, reflecting remarkable similarity in displacement trends. Generally, the correlation decreases with increasing distance. However, for stations located at different blocks (e.g., ZG92 and ZG93), the monitoring displacements still show similarity in local features (Figure 8d), with the value of GRD equalling 0.54 (Table 2).

Table 2. Grey relation analysis results of the Baishuihe landslide.

Point	ZG91	ZG92	ZG93	ZG94	ZG118	ZG119	ZG120
ZG91	1.00	0.69	0.51	0.76	0.52	0.79	0.77
ZG92	0.69	1.00	0.54	0.70	0.57	0.65	0.65
ZG93	0.51	0.54	1.00	0.59	0.74	0.54	0.59
ZG94	0.76	0.70	0.59	1.00	0.59	0.72	0.71
ZG118	0.52	0.57	0.74	0.59	1.00	0.55	0.54
ZG119	0.79	0.65	0.54	0.72	0.55	1.00	0.68
ZG120	0.77	0.65	0.59	0.71	0.54	0.68	1.00

In summary, the spatiotemporal correlation of monitoring sites in the GNSS network shows medium to strong relations. Grey relational analysis (GRA) results also show a strong location dependence consistent with the results calculated by Gaussian similarity functions. It confirms that the landslide displacement prediction should consider the spatiotemporal relationship between monitoring points.

3.2. Model and Parameter Setting

3.2.1. Model Inputs

In the experiments, GNSS measured displacements acquired monthly (from July 2003 to March 2013 for Baishuihe landslide, from September 2007 to May 2015 for Shuping landslide), the associated daily reservoir water level and precipitation are used, all of which are normalized to the interval from 0 to 1 using the max–min normalization. These datasets are further divided into a training dataset and a test dataset. The training dataset (from July 2003 to August 2011 for Baishuihe landslide, from September 2007 to March 2014 for Shuping landslide) of the input layer are taken as inputs in the training process, and the remaining dataset is used to evaluate the performance of our proposed method.

For the input layer, the training dataset of single GNSS displacement time series can be denoted as $D_m = \{d_1, d_2, \ldots, d_m\}$, m represents the length of the time series. GNSS observations from July 2003 to August 2011 for the Baishuihe landslide and from September 2007 to March 2014 for the Shuping landslide are taken separately as training datasets in the training process. The remaining GNSS observations are used to evaluate the performance of our proposed method.

The way of sample division from each training dataset is shown in Figure 9. A sliding window with window length equals l and step size equal n is used. Thus, the length of each obtained sample is l $(2 \leq l < m)$, represented by $D_{train} = \{d_{m-l}, d_{m-l+1}, \ldots, d_m-_1\}$ with the last y $(1 \leq y < l)$ serving as label sample and the other values $(1 - y)$ used as the sample input.

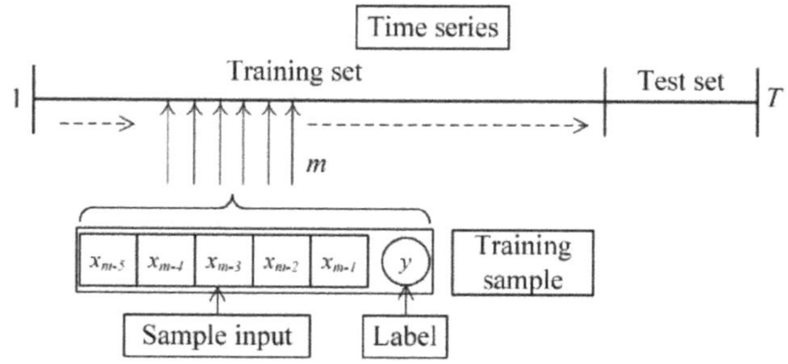

Figure 9. The sample division of a single time series.

In this paper, the sliding window size is set to 6 by taking account of the typicality and quantity of the training samples, which stands for the half-cycle of the external factors to facilitate the recurrent layer to capture the temporal dynamics. The first five are input samples, the last 1 marked as the label. As suggested by [32], the model error increase as the value of y increases; thus, we set it to 1. Therefore, the dimension of the training sample is 93 and 74, separately. The test datasets are treated in the same way.

For the Baishuihe landslide, a 7×7 weighted adjacency matrix A_w is first construed using the Gaussian similarity function based on the spatial proximity of the deployed GNSS monitoring stations (Section 2.2.1). Then a feature matrix X with a size of 7×117 is constructed to represent the temporal displacement of each station. The number of rows equals the number of stations; the number of columns equals the measured displacement time series. Thus, in the same way, the dimension of A_w and X for the Shuping landslide is 6×6 and 6×93, respectively.

As illustrated in Section 2.1, for both landslides, the fluctuation of the reservoir water level and varying precipitation are two main external factors influencing landslides behaviours. We introduce an attribute-augmented unit that integrates features of the displacements, the seasonal rainfall, and the water level fluctuation to represent the effect of external influencing factors on landslide deformation. The augmented matrix with weighted adjacency matrices is incorporated into the forecast model to enhance the extraction of realistic spatial–temporal dependency, and the derived results will be used as the model inputs.

3.2.2. Model Parameters and Settings

Our proposed model has four hyper-parameters: the learning rate, the number of training iteration epochs, the number of hidden units, and the batch size. In the experiment, we empirically set the learning rate to 0.001 and the batch size to 32 [33]. However, the numbers of training iteration epochs and hidden units are two crucial parameters that may affect the prediction precision and, therefore, should be determined by designed comparison experiments. The ReLU is employed as the activation for each convolutional layer and the Adam optimizer for minimizing the loss function (Equation (13)).

$$loss = \sum_{t=1}^{n} \left(Y_t - \hat{Y}_t\right)^2 / n \tag{13}$$

where n is the time series length, Y_t represents the actual measurement, $\hat{Y}_t$ denotes the predicted value.

Comparison experiments for selecting the optimal hyper-parameters are performed by setting the number of hidden units to 64 first to analyze the changes of the prediction precision with a varying number of training epochs designed to be {100, 250, 500, 1000, 1500, 2000}. Figure 10 shows the variation of metrics with different training epochs. The horizontal axis represents the number of training epochs, and the vertical axis represents the variation of the metrics; it can be seen that when the training epochs equals 1000, the metrics obtain a minimum value. Thus, the model reaches its optimal performance. Accordingly, in the following comparison experiments, we set the training epochs value to 1000 to analyze the changes of the prediction precision with varying numbers of hidden units; these numbers are designed to be {8, 16, 32, 64, 100, 128}. Figure 11 gives the variation of metrics with different hidden units. The horizontal axis represents the number of hidden units, and the vertical axis represents the variation of the metrics. It can be seen that when the hidden units equal 64, the metrics obtain a minimum value. Thus, the model reaches its optimal performance. Consequently, in the following experiment, the number of training epochs and hidden units is set to be 1000 and 64, respectively.

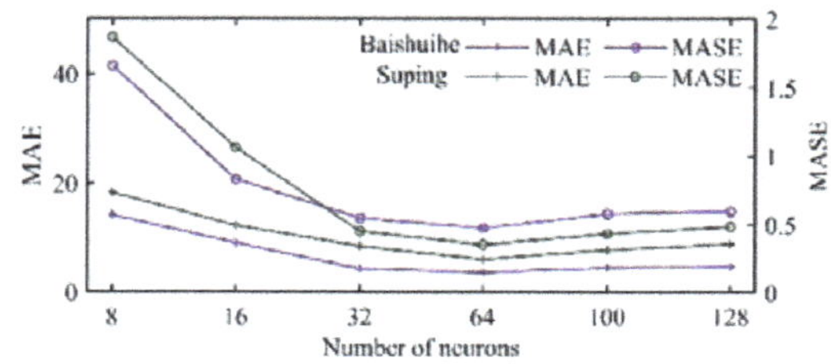

Figure 10. The variation of metrics with different training epochs.

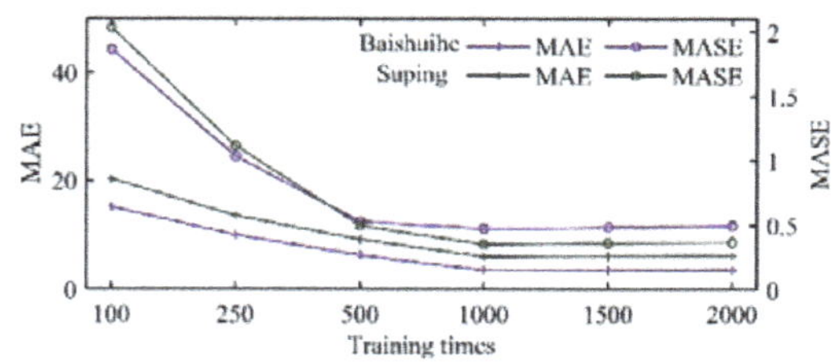

Figure 11. The variation of metrics with different hidden units.

3.3. Predicted Results and Analysis

3.3.1. Predicted Results Using the GC-GRU-N

To prove the effectiveness of the proposed model, we use four classical prediction models, which are MLR, ARIMA, SVR, and LSTM, to compare with the GC-GRU-N model for the two study sites. This section also conducts comparative analysis using the temporal graph convolutional network (T-GCN) without attribute augmentation to verify the model enhancement using the attribute-augmented graph convolution (GC) operations. We evaluate the effect of the GC-GRU-N model from two aspects: prediction performance and modelling time.

As the above-mentioned classical prediction models can only realize single time-series prediction, the model predictions only reflect the displacement behaviour of a single monitoring station. Thus, for a GNSS network with m stations, classic models need to calculate m times separately to obtain the displacement forecasts of all stations. The GC-GRU-N utilizes a feature matrix $X_{m \times n}$ (Section 3.2.1) to represent the displacement over time of each station, predicting the displacement of the entire monitoring system.

The predicted results of the Baishuihe landslide and the Shuping landslide by the proposed model are shown in Figures 12 and 13, respectively. The predictions of each monitoring station are consistent well with the actual observations as a whole. According to Figures 6 and 7, measurements of several monitoring stations show mutational transition appeared in September each year; a more significant prediction error arises at this abrupt state with a maximum of 16.66 mm and 30.35 mm, respectively. The maximum error does not exceed 10 mm for the rest of the year. It could be due to fewer samples being available for the mutation state than for the other states because a monthly prediction time scale is used due to data acquisition limitations. Generally speaking, as the number of samples for mutation state increases, e.g., with daily-scale displacement, the model's errors gradually decrease. In addition, since the GCN captures spatial features by constantly moving a smooth filter in the Fourier domain, it might also lead to the peaks being smoother.

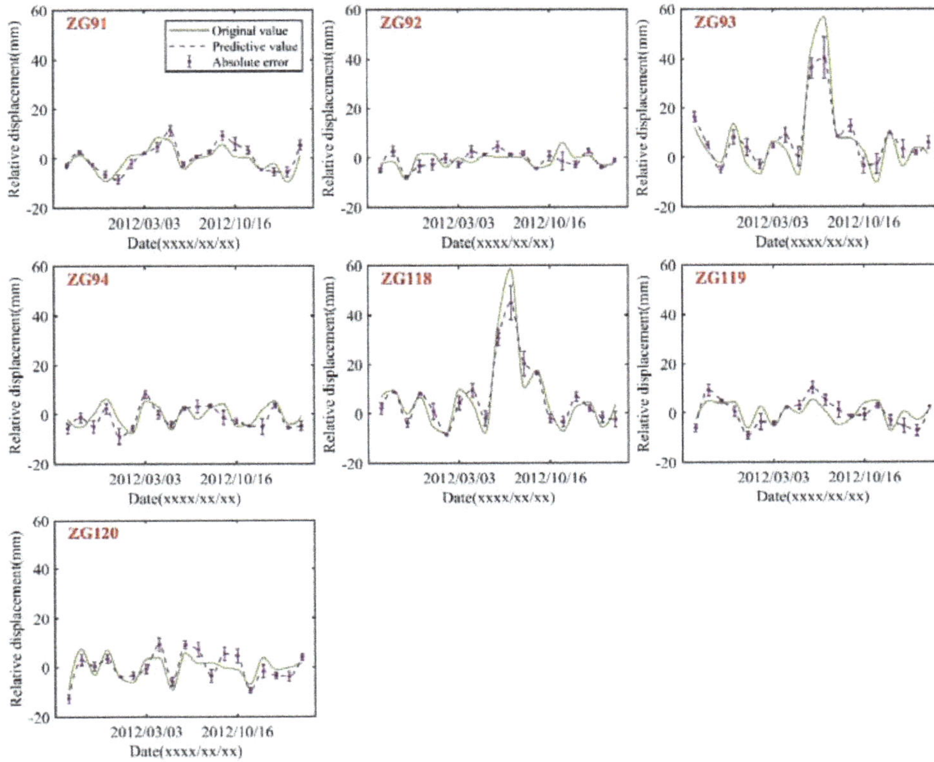

Figure 12. Comparison diagram of the overall prediction of landslide displacement; the vertical purple line represents the absolute error.

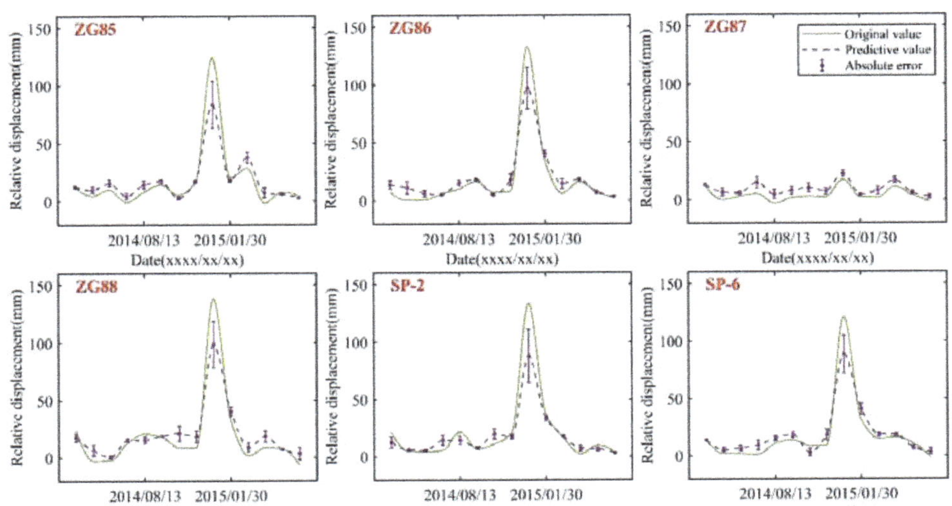

Figure 13. Comparison diagram of the overall prediction of landslide displacement; the vertical purple line represents the absolute error.

3.3.2. Comparative Experiments

The performance of the forecast models is shown in Table 3. Our proposed model has outperformed the other five models in terms of three evaluation indicators in two study areas (Table 3). The errors of models such as MLR, ARIMA, and SVR are relatively large, resulting in poor prediction performance. However, the LSTM as a neural network model is better than traditional machine learning models (SVR) and time series models (MLR and ARIMA). Compared with the LSTM model, the GC-GRU-N and the T-GCN model can better describe the displacement trend because the model structure captures the spatial feature of the monitoring network. Consequently, the prediction accuracy of the GC-GRU-N is effectively enhanced. For the Baishuihe landslide, the three metrics of GC-GRU-N are 3.682 mm of *MAE*, 0.477 of *MASE*, and 4.429 mm of *RMSE*, respectively.

Table 3. The performance of different forecast models.

Model	Evaluation Index						Average Time
	Baishuihe			Shuping			
	MAE/mm	*MASE*	*RMSE*/mm	*MAE*/mm	*MASE*	*RMSE*/mm	
The proposed	3.682	0.477	4.429	6.123	0.353	8.321	44.88 s
T-GCN	4.707	0.61	6.183	7.071	0.401	9.796	19.93 s
MLR	7.514	0.974	12.319	13.548	0.782	17.566	986.435 s
ARIMA	6.718	0.87	10.041	10.953	0.632	13.917	0.534 h
SVR	6.765	0.877	10.512	13.936	0.804	16.734	349.971 s
LSTM	5.981	0.727	8.401	8.825	0.509	12.788	229.936 s

In the following discussion, we use the *RMSE* as the primary metric to represent the model's performance. The *RMSE* of the GC-GRU-N model and T-GCN model are reduced by approximately 64% and 50% compared with the MLR model. In comparison, the *RMSE* of the T-GC-GRU-N model and T-GCN model are around 56% and 38% lower than that of the ARIMA model. Compared with the SVR model, the RMSEs of the GC-GRU-N model and T-GCN model are reduced by 58% and 41%, respectively. In contrast, the RMSEs of the T-GC-GRU-N model and T-GCN model are decreased by 47% and 41% compared with the LSTM. Compared with the GC-GRU-N model, the T-GCN model is less effective because the T-GCN considers the spatial features and ignores the impact of the external factors on landslide displacement.

In terms of computation time, T-GCN is the most efficient model amongst all tested models, only requiring 19.93 s (Table 3). The proposed GC-GRU-N achieves competitive training efficiency ranking top two, taking 44.88 s, followed by the LSTM model costing 229.936 s. The GC-GRU-N is slower than T-GCN because the method needs to develop a unit to represent the effects of the triggering factors during convolution operation. The SVR takes 349.971 s, slightly higher than LSTM. In contrast, the modelling time of the MLR model and the ARIMA model is much longer than other methods presented in this paper, requiring 986.435 s and 0.534 h, respectively. In summary, the GC-GRU-N is significantly efficient considering its high accuracy among other advanced models.

Results of the ZG93 station installed on the Baishuihe landslide and the ZG85 station deployed on the Shuping landslide are depicted in Figure 14. The predictions of the proposed model are consistently well with the actual deformation trend and superior to other methods as a whole. Despite sometimes overreacting to rapid decreases and producing underestimated results at abrupt increases, our model outperforms all other time series forecast models at both landslides. This result indicates that the graph convolution with spatial correlation consideration scheme can efficiently capture the dynamics in the landslide monitoring.

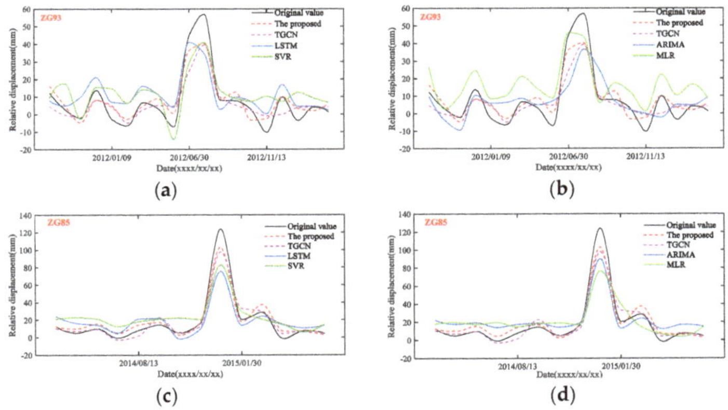

Figure 14. Comparison results of the proposed model and five other models. Results of ZG93 installed on the Baishuihe landslide are shown in figure (**a**,**b**), and that of ZG85 deployed on the Shuping landslide are shown in figure (**c**,**d**).

Specifically, MLR and ARIMA as statistics methods can also depict the variation trend of landslide displacement, but with more significant overestimated or underestimated errors. The LSTM model is more efficient and shows more promising results than the SVR model in the machine-learning-based models, especially in predicting displacement around transition states. However, sudden rapid changes in the evolution may increase the model's errors. This could be due to fewer samples being available for the mutation state.

The T-GCN and the GC-GRU-N models capturing spatial and temporal features have achieved more promising time-series forecasts. The T-GCN model gives a lower prediction accuracy. This is because the T-GCN model only considers the spatial features, and ignores the external factors impacting landslide displacement. In summary, the GC-GRU-N as a spatial and temporal mode is significantly efficient with high accuracy amongst other models in landslide displacement forecasting.

3.3.3. Ablation Experiment and Analysis

Ablation is utilized to demonstrate the importance of attribute enhancement to improve model performance. It refers to an attribute-augmented unit in the forecast model. We design the ablation experiment as the following: only consider the rainfall factor or the reservoir water effect, and both factors together (Figure 15). Table 4 shows the results for the Baishuihe landslide, with the T-GCN representing a model without an attribute-augmented unit. According to Table 4, the performance gains of using an attribute-augmented unit is apparent.

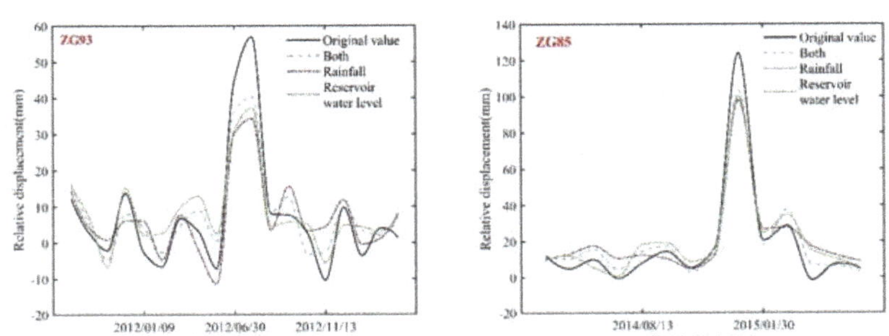

Figure 15. Comparison results of the ablation experiments considering only the rainfall factor or the reservoir water effect, and both factors together. The solid black line represents the original value.

Table 4. Ablation Experiments with different settings.

Evaluation Index	T-GCN	The Proposed Model (the Baishuihe Landslide)		
		Rainfall	R.w.l	Both Factors
MAE/mm	4.707	3.724	3.704	3.682
$MASE$	0.610	0.491	0.489	0.477
$RMSE$/mm	6.183	4.442	4.434	4.429

Note: R.w.l is the reservoir water level.

We also use the *RMSE* as the primary metric to represent the model's performance. As the experiment considers the rainfall factor alone, the reservoir water level factor alone, and both factors together, the *RMSE* values of the proposed model are 4.442 mm, 4.434 mm, and 4.429 mm, respectively, all of which are lower than that of the T-GCN (6.183 mm). Specifically, the ablation experiment demonstrates the effectiveness of assembling the external inducing factors in graph convolutional network, and the best performance in all indicators is achieved when both factors are considered simultaneously. As shown in Figure 15, the predictions of considering both trigging factors are consistent well with the actual deformation trend, which is superior to the other two scenarios and is still valid around transition states, including rapid decrease and abrupt increase conditions.

4. Discussion

4.1. Advantage of the Proposed Method

Unlike the time-series forecast models that only explore temporal features and focus on a single point, this paper presents a new deep learning architecture that considers the spatial and temporal correlation for landslide displacement prediction. More specifically, the spatial correlation of the entire monitoring system and the temporal dependency of the monitoring time series are explored to establish the forecast model predicting the displacement of the monitoring network instead of a specific station. Considering the displacement prediction of a landslide relies not only on historical GNSS measurements and the spatial correlations of the monitoring network but also on various external incentive factors. An attribute-augmented unit is designed to integrate weighted adjacency matrix, displacements, and triggering factors to enhance the capture of spatial–temporal dependency serving as the model inputs.

To the best of the authors' knowledge, there is currently no related work focusing on addressing the prediction of rainfall reservoir-induced landslide displacement from a holistic perspective combining the external incentive factors. This paper presents a new deep learning GC-GCN-N model based on the GCN and GRU models, which effectively utilizes the spatial and temporal features contained in the model input data. The results show that the proposed model outperforms comparative models in both landslides over our study site in China's Three Gorge Reservoir (TGR).

4.2. Shortcoming and Outlook of the Proposed Method

As shown in Figures 6 and 7, several GNSS-monitored displacements show mutational transitions in September. Accordingly, significant prediction error often appears at this abrupt state (Figures 12–14), which is true to other forecast models. Considering the monthly data-acquisition limitation, this could be due to fewer samples available for the mutation state than for the other states. Thus, the model's errors probably gradually decrease as the number of samples for the mutation state increases. In addition, the GCN captures spatial features by constantly moving a smooth filter in the Fourier domain, which might also lead to the peaks being smoother.

Limited datasets in geohazard domains might be a prevalent phenomenon. Results of the Shuping landslide and the Baishuihe landslide also show that the number of motoring stations in a GNSS network also affects the prediction result. As monitoring equipment and data transmission technology advance, daily, hourly, and even minute-scale displacements could be collected and predicted in real time. Additionally, several other solutions have emerged in different domains for handling dataset limitations, including data augmentation [34], synthetic data [35], and transfer learning [36].

Data augmentation refers to increasing the number of data points without changing the label. For example, variable factors include random noise, and adequate time characteristics can be employed to enlarge the time-series data [34]. Although not real data, synthetic data contain the same patterns and statistical properties as actual data, generated by a deep-learning model called generative adversarial networks (GAN) [35]. Transfer learning uses knowledge from other relevant datasets or an existing model to construct new models that lack enough training data to provide an alternative solution [23,36].

In this study, periodic rainfall and reservoir water level fluctuations are the main factors triggering landslide kinematic evolution in the TGR area. Therefore, we consider only these two trigging factors. Subsequent studies might include more complicated datasets to establish a more comprehensive model. For example, factors affecting landslide motion can consist of other essential characteristics of landslides, such as strata lithology, slope aspect and angle, etc.

5. Conclusions

This research develops a new deep-learning approach for landslide displacement forecasting called GC-GCN-N, which combines the GCN and the GRU. The architecture inherits the merits from both GCN in extracting spatial dependencies and GRU in capturing temporal correlation features to tackle the spatiotemporal landslide displacement forecast. In the proposed model, (1) a weighted adjacency matrix is built to interpret the spatial correlations between all monitoring stations, (2) a feature matrix is assembled to handle the time-series measurements of all monitoring stations, (3) an attribute-augmented unit is designed to represent the effects of the triggering factors and integrate the matrix mentioned above into a single graph convolutional network, and (4) a novel neural network-based approach is developed to enable to process the above graph-structured data. Experiments have been carried out on two landslides in Three George Reservoir, China. Compared with the MLR model, the ARIMA model, the SVR model, the LSTM model, and the T-GCN model, the GC-GCN-N model outperforms other forecasting models at both landslide sites. In summary, the GC-GCN-N model successfully captures the spatial and temporal features from the landslide monitoring dataset, showing great potential for other spatiotemporal forecast tasks.

Author Contributions: Conceptualization, Q.X. and Y.J.; methodology and validation, Y.J. and H.L. (Huiyuan Luo); formal analysis and investigation, L.L. resources, Q.X. and H.L.(Huajin Li); writing—original draft preparation, Y.J.; writing—review and editing, Y.J. and Z.L.; visualization, H.L. (Huiyuan Luo); funding acquisition, Q.X., L.H. and Y.J. All authors have read and agreed to the published version of the manuscript.

Funding: This research is financially supported by the National Key Research and Development Program of China (2018YFC1505101, 2021YFC3000401). The Key Research and Development Program of Sichuan Province (2020YFS0353) and the Open Fund of State Key Laboratory of Geohazard Prevention and Geoenviroment Protection (SKLGP2017K016).

Institutional Review Board Statement: Not applicable.

Informed Consent Statement: Not applicable.

Data Availability Statement: The raw data used in this study can be downloaded for scientific research after approval at http://www.crensed.ac.cn/, accessed on 14 February 2022.

Acknowledgments: The authors would like to thank the National Service Center for Specialty Environmental Observation Stations for providing the in situ monitoring data set.

Conflicts of Interest: The authors declare no conflict of interest.

References

1. Runqiu, H. Some Catastrophic Landslides since the Twentieth Century in the Southwest of China. *Landslides* **2009**, *6*, 69–81. [CrossRef]
2. Hu, X.; Bürgmann, R.; Schulz, W.H.; Fielding, E.J. Fielding. Four-Dimensional Surface Motions of the Slumgullion Landslide and Quantification of Hydrometeorological Forcing. *Nat. Commun.* **2020**, *11*, 2792–2801. [CrossRef] [PubMed]
3. Fan, X.; Scaringi, G.; Korup, O.; West, A.J.; van Westen, C.J.; Tanyas, H.; Hovius, N.; Hales, T.C.; Jibson, R.W.; Allstadt, K.E.; et al. Earthquake-Induced Chains of Geologic Hazards: Patterns, Mechanisms, and Impacts. *Rev. Geophys.* **2019**, *57*, 421–503. [CrossRef]
4. Jiang, Y.; Xu, Q.; Lu, Z.; Luo, H.; Liao, L.; Dong, X. Modelling and Predicting Landslide Displacements and Uncertainties by Multiple Machine-Learning Algorithms: Application to Baishuihe Landslide in Three Gorges Reservoir, China. *Geomat. Nat. Hazards Risk* **2021**, *12*, 741–762. [CrossRef]
5. LeCun, Y.; Bengio, Y.; Hinton, G. Deep Learning. *Nature* **2015**, *521*, 436–444. [CrossRef] [PubMed]
6. Ma, Z.; Mei, G. Deep Learning for Geological Hazards Analysis: Data, Models, Applications, and Opportunities. *Earth Sci. Rev.* **2021**, *223*, 103858. [CrossRef]
7. Krkač, M.; Bernat Gazibara, S.; Arbanas, Ž.; Sečanj, M.; Mihalić Arbanas, S. A Comparative Study of Random Forests and Multiple Linear Regression in the Prediction of Landslide Velocity. *Landslides* **2020**, *17*, 2515–2531. [CrossRef]
8. Zhang, G.P. Time Series Forecasting Using a Hybrid ARIMA and Neural Network Model. *Neurocomputing* **2003**, *50*, 159–175. [CrossRef]
9. Zhang, Y.; Tang, J.; He, Z.; Tan, J.; Li, C. A Novel Displacement Prediction Method Using Gated Recurrent Unit Model with Time Series Analysis in the Erdaohe Landslide. *Nat. Hazards* **2021**, *105*, 783–813. [CrossRef]
10. Xu, S.; Niu, R. Displacement Prediction of Baijiabao Landslide Based on Empirical Mode Decomposition and Long Short-Term Memory Neural Network in Three Gorges Area, China. *Comput. Geosci.* **2018**, *111*, 87–96. [CrossRef]
11. Li, H.; Xu, Q.; He, Y.; Fan, X.; Li, S. Modeling and Predicting Reservoir Landslide Displacement with Deep Belief Network and EWMA Control Charts: A Case Study in Three Gorges Reservoir. *Landslides* **2020**, *17*, 693–707. [CrossRef]
12. Han, H.; Shi, B.; Zhang, L. Prediction of Landslide Sharp Increase Displacement by SVM with Considering Hysteresis of Groundwater Change. *Eng. Geol.* **2021**, *280*, 105876. [CrossRef]
13. Lv, M.; Hong, Z.; Chen, L.; Chen, T.; Zhu, T.; Ji, S. Temporal Multi-Graph Convolutional Network for Traffic Flow Prediction. *IEEE Trans. Intell. Transp. Syst.* **2021**, *22*, 3337–3348. [CrossRef]
14. Huang, Y.; Han, X.; Zhao, L. Recurrent Neural Networks for Complicated Seismic Dynamic Response Prediction of a Slope System. *Eng. Geol.* **2021**, *289*, 106198. [CrossRef]
15. Chen, P.; Fu, X.; Wang, X. A Graph Convolutional Stacked Bidirectional Unidirectional-LSTM Neural Network for Metro Ridership Prediction. *IEEE Trans. Intell. Transp. Syst.* **2021**, 1–13. [CrossRef]
16. Navarin, N.; Van Tran, D.; Sperduti, A. Universal Readout for Graph Convolutional Neural Networks. In Proceedings of the 2019 International Joint Conference on Neural Networks (IJCNN), Budapest, Hungary, 14–19 July 2019; IEEE: Piscataway, NJ, USA, 2019; pp. 1–7.
17. Panahi, M.; Sadhasivam, N.; Pourghasemi, H.R.; Rezaie, F.; Lee, S. Spatial Prediction of Groundwater Potential Mapping Based on Convolutional Neural Network (CNN) and Support Vector Regression (SVR). *J. Hydrol.* **2020**, *588*, 125033. [CrossRef]
18. Jepsen, T.S.; Jensen, C.S.; Nielsen, T.D. Relational Fusion Networks: Graph Convolutional Networks for Road Networks. *IEEE Trans. Intell. Transp. Syst.* **2020**, *23*, 418–429. [CrossRef]
19. Jiang, Y.; Liao, M.; Zhou, Z.; Shi, X.; Zhang, L.; Balz, T. Landslide Deformation Analysis by Coupling Deformation Time Series from SAR Data with Hydrological Factors through Data Assimilation. *Remote Sens.* **2016**, *8*, 179. [CrossRef]
20. Jiang, C.; Fan, W.; Yu, N.; Liu, E. Spatial Modeling of Gully Head Erosion on the Loess Plateau Using a Certainty Factor and Random Forest Model. *Sci. Total Environ.* **2021**, *783*, 147040. [CrossRef]
21. Deng, L.; Smith, A.; Dixon, N.; Yuan, H. Machine Learning Prediction of Landslide Deformation Behaviour Using Acoustic Emission and Rainfall Measurements. *Eng. Geol.* **2021**, 106315. [CrossRef]
22. Hu, X.; Wu, S.; Zhang, G.; Zheng, W.; Liu, C.; He, C.; Liu, Z.; Guo, X.; Zhang, H. Landslide Displacement Prediction Using Kinematics-Based Random Forests Method: A Case Study in Jinping Reservoir Area, China. *Eng. Geol.* **2021**, *283*, 105975. [CrossRef]
23. Long, J.; Li, C.; Liu, Y.; Feng, P.; Zuo, Q. A Multi-Feature Fusion Transfer Learning Method for Displacement Prediction of Rainfall Reservoir-Induced Landslide with Step-like Deformation Characteristics. *Eng. Geol.* **2022**, *297*, 106494. [CrossRef]
24. Shin, Y.; Yoon, Y. Incorporating Dynamicity of Transportation Network With Multi-Weight Traffic Graph Convolutional Network for Traffic Forecasting. *IEEE Trans. Intell. Transp. Syst.* **2020**, 1–11. [CrossRef]
25. Ma, Z.; Mei, G.; Prezioso, E.; Zhang, Z.; Xu, N. A Deep Learning Approach Using Graph Convolutional Networks for Slope Deformation Prediction Based on Time-Series Displacement Data. *Neural Comput. Appl.* **2021**, *33*, 14441–14457. [CrossRef]
26. Yao, X.; Gao, Y.; Zhu, D.; Manley, E.; Wang, J.; Liu, Y. Spatial Origin-Destination Flow Imputation Using Graph Convolutional Networks. *IEEE Trans. Intell. Transp. Syst.* **2020**, *22*, 7474–7484. [CrossRef]
27. Bengio, Y.; Simard, P.; Frasconi, P. Learning Long-Term Dependencies with Gradient Descent Is Difficult. *IEEE Trans. Neural Netw.* **1994**, *5*, 157–166. [CrossRef]
28. Hochreiter, S.; Schmidhuber, J. Long Short-Term Memory. *Neural Comput.* **1997**, *9*, 1735–1780. [CrossRef]

29. Cho, K.; van Merrienboer, B.; Bahdanau, D.; Bengio, Y. On the Properties of Neural Machine Translation: Encoder-Decoder Approaches. *arXiv* **2014**, arXiv:1409.1259.
30. Man, J.; Dong, H.; Yang, X.; Meng, Z.; Jia, L.; Qin, Y.; Xin, G. GCG: Graph Convolutional Network and Gated Recurrent Unit Method for High-Speed Train Axle Temperature Forecasting. *Mech. Syst. Signal Process.* **2022**, *163*, 108102. [CrossRef]
31. Hyndman, R.J.; Koehler, A.B. Another Look at Measures of Forecast Accuracy. *Int. J. Forecast.* **2006**, *22*, 679–688. [CrossRef]
32. Qinghao, L.I.U.; Yonghong, Z.; Min, D.; Hongan, W.U.; Yonghui, K.; Jujie, W.E.I. Time Series Prediction Method of Large-Scale Surface Subsidence Based on Deep Learning. *Acta Geod. Cartogr. Sin.* **2021**, *50*, 396. [CrossRef]
33. Zhao, L.; Song, Y.; Zhang, C.; Liu, Y.; Wang, P.; Lin, T.; Deng, M.; Li, H. T-GCN: A Temporal Graph Convolutional Network for Traffic Prediction. *IEEE Trans. Intell. Transp. Syst.* **2020**, *21*, 3848–3858. [CrossRef]
34. Rashid, K.M.; Louis, J. Times-Series Data Augmentation and Deep Learning for Construction Equipment Activity Recognition. *Adv. Eng. Inform.* **2019**, *42*, 100944. [CrossRef]
35. Creswell, A.; White, T.; Dumoulin, V.; Arulkumaran, K.; Sengupta, B.; Bharath, A.A. Generative Adversarial Networks: An Overview. *IEEE Signal Process. Mag.* **2018**, *35*, 53–65. [CrossRef]
36. Li, N.; Hao, H.; Gu, Q.; Wang, D.; Hu, X. A Transfer Learning Method for Automatic Identification of Sandstone Microscopic Images. *Comput. Geosci.* **2017**, *103*, 111–121. [CrossRef]

remote sensing

MDPI

Article

Seasonal and Interhemispheric Effects on the Diurnal Evolution of EIA: Assessed by IGS TEC and IRI-2016 over Peruvian and Indian Sectors

Xin Wan [1,2], Jiahao Zhong [1,2,*], Chao Xiong [3], Hui Wang [3], Yiwen Liu [4], Qiaoling Li [1,2,5], Jiawei Kuai [6] and Jun Cui [1]

1 Planetary Environmental and Astrobiological Research Laboratory (PEARL), School of Atmospheric Sciences, Sun Yat-sen University, Zhuhai 519082, China; wanx7@mail.sysu.edu.cn (X.W.); liqling6@mail.sysu.edu.cn (Q.L.); cuijun7@mail.sysu.edu.cn (J.C.)
2 Key Laboratory of Tropical Atmosphere-Ocean System, Ministry of Education, Zhuhai 519082, China
3 Department of Space Physics, College of Electronic Information, Wuhan University, Wuhan 430079, China; xiongchao@whu.edu.cn (C.X.); h.wang@whu.edu.cn (H.W.)
4 School of Physics and Electronic Information, Shangrao Normal University, Shangrao 334000, China; liuyiwen@whu.edu.cn
5 Mengcheng National Geophysical Observatory, University of Science and Technology of China, Hefei 230026, China
6 College of Astronautics, Nanjing University of Aeronautics and Astronautics, Nanjing 210016, China; jwkuai@nuaa.edu.cn
* Correspondence: zhongjh55@mail.sysu.edu.cn

Abstract: The global total electron content (TEC) map in 2013, retrieved from the International Global Navigation Satellite Systems (GNSS) Service (IGS), and the International Reference Ionosphere (IRI-2016) model are used to monitor the diurnal evolution of the equatorial ionization anomaly (EIA). The statistics are conducted during geomagnetic quiet periods in the Peruvian and Indian sectors, where the equatorial electrojet (EEJ) data and reliable TEC are available. The EEJ is used as a proxy to determine whether the EIA structure is fully developed. Most of the previous studies focused on the period in which the EIA is well developed, while the period before EIA emergence is usually neglected. To characterize dynamics accounting for the full development of EIA, we defined and statistically analyzed the onset, first emergence, and the peaks of the northern crest and southern crest based on the proposed crest-to-trough difference (CTD) profiles. These time points extracted from IGS TEC show typical annual cycles in the Indian sector which can be summarized as winter hemispheric priority, i.e., the development of EIA in the winter hemisphere is ahead of that in the summer hemisphere. However, these same time points show abnormal semiannual cycles in the Peruvian sector, that is, EIA develops earlier during two equinoxes/solstices in the northern/southern hemisphere. We suggest that the onset of EIA is a consequence of the equilibrium between sunlight ionization and ambipolar diffusion. However, the latter term is not considered in modeling the topside ionosphere in IRI-2016, which results in a poor capacity in IRI to describe the diurnal evolution of EIA. Meridional neutral wind's modulation on the ambipolar diffusion can explain the annual cycle observed in the Indian sector, while the semiannual variation seen in the Peruvian sector might be due to additional competing effects induced by the F region height changes.

Keywords: GPS TEC; EIA; diurnal evolution; seasonal variation

Citation: Wan, X.; Zhong, J.; Xiong, C.; Wang, H.; Liu, Y.; Li, Q.; Kuai, J.; Cui, J. Seasonal and Interhemispheric Effects on the Diurnal Evolution of EIA: Assessed by IGS TEC and IRI-2016 over Peruvian and Indian Sectors. *Remote Sens.* **2022**, *14*, 107. https://doi.org/10.3390/rs14010107

Academic Editors: Serdjo Kos, José Fernández and Juan F. Prieto

Received: 6 November 2021
Accepted: 24 December 2021
Published: 27 December 2021

Publisher's Note: MDPI stays neutral with regard to jurisdictional claims in published maps and institutional affiliations.

1. Introduction

The equatorial ionization anomaly (EIA) is the result of ionization and electrodynamic processes in the ionosphere, representing a plasma density trough at the equator and two crests approximately 15° to the north and south. The formation process of EIA can be characterized as a fountain effect. That is, the daytime E region wind dynamo produces an eastward electric field to move the plasma upward via E×B force; the increasing gravitational potential and pressure gradient would finally result in ambipolar diffusion, which transports the equatorial source plasma downward/poleward to the north and south, to create double plasma density crests [1–10].

Extensive studies have investigated the development of the EIA under various conditions configured by solar radiation, season, local time (LT), longitude, and geomagnetic disturbances, to reveal the impacts of those parameters on the ionizations, electrodynamics, and neutral dynamics behind the fountain effects. Yeh et al. [1] found that EIA crests generally begin to appear at 0900 LT, and the farthest latitude of the crests during daytime is correlated with the level of the fountain effect and the local ionospheric total electron content (TEC). Liu et al. [2] reported that the crest-to-trough ratio (CTR) of EIAs observed by the CHAMP satellite gradually increases from morning to noon, reaching its maximum value between 1800 and 1900 LT as a result of pre-reversal enhancement (PRE). Xiong et al. [3] reported that the electron density and magnetic latitudes of both EIA crests peak at approximately 1400 LT. The local time variation of the electron density crest during daylight hours is similar to that of the trough but with a 1–2 h delay [4].

Another interesting aspect of the EIA is the interhemispheric asymmetry, which is most pronounced during solstitial seasons. In the morning hours, the crest in the winter hemisphere generally forms earlier and has a greater magnitude than that in the summer hemisphere; during the afternoon, the summer hemisphere features the larger EIA crest [3,6–10]. The transition time of this interhemispheric EIA asymmetry occurs at around 1200–1400 LT, depending on solar activity levels [6,7]. In addition, this transition time is a function of observed altitude, since the time lag of fountain effects varies with height [3]. Interhemispheric EIA asymmetry has also been reported during equinoxes [7,10,11]. Neutral wind variations associated with displacements of the geographic and geomagnetic equators as well as magnetic declination angles have been proposed as the main factors affecting EIA asymmetry [10]. Balan et al. [11] argued that the displacement of the geographic and geomagnetic equators is a more significant factor than the declination angle.

The above-mentioned studies mainly focus on the EIA that exhibits clear double crests signature. However, before the first emergence of EIA double crests, the fountain-like processes had already been launched, but this stage receives much less attention in the research community. In detail, the sunlit ionization process is instantaneous, while the mechanical transportation process is much slower. Thus, at the beginning, stronger sunlit ionization actually would cause faster plasma accumulation at the subsolar position near the equator. The fountain effects take time to compete against the uneven ionization, to form EIA. The contributions of various physical processes during different stages of the EIA diurnal evolution are still not well known. In particular, the dynamic process before the emergence of the EIA crest is usually neglected and had not been investigated yet.

Note that the sunlit ionization and electrodynamical transportation have different reference equators, i.e., the geographic and geomagnetic equators. The displacement between the two equators varies at different longitudes, which would significantly impact the EIA evolution. Moreover, meridional thermospheric wind, which drags the ion along with it, is reported to impact the EIA evolution in two opposite ways. On the one hand, the transequatorial thermospheric wind pushes the plasma along the field line to contribute/counter the ambipolar diffusion in the winter/summer hemisphere, leading to a winter hemispheric priority during EIA's development [6,7]. On the other hand, the transequatorial thermospheric wind would lift/lower the F region height in the summer/winter hemisphere, leading to stronger/weaker intensity (i.e., TEC) of EIA crest [11–13]. It can be seen that the inclusion of thermospheric neutral wind effects could make the EIA evolution more complicated, the dominant seasonal cycles at different longitudes are still not well understood, and the physical processes and mechanisms involved are still in debate [6,7,12–14].

We dedicate this study not only to monitoring the interhemispheric asymmetry in a traditional way that focuses on the intensity of EIA, but also to trying to characterize the detailed time evolution to clarify the dynamical competition and cooperation between the sunlit ionization, ambipolar diffusion, and neutral wind drag. The TEC maps provided by IGS [15] are used under the geomagnetic quiet condition in 2013 in the Peruvian and Indian sectors. At low geomagnetic latitudes, these two sectors are both deployed with magnetometers to retrieve an equatorial electrojet (EEJ) that can be used as a proxy to select days with developed EIA, and GNSS receivers that provide reliable TEC product to investigate the seasonal and interhemispheric effects on the diurnal evolution of EIA. The International Reference Ionosphere (IRI-2016) model, widely recognized as a powerful tool to represent the climatological behavior of the ionosphere [16–19], is also adopted to check whether the empirical model can capture the real features of the EIA evolution.

In Section 2, we provide descriptions of the datasets. In Section 3, we introduce how the geomagnetic quiet days and days with well-developed EIAs are sorted out. Section 4 presents the seasonal/local variation of EIA, followed by detailed statistics of crests and trough variations. The crest-to-trough difference (CTD), is defined and described in Section 4.2. The EIA onset time, first emergence and peaks are further derived from CTD profiles and we presented the statistical results. In Section 5, we discuss the physical mechanisms involved. Conclusions of this study are provided in Section 6.

2. Dataset

2.1. IGS TEC Maps and the IRI-2016 Model

The IGS TEC data is an interpolated data product based on the TEC measurements from ground-based GNSS receivers that are distributed mainly over the continents [15]. Thus, the IGS TEC should provide trustable EIA observation over Indian and Peruvian sectors. The dataset has a time resolution of 15 min and spatial resolution of 2.5° × 5° in geographic latitude and longitude.

The distribution of GPS receivers over the oceanic region is much sparser than the continents, IGS TEC was found to overestimate Jason-2/3 derived TEC by more than 5 TECU [20]. This overestimation would possibly impact the climatological behavior of the retrieved EIA features. Thus, whether the IGS TEC is suitable to be extendedly applied to the longitudes over the oceanic region remains unclear. We choose another candidate data source from the IRI-2016 model, a widely used empirical ionospheric model, and it was recently improved with a new hmF2 model based on a new database from the worldwide network of ionosondes [17,19]. The IRI-2016 is used to retrieve the EIA feature and compare with IGS TEC data. This comparison would not only help to assess the performance of the empirical model in describing the regional EIA evolution, but also evaluate the feasibility of whether the empirical model can be applied to extend EIA study over the oceanic region.

For a given location, time, and date, the IRI model provides monthly averages of the ionospheric parameters, including electron density, electron temperature, ion temperature, ion composition, and TEC from an altitude range of 50–2000 km [17–19]. Options were set to calculate the TEC from the IRI-2016 model: The fof2 storm model was switched off, the Shubin-cosmic option was used for the hmf2 model, and NeQuick was used as the topside model. The maximum height of the TEC calculation was set to 2000 km.

2.2. The EEJ Derived from Ground-Based Magnetometers

Ground-based paired magnetometer measurements over the Peruvian and Indian sectors in 2013 were utilized to estimate the equatorial electrojets (EEJ). As a narrow current that flows in the E region above the magnetic equator, EEJ can be extracted by removing the solar quiet (Sq) current that barely shows latitudinal dependence [21]. To estimate the EEJ, we calculated the differences of the horizontal (H) component of the geomagnetic field between the paired magnetometer, as the residual horizontal magnetic field is recognized to be caused by the EEJ [22]. The measurements of Huancayo (HUA, −12.05° N, −75.33° E, 0.59° dip latitude) and Fuquene, (FUQ,18.11° N, −66.15° E, 17.06° dip latitude) are used for the Peruvian sector, and Tirunelveli (TIR, 8.7° N, 77.8° E, 0.59° dip latitude) versus Alibag (ABG, 18.6° N, 72.9° S, 13.67° dip latitude) are used for the Indian sector [19].

2.3. Horizontal Wind Simulated by TIEGCM

To estimate the neutral wind effects on the development of EIA, the horizontal wind simulated from the Thermosphere Ionosphere Electrodynamics General Circulation Model (TIEGCM) is adopted. The TIEGCM is a first principle and physics-based model driven by a high-latitude electric field [23], solar EUV, and UV spectral fluxes parameterized by the F10.7 index [24].

3. Methodology and Methods

3.1. Sorting Geomagnetic Quiet Days Using Kp Index

Kp index is a quasi-logarithmic index, ranging in steps of 1/3 from 0 to 9, to quantify the level of the geomagnetic disturbance on a global scale [25,26]. All the data used in this study were firstly sorted under geomagnetic quiet conditions when the daily mean *Kp* values were less than 3.

3.2. Sorting Developed EIA Using EEJ as a Proxy

Note that sometimes the EIA structure is dismissed, which could not be used to monitor the time evolution of EIA. Thus, we firstly determine whether the EIA appears for further statistics. As mentioned previously, the formation of EIA is a product of the eastward zonal electric field; thus, the intensity of the zonal electric field could be a good indicator for the EIA development. However, the direct measurement of the electric field is rare; an alternate option is EEJ, which serves as a proxy to quantify the daytime zonal electric field. The EEJ refers to a narrow band of intense electric current flowing above the equatorial dip in the daytime E-region driven by the E-region electric field and conductivity. The EEJ mainly flows eastward corresponding to an eastward electric field, the infrequent westward flow of EEJ is called counter electrojet (CEJ) corresponding to a westward electric field. The EEJ is considered to be a suitable proxy for EIA intensity. Stolle et al. [27] found correlation coefficients greater than 0.8 between EEJ strength and EIA intensity. Venkatesh et al. [28] discovered that the daily summed EEJ strength had correlations of 0.62 and 0.72 with the EIA crest amplitude and latitude, respectively.

Figure 1 shows examples of a weak EEJ profile (Figure 1a) and a CEJ profile (Figure 1b), with associated TEC maps for the day. The double-crest structure barely formed as the EEJ intensity was relatively weak (Figure 1a), while the CEJ resulted in a single peak at the equator, meaning that the EIA was inhibited completely.

It is reported that the strength of the EIA shows a better correlation with integrated EEJ values than the daily maximum of EEJ [28]. In addition, the time delay of the EIA response to EEJ strength is 2–3 h [27,28]. To identify the days with weak EEJ or CEJ, the averaged ΔH at an LT bin of 0800–1200 is calculated first. Please note that the Weak EEJ or CEJ were combined to be referred to as WEC in this study. Days with WEC were then determined when the averaged ΔH components were less than the threshold value; the remainder of the quiet days were recognized as EEJ days. Our experiments (not shown here) showed that the EEJ intensity was generally higher in the Peruvian sector than in the Indian sector, probably as a result of tidal effects. Thus, we chose two thresholds, ~70 nT and ~20 nT of averaged EEJ intensity during 0800–1200 LT for the Peruvian and Indian sectors, respectively.

Figure 2 shows the day numbers of the WEC and EEJ cases at longitudes of $-75°$ E and $75°$ E, which represent the Peruvian and Indian sectors, respectively. The WEC case showed a preferential occurrence during the two solstices at both longitudes, consistent with previous studies which had found that the EEJ is characterized by solstitial minima [29–31]. However, for a given longitude, the WEC showed specific seasonal preferences as well. There were more WEC days around the December Solstice (Nov, Dec, Jan, Feb) than the June Solstice (May, Jun, Jul, Aug) in the Peruvian sector, with this preference being reversed in the Indian sector. Considering the northward–southward deviation of the dip equator from the geographic equator at both the Indian and Peruvian longitudes, this implied that more WEC occurred when the dip equator is in the summer hemisphere. One may also note that the WEC day is missed in Mar, Apr, Aug, Sep, and Oct in the Indian sector (Figure 2b).

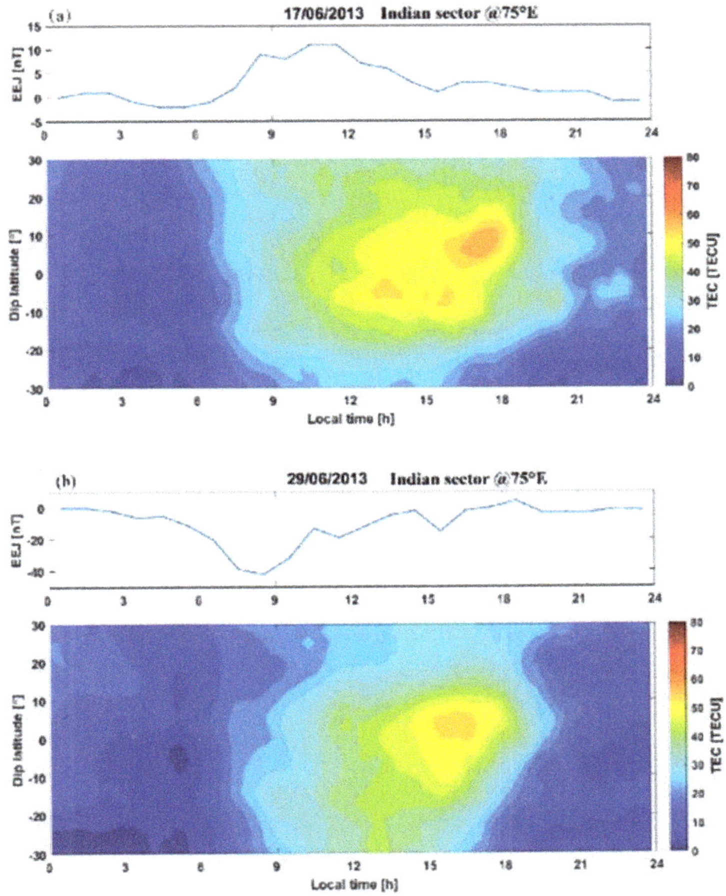

Figure 1. Two examples of the GPS-measured TEC map at Indian sector during (**a**) a weak EEJ day and (**b**) a CEJ day.

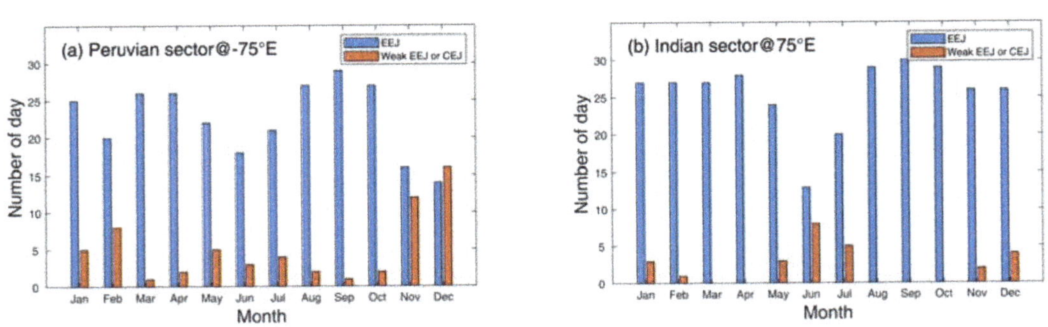

Figure 2. Number of days with EEJ signatures and weak EEJ or CEJ signatures at (**a**) Peruvian sector and (**b**) Indian sector.

4. Results

4.1. Overview of EIA during EEJ and Weak EEJ/CEJ Days

Figure 3 shows the TEC maps during EEJ days in the Peruvian and Indian sectors. The data shown in the left, middle, and right columns represent different seasons: December Solstice (Dec. S., which includes Nov, Dec, Jan, and Feb), June Solstice (June. S., which includes May, Jun, Jul, and Aug), and equinoxes (March, April, September, October), respectively. The daily EEJ was first plotted in Figure 3a,d; solid black lines represent four-month averages.

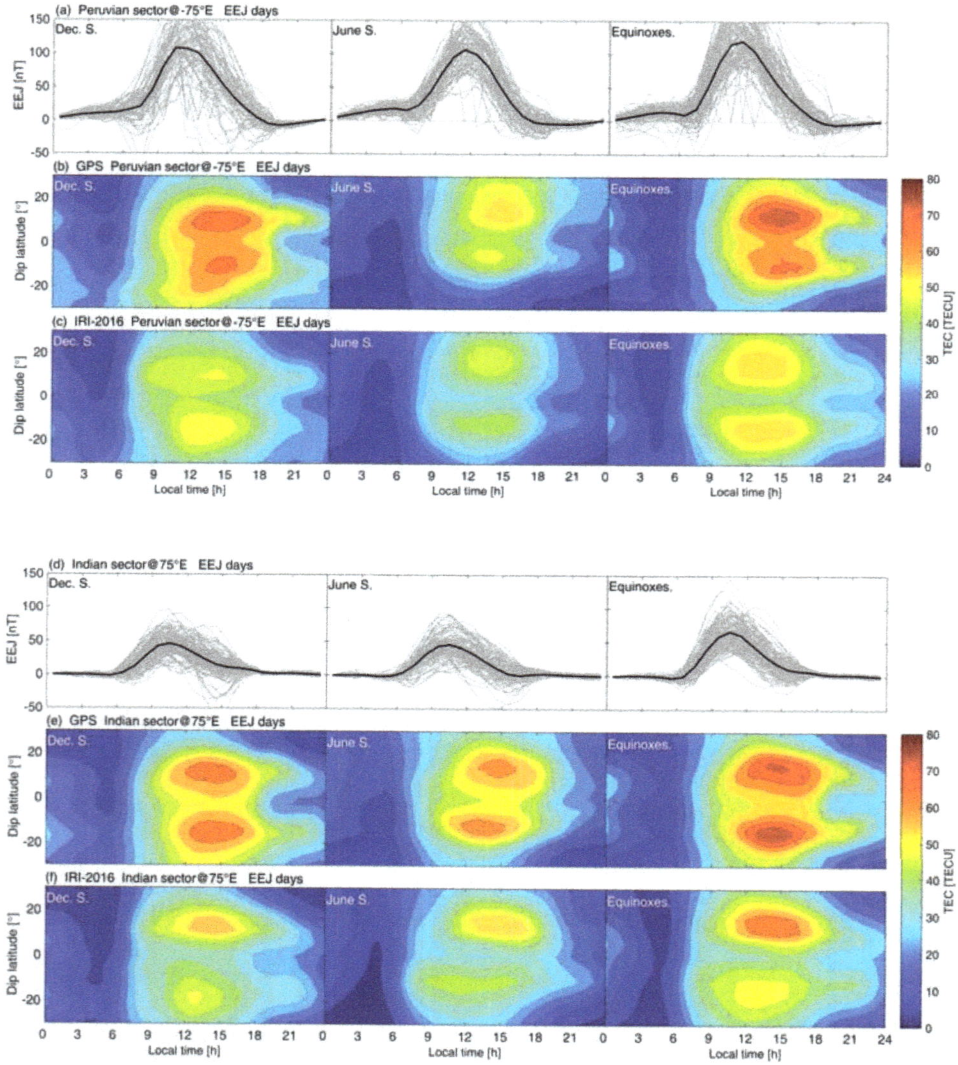

Figure 3. Seasonal averaged TEC maps over the Peruvian and Indian sectors during EEJ days when the averaged EEJ intensity at 0800–1200 UT was relatively high during three seasons of December Solstice (left column), June Solstice (middle column) and equinoxes (right column). (**a**) EEJ profiles of Peruvian sector; (**b**) GPS TEC at Peruvian sector; (**c**) IRI-2016 derived TEC at Peruvian sector; (**d–f**) are organized in a same format as (**a–c**) but for the Indian sector.

For GPS TEC observations at both Peruvian and Indian sectors, the most prominent feature of EIA is the interhemispheric asymmetry. During two solstices, the crests in the winter/summer hemisphere are stronger during morning/afternoon hours. Moreover, the stronger crest generally resides at a higher latitude. An exception is that at the Peruvian during Dec. S., the southern crest (in the summer hemisphere) was stronger first, which lasted an interval of 0800–0900 LT; afterward, the northern crest (in the winter hemisphere) turned out to be dominant. During equinoxes, interhemispheric asymmetry still exists in that a stronger northern/southern appeared in the Peruvian/Indian sector. Similar seasonal and longitudinal variations of the interhemispheric asymmetry have been investigated by many studies. Those studies generally accepted that the displacements of the geographic and geomagnetic equators, the different geomagnetic declinations, and the meridional neutral wind show clear seasonal and longitudinal variations, which affects the diurnal evolution of EIA [3,6–9,11].

The developed EIA signatures can also be characterized by the IRI-2016, nevertheless, with a lower absolute TEC level by approximately 20 TECU, compared with the GPS TEC. In the Peruvian sector, IRI-2016 performed stronger crests in the summer hemisphere. However, in the Indian sector, the northern crest is persistently stronger than the southern crests. Hence, it can be noticed that the IRI-2016 performed quite a different interhemispheric asymmetry pattern compared with that of the GPS observations, especially during equinoxes.

On WEC days (Figure 4), as expected, the EIA barely formed as observed by the GPS TEC. The Peruvian sector even showed a single-peak structure over the equator during the December Solstice and two equinoxes, indicating that the fountain effect was completely inhibited. For the developed EIA, the intensity is weaker than that during EEJ days; that is, the double crest was narrower, and the TEC values were reduced by approximately 15 TECU and 5 TECU in the Peruvian and Indian sectors, respectively. Note that during the equinoxes there are no WEC days for the Indian sector as displayed in the statistical day numbers in Figure 2b, thus the corresponding TEC data were not available (Figure 4d–f).

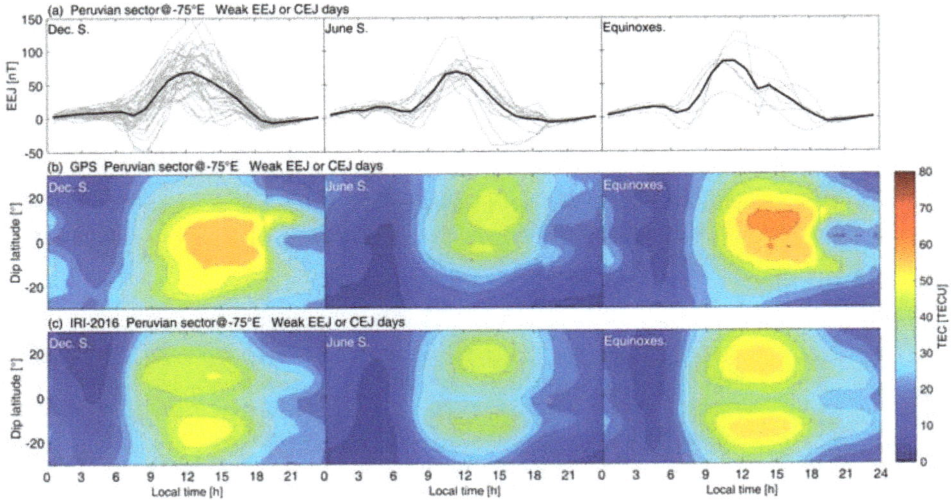

Figure 4. *Cont.*

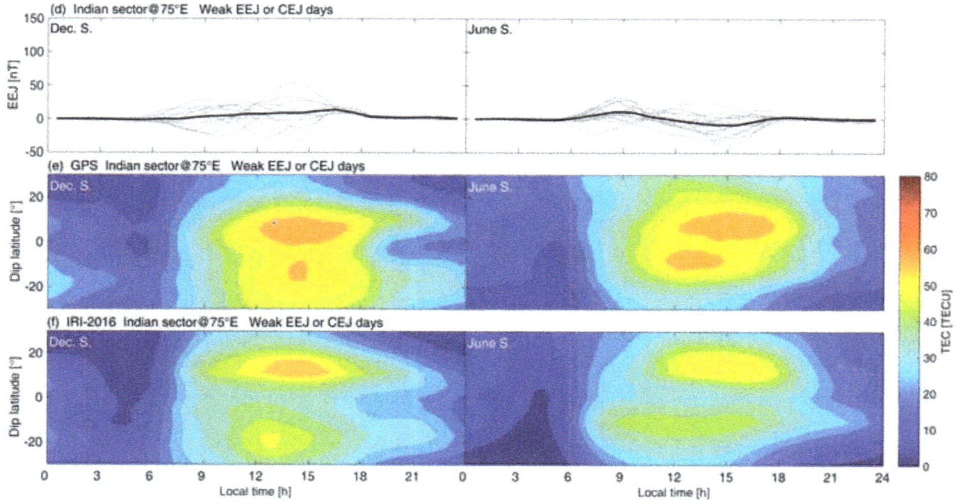

Figure 4. Similar to Figure 3, but during days that the average EEJ intensity at 0800–1200 UT was relatively low (refer to WEC days) at Peruvian sector (**a**–**c**) and Indian sector (**d**–**f**). The lack of equinox results in (**d**–**f**) is due to that the WEC days are not found.

However, TEC values predicted by the IRI-2016 model on WEC days remained at nearly the same levels as those during EEJ days, indicating that the model cannot capture an inhibited EIA signature. For a given time, the IRI-2016 predicted the result in terms of the monthly averaged value. Hence, this inherent characteristic of the empirical model would result in unreliable predictions for days with unusual space weather occurring during geomagnetic quiet times such as CEJs. To successfully ascertain the characteristics of daytime EIAs, we excluded those types of WEC days in the following analysis.

4.2. Crest-to-Trough Differences (CTD)

Despite the EIA intensity, it is easy to notice in Figures 3 and 4 that the local time of the first appearance of the crests also showed interhemispheric asymmetry. That is, a stronger crest corresponding to earlier emergence. Extensive studies have investigated the interhemispheric asymmetry of the EIA intensity; hence, in the following, we provide another perspective to investigate asymmetrical behavior concerning the time evolution of EIA. In particular, to assess the time evolution of EIA, we defined crest-to-trough difference (CTD):

$$CTD = TEC_{\text{off-equator}} - TEC_{\text{equator}} \tag{1}$$

That is, the difference between the TEC at off-equator (set as fixed dip latitude bin of 10°–15° N/S) and the TEC at the equator (set as fixed dip latitude bin of 2.5° S–2.5° N), where the EIA crest and trough normally reside, respectively. Note that the 'crest' location we defined here is not the accurate position of the EIA crest; thus, the calculation of CTD does not require the presence of a clear EIA crest feature.

Another criterion used to quantify the developed EIA intensity is the crest-to-trough ratio, defined as the ratio of the mean of the northern and southern EIA crest peak value to the minimum TEC in the EIA trough:

$$CTR = (TEC_{\text{ncrest}} + TEC_{\text{screst}})/(2 \times TEC_{\text{trough}}) \tag{2}$$

CTR is much more extensively used in other studies [2,3,27,32,33]. The CTR provides information on the overall intensity of the developed EIA normalized by the background TEC at the EIA trough. As we intend to monitor the full development of the EIA throughout the daytime, even when the EIA crests have not been developed, the defined CTD is a more suitable parameter to achieve the goal.

Figure 5 shows an example of the northern crest-to-trough difference (NCTD) and southern crest-to-trough difference (SCTD), extracted from GPS TEC and IRI TEC maps (Figure 5a,b). GPS TEC measurements (Figure 5a) show that at 0600–1000 LT the double crest had not been developed and CTD experienced a falling and then a rising (Figure 5c). The falling occurred right after the sunrise, due to that the ionization creates more plasma around the equator. The rising should be related to the ambipolar diffusion that transports equatorial plasma to higher latitudes. Thus, the transition of the falling and rising marks the time that the transportation term dominates, i.e., the net accumulation of plasma at the off-equator exceeds that at the equator. This transition is marked as the inflection points on the CTD curved in the morning hours and is defined as the onset of EIA.

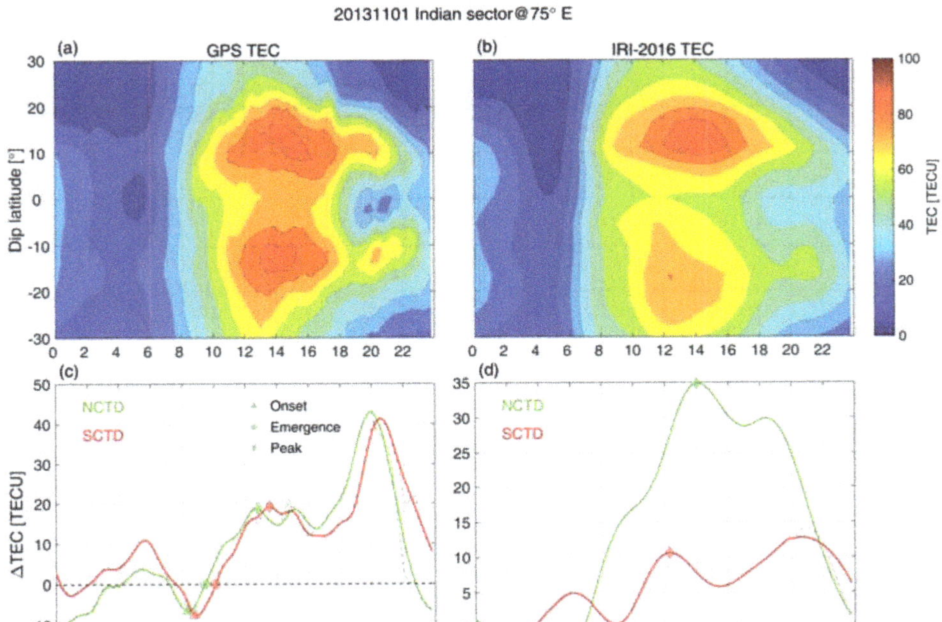

Figure 5. Top panels: TEC at Indian sector as a function of local time and dip latitude, using data from (**a**) IGS GPS TEC and (**b**) IRI-2016. Bottom panels: Local time evolution of the extracted NCTD and SCTD from (**c**) IGS GPS TEC and (**d**) IRI-2016. Dash curves are the original calculation from TEC data; solid curve is the smoothed results. The triangles, circles, and diamonds mark the time of the onsets, first emergences, and the peaks of the EIA crests during the evolution.

In the GPS TEC map (Figure 5a), the EIA northern crest appears earlier than the southern crest, at a local time near 0900. The emergence of the EIA has been marked as circles on the CTD (Figure 5c) curves as the CTD equals 0, which also exhibit northern crest priority. However, the local time of the marked emergence lags behind the crest's appearance on the TEC map (Figure 5a). This is because the calculation of CTD set the crest at fixed latitudes of 10°–15° N/S, while the real emergence of the EIA crest appeared at latitudes closer to the magnetic equator where the fountain effects launch. After the first emergence, CTD grows continuously, representing the development of EIA. The peaks of the CTD mark that the EIA is fully developed and the TEC at the EIA crest reaches its peak value. Thus, we mark the peak CTD in the afternoon as the peaks of EIA.

Figure 5b,d show the IRI-2016 predicted TEC map and the corresponding CTD profiles, respectively. Besides a clearly different morphology of EIA structure (Figure 5b) compared with the GPS TEC measurements (Figure 5a), the IRI-2016 derived CTD also exhibited abnormal local time evolution. In detail (Figure 5d), there is no post-sunrise falling that ends in negative vales for NCTD; SCTD showed persistent positive value during 0700–2300 LT. Thus, the onset, as well as the first emergence of the EIA crest, can neither be identified on the IRI-2016 CTD. Only the peaks of the EIA crest can be marked as the peaks of CTD profiles. Note that a similar situation is a common feature for IRI-2016, though the three time points can be occasionally identified.

In summary, the local time of the onset, first emergence, and peak of EIA crest can be identified from GPS TEC measurement, while IRI-2016 cannot regularly capture the real time evolution process of the EIA development.

4.3. Time Evolution of EIA: Seasonal and Longitudinal Effects

In the last subsection, the time points of the onset, first emergence, and peak during the evolution of EIA can be marked on the CTD profiles. Figure 6 presents the variations of these time points as a function of months in 2013.

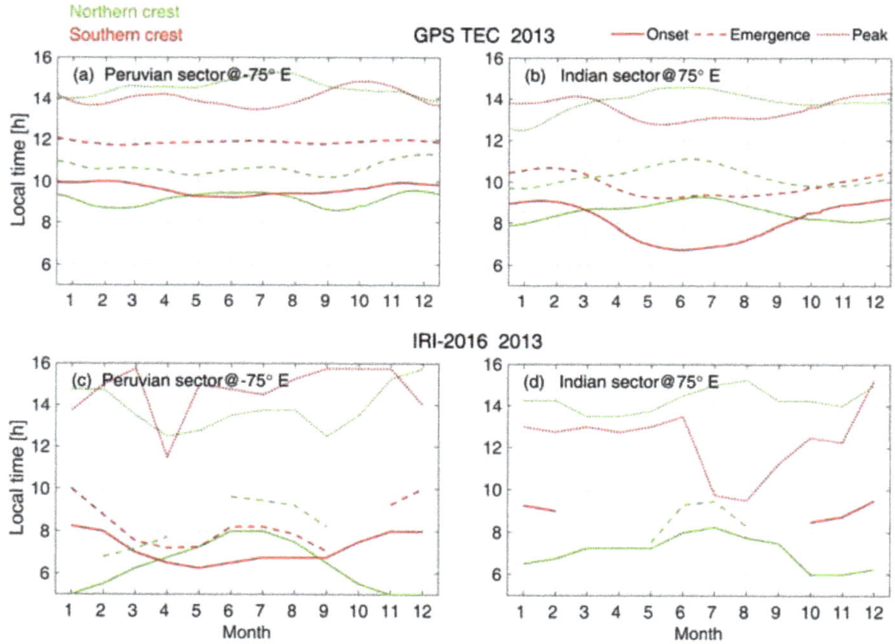

Figure 6. Local time of the onsets (solid lines), emergence (dash lines), peaks (dotted lines) of the EIA northern crest (green) and southern crest (red), at Peruvian (**a–c**) and Indian sector (**b–d**), extracted from GPS TEC (top panels) and IRI-2016 (bottom panels). The data gap in Figure 6c,d is due to that the time points sometimes cannot be identified in the IRI model.

Results from GPS TEC (Figure 6a,b) revealed that the patterned annual variations were generally shared by the three time points. The time lag between the first emergence of EIA crests with the onset is ~2 h, and it takes another ~4 h for the EIA crest to reach peak values. Specifically, semiannual and annual cycle patterns in the Peruvian and Indian sectors, respectively, can be captured. Take the onset of the northern EIA crest as an example, the semi-annual cycle (Figure 6a) is characterized by an earlier onset time near two equinoxes and a later onset time near two solstices; the annual cycle (Figure 6b) is characterized by an earlier onset time in the winter and a later onset time in the summer. As for the southern crests, the time points generally exhibit reversed seasonal patterns compared with those of the northern crests, except that the emergence time at the Peruvian sector (Figure 6a) stays constantly near 1200 LT throughout the year 2013. Besides, the semiannual variation of the onset of the southern crest at the Peruvian sector is not as prominent as the northern crest, resulting in an earlier onset of southern crest than northern crest during winter seasons, which is similar to that at the Indian sector.

Figure 6c,d show the IRI-2016 predictions. As has been mentioned above, the onset and the emergence of EIA sometimes cannot be identified from IRI-2016 data. It can be witnessed that the onset and emergence data are sparse, due to the incapability for identifying those time points on CTD profiles. Nevertheless, the available time points generally predicted earlier onset and emergence, compared with those of the GPS observations. Interestingly, the annual cycle of EIA onset time is a prevailing phenomenon in both Peruvian and Indian sectors, while GPS revealed a semiannual cycle seen in the Peruvian sector. However, a similar semiannual cycle seen in GPS observation can be found on the peak time of the northern crest in the Peruvian sector (Figure 6c). Besides, the emerging time of the southern crest also exhibits semiannual variation (Figure 6c), but in the same phase of the time points of the northern crest seen in GPS observations (Figure 6a). It can be concluded that, although the semiannual cycle can be occasionally found on the emerging and peak time of EIA crest from IRI-2016, the onset time of EIA still shows the classic picture that the crest in the winter hemisphere develops earlier than that in the summer hemisphere. Note that the IRI-2016 derived peak time of EIA sometimes shows abnormal values (e.g., April at Peruvian sector, July and August at Indian sector, as shown in Figure 6c,d); this also indicated the inaccuracy of the IRI-2016 in retrieving EIA evolution.

The annual cycle shown in the Indian sector represents a classical picture of the neutral wind modulated ambipolar diffusion during the development of EIA. That is, the transequatorial wind blows from the summer hemisphere to the winter hemisphere, which pushes the plasma equatorward and poleward along the field line (referred to as pile-up effects) in the summer and winter hemispheres, respectively. These neutral wind effects contradict/favor the ambipolar diffusion in the summer/winter hemisphere; hence, the development of EIA crest in the summer/winter hemisphere is inhibited/promoted, resulting in that the time points of EIA's development show clear winter hemispheric priority.

Take the onset of the northern crest during the winter season as an example, the essential difference between the annual cycle (Indian sector, Figure 6b) and the semiannual cycle (Peruvian sector, Figure 6a) occurs during winter seasons. That is, the semiannual cycle consists of a delayed northern crest onset during northern winter (Figure 6a), which ought to occur earlier (Figure 6b) to exhibit the annual cycle. Thus, the fundamental question is what causes the retarded EIA northern crest development that is supposed to be advanced.

5. Discussion

The semi-annual cycle is a common feature on the overall intensity of plasma density at low and middle latitudes, which appears as density maximums at two equinoxes. The causes of the semi-annual cycle are discussed as follows. There is the ratio of atomic oxygen to molecular nitrogen O/N_2, which also shows similar semiannual variation, especially at middle latitudes [34]. In detail, during high O/N_2 ratio seasons (i.e., equinoxes), the chemical recombination rate is low, which causes a relatively high plasma density. He O/N_2 ratio is higher in the winter hemisphere, leading to a higher ionization rate [35]. However, this effect is more suitable to explain the middle latitudes winter anomaly [36] rather than that at low latitudes, since the hemispheric difference in the O/N_2 ratio is smaller in the low latitudes during the daytime. Thus, the O/N_2 ratio is less likely to contribute to the interhemispheric asymmetry of EIA's evolution seen in this study.

In addition to the O/N_2 ratio, the semi-annual cycle is also the property possessed by the daytime equatorial E×B drift that exhibits two equinoctial maximums and two solstitial minimums [37,38], via atmospheric tides modulated E-region dynamo [39]. Hence, the E×B drift is widely recognized as a major mechanism for the appearance of semiannual variation of the general EIA intensity [33]. However, there exist arguments regarding the E×B drift effects on the detailed EIA morphology. Wu et al. [40] attributed the semiannual variation of the northern EIA crest location to the E×B drift, while Liu et al. [14] doubt the mechanism as the daytime EEJ shows poor correlation with the EIA crest location.

The semiannual cycle of either the O/N_2 ratio or the E×B drift is a global phenomenon, if the O/N_2 ratio or the E×B drift is indeed involved in altering the efficiency of the EIA development, the earlier onset during equinoxes should also be witnessed in the southern hemisphere, and in other longitudes (i.e., Indian sector). However, this is not supported by the current observations. Hence, neither the O/N_2 ratio nor the E×B drift is responsible for the abnormal semiannual variation of the onset, emergence, and peaks of the EIA at the Peruvian sector, at least in a sole way.

The meridional neutral wind is recognized as the major impact of the asymmetric development of EIA. Thus, we raise a question of whether the semiannual cycle seen in the Peruvian sector is due to the local abnormal neutral wind configurations. To answer the question, the neutral wind result from TIEGCM simulation during two solstices and two equinoxes is adopted, during the geomagnetic quiet periods. Note that when discussing the pushing effects from neutral wind to plasma, the zonal wind contribution under the presence of magnetic declination should be considered. Hence, the effective magnetic meridional neutral wind velocity (U) could be written as:

$$U = U_\theta \cos D + U_\varphi \sin D \quad (3)$$

where U_θ (positive southward) and U_φ (positive eastward) are the meridional and zonal wind velocities, and D (positive eastward) is the declination angle [11]. Figure 7 presented the effective magnetic meridional wind as a function of local time and dip latitudes, in the Peruvian sector and Indian sector during March Equinox, June Solstice, September Equinox, and December Solstice. The dashed lines mark the dip equator; the two shaded bars in each plot mark the 'fixed crest' defined in this study at dip latitude of 10°~15° S/N.

To illustrate the wind's effects on the ambipolar diffusion during the initial stage of the EIA development, we focus on regions of the equatorward side of the two 'fixed crests' (regions between the dip equator and the crests, 0°~10° S/N in dip latitude). All three time points shown in Figure 6 are generally within 0600~1500 LT, so we focus on this local time bin.

In the Peruvian sector, the southward winds prevail through geomagnetic low latitude regions (i.e., regions between two shaded bars), except for the region near the northern crest (still in the geographic southern hemisphere) during December Solstice. For the southern crest, the southward wind is strongest/weakest at June Solstice and December Solstice, respectively. Thus, the promotion of the ambipolar diffusion to the southern crest is most/least prominent at June/December Solstice, which should result in the earliest/latest onset of EIA southern crest, consistent with the results shown in Figure 6a (red solid line). For the northern crest, the strongest southward wind occurs during June Solstice, while the northward wind appears during December Solstice. Thus, the ambipolar diffusion is inhibited/promoted during June/December Solstice that would result in an earlier/latest onset of northern EIA crest, which is inconsistent with the observed semiannual cycle (Figure 6a).

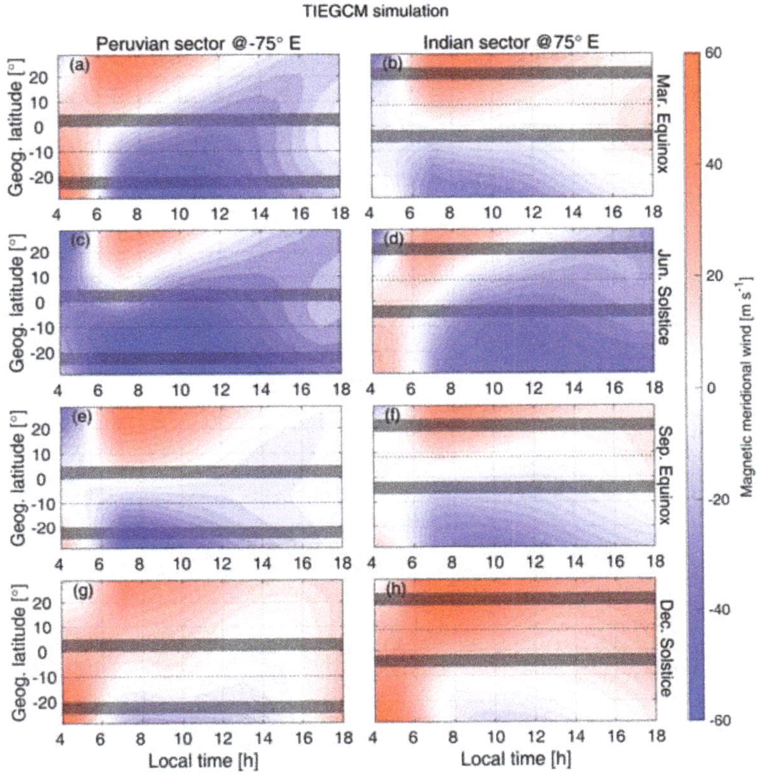

Figure 7. Effective magnetic meridional winds (positive northward). Avergaed winds centered on the days of (**a**,**b**) March Equinox, (**c**,**d**) June Solstice, (**e**,**f**) September Equinoxes, and (**g**,**h**) December Solstice, at Peruvian (left column) and Indian sector (right column). The dash line marks the dip equator, the two shaded bars in each plot mark the 'fixed crest' defined in this study at dip latitude of 10°~15° S/N.

In the Indian sector, the locations at both the northern and southern crests exhibit the strongest northward wind during December Solstice and the strongest southward wind during June Solstice. Hence, the ambipolar diffusion is severely retarded/promoted near southern/northern crest during December Solstice, and vice versa for June Solstice, resulting in an annual cycle on the onset of EIA, which is consistent with the observations (Figure 6b). We conclude that, under the assumption that the meridional wind only takes direct effects on the ambipolar diffusion, the wind configuration simulated by the TIEGCM would result in the typical annual cycle (winter crest priority) of the EIA evolution in both Peruvian and Indian sectors.

Note that the classical scenarios of the transequatorial wind effect on EIA have been challenged in both simulation [12,13] and observations [14]. Abdu [12] showed that the northward thermospheric wind would drive both the southern and northern EIA crest to move southward (upwind direction) in simulations, while the height of the F region is lifted/lowered near the northern/southern crest. Liu et al. [14] found an abnormal annual variation of EIA in the American sector; that is, the northern EIA crest resides at the highest/lowest latitude during local summer/winter when the southward/northward winds prevail, consistent with upwind seasonal movement of EIA crest proposed by Abdu [12], but contradicting the classical scenarios of the transequatorial wind effect.

So far, the physical mechanism of the tilt (in upwind direction) development of EIA has not been addressed either in the aforementioned simulation studies [12] or observational study [13]. Nevertheless, the mechanical effects of neutral wind would push the plasma not only equatorward/poleward but also upward/downward. The uplifting of the ionosphere leaves a depleted bottom side which encourages more ionizations; hence, the TEC should increase, and vice versa. In other words, though the equatorward wind retards the ambipolar diffusion, the accumulation of TEC near the EIA crest can still be accelerated by the uplifting of the ionosphere. In the same manner, the poleward wind that lowers the ionosphere would retard the speed of the growth of TEC near the EIA crest. Near the northern EIA crest at the Peruvian sector, the northward winds only occur at December Solstice (Figure 7h) when the F region resides at lower altitudes [41]. Thus, during the development of EIA, the TEC growth of the northern crest might be retarded, resulting in a delayed onset.

We emphasized that the above-proposed opposite neutral wind effects on the EIA evolution would depend on the longitude (magnetic declination) seasons, and possibly the solar activities. Hence, it is necessary to extend this study to a longer interval and more longitudes to further validate the proposed scenario.

6. Conclusions and Future Work Remarks

The study adopted the IGS TEC map and IRI-2016 to investigate the seasonal, interhemispheric variations on the time evolution (onset, first emergence, and peak) of EIA at Peruvian and Indian sectors. Major findings are listed below:

1. Three time points can be concluded as: The onset occurs at 0600–1000 LT; the first emergence occurs at 0900–1200 LT; the peak occurs at 1200–1500 LT.
2. The onset, first emergence, and peak of EIA show semiannual/annual cycle at the Peruvian/Indian sector. The annual cycle is characterized by a winter priority; that is, the EIA crest during local winter/summer develops earliest/latest. The semiannual is characterized as the northern/southern crest developing earlier during two equinoxes/solstices.
3. The winter priority of the annual cycle can be explained by the transequatorial neutral wind that pushes the plasma along the field line to suppress/promote the EIA development in the summer/winter hemisphere. The semi-annual cycle might be associated with the effect of the neutral wind on the modulation of the F region height, which significantly alters the TEC.
4. We suggest that the transequatorial wind would not only influence the EIA development via the modulation of ambipolar diffusion but also alter the F region height to further modulate the TEC growth speed. The two effects could be in a completive relationship, which causes complex seasonal variations of the EIA development. More studies are needed to further validate this mechanism.
5. The IRI-2016 outputs generally underestimated the TEC value and showed abnormal interhemispheric asymmetry, and sometimes cannot correctly characterize the different stages of the EIA evolution, while the IGS TEC presented a more convincing pattern of the EIA evolution. We suggest that the lack of zonal electric field data that launches the ambipolar diffusion results in IRI's poor ability to describe the diurnal EIA evolution, and we indicate that the empirical model needs to be further improved. Thus, the I[2]RI-2016 model is not a good candidate to extend this study to longitudes where the GNSS observation is inadequate.

An interesting question remains: Why the F region height effect seems invalid in the Indian sector where the development of EIA showed typical winter crest priority. A possible explanation is that this competition varies at different longitudes/seasons/solar activity levels. In the future, by including the ionosonde data, we intend to extend this study to more longitude sectors and longer time intervals to further address the competition between the modulation of ambipolar diffusion and F region height variations.

Author Contributions: Conceptualization, X.W., C.X. and J.Z.; methodology, X.W. and J.Z.; software, X.W.; validation, X.W., H.W., C.X. and J.Z.; formal analysis, X.W.; investigation, X.W.; resources, J.C. and J.Z.; data curation, X.W.; writing—original draft preparation, X.W.; writing—review and editing, X.W., J.Z., C.X., H.W., Y.L., Q.L. and J.K.; visualization, X.W.; supervision, J.Z. and J.C.; project administration, X.W., J.Z., J.C. and J.K.; funding acquisition, X.W., J.Z., J.C. and J.K. All authors have read and agreed to the published version of the manuscript.

Funding: This research was funded by the Strategic Priority Research Program of the Chinese Academy of Sciences, grant number: XDB41000000; the National Natural Science Foundation of China, grant number: 41804150, 42104169, 42104147, 41804153, 41431073, 41521063, and 41674153; Guangdong Basic and Applied Basic Research Foundation, grant number: 2021A1515011216, 2020A1515110242; China Postdoctoral Science Foundation grant number: 2020M6830265, the Fundamental Research Funds for the Central Universities, Sun Yat-sen University, grant number: 2021qntd29; the Natural Science Fundation of Jiangsu Province, grant number: BK20180445; the Joint Open Fund of Mengcheng National Geophysical Observatory, grant number: MENGO-202018; the Opening Funding of the Chinese Academy of Sciences dedicated for the Chinese Meridian Project; the Open Research Project of Large Research Infrastructures of CAS—"Study on the interaction between low/mid-latitude atmosphere and ionosphere based on the Chinese Meridian Project".

Data Availability Statement: The International Global Navigation Satellite Systems Service (IGS) provide the GPS TEC map; data are available on the Coordinated Data Analysis Web (CDAWeb): Ftp://Cdaweb.Gsfc.Nasa.Gov/Pub/Data/Gps/Tec15min_Is/. The Kp index can be accessed on http://wdc.kugi.kyoto-u.ac.jp/kp/index.html#LIST.

Acknowledgments: Geomagnetic field measurements were collected at Fuquene, Huancayo, Tirunelveli, and Alibag. The author thanks T. A. Siddiqui for providing the geomagnetic field data from Huancayo and Fuquene; Virendra Yadav for providing the geomagnetic field data from Tirunelveli and Alibag. We thank the Colombian Instituto Geográfico Agustin Codazzi, the Instituto Geofisico del Peru, and the Indian Institute of Geomagnetism for supporting geomagnetic observatory operations. We also thank the World Data Center for Geomagnetism at Kyoto for providing the Kp index product.

Conflicts of Interest: The authors declare no conflict of interest.

References

1. Yeh, K.; Franke, S.J.; Andreeva, E.; Kunitsyn, V. An investigation of motions of the equatorial anomaly crest. *Geophys. Res. Lett.* **2001**, *28*, 4517–4520. [CrossRef]
2. Liu, H.; Stolle, C.; Förster, M.; Watanabe, S. Solar activity dependence of the electron density in the equatorial anomaly regions observed by CHAMP. *J. Geophys. Res. Space Phys.* **2007**, *112*. [CrossRef]
3. Xiong, C.; Lühr, H.; Ma, S. The magnitude and inter-hemispheric asymmetry of equatorial ionization anomaly-based on CHAMP and GRACE observations. *J. Atmos. Sol.-Terr. Phys.* **2013**, *105*, 160–169. [CrossRef]
4. Xiong, C.; Zhou, Y.L.; Lühr, H.; Ma, S.Y. Diurnal evolution of the F region electron density local time gradient at low and middle latitudes resolved by the Swarm constellation. *J. Geophys. Res. Space Phys.* **2016**, *121*, 9075–9089. [CrossRef]
5. Chen, Y.; Liu, L.; Le, H.; Wan, W.; Zhang, H. Equatorial ionization anomaly in the low-latitude topside ionosphere: Local time evolution and longitudinal difference. *J. Geophys. Res. Space Phys.* **2016**, *121*, 7166–7182. [CrossRef]
6. Lin, C.; Liu, J.; Fang, T.; Chang, P.; Tsai, H.; Chen, C.; Hsiao, C. Motions of the equatorial ionization anomaly crests imaged by FORMOSAT-3/COSMIC. *Geophys. Res. Lett.* **2007**, *34*, L19101. [CrossRef]
7. Tulasi Ram, S.; Su, S.Y.; Liu, C. FORMOSAT-3/COSMIC observations of seasonal and longitudinal variations of equatorial ionization anomaly and its interhemispheric asymmetry during the solar minimum period. *J. Geophys. Res. Space Phys.* **2009**, *114*, A06311. [CrossRef]
8. Luan, X.; Wang, P.; Dou, X.; Liu, Y.C.M. Interhemispheric asymmetry of the equatorial ionization anomaly in solstices observed by COSMIC during 2007–2012. *J. Geophys. Res. Space Phys.* **2015**, *120*, 3059–3073. [CrossRef]
9. Dang, T.; Luan, X.; Lei, J.; Dou, X.; Wan, W. A numerical study of the interhemispheric asymmetry of the equatorial ionization anomaly in solstice at solar minimum. *J. Geophys. Res. Space Phys.* **2016**, *121*, 9099–9110. [CrossRef]
10. Su, Y.; Bailey, G.; Oyama, K.; Balan, N. A modelling study of the longitudinal variations in the north-south asymmetries of the ionospheric equatorial anomaly. *J. Atmos. Sol.-Terr. Phys.* **1997**, *59*, 1299–1310. [CrossRef]
11. Balan, N.; Rajesh, P.; Sripathi, S.; Tulasiram, S.; Liu, J.; Bailey, G. Modeling and observations of the north–south ionospheric asymmetry at low latitudes at long deep solar minimum. *Adv. Space Res.* **2013**, *52*, 375–382. [CrossRef]
12. Abdu, M. Equatorial ionosphere–thermosphere system: Electrodynamics and irregularities. *Adv. Space Res.* **2005**, *35*, 771–787. [CrossRef]

13. Nanan, B.; Chen, C.; Rajesh, P.; Liu, J.; Bailey, G. Modeling and observations of the low latitude ionosphere-plasmasphere system at long deep solar minimum. *J. Geophys. Res. Space Phys.* **2012**, *117*, 8316. [CrossRef]
14. Liu, J.; Zhang, D.; Mo, X.; Xiong, C.; Hao, Y.; Xiao, Z. Morphological Differences of the Northern Equatorial Ionization Anomaly Between the Eastern Asian and American Sectors. *J. Geophys. Res. Space Phys.* **2020**, *125*, e2019JA027506. [CrossRef]
15. Hernández-Pajares, M.; Juan, J.; Sanz, J.; Orus, R.; Garcia-Rigo, A.; Feltens, J.; Komjathy, A.; Schaer, S.; Krankowski, A. The IGS VTEC maps: A reliable source of ionospheric information since 1998. *J. Geod.* **2009**, *83*, 263–275. [CrossRef]
16. Lühr, H.; Xiong, C. IRI-2007 model overestimates electron density during the 23/24 solar minimum. *Geophys. Res. Lett.* **2010**, *37*, L23101. [CrossRef]
17. Bilitza, D.; Altadill, D.; Reinisch, B.; Galkin, I.; Shubin, V.; Truhlik, V. The international reference ionosphere: Model update 2016. In Proceedings of the EGU General Assembly Conference Abstracts, Vienna, Austria, 17–22 April 2016; p. EPSC2016-9671.
18. Bilitza, D.; Altadill, D.; Zhang, Y.; Mertens, C.; Truhlik, V.; Richards, P.; McKinnell, L.-A.; Reinisch, B. The International Reference Ionosphere 2012–a model of international collaboration. *J. Space Weather. Space Clim.* **2014**, *4*, A07. [CrossRef]
19. Bilitza, D.; Altadill, D.; Truhlik, V.; Shubin, V.; Galkin, I.; Reinisch, B.; Huang, X. International Reference Ionosphere 2016: From ionospheric climate to real-time weather predictions. *Space Weather* **2017**, *15*, 418–429. [CrossRef]
20. Ren, X.; Chen, J.; Li, X.; Zhang, X.; Freeshah, M. Performance evaluation of real-time global ionospheric maps provided by different IGS analysis centers. *GPS Solut.* **2019**, *23*, 113. [CrossRef]
21. Manoj, C.; Lühr, H.; Maus, S.; Nagarajan, N. Evidence for short spatial correlation lengths of the noontime equatorial electrojet inferred from a comparison of satellite and ground magnetic data. *J. Geophys. Res. Space Phys.* **2006**, *111*, A11312. [CrossRef]
22. Siddiqui, T. Long-Term Investigation of the Lunar Tide in the Equatorial Electrojet during Stratospheric Sudden Warmings. Ph.D. Thesis, Universität Potsdam Potsdam, Potsdam, Germany, 2017.
23. Heelis, R.; Lowell, J.K.; Spiro, R.W. A model of the high-latitude ionospheric convection pattern. *J. Geophys. Res. Space Phys.* **1982**, *87*, 6339–6345. [CrossRef]
24. Richards, P.; Fennelly, J.; Torr, D. EUVAC: A solar EUV flux model for aeronomic calculations. *J. Geophys. Res. Space Phys.* **1994**, *99*, 8981–8992. [CrossRef]
25. Bartels, J. *The standardized Index Ks, and the Planetary index Kp, IATME Bull., 12 (b), 97*; IUGG Publ. Office: Paris, France, 1949.
26. Matzka, J.; Stolle, C.; Yamazaki, Y.; Bronkalla, O.; Morschhauser, A. The Geomagnetic Kp Index and Derived Indices of Geomagnetic Activity. *Space Weather* **2021**, *19*, e2020SW002641. [CrossRef]
27. Stolle, C.; Manoj, C.; Lühr, H.; Maus, S.; Alken, P. Estimating the daytime equatorial ionization anomaly strength from electric field proxies. *J. Geophys. Res. Space Phys.* **2008**, *113*, A09310. [CrossRef]
28. Venkatesh, K.; Fagundes, P.R.; Prasad, D.V.; Denardini, C.M.; De Abreu, A.; De Jesus, R.; Gende, M. Day-to-day variability of equatorial electrojet and its role on the day-to-day characteristics of the equatorial ionization anomaly over the Indian and Brazilian sectors. *J. Geophys. Res. Space Phys.* **2015**, *120*, 9117–9131. [CrossRef]
29. RG, R.; Alex, S.; Patil, A. Seasonal variations of geomagnetic D, H and Z fields at low latitudes. *J. Geomagn. Geoelectr.* **1994**, *46*, 115–126. [CrossRef]
30. Alken, P.; Maus, S. Spatio-temporal characterization of the equatorial electrojet from CHAMP, Ørsted, and SAC-C satellite magnetic measurements. *J. Geophys. Res. Space Phys.* **2007**, *112*, A09305. [CrossRef]
31. Doumouya, V.; Vassal, J.; Cohen, Y.; Fambitakoye, O.; Menvielle, M. Equatorial electrojet at African longitudes: First results from magnetic measurements. *Proc. Ann. Geophys.* **1998**, *16*, 658–676. [CrossRef]
32. Mendillo, M.; Lin, B.; Aarons, J. The application of GPS observations to equatorial aeronomy. *Radio Sci.* **2000**, *35*, 885–904. [CrossRef]
33. Zhao, B.; Wan, W.; Liu, L.; Ren, Z. Characteristics of the ionospheric total electron content of the equatorial ionization anomaly in the Asian-Australian region during 1996–2004. *Ann. Geophys.* **2009**, *27*, 3861–3873. [CrossRef]
34. Qian, L.; Burns, A.G.; Solomon, S.C.; Wang, W. Annual/semiannual variation of the ionosphere. *Geophys. Res. Lett.* **2013**, *40*, 1928–1933. [CrossRef]
35. Rishbeth, H. How the thermospheric circulation affects the ionospheric F2-layer. *J. Atmos. Sol.-Terr. Phys.* **1998**, *60*, 1385–1402. [CrossRef]
36. Zhao, B.; Wan, W.; Liu, L.; Mao, T.; Ren, Z.; Wang, M.; Christensen, A. Features of annual and semiannual variations derived from the global ionospheric maps of total electron content. *Ann. Geophys.* **2007**, *25*, 2513–2527. [CrossRef]
37. Alken, P. A quiet time empirical model of equatorial vertical plasma drift in the Peruvian sector based on 150 km echoes. *J. Geophys. Res. Space Phys.* **2009**, *114*, A02308. [CrossRef]
38. Chaitanya, P.P.; Patra, A. A neural network-based model for daytime vertical E× B drift in the Indian sector. *J. Geophys. Res. Space Phys.* **2020**, *125*, e2020JA027832. [CrossRef]
39. Fejer, B.G. Low latitude electrodynamic plasma drifts: A review. *J. Atmos. Terr. Phys.* **1991**, *53*, 677–693. [CrossRef]
40. Wu, C.-C.; Liou, K.; Shan, S.-J.; Tseng, C.-L. Variation of ionospheric total electron content in Taiwan region of the equatorial anomaly from 1994 to 2003. *Adv. Space Res.* **2008**, *41*, 611–616. [CrossRef]
41. Bello, S.; Abdullah, M.; Hamid, N.A.; Reinisch, B.; Yoshikawa, A.; Fujimoto, A. Response of ionospheric profile parameters to equatorial electrojet over Peruvian station. *Earth Space Sci.* **2019**, *6*, 617–628. [CrossRef]

remote sensing

MDPI

Article

Evaluating Total Electron Content (TEC) Detrending Techniques in Determining Ionospheric Disturbances during Lightning Events in A Low Latitude Region

Louis Osei-Poku [1], Long Tang [1,2], Wu Chen [1,3,*] and Chen Mingli [4]

1 Department of Land Surveying and Geo-Informatics, The Hong Kong Polytechnic University, Kowloon, Hong Kong 999077, China; louis.oseipoku@connect.polyu.hk (L.O.-P.); ltang@gdut.edu.cn (T.L.)
2 School of Civil and Transportation Engineering, Guangdong University of Technology, Guangzhou 510090, China
3 Shenzhen Research Institute, The Hong Kong Polytechnic University, Shenzhen 518172, China
4 Department of Building Environment and Energy Engineering, The Hong Kong Polytechnic University, Kowloon, Hong Kong 999077, China; mingli.chen@polyu.edu.hk
* Correspondence: wu.chen@polyu.edu.hk

Abstract: Total Electron Content (TEC) from Global Navigation Satellite Systems (GNSS) is used to ascertain the impact of space weather events on navigation and communication systems. TEC is detrended by several methods to show this impact. Information from the detrended TEC may or may not necessarily represent a geophysical parameter. In this study, two commonly used detrending methods, Savitzky–Golay filter and polynomial fitting, are evaluated during thunderstorm events in Hong Kong. A two-step approach of detection and distinguishing is introduced alongside linear correlation in order to determine the best detrending model. Savitzky–Golay filter on order six and with a time window length of 120 min performed the best in detecting lightning events, and had the highest moderate positive correlation of 0.4. That the best time frame was 120 min suggests that the observed disturbances could be travelling ionospheric disturbance (TID), with lightning as the potential source.

Keywords: TEC; detrending; Savitzky–Golay; polynomial; lightning

Citation: Osei-Poku, L.; Tang, L.; Chen, W.; Mingli, C. Evaluating Total Electron Content (TEC) Detrending Techniques in Determining Ionospheric Disturbances during Lightning Events in A Low Latitude Region. *Remote Sens.* **2021**, *13*, 4753. https://doi.org/10.3390/rs13234753

Academic Editors: Serdjo Kos, José Fernández and Juan F. Prieto

Received: 18 October 2021
Accepted: 22 November 2021
Published: 24 November 2021

1. Introduction

Total electron content (TEC) obtained from Global Navigation Satellite Systems (GNSS) can be used to ascertain the impact of space weather events on communication and navigation systems. The spatio-temporally wide range of capabilities of GNSS make it useful in studying these impacts. One technique in using GNSS to reveal weather impacts is by detrending the original TEC [1]. That is, a best fit model or method is first fitted or applied to the original TEC, then the difference between the original TEC and the best fit model is computed. The difference obtained can determine amplitude, frequency, and other signal changes in case of a weather event.

Over time, different best fit models have been used. Each best fit method produces different results, which may or may not necessarily detect the impact a weather event makes on the signal or adequately show the occurrence of the event. In using GNSS for space weather event studies, extensive studies have been done on geomagnetic storms [2–4], earthquakes [5,6], and typhoons [7–9]. The fitting methods have been mostly used in these studies to detect travelling ionosphere disturbance (TID) and other ionosphere irregularities.

Thunderstorms/lightning, a troposphere weather event, have recently garnered interest from the scientific community in the context of harnessing the spatio-temporal capabilities of GNSS to help understand some characteristics of this weather event. Thunderstorm studies in the mid-latitude US plains by Lay, et al. [10] pointed out that ionosphere gravity waves (IGW) damp out before a thunderstorm dies out, contrary to a recent study by Rahmani, et al. [11] in the same study area but at a different time period. Ogunsua, et al. [12] also reported on the ionosphere's response to thunderstorms in the West African and Congo sector of the equatorial region, stating that thunderstorm impact on the ionosphere at nighttime is negligible compared to the daytime. Mahmud M [13], who also studied thunderstorm events over Southern Africa, showed that a strong correlation exists between hourly lightning and ionosphere irregularity event occurrence. Recent studies by Tang, et al. [14] and Liu, et al. [15] showed the characteristics of thunderstorm generated ionosphere gravity waves (IGW) in the Southern China region.

In contributing to the GNSS and thunderstorm related studies, this study assesses two commonly used fitting methods, the Savitzky–Golay and polynomial methods, to detect lightning events as they have been used to detect the presence of other space events. These two methods have been demonstrated in the existing literature to best indicate the incidence of weather events. The study area is Hong Kong, a low latitude region in the southern China coastal region. In the following sections, a description of the detrending methods is briefly discussed. A two-step approach, used to detect and distinguish lightning from non-lightning events using the detrending methods, is introduced, followed by the results. Discussions on the results and derived conclusions are then presented. Aside from the obvious visual changes in amplitude, statistical means were also used to choose the best fitting method.

2. Materials and Methods

2.1. Detrending Methods

The detrending or best fitting methods adapted in this study are briefly discussed below.

2.1.1. Savitzky–Golay Filter

This filter is named after Abraham Savitzky and Marcel Golay, who first made it known as a solution to smooth out noise in data from a chemical spectrum analyzer. The filter falls into category of low-pass time domain filters that smooth out high data variability [16] and is used in many applications, such as electrocardiogram denoising [17], vegetation monitoring and GNSS-TEC changes. The filter operates by the convolution process, with least squares fitting of successive subsets in a given time window [18]. The formulae and detailed explanation are found in the original work of Savitzky, et al. [19].

2.1.2. Polynomial Filtering

This kind of filter first approximates the entire data set by repeatedly evaluations at a given order. Given a time series F(t) function measured from series $x_1, x_2, x_3, \ldots .x_i$, the polynomial P is obtained in Equation (1) as

$$P_n(X) = q_1(X^n) + q_2\left(X^{n-1}\right) + \ldots . + q_n(X) + q_{n+1} \quad (1)$$

where q_n are quantities derived based on P using least squares and n is the order. The residuals, or the difference between F(t) and $P(X)$, are computed to filter out gross effects from F(t), from which some useful information can be obtained. An example using TEC time series can be found in Rahmani, et al. [11].

2.2. GNSS Data

The local GNSS network was used to characterize the TEC changes in Hong Kong. The network referred to as Hong Kong Satellite Reference (HK SatRef) covers the entire Hong Kong area. Information on the network is given in the work of Ji, et al. [20]. More details can be obtained from the Hong Kong Survey Department website (https://www.geodetic.gov.hk/en/rinex/downv.aspx (accessed on 14 June 2019)). The American Global Positioning System (GPS) constellation comprising thirty-two satellites was used. Figure 1 shows the study area and GNSS network.

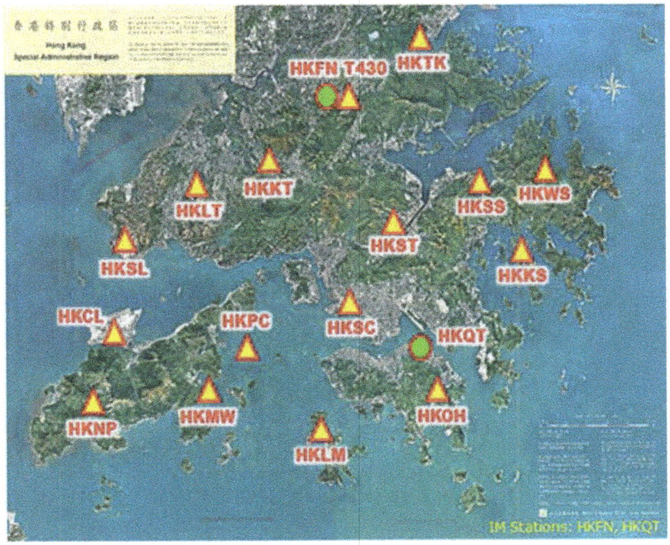

Figure 1. Study area showing the Hong Kong GNSS network. GNSS receivers are colored triangles and circles with their names written beside them. The image shown was obtained from the Hong Kong Geodetic Survey Department website. (https://www.geodetic.gov.hk/en/satref/satref.htm, accessed on 14 June 2019).

In order to detect ionosphere irregularities caused by lightning, the well-known geometry free linear combination of pseudo- and carrier-phase signals was first used to compute the Slant TEC (STEC) from observations at a sampling interval of 30 s. An elevation cut-off angle of 15° was set to eliminate the multi-path effect [9]. The computed STEC was converted to Vertical TEC (VTEC) by applying a mapping function in Equation (2) below, where R_e is the earth's radius, θ is the elevation angle at the ionosphere pierce point (IPP) of the signal–receiver path, and h_i is the ionospheric single layer, approximated at 350 km.

$$\text{VTEC} = \sqrt{1 - \left(\frac{R_e cos\theta}{R_e + h_i}\right)^2} * \text{STEC} \tag{2}$$

VTEC was then detrended using the two detrending methods stated above to get detrended TEC (DTEC), using Equation (3):

$$\text{DTEC}_{model} = \text{VTEC}_{org} - \text{VTEC}_{model} \tag{3}$$

where VTEC_{org} is the original VTEC, VTEC_{model} is the VTEC obtained from the fitting model and DTEC_{model} is the detrended TEC derived according to Equation (3). The unit of VTEC and DTEC is Total Electron Content Units (TECU; 1 TECU = 10^{16} e/m^2). Orders of 3 and 6 and time window lengths 30, 60, 90 and 120 min [13,21,22] were selected for Savitzky–Golay, and orders of 3, 5 [23,24], 6, [1,12,25], and 10 [11] were used for polynomial fitting. The selected parameters for detrending are summarized below in Table 1.

2.3. Thunderstorm/Lightning Data

At very low frequency and low frequency (VLF/LF), lightning discharges produce electric current in the lower D layer of the ionosphere [26]. Lightning data were obtained from a local VLF/LF network in the low-latitude region of Southern China. Total current generated is strongly correlated with lightning activity [27]; a day with a lightning count greater than 10,000 was deemed as a "lightning day".

Table 1. Detrending methods and their selected parameters.

Savitzky–Golay				
Order	**Window (min)**			
3	30	60	90	120
6	30	60	90	120
Polynomial				
Order	3	5	6	10

2.4. Selection Criteria

A total of nine days in the months of July and August 2015 were used in this study. The days were grouped into three sets of three. The first (9th to 11th July) and third (1st to 3rd August) sets comprise three continuous non-lightning days before and after the second (17th to 19th July) set of lightning days respectively. The lightning counts for the days are as follows: 319, 1277 and 91 for 9th to 11th July; 200,435, 75,078, and 33,709 for 17th to 19th July; and 6875, 1775 and 1589 for 1st to 3rd August. All days were void of geomagnetic storm or solar condition events. The disturbance storm time (Dst) and solar condition index (F10.7 index) were less than −30 nT [28] and 150 [8], respectively. Figure 2 shows the Dst and F10.7 indices for the set of days.

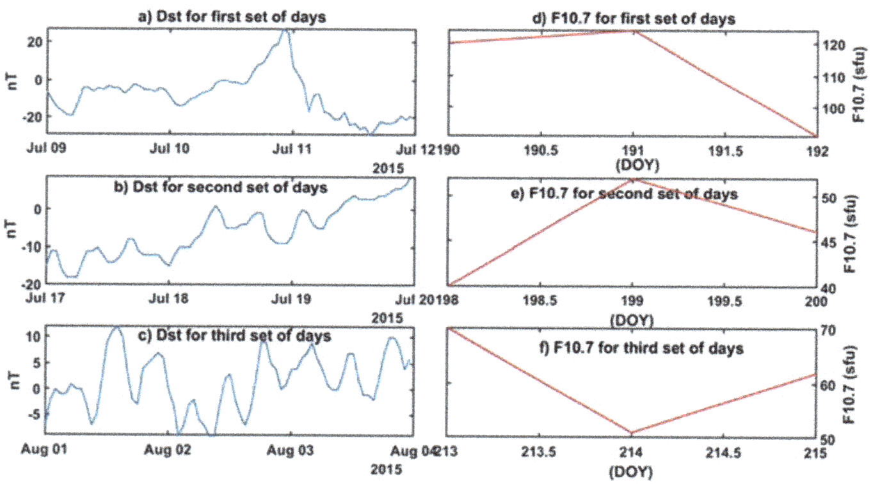

Figure 2. Dst (panels **a–c**) and F10.7 (panels **d–f**) indices for the set of days. Dst index is greater than −30 nT and F10.7 index is less than 150 sfu, indicating the days were void of geomagnetic activity and solar condition.

In determining which detrending method was most suitable for detecting and distinguishing lightning days from non-lightning days, the detection and distinguishing conditions (2DC) approach was used. For the detection condition, because electrical discharge from lightning takes about three hours to travel to higher ionosphere heights, the changes in DTEC amplitude, mostly an increment, may be observed after 3 h of lightning occurrence and last 1–2 h. Being able to show this change using the DTEC method indicates that the DTEC method can detect lightning activity. For the distinguishing condition, anomalous behaviour of DTEC amplitude was checked on non-lightning days against that of lightning days. A non-lightning event day is expected to have one absolute maxima constant as the DTEC amplitude or value throughout, as there is no or little lightning or other space weather events to cause such changes. This constant value is then set as the threshold with which to assess lightning days. As indicated under the detection condition, lightning is expected to cause changes to

DTEC amplitude or value. The absolute maximum value of lightning days is compared to the threshold from a non-lightning day. When the increased value is greater than the threshold, a lightning day has either been distinguished from non-lightning days or not. A DTEC method achieving this is said to have both detected lightning activity and distinguished lightning days from non-lightning. Furthermore, 2DC is explained using the following example. Day 1 is a non-lightning day, Day 2 is a lightning day, and the detrending method is DM. First, DM is used to detrend TEC on Day 1. The DTEC on Day 1 was mostly between ± 0.5TECu. Next, TEC on Day 2 is detrended. DTEC on Day 2 was initially ±0.1TECu but at the time of lightning increased to ±0.5TECu. At this point, DM has met the detection condition, and hence is able to detect lightning, but not the distinguish condition, as the DTEC maxima for both days are the same. If Day 2 DTEC at time of lightning increased to, for instance, ±1TECu, DM would have successfully detected and distinguished Day 1 from Day 2. On the other hand, should Day 2 DTEC remain at ±0.1 throughout the entire period, DM could neither detect nor distinguish lightning events from non-lightning event. Figure 3 shows the flow chart of 2DC.

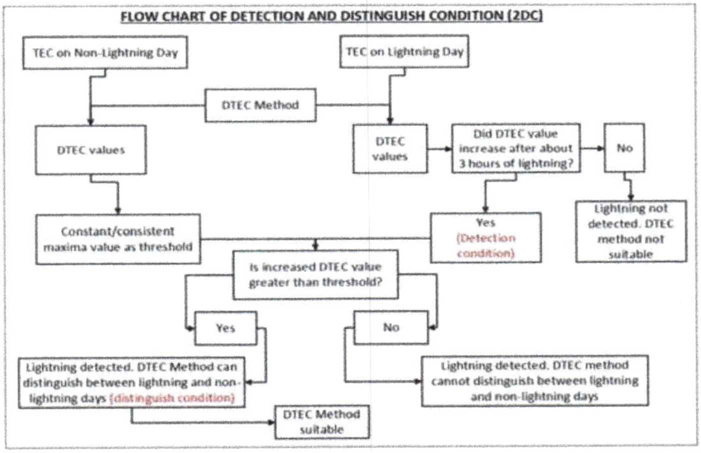

Figure 3. Flow Chart .showing the Detection and Distinguish Condition (2DC).

The evaluation was done on the basis of each satellite–receiver pair rather than an average of TEC over a station, in order to obtain greater detail. Satellites passing from the time of lightning occurrence to about 3 h afterwards were investigated.

3. Results

The results are presented in ascending order of set of days. The days in each set are also presented in chronological order. For brevity, the parameters in Table 1 are shortened as follows: polynomials are prefixed by the letter P, followed by the order. For instance, the polynomial of order 3 is represented as P3. That of Savitzky–Golay is prefixed as sgf, followed by the order and window length. For example, the Savitzky–Golay of order 3 with a window length of 30 min is represented as sgf_3_30. Figures showing the lightning count and current are also presented. With TEC and DTEC, the stations of the HK SatRef are quite close; hence, similar observations are made by most stations at a given time [14]. Only observations from station HKOH meeting the criteria in Section 2.4 are presented. A plot of TEC and DTEC for a given satellite in a day has six rows. The first row comprises the original and fitted TEC of P3, P5, P6 and P10 while the second row comprises their respective DTEC. The third row also comprises the original and fitted TEC of sgf_3_30 to sgf_3_120, with the fourth row comprising their respective DTEC. The fifth and last rows follow the same order as the third and fourth rows for sgf_6 parameters. All of the subsequent figures presented showing TEC and DTEC in this section follow the same format.

3.1. First Set of Days (Quiet Days before Lightning Events)

3.1.1. 9th July

This day was the first day of non-lightning days before lightning days. The upper and lower panels of Figure 4 show the lightning count and current, respectively. Few counts are seen from 00–01 local time (LT: UT + 8). Figures 5 and 6 show the TEC-DTEC for satellites of pseudorandom noise code (PRN) 1 and 13. PRN 1 was chosen as it covered the time with no lightning (09–20LT), while PRN 13 was specifically chosen as it covers the time of the few lightning counts.

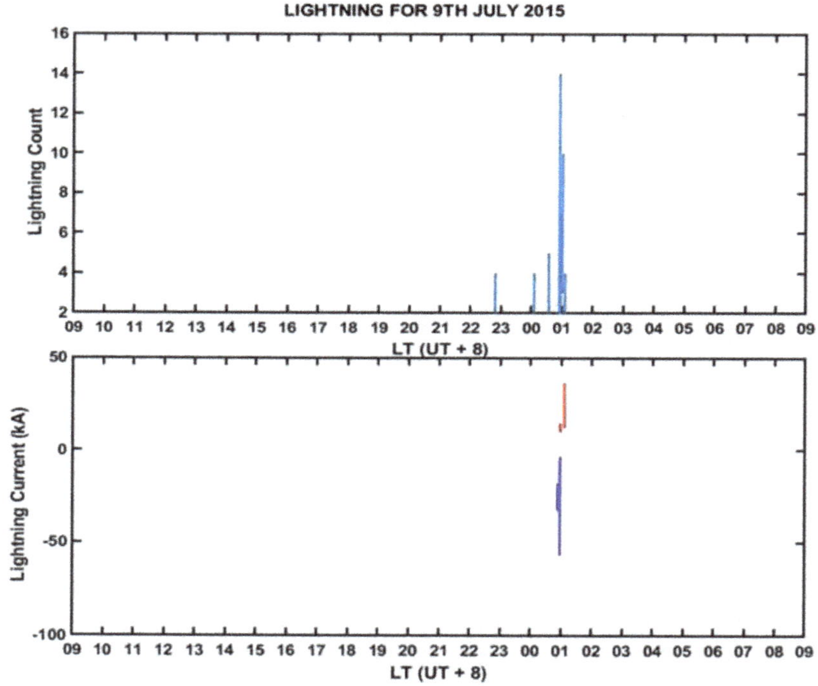

Figure 4. Lightning activity on 9th July. The (**upper panel**) shows the counts; the (**lower panel**) is the current.

From panel e of Figure 5, P3 recorded DTEC values between ±3TECu for PRN 1. The amplitudes remained same during the time the satellite was available. For P5, PRN 1 had an amplitude of DTEC of about ±0.5TECu, which remained same throughout the periods. On PRN 13, P5 and P6 had similar characteristics to P3. For P10 the amplitude of DTEC was about ±0.3TECu, which is 0.2 less than that of P5 and P6 on PRN 1. Similar observations to P5 on PRN 13 were made by P10.

Observations on PRN 1 reveal the quiet day phenomena, as no change in amplitude was seen. Increased amplitude on PRN 13 also reveals that P3, P5, P6 and P10 could detect low lightning counts, around 00–01 LT.

sgf_3_30 had an amplitude of DTEC of about ±0.03TECu on PRN 1, shown in panel m of Figure 5, which remained same throughout the day. For sgf_3_60, the amplitude of DTEC was about ±0.1TECu compared to sgf_3_30. sgf_3_90 had an amplitude of DTEC of ±0.5TECu, which was constant for the whole day for PRN 1. sgf_3_120 also had a DTEC amplitude of ±0.2TECu on PRN 1. With PRN 13, sgf_3 parameters saw an increase in amplitude about the same time as lightning. Sgf_3_30 saw an increase to ±0.1TECu, and that of sgf_3_60 to ±0.6TECu. The increases in amplitude for sgf_3_90 and 120 are no different than the amplitude recorded on PRN 1.

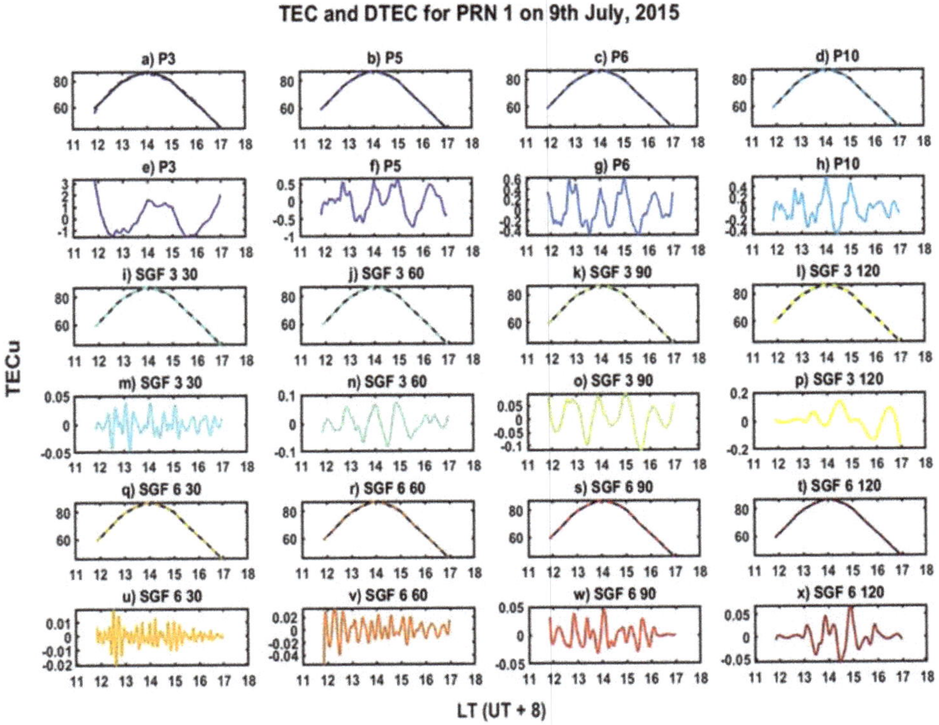

Figure 5. TEC and DTEC for PRN 1 on 9th July, 2015. Original TEC are in black solid lines and modelled TEC are in dashed colored lines. In panels (**a–d**) are the original TEC and modelled TEC of P3, P5, P6 and P10, respectively. Their respective DTECs are in panels (**e–h**) with colors corresponding to their modelled TEC. Panels (**i–l**) show the original TEC and modelled TEC of sgf_3_30, sgf_3_60, sgf_3_90, and sgf_3_120. Their respective DTECs are in panels (**m–p**), with colors corresponding to their modelled TEC. Original and modelled TEC of sgf_6_30, sgf_6_60, sgf_6_90, and sgf_6_120 are in panels (**q–t**), with their respective DTEC in corresponding colors in panels (**u–x**). The horizontal axis is hours in local time (LT: UT + 8). The vertical axis is in TEC units.

The DTEC amplitude of sgf_6_30 was mostly between ±0.01TECu, and remained constant for the whole day on PRN 1. That of sgf_6_60 was between ±0.03TECu compared to sgf_6_30. For sgf_6_90 and 120, the amplitude was about ±0.05TECu compared to sgf_6_30 and sgf_6_60. On PRN 13, sgf_6_30 saw an increase in amplitude of about ±0.05TECu, while sgf_6 30, 90 and 120 saw an increase of about ±0.2TECu during the time of the lightning count.

In all, most parameters had their DTEC amplitude constant throughout the day, and were thus able to represent a quiet day. Sgf parameters had lower amplitudes compared to polynomials.

3.1.2. 10th July

This day was the second of the non-lightning days before lightning days. Figure 7 shows the lightning count and current. Few lightning counts were seen at 08–09 and 18–20 LT. Figure 8 shows the TEC-DTECs for PRN 13.

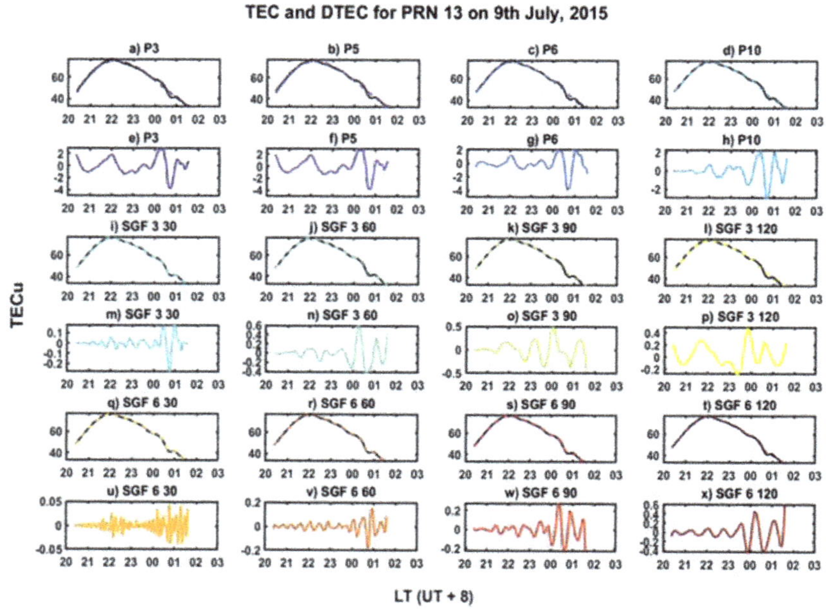

Figure 6. TEC and DTEC for PRN 13 on 9 July 2015. Original TEC are in black solid lines and modelled TEC are in dashed colored lines. In panels (**a–d**) are the original TEC and modelled TEC of P3, P5, P6 and P10, respectively. Their respective DTECs are in panels (**e–h**), with colors corresponding to their modelled TEC. Panels (**i–l**) show the original TEC and modelled TEC of sgf_3_30, sgf_3_60, sgf_3_90, and sgf_3_120. Their respective DTECs are in panels (**m–p**), with colors corresponding to their modelled TEC. Original and modelled TEC of sgf_6_30, sgf_6_60, sgf_6_90, and sgf_6_120 are in panels (**q–t**), with their respective DTECs in corresponding colors in panels (**u–x**). The horizontal axis is hours in local time (LT: UT + 8). The vertical axis is in TEC units.

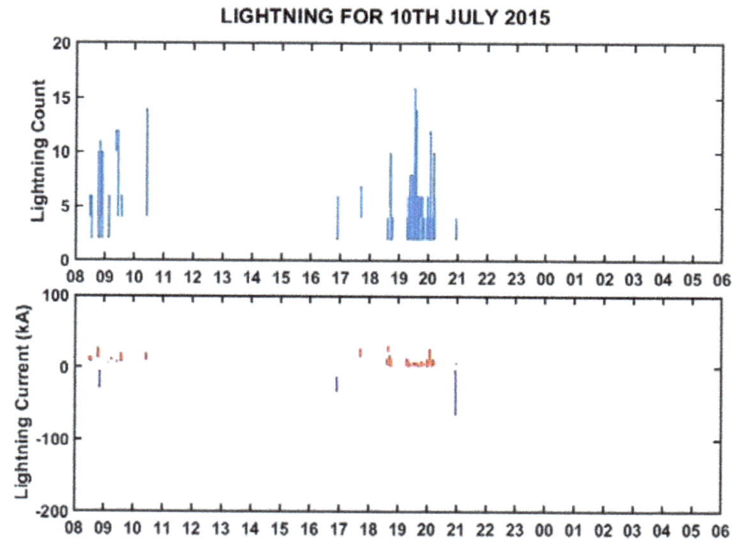

Figure 7. Lightning activity on 10th July. The (**upper panel**) shows the counts; the (**lower panel**) is the current.

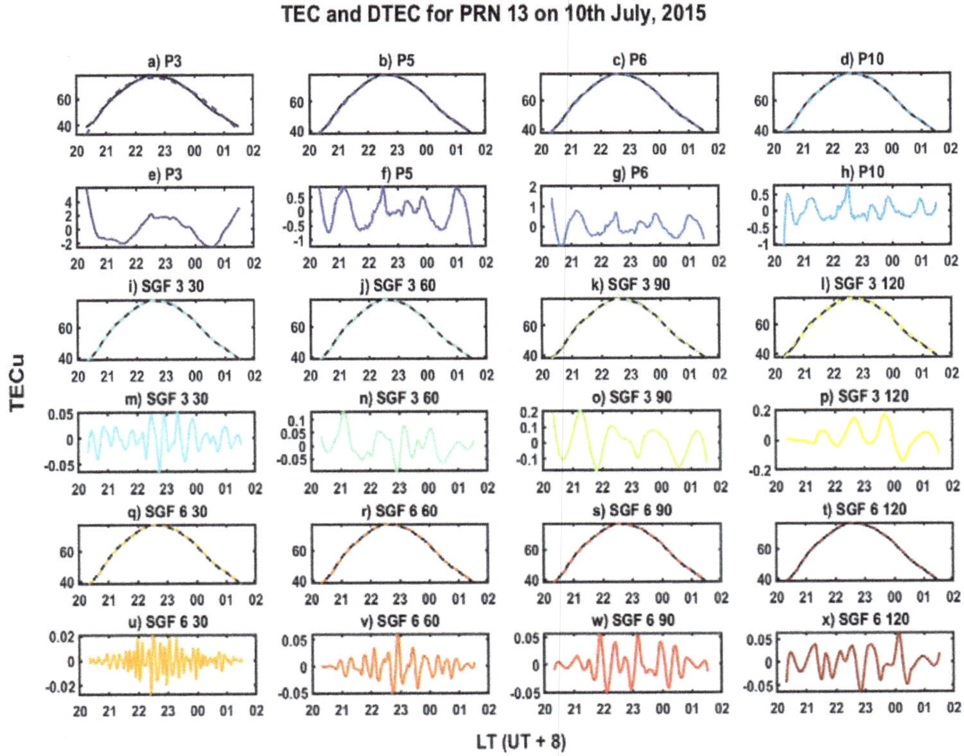

Figure 8. TEC and DTEC for PRN 13 on 10th July 2015. Information about figure panels is same as that of Figures 5 and 6.

From panels f–h of Figure 8, P5, P6 and P10 all had a constant amplitude of ±0.5TECu. P3 had the highest DTEC value at ±4TECu. The amplitudes remained the same for the whole period.

Sgf_3_30 had an amplitude of DTEC of about ±0.05TECu on PRN 13 in Figure 8, and remained constant. Panels n–p in Figure 8 show that sgf_3_60, 90 and 120 had a constant amplitude of ±0.2TECu on PRN 13.

For sgf_6_30, the DTEC amplitude was ±0.02TECu and remained constant for the whole period on PRN 13 (panel u of Figure 8). Sgf_6_60, 90 and 120 recorded DTEC amplitudes of ±0.05TECu throughout the period of PRN 13

Similar to 9th July, most parameters had their DTEC amplitude constant throughout the day, and were thus ably to represent a quiet day. Sgf parameters had lower amplitudes compared to polynomials.

3.1.3. 11th July

This day was the last of the non-lightning days before lightning days. Compared to the previous days in this set of days, 11th July virtually had no lightning count, making it an ideal example of a truly quiet day. Figure 9 shows the lightning count and current. Figure 10 shows the TEC-DTECs for PRN 2. Most satellites had similar observations to PRN 2.

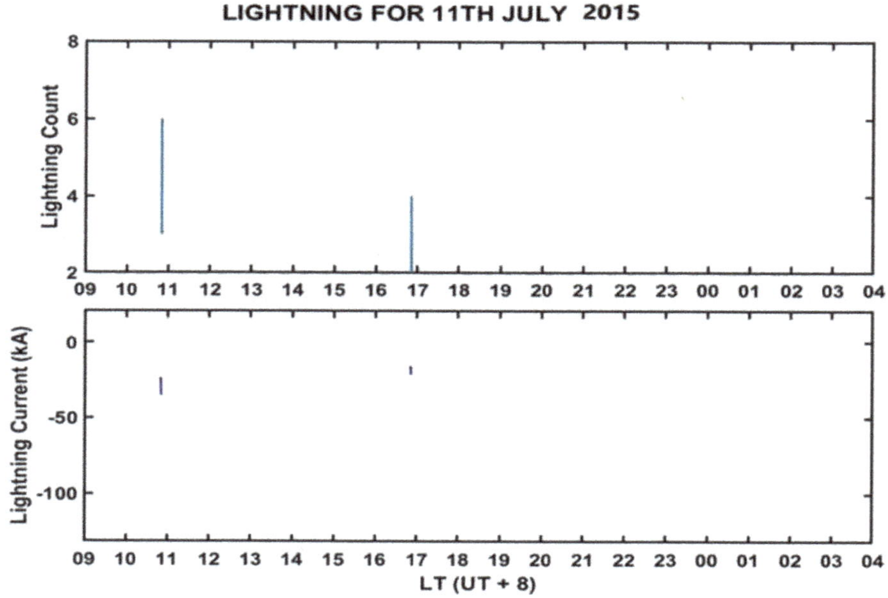

Figure 9. Lightning activity on 11th July. The (**upper panel**) shows the counts; the (**lower panel**) is the current.

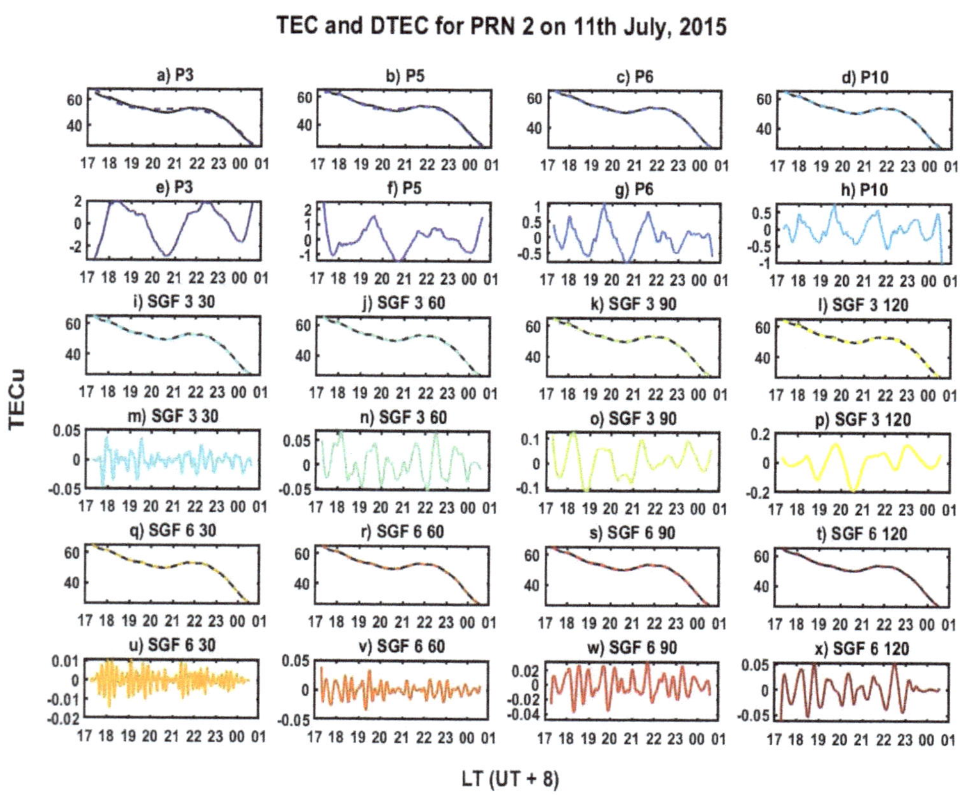

Figure 10. TEC and DTEC for PRN 2 on 11th July 2015. Information about figure panels is same as that of Figures 5 and 6.

From Figure 10, the polynomial and sgf parameters had constant amplitudes throughout the period of PRN 2. DTECs for all the parameters had similar characteristics to those of 9th and 10th July; this example thus represents a typical quiet day.

3.2. Second Set of Days (Lightning Days)

3.2.1. 17th July

This was the first of the lightning days. Lightning happened for the whole day, with a peak count around 15–17 LT. After 22 LT, the lightning activity reduced to a minimum. Lightning activity is shown in Figure 11. TEC and DTEC of PRN 13 and 21 are shown in Figures 12 and 13 to check the changes in DTEC for the first peak expected at 18LT and second peak expected at 01LT, respectively.

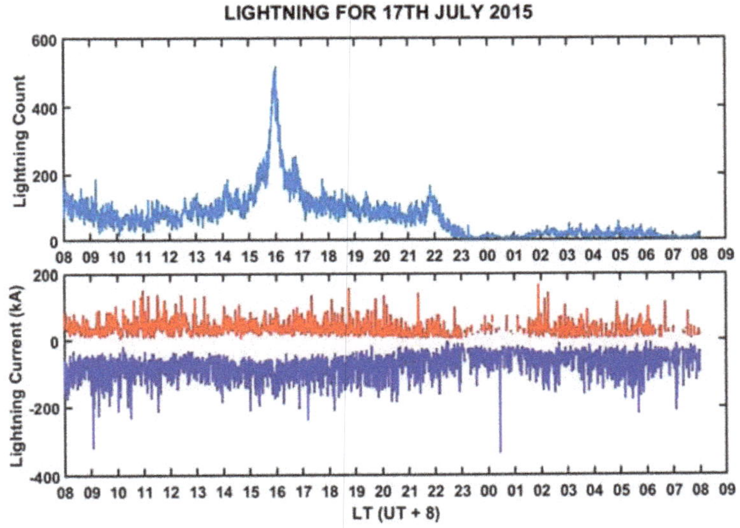

Figure 11. Lightning activity on 17th July. (**Upper panel**) shows the counts; (**lower panel**) is the current.

Since it takes about three hours for lightning to effect changes in TEC, and the change lasts for about three hours, amplitude changes for the first peak (15–17LT) can be expected to be around 20–23LT, while for the second peak (22LT) they can be expected to show changes of around 00–02LT. In panels e–g in Figure 12, P3, P5 and P6 show the same amplitude values throughout. No increase is seen in the DTEC amplitude values. For P10 in panel f, DTEC amplitude increases to ±0.5TECu at the expected time. P10 was thus able to detect the lightning event on PRN 13. In Figure 13, the DTEC amplitude drops sharply at the expected hour of 01LT from 1 to −2TECu for P5, P6 and P10, signifying a detection. P3 also records a similar observation, though this is not as obvious as P5, P6 and P10.

For the sgf_3 parameters, only sgf_3_30 and 60 could detect the lightning event on PRN 13 in Figure 12. Their amplitudes increased to ±0.04 within the expected time range. Sgf_3_90 and 120, on the other hand, had constant DTEC amplitude. For the second lightning peak at 22LT, all of the sgf_3 parameters could detect it within the expected time range. Their amplitudes increased to between ±0.2–0.4TECu, as seen in panels m to p in Figure 13.

Sgf_6_30 could not detect the first lightning peak but did detect the second. Sgf_6_60, 90 and 120 detected both lightning peaks as increase in amplitude, as seen in panels v to x in Figures 12 and 13.

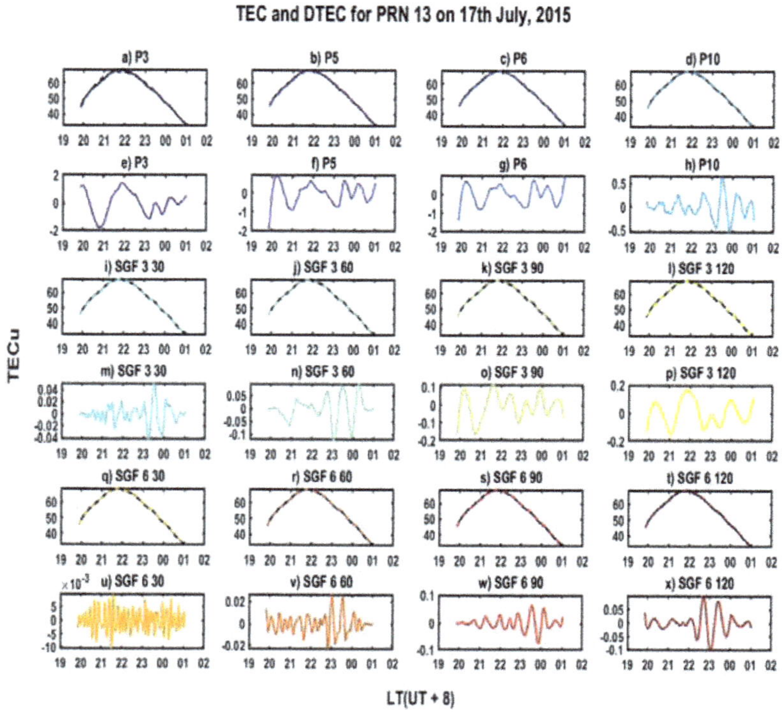

Figure 12. TEC and DTEC for PRN 13 on 17th July 2015. Information about figure panels is same as that of Figures 5 and 6.

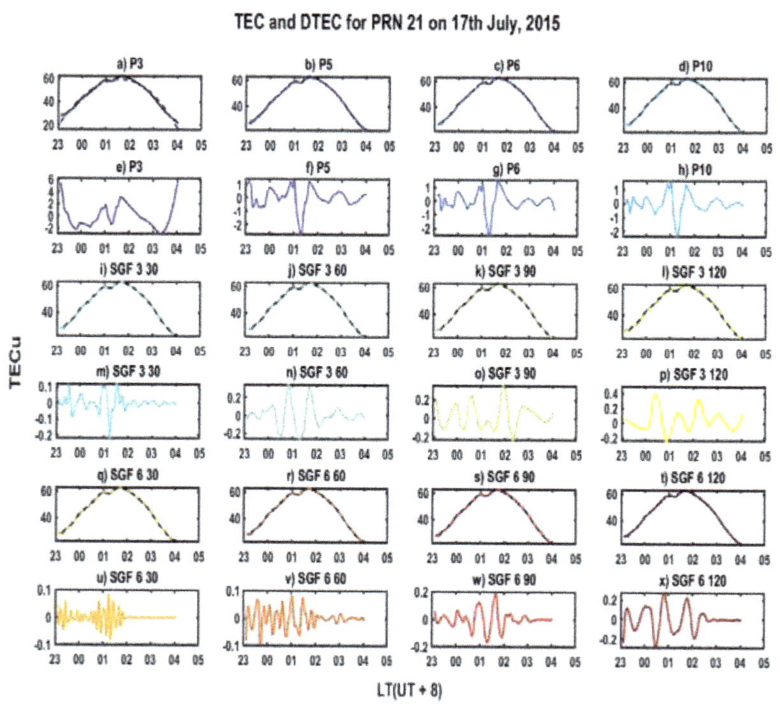

Figure 13. TEC and DTEC for PRN 21 on 17th July 2015. Information about figure panels is same as that of Figures 5 and 6.

P5, P6 and P10 could detect lightning events; although the change in amplitude for lightning and non-lightning days were similar, the DTEC amplitude values of sgf parameters indicating the presence of lightning were higher compared to the quiet non-lightning days.

3.2.2. 18th July

This was the second of the lightning days. Lightning happened for the whole day, with peaks around 15LT. At 21LT, a small peak was also seen. Figure 14 shows the lightning event. DTEC-TEC are shown for PRN 5 in Figure 15, as its time of availability covered the expected time of the amplitude changes of the two peaks, 18LT for the peak at 15LT and 23LT for that at 21LT.

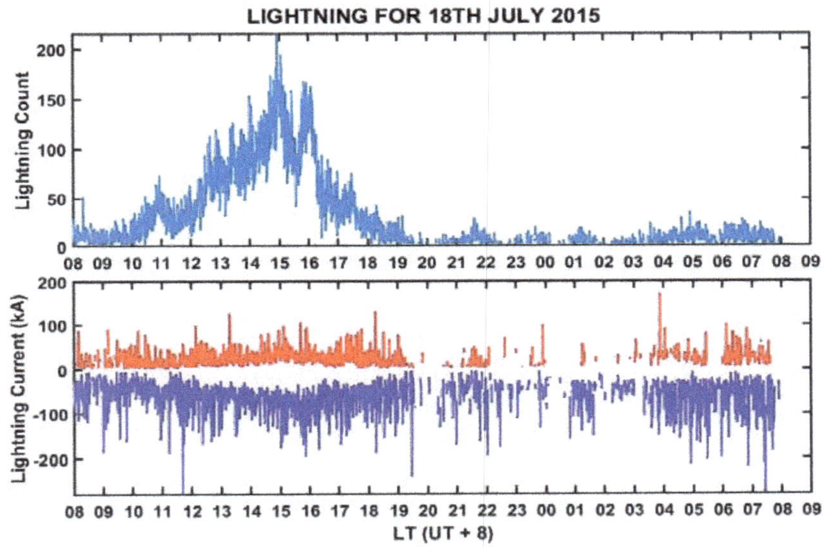

Figure 14. Lightning activity on 18th July. (**Upper panel**) shows the counts; (**lower panel**) is the current.

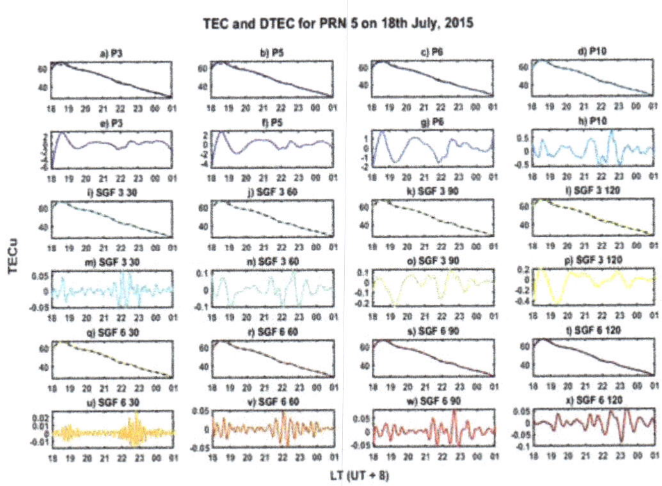

Figure 15. TEC and DTEC for PRN 5 on 18th July 2015. Information about figure panels is the same as that of Figures 5 and 6.

With peak counts around 15 and 21LT, changes in DTEC amplitude were expected around 18 and 23LT. P3, P5 and P6 had constant amplitudes of ±3TECu, and hence were not able to detect the lightning event. P10 first saw an increase in amplitude of about ± 0.5TECu, then a reduced amplitude of ±0.2TECu, and finally another increase to ±0.5TECu. These increments all happened at the expected time of amplitude change, indicating that P10 could detect the lightning event. From panels m to p in Figure 15, sgf_3 parameters all saw amplitude changes at the expected time. Sgf_3_30 recorded ±0.06TECu, while sgf_3_60, 90 and 120 all recorded ±0.2TECu at the expected time of DTEC changes. Sgf 6 parameters also showed an increase in DTEC amplitude at the expected time. Sgf_6_30 recorded ±0.03TECu and sgfs 6_60, 90 and 120 all recorded ±0.07TECu.

P10 and all of the sgf parameters therefore detected the lightning activity on this day.

3.2.3. 19th July

This was the last of the lightning days. Lightning happened from 08–09LT and reoccurred from 16–07LT. The peak periods were 16–17, 19–20 and 23–03LT, with DTEC amplitude changes expected at 11, 23 and 02LT, respectively. PRN 15 was available at 20–02LT, covering the expected time changes of the second peak period and hence able to be investigated. Figures 16 and 17 show the lightning activity and DTEC for PRN 15, respectively.

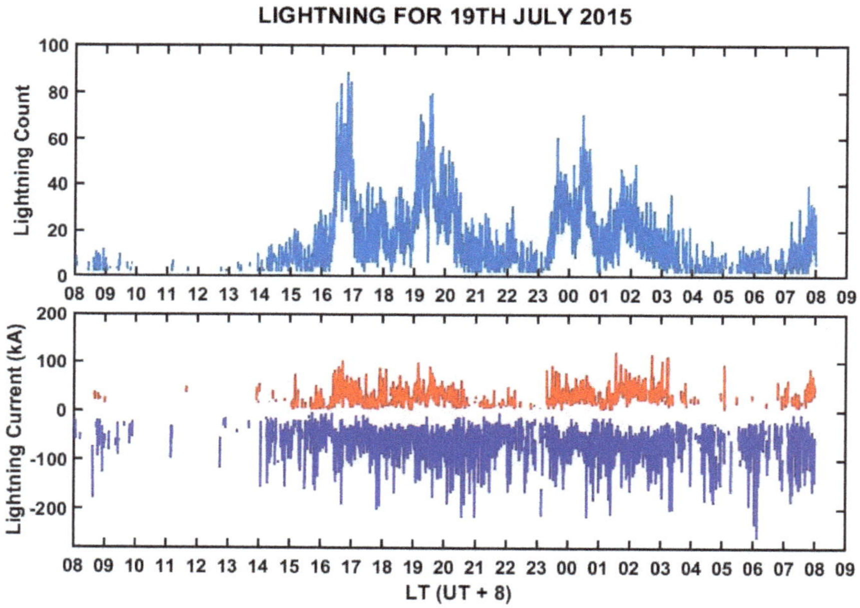

Figure 16. Lightning activity on 19th July. (**Upper panel**) shows the counts; (**Lower panel**) is the current.

Less useful information could be derived from the DTEC of P3 (panel e of Figures 17 and 18). P5, P6 and P10 made similar observations to PRN 15. An increase in amplitude of ±2TECu was seen around 22–23LT, and the amplitude changes occur at the expected time.

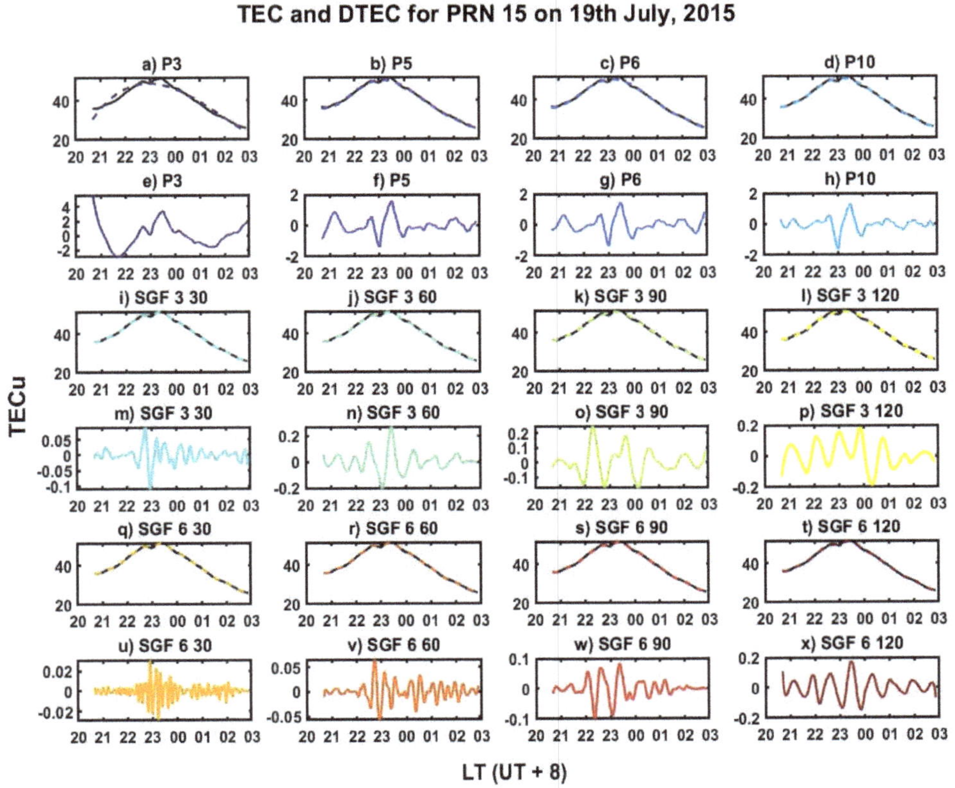

Figure 17. TEC and DTEC for PRN 15 on 19th July 2015. Information about figure panels is the same as that of Figures 5 and 6.

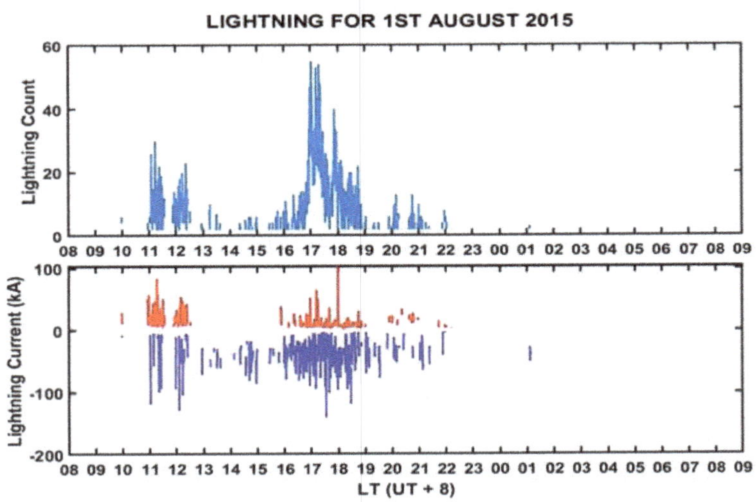

Figure 18. Lightning activity on 1st August. (**Upper panel**) shows the counts; (**lower panel**) is the current.

Sgf_3_30, 60, 90 and 120 all saw an increase in DTEC amplitude at the expected time. Similar observations were also made by the sgf_6 parameters. Sgf_3_30 recorded ±0.06TECu, while sgf_3_60, 90 and 120 all recorded ±0.2TECu at the expected time of DTEC changes. Sgf 6 parameters also showed an increase in DTEC amplitude at the expected time. Sgf_6_30 recorded ±0.03TECu, sgf 6_60 and 90 recorded ±0.1TECu, and 6_120 recorded ±0.2TECu.

P5, P6,P5, P10 and all the sgf parameters were able to detect the lightning activity on this day.

3.3. Third Set of Days (Quiet Days after Lightning Events)

3.3.1. 1st August

This day was the first of the non-lightning days after lightning days. Few lightning counts were seen at 11–12, 16–18 and 21–22 LT. Figures 18 and 19 show the lightning activity, TEC and DTEC of PRN 11.

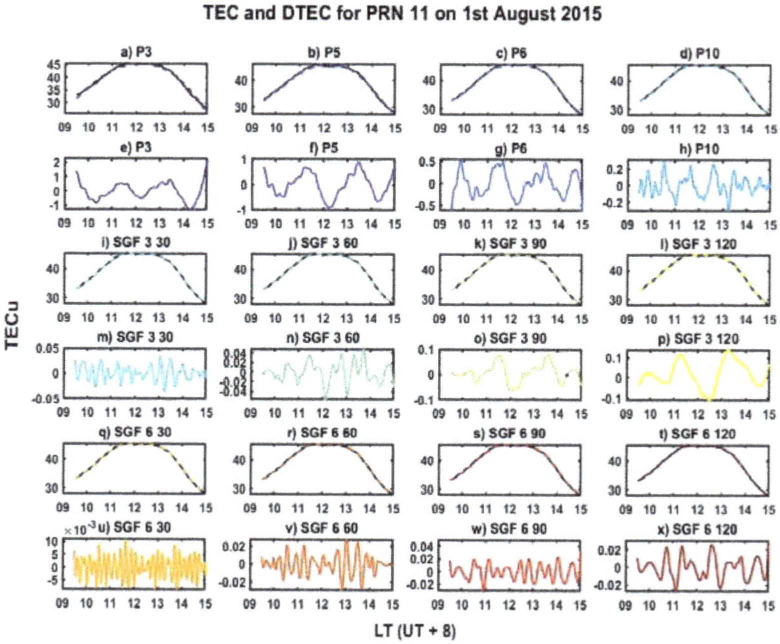

Figure 19. TEC and DTEC for PRN 11 on 1st August 2015. Information about figure panels is the same as that of Figures 5 and 6.

P3, P5, P6 and P10 all had constant DTEC amplitude of ±0.2 on PRN 7, as seen in panels e to h of Figures 20 and 21. The Sgf_3_and 6 parameters all had constant DTEC amplitudes. Sgfs 3_30, 3_60, 6_30, 6_90 and 6_120 had amplitudes of about ±0.02TECu, while 3_90 and 3_120 all had ±0.1TECu. All of the DTEC techniques used were able to show that this day was a quiet day, having constant DTEC amplitudes.

3.3.2. 2nd August

This day was the second non-lightning day after a lightning day. Figure 20 shows the lightning activity in terms of current and count. Few lightning counts are seen at 14–19 or 03–04 LT. Figure 21 shows TEC and DTEC for PRN 5.

Less useful information could be derived from the DTEC of P3 in panel e of Figure 21. P5, P6 and P10 had a constant DTEC amplitude of about ± 0.1TECu. Sgf parameters also had constant amplitude throughout. Sgf_6_30 recorded ±0.02TECu. Sgfs 3_30, 6_60, 6_90 and 6_120 recorded ±0.05TECu, while 3_90 and 3_120 had ±0.1TECu.

The constant amplitude shown by the detrending techniques indicate that this day was a quiet day.

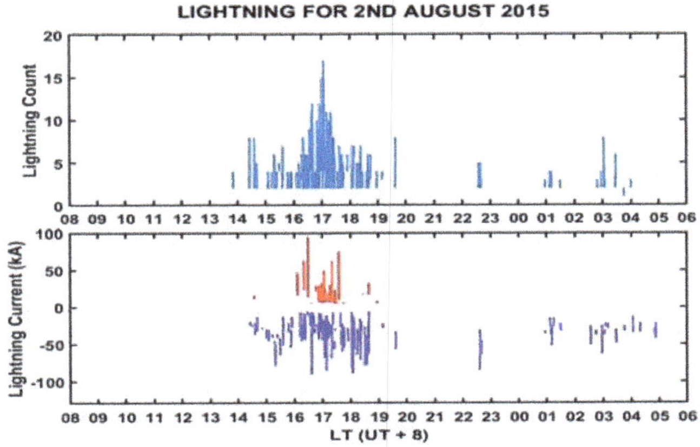

Figure 20. Lightning activity on 2nd August. (**Upper panel**) shows the counts; (**lower panel**) is the current.

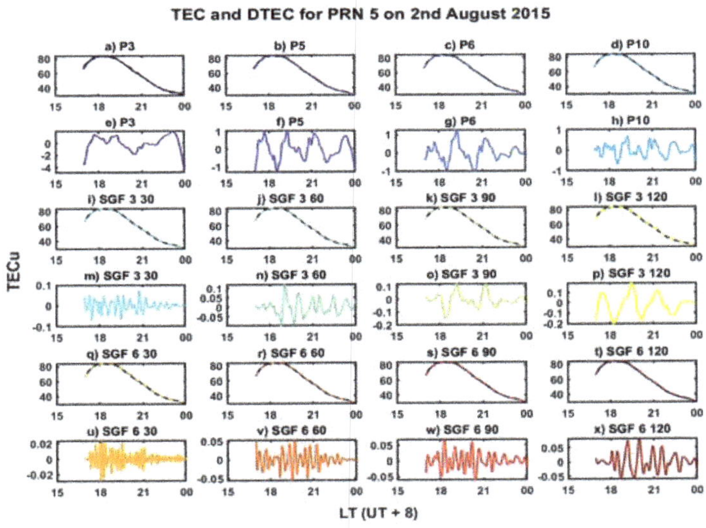

Figure 21. TEC and DTEC for PRN 5 on 2nd August 2015. Information about figure panels is the same as that of Figures 5 and 6.

3.3.3. 3rd August

This was the last non-lightning day in the third set of days. Figure 22 shows the lightning activity for this day. Few lightning counts were seen, mostly at 14–19LT. Figure 23 show the time of passage of PRN 4 and its respective TEC and DTECs.

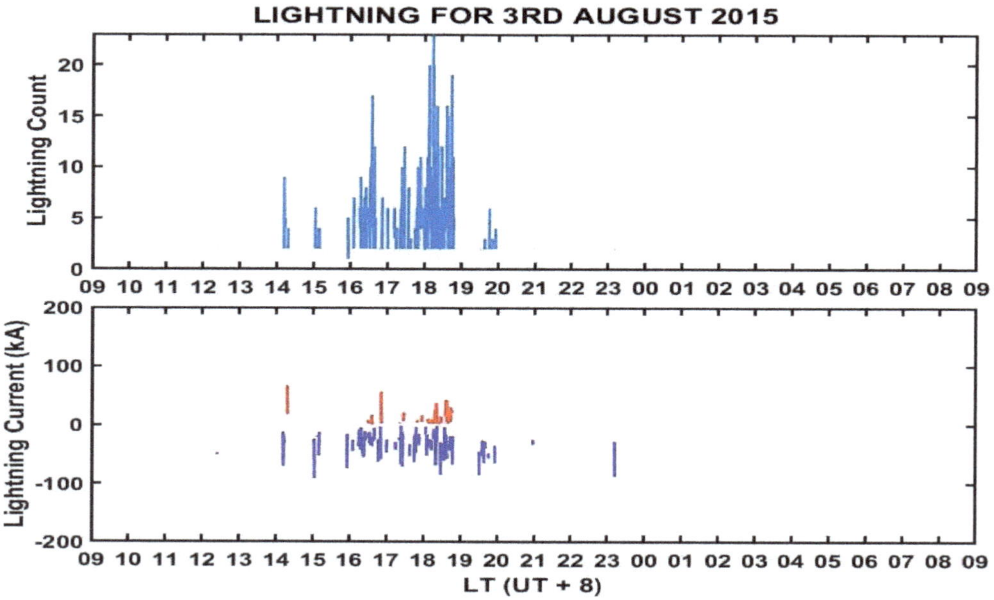

Figure 22. Lightning activity on 3rd August. (**Upper panel**) shows the counts; (**lower panel**) is the current.

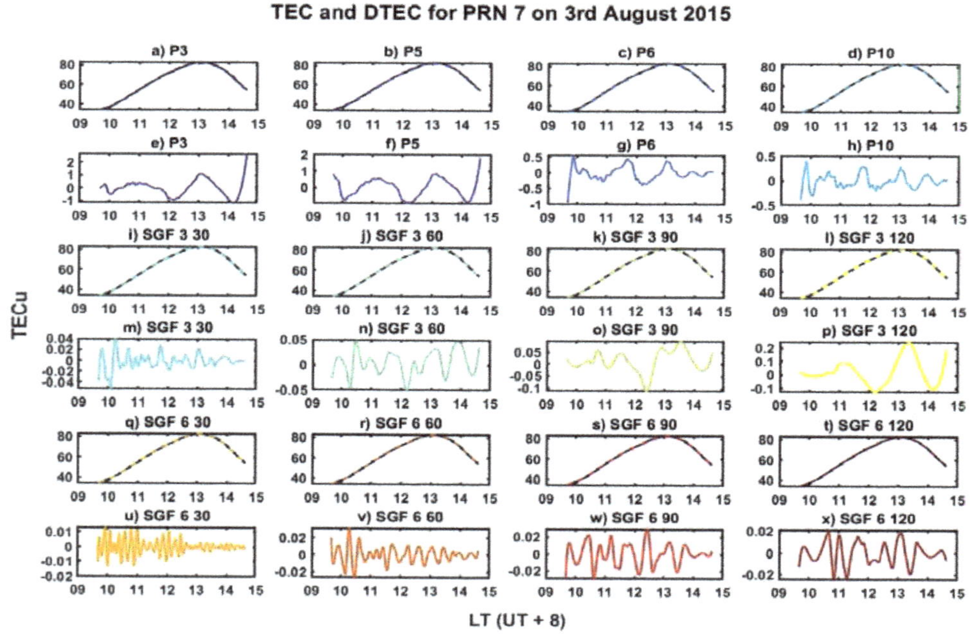

Figure 23. TEC and DTEC for PRN 7 on 3rd August 2015. Information about figure panels is the same as that of Figures 5 and 6.

All detrending parameters for PRN 4 had constant DTEC amplitudes at the time of its passage. The polynomials had higher values compared to the those of sgf. The constant amplitude once again indicates that this day was a quiet a day. The amplitudes recorded by the detrending parameters were similar to those of other non-lightning days.

Table 2 provides a summary of which DTEC method and parameters were able to detect lightning events on lightning days and to represent the quiet nature of non-lightning days according to the 2DC approach for the results enumerated above.

Table 2. Summary of which DTEC parameters could indicate non-lightning and lightning events. R means represent, D means detected, ND means not detected.

Day/DTEC Method		P3	P5	P6	P10	sgf_3_30	sgf_3_60	sgf_3_90	sgf_3_120	sgf_6_30	sgf_6_60	sgf_6_90	sgf_6_120
First set of days	9th July	R	R	R	R	R	R	R	R	R	R	R	R
	10th July	R	R	R	R	R	R	R	R	R	R	R	R
	11th July	R	R	R	R	R	R	R	R	R	R	R	R
Second set of days	17th July	D	D	D	D	D	D	D	D	D	D	D	D
	18th July	ND	ND	ND	D	D	D	D	D	D	D	D	D
	19th July	-	D	D	D	D	D	D	D	D	D	D	D
Third set of days	1st August	R	R	R	R	R	R	R	R	R	R	R	R
	2nd August	R	R	R	R	R	R	R	R	R	R	R	R
	3rd August	R	R	R	R	R	R	R	R	R	R	R	R

4. Discussion

From the results in Section 3 and Table 2, all of the detrending methods had individual DTEC amplitudes that were mostly the same or constant during both sets of non-lightning days at the time of passage of the satellites (panels e–h, m–p, and w–x of Figures 6, 8, 10, 19, 21 and 23). Days without changes in amplitude show that the ionosphere was quiet, which truly reflects the weather events, as there were no geomagnetic storms, lightning events, or sunspots.

During the lightning days, P5, P6 and P10 could detect lightning events using the 2DC approach. DTEC amplitude increased, as seen in panels e to h of Figures 13 and 17. This observation agrees with Rahmani, et al. [11] and Ogunsua, et al. [12], who used P10 and P6 respectively to detect the occurrence of lightning. The amplitude values, however, were no different from those of non-lightning days. The amplitude for non-lightning days was on average between ±0.5–5TECu. Lightning days at the time of expected DTEC changes also recorded increases in amplitude values between ±0.5–5TECu. This does not show a clear distinction between the lightning days and non-lightning days. Thus, the distinguishing condition of 2DC was not met. This non-distinction could be the reason Kumar, et al. [29] reported no difference between lightning and non-lightning days, and makes polynomials less suitable for distinguishing lightning days.

Coster, et al. [30], suggests the accuracy of DTEC is about ±0.05TECu. Any fluctuations above this could be a disturbance. The Savitzky–Golay parameters mostly had ±0.05TECu on non-lightning days and saw an increase to about ±0.06–2TECu on lightning days (panels m–p and w–x of Figures 12, 13, 15 and 17), in agreement with this suggestion. This further suggests that the Savitzky–Golay filters were better at detecting lightning activity and representing the quiet activity of non-lightning days. Sgf_3_30, 6_30, 6_60 and 6_120 mostly had amplitudes of ±0.05TECu on non-lightning days and saw an increase to ±0.06–0.2TECu at the time of expected DTEC amplitude changes on lightning days. These parameters were therefore able to detect lightning events and distinguish lightning days from non-lightning days using 2DC. Sgf_3_90 and 3_120 had amplitudes of ±0.1TECu on non-lightning days (panels o and p of Figures 6, 8, 10, 19 and 21) and an increase to the same value at the time of expected DTEC changes on lightning days (panel o and p of Figures 12, 15 and 17). These two parameters, like the polynomials, could only detect lightning activity, not distinguish between lightning and non-lightning days. The time window of 90 to 120 min is the typical period of TIDs. Sgfs 3_90, 3_120, 6_90 and 6_120 being able to detect DTEC changes affirms that lightning can induce TIDs as, suggested by Mahmud M [13]. Another interesting observation can be seen in Figure 5: DTEC amplitude changes are observed on PRN 13, although 9th July was a non-lightning day. PRN 13 passed at a time about 1–2 h after the few lightning counts on 9th July, as seen in Figure 8. It could be that PRN 13 passed directly over the location of the lightning strokes, and was therefore able to detect them. This also confirms the observation by Qin, et al. [31] that even a small lightning stroke can effect changes in the ionosphere.

As the Savitzky–Golay parameters sgf_3_30, 6_30, 6_60, 6_90 and 6_120 met all the conditions in 2DC, further evaluation through statistical means was deployed to choose the most suitable parameter. Linear correlations between lightning count and DTEC on lightning days for the PRNs presented in Section 3 above were conducted. The significant level (α) for accuracy assessment was 0.05, and the correlation coefficients and p-values for each PRN are presented in Table 3. Figure 24 shows the respective scatter diagrams of the correlations for these parameters.

Table 3. Correlation coefficients between lightning counts and DTEC values from the various parameters during lightning days. PRNs are placed in brackets. E is the scientific notation for base ten.

	CORRELATION COEFFICIENT				P-VALUE			
DTEC Method/Satellite	17th July (13)	17th July (21)	18th July (5)	19th July (15)	17th July (13)	17th July (21)	18th July (5)	19th July (15)
sgf_3_30	−0.0532432	0.212518	0.1380252	0.2172065	0.1844358	1.837E-07	6.041E-05	2.313E-09
sgf_6_30	0.0205577	0.2196998	0.3005218	0.3456595	0.6085514	6.823E-08	5.672E-19	3.204E-22
sgf_6_60	−0.0195246	0.0597396	−0.0388213	0.1613585	0.6266822	0.1469133	0.2613414	1.015E-05
sgf_6_90	0.0040953	0.5853365	0.465207	0.4095069	0.9187456	1.245E-55	2.789E-46	2.492E-31
sgf_6_120	0.34839	0.2449198	0.3050079	0.4778427	3.23E-19	1.603E-09	1.599E-19	1.554E-43

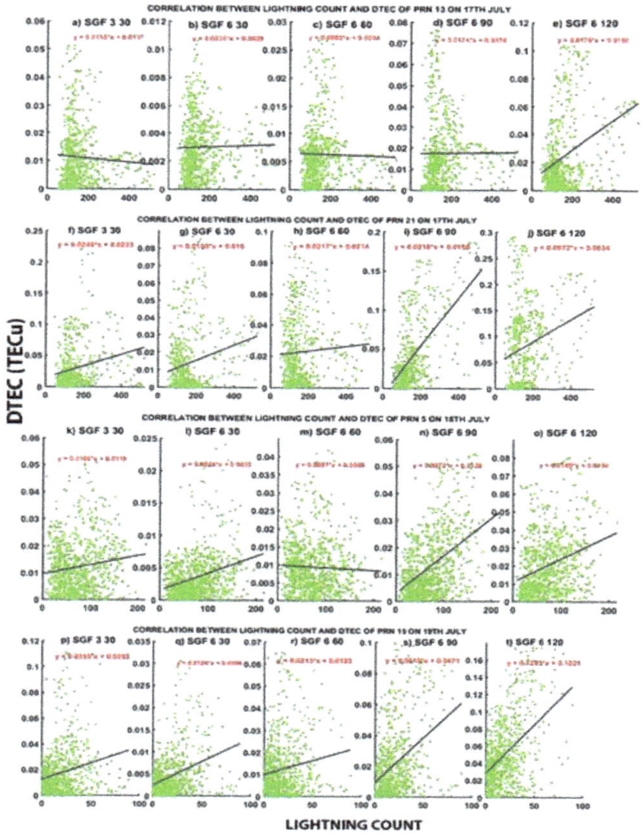

Figure 24. Scatter diagram for correlation between lightning count and DTEC amplitude in TECu on lightning days. Columns 1 to 4 are for the parameters sgf_3_30, 6_30, 6_60, 6_90 and 6_120, respectively. Row 1 (panels **a–e**) is for PRN 13 on 17th July. Row 2 (panels **f–j**) is for PRN 21 on 17th July. Row 3 (panels **k–o**) is for PRN 5 on 18th July. Row 4 (panels **p–t**) is for PRN 15 on 19th July.

From Table 3 and Figure 24, the DTEC amplitudes were mostly moderately positively correlated to lightning. The sgf_6_120 parameter was the most consistent with lightning count for all the days and PRNs, with an average moderate positive correlation of about 0.5. With such consistency and moderate positive correlation, sgf_6_120 was selected as the most suitable after meeting the conditions of 2DC. The time frame of 120 min further suggests that the disturbances could be TIDs, with lightning as the potential source. The coefficient of 0.4–0.5, though moderate, could be deemed as significant. Gravity waves in equatorial regions correlated with equatorial plasma bubbles (EPB) was 0.2. Though a weak correlation, this was consistent over a long period of time, and given the multiple sources of EPBs it should not be neglected. [32] Lightning is also a source of gravity waves [14], and a relative higher correlation of 0.4 could equally be deemed significant. The findings from the sgf confirms studies in other disciplines, such as medicine, that a Savitzky–Golay filter can represent physical parameters and events and provide detail which could otherwise be missed.

5. Conclusions

In this study, two commonly used TEC detrending methods were evaluated for the detection of lightning events in a low latitude region. The results show that a Savitzky–Golay filter of order 6 with a time window length of 120 min best detected the occurrence of lightning compared to the other parameters used in the study. The evaluation was done on the individual satellites rather than the average TEC over a station, as in some studies cited [13]. The individual satellite approach offers a better view than the average approach, as some minute details may be missed or cancelled out during the averaging. In addition, the best time frame being 120 min shows the potential of lightning to inducing TID. Further investigation is suggested in order to explore this observation.

Author Contributions: L.O.-P. conceptualized the work. Research design was done by L.O.-P. and W.C., L.O.-P., L.T. and C.M. contributed to data processing and analysis. W.C. and C.M. supervised the work. Manuscript writing and editing was done by all authors. All authors have read and agreed to the published version of the manuscript.

Funding: This research was funded by the Shenzhen Science and Technology Innovation Commission (Project No. JCYJ20170818104822282), the Hong Kong RGC Joint Research Scheme (E-PolyU501/16), the Smart City Research Institute, Hong Kong Polytechnic University and National Natural Science Foundation of China (Grant number 41804021).

Data Availability Statement: The Dst-index data (IAGA 2002-like format) can be obtained from the Data Analysis Center for Geomagnetism and Space Magnetism, Kyoto University, operating WDC for Geomagnetism, Kyoto (http://wdc.kugi.kyoto-u.ac.jp/Dstae/index.html, accessed on 14 June 2019). The Dst indices were obtained from the GSFC/SPDF OMNIWeb interface at http://omniweb.gsfc.nasa.gov, accessed on 14 June 2019. The GNSS data can be obtained from Hong Kong SatRef of the Lands Department of the Hong Kong Government. (https://www.geodetic.gov.hk/en/rinex/downv.aspx, accessed on 14 June 2019).

Acknowledgments: The authors thank the anonymous reviewers for their useful and insightful comments in improving the manuscript.

Conflicts of Interest: The authors declare no conflict of interest.

References

1. Maletckii, B.; Yasyukevich, Y.; Vesnin, A. Wave signatures in total electron content variations: Filtering problems. *Remote. Sens.* **2020**, *12*, 1340. [CrossRef]
2. Cherniak, I.; Zakharenkova, I. Large-scale traveling ionospheric disturbances origin and propagation: Case study of the december 2015 geomagnetic storm. *Space Weather* **2018**, *16*, 1377–1395. [CrossRef]
3. Ray, S.; Roy, B.; Paul, K.S.; Goswami, S.; Oikonomou, C.; Haralambous, H.; Chandel, B.; Paul, A. Study of the effect of 17–18 March 2015 geomagnetic storm on the Indian longitudes using GPS and C/NOFS. *J. Geophys. Res. Space Phys.* **2017**, *122*, 2551–2563. [CrossRef]

4. Belehaki, A.; Kutiev, I.; Marinov, P.; Tsagouri, I.; Koutroumbas, K.; Elias, P. Ionospheric electron density perturbations during the 7–10 March 2012 geomagnetic storm period. *Adv. Space Res.* **2017**, *59*, 1041–1056. [CrossRef]
5. Kong, J.; Yao, Y.; Zhou, C.; Liu, Y.; Zhai, C.; Wang, Z.; Liu, L. Tridimensional reconstruction of the Co-Seismic Ionospheric Disturbance around the time of 2015 Nepal earthquake. *J. Geod.* **2018**, *92*, 1255–1266. [CrossRef]
6. Liu, Y.; Jin, S. Ionospheric Rayleigh wave disturbances following the 2018 Alaska earthquake from GPS observations. *Remote. Sens.* **2019**, *11*, 901. [CrossRef]
7. Polyakova, A.S.; Perevalova, N.P. Investigation into impact of tropical cyclones on the ionosphere using GPS sounding and NCEP/NCAR Reanalysis data. *Adv. Space Res.* **2011**, *48*, 1196–1210. [CrossRef]
8. Yu, S.; Liu, Z. The ionospheric condition and GPS positioning performance during the 2013 tropical cyclone Usagi event in the Hong Kong region. *Earth Planets Space* **2021**, *73*, 1–16. [CrossRef]
9. Yang, Z.; Liu, Z. Observational study of ionospheric irregularities and GPS scintillations associated with the 2012 tropical cyclone Tembin passing Hong Kong. *J. Geophys. Res. Space Phys.* **2016**, *121*, 1–13. [CrossRef]
10. Lay, E.H.; Shao, X.M.; Kendrick, A.K.; Carrano, C.S. Ionospheric acoustic and gravity waves associated with midlatitude thunderstorms. *J. Geophys. Res. Space Phys.* **2015**, *120*, 6010–6020. [CrossRef]
11. Rahmani, Y.; Alizadeh, M.M.; Schuh, H.; Wickert, J.; Tsai, L.-C. Probing vertical coupling effects of thunderstorms on lower ionosphere using GNSS data. *Adv. Space Res.* **2020**, *66*, 1967–1976. [CrossRef]
12. Ogunsua, B.O.; Srivastava, A.; Bian, J.; Qie, X.; Wang, D.; Jiang, R.; Yang, J. Significant day-time ionospheric perturbation by thunderstorms along the West African and Congo sector of equatorial region. *Sci. Rep.* **2020**, *10*, 8466. [CrossRef]
13. Amin, M.M. Influence of Lightning on Electron Density Variation in the Ionosphere Using WWLLN Lightning Data and GPS Data. Master's Thesis, University of Cape Town, Cape Town, South Africa, 2015.
14. Tang, L.; Chen, W.; Chen, M.; Osei-Poku, L. Statistical observation of thunderstorm-induced ionospheric gravity waves above low-latitude areas in the northern hemisphere. *Remote. Sens.* **2019**, *11*, 2732. [CrossRef]
15. Liu, T.; Yu, Z.; Ding, Z.; Nie, W.; Xu, G. Observation of Ionospheric Gravity Waves Introduced by Thunderstorms in Low Latitudes China by GNSS. *Remote. Sens.* **2021**, *13*, 4131. [CrossRef]
16. Jardim, R.; Morgado-Dias, F. Savitzky–Golay filtering as image noise reduction with sharp color reset. *Microprocess. Microsyst.* **2020**, *74*, 103006. [CrossRef]
17. Chakraborty, M.; Das, S. Determination of signal to noise ratio of electrocardiograms filtered by band pass and Savitzky-Golay filters. *Procedia Technol.* **2012**, *4*, 830–833. [CrossRef]
18. Coster, A.J.; Goncharenko, L.; Zhang, S.R.; Erickson, P.J.; Rideout, W.; Vierinen, J. GNSS observations of ionospheric variations during the 21 August 2017 solar eclipse. *Geophys. Res. Lett.* **2017**, *44*. [CrossRef]
19. Savitzky, A.; Golay, M.J.E. Smoothing and differentiation of data by simplified least squares procedures. *Anal. Chem.* **1964**, *36*, 1627–1639. [CrossRef]
20. Ji, S.; Chen, W.; Wang, Z.; Xu, Y.; Weng, D.; Wan, J.; Fan, Y.; Huang, B.; Fan, S.; Sun, G.; et al. A study of occurrence characteristics of plasma bubbles over Hong Kong area. *Adv. Space Res.* **2013**, *52*, 1949–1958. [CrossRef]
21. Tang, L.; Chen, W.; Louis, O.-P.; Chen, M. Study on seasonal variations of plasma bubble occurrence over Hong Kong area using GNSS observations. *Remote Sens.* **2020**, *12*, 2423. [CrossRef]
22. Zhang, S.-R.; Coster, A.J.; Erickson, P.J.; Goncharenko, L.P.; Rideout, W.; Vierinen, J. Traveling ionospheric disturbances and ionospheric perturbations associated with solar flares in September 2017. *J. Geophys. Res. Space Phys.* **2019**, *124*, 5894–5917. [CrossRef]
23. Afraimovich, E.L.; Kosogorov, E.A.; Leonovich, L.A.; Palamartchouk, K.S.; Perevalova, N.P.; Pirog, O.M. Determining parameters of large-scale traveling ionospheric disturbances of auroral origin using GPS-arrays. *J. Atmos. Sol. Terr. Phys.* **2000**, *62*, 553–565. [CrossRef]
24. Liu, J.; Zhang, D.-H.; Coster, A.J.; Zhang, S.-R.; Ma, G.-Y.; Hao, Y.-Q.; Xiao, Z. A case study of the large-scale traveling ionospheric disturbances in the eastern Asian sector during the 2015 St. Patrick's Day geomagnetic storm. *Annales Geophysicae* **2019**, *37*, 673–687. [CrossRef]
25. Freeshah, M.; Zhang, X.; Chen, J.; Zhao, Z.; Osama, N.; Sadek, M.; Twumasi, N. Detecting ionospheric TEC disturbances by three methods of detrending through dense CORS during a strong thunderstorm. *Ann. Geophys.* **2020**, *63*, 667. [CrossRef]
26. Shao, X.M.; Lay, E.H.; Jacobson, A.R. Reduction of electron density in the night-time lower ionosphere in response to a thunderstorm. *Nat. Geosci.* **2013**, *6*, 29–33. [CrossRef]
27. Salem, M.A.; Liu, N.; Rassoul, H.K. Modification of the lower ionospheric conductivity by thunderstorm electrostatic fields. *Geophys. Res. Lett.* **2016**, *43*, 5–12. [CrossRef]
28. Loewe, C.A. Classification and mean behavior of magnetic storms. *J. Geophys. Res. Space Phys.* **1997**, *102*, 14209–14213. [CrossRef]
29. Kumar, S.; Chen, W.; Chen, M.; Liu, Z.; Singh, R.P. Thunderstorm-/lightning-induced ionospheric perturbation: An observation from equatorial and low-latitude stations around Hong Kong. *J. Geophys. Res. Space Phys.* **2017**, 9032–9044. [CrossRef]
30. Coster, A.; Herne, D.; Erickson, P.; Oberoi, D. Using the Murchison Widefield Array to observe midlatitude space weather. *Radio Sci.* **2012**, *47*, 1–8. [CrossRef]

31. Qin, Z.; Chen, M.; Lyu, F.; Cummer, S.A.; Gao, Y.; Liu, F.; Zhu, B.; Du, Y.p.; Usmani, A.; Qiu, Z. Prima facie evidence of the fast impact of a lightning stroke on the lower ionosphere. *Geophys. Res. Lett.* **2020**, *47*. [CrossRef]
32. Wu, Q.; Chou, M.-Y.; Schreiner, W.; Braun, J.; Pedatella, N.; Cherniak, I. COSMIC observation of stratospheric gravity wave and ionospheric scintillation correlation. *J. Atmos. Sol. Terr. Phys.* **2021**, *217*, 105598. [CrossRef]

remote sensing MDPI

Article

Determination of Navigation System Positioning Accuracy Using the Reliability Method Based on Real Measurements

Mariusz Specht

Department of Transport and Logistics, Gdynia Maritime University, 81-225 Gdynia, Poland; m.specht@wn.umg.edu.pl

Abstract: In navigation, the Twice the Distance Root Mean Square (2DRMS) is commonly used as a position accuracy measure. Its determination, based on statistical methods, assumes that the position errors are normally distributed and are often not reflected in actual measurements. As a result of the widespread adoption of this measure, the positioning accuracy of navigation systems is overestimated by 10–15%. In this paper, a new method is presented for determining the navigation system positioning accuracy based on a reliability model where the system's operation and failure statistics are referred to as life and failure times. Based on real measurements, the method proposed in this article will be compared with the classical method (based on the 2DRMS measure). Real (empirical) measurements made by the principal modern navigation positioning systems were used in the analyses: Global Positioning System (GPS) (168'286 fixes), Differential Global Positioning System (DGPS) (900'000 fixes) and European Geostationary Navigation Overlay Service (EGNOS) (900'000 fixes). Research performed on real data, many of which can be considered representative, have shown that the reliability method provides a better (compared to the 2DRMS measure) estimate of navigation system positioning accuracy. Thanks to its application, it is possible to determine the position error distribution of the navigation system more precisely when compared to the classical method, as well as to indicate those applications that can be used by this system, ensuring the safety of the navigation process.

Keywords: navigation positioning system; positioning accuracy; reliability method; Global Positioning System (GPS); Differential Global Positioning System (DGPS); European Geostationary Navigation Overlay Service (EGNOS)

Citation: Specht, M. Determination of Navigation System Positioning Accuracy Using the Reliability Method Based on Real Measurements. *Remote Sens.* **2021**, *13*, 4424. https://doi.org/10.3390/rs13214424

Academic Editor: Serdjo Kos

Received: 30 September 2021
Accepted: 2 November 2021
Published: 3 November 2021

1. Introduction

The main goal behind positioning systems is to provide air [1–5], land [6–10] and marine navigation [11–15] applications with such accuracy that the process is carried out safely. Such applications may include the following: a ship entering a port, driving a car along a route, landing an aircraft at an airport, stopping a tram on a platform, etc. The decision to qualify a positioning system as safe for a given navigation application is made based on a comparison of the position error characterizing the system with the minimum navigation requirements for a specific application specified in normative documents, such as radio navigation plans [16–19] and other recommendations or regulations [20–24]. These requirements most often include the following: positioning accuracy, availability, continuity, fix rate, integrity, operation range, reliability [25].

There is no doubt that, for decades of navigation development, it is the positioning accuracy that has been, and still is, the decisive factor for the use of a system for a specific navigation task, while the availability level of a specific position error is related to the threat that a positioning failure may pose to the safety of an object. Hence, air navigation requires top positioning availability. Figure 1 presents a synthesis of the requirements for different navigation applications in terms of the maximum permissible position error and its availability. The requirements found in various normative documents were used for the analysis.

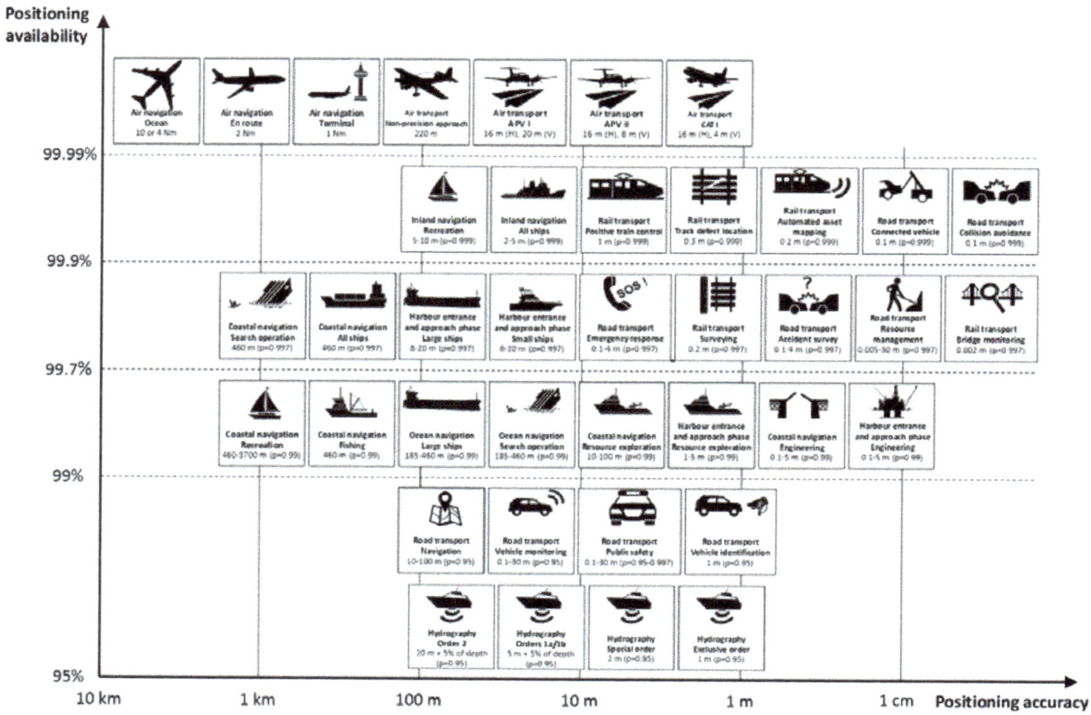

Figure 1. Requirements for navigation positioning systems. Own study based on [16–24,26].

Figure 1 shows that each of the navigation applications should use a system with adequate positioning accuracy. This means, among other things, that a system with a positioning accuracy of 100 m cannot be used in hydrography as it requires a system with a positioning accuracy of less than 20 m for order 2, 5 m for orders 1a/1b, 2 m for special order and 1 m for exclusive order.

The navigation system positioning accuracy refers to the overlap between the statistics of the measured position coordinates and their real values or values assumed to be real (most often the average value from latitude (φ) and longitude (λ) measurements) [19]. A position accuracy measure is its error, which can be evaluated with reference to any dimension: plane (2D) or space (3D). In navigation, a position error is defined as the radius of the circle or sphere within which a certain proportion of position determinations should statistically fall. It is commonly accepted in navigation that position error statistics refer to 95% of the population, as the safety of the navigation process depends largely on the navigation system positioning accuracy. The Twice the Distance Root Mean Square (2DRMS(2D)) is used as a position accuracy measure. Its determination starts with the calculation of the DRMS(2D), which is the square root from the sum of the squares of the standard deviations of the position coordinates relative to φ and λ, as per the following relationship:

$$DRMS(2D) = \sqrt{(s_\varphi)^2 + (s_\lambda)^2}, \quad (1)$$

where: s_φ—standard deviation of the geodetic (geographic) latitude; s_λ—standard deviation of the geodetic (geographic) longitude.

The probability of the DRMS(2D) lies in the 63.2–68.3% range and depends on the relationship between the standard deviations. For $s_\varphi = s_\lambda$, $p = 0.63$, while for the relation $s_\varphi = 10 \cdot s_\lambda$, $p = 0.68$.

To provide greater statistical reliability of the DRMS(2D) in navigation, the 2DRMS(2D) measure is commonly used, taking the following form:

$$2DRMS(2D) = 2\sqrt{(s_{\varphi})^2 + (s_{\lambda})^2}. \tag{2}$$

In navigation literature, the 2DRMS corresponds to a probability lying in the range 95.4–98.2% and is related to the relationship between the standard deviations determined with the two coordinates. Figure 2 shows the geometric interpretation of the position error.

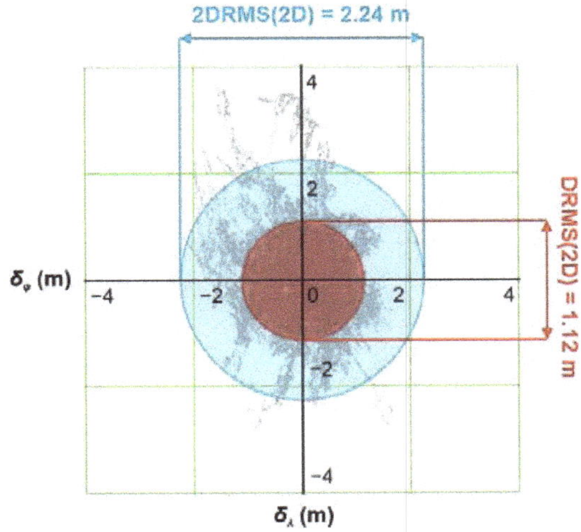

Figure 2. Geometric interpretation of the concept of navigation system position error in 2D plane using DRMS and 2DRMS values.

Taking the 2DRMS value as the primary position accuracy measure by a navigation system is based on the assumption that φ and λ errors are normally distributed [27,28]. However, this belief has been questioned in several publications. The most important standard describing Global Positioning System (GPS) accuracy characteristics [28] states that the difference between the empirical value (64 m) and the theoretical value (83 m), as determined by the 2DRMS measure, was as much as 19 m. Similar conclusions concerning the inconsistency of the statistical distributions of Differential Global Positioning System (DGPS) and GPS position errors were raised by Frank van Diggelen, but with much smaller discrepancies [29].

The author's research, conducted on various navigation positioning systems, has repeatedly confirmed the existence of such discrepancies. They related to systems such as DGPS and the European Geostationary Navigation Overlay Service (EGNOS) [30], the Global Navigation Satellite System (GNSS) and geodetic networks and multi-GNSS solutions [31–33].

Questioning normal distribution as a model for navigation positioning system errors has prompted the search for other methods to determine the value of the position error with 95% probability, as commonly used in navigation [27,28]. One of the methods based on reliability theory has already been proposed in [25]. This method allows the navigation system positioning availability to be determined for a specific (given) value of the position error based on life and failure times, and not based on measurement errors. The positioning system is fit when the position error does not exceed the allowable error. The failure period is defined as the logical negation of the fit period. A comparison of both methods (classical and reliability) is presented in Figure 3.

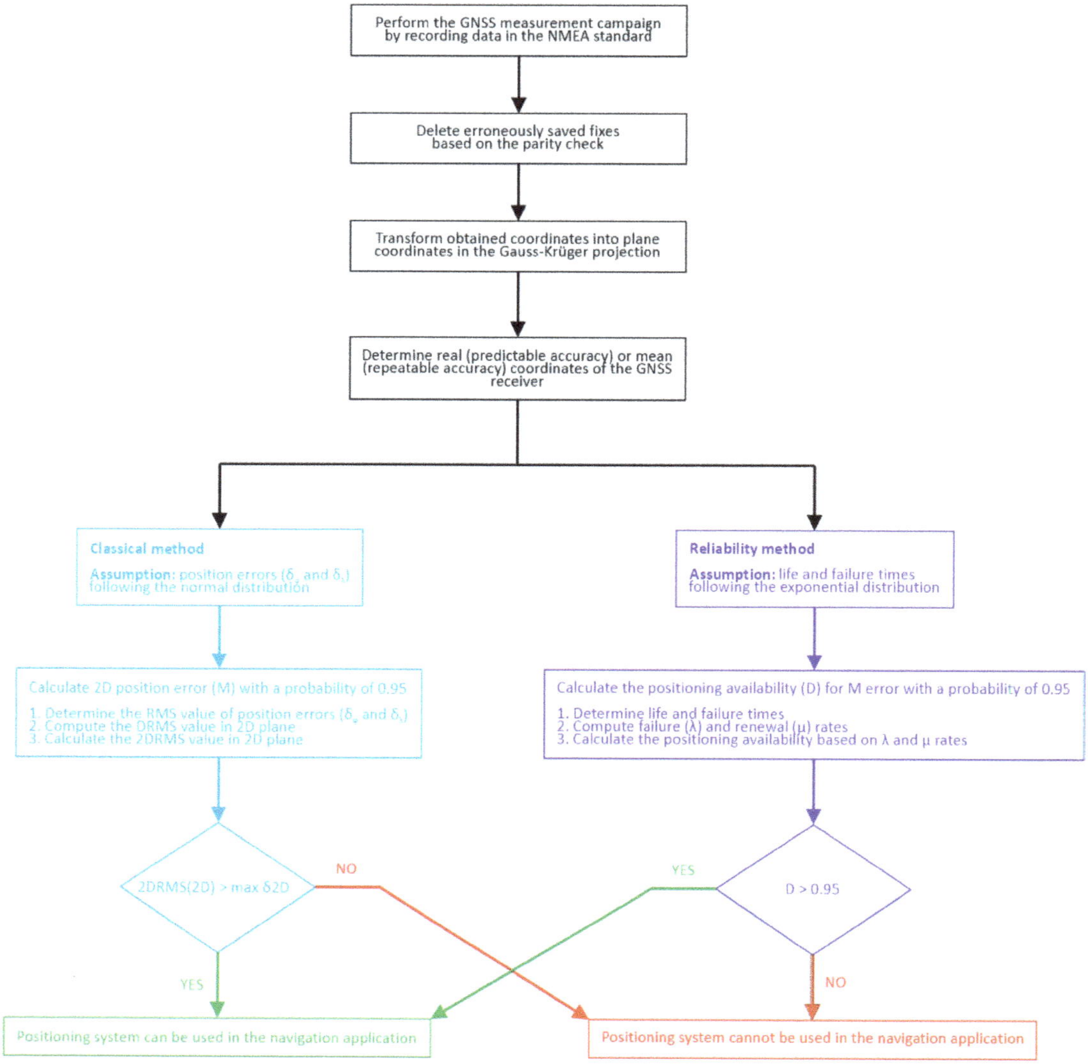

Figure 3. Comparison of the classical and reliability methods for assessing the positioning system's ability to meet the accuracy requirements for a navigation application.

In the classical approach, for a fixed probability (usually amounting to 95%), the position error (value of the 2DRMS) is calculated, whereas in the reliability method an acceptable error value is determined first, and only then is its probability is calculated. In the first method, the position error is a random variable and in the second method the random variable is the life or failure time. In the first method, 1D errors are assumed to be normally distributed and 2D errors are assumed to follow a chi-square distribution, while in the second method, exponential distributions of life and failure times are assumed.

The characteristics of the classical (based on the normal distribution) approach are as follows:

- The calculations are based on simple Root Mean Square (RMS) determination relationships;

- Gross errors and outliers significantly affect RMS (φ) and RMS (λ), causing a change in the 2DRMS measure;
- Errors are analysed, not as a function of time, but as a function of the subsequent measurement error. The navigation process runs as a function of time. The problem of missing synchronisation with time will emerge in the case of erroneous measurements (recording errors, which have to be removed from the dataset).

The reliability approach has the following characteristics:

- The calculations are quite complex;
- Gross errors and outliers affect the life and failure times in the same way as the other measurements;
- The analysis is carried out as a function of time, similar to the navigation process.

This paper aims to compare both methods and to evaluate the positioning system in terms of its fitness for a specific navigation application, based on real measurement data from the positioning system (long sessions). Therefore, the scientific purpose of this article are as follows:

1. To propose a new (reliability-based) method to calculate position error values for a navigation system with a probability of 95%;
2. To verify which method (classical or reliability) produces results closer to empirical data;
3. To check, based on empirical data, the actual measurements of GPS, DGPS and EGNOS systems, whether the distributions of life and failure times for position errors are, in fact, exponential. The other distributions most commonly used in statistics will be tested: beta, Cauchy, chi-square, exponential, gamma, Laplace, logistic, lognormal, normal, Pareto, Rayleigh, Student's and Weibull.

To assess which of the methods allows for more precise determination of the statistical value of the position errors, the measured position errors were sorted from the smallest to the largest, and based on this the value of the error that is greater exactly than 95% of the error population will be determined. In the navigation literature [34], this value is referred to as the R95 measure.

This is the third article in a series of monothematic publications "Research on empirical (actual) statistical distributions of navigation system position errors" [35–37]. The main scientific aim of this series is to answer the question of what statistical distributions follow the position errors of navigation systems such as GPS, Global Navigation Satellite System (GLONASS), BeiDou Navigation Satellite System (BDS), Galileo, DGPS, EGNOS and others. It must be emphasised that the purpose of both this paper and the whole series of publications is not to analyse the causes of Position Random Walk (PRW), such as ionospheric and tropospheric effects, multipath, noise, etc. This article rather analyses the statistical distributions of 1D and 2D position errors resulting from PRW. The causes might be very complex and probably deserve a separate series of publications.

2. Materials and Methods

2.1. Classical Method for Determining the Positioning Accuracy of a Navigation System with 95% Probability

The classical method for determining the positioning accuracy of a navigation system is based on a statistical approach. The works of Gerolamo Cardano [38], Pierre de Fermat and Blaise Pascal [39] on gambling, as well as the works of Christiaan Huygens [40] constitute the foundation of modern statistics. In 1812, Pierre-Simon Laplace formulated the classic definition of probability [41], which was mathematically formalised in 1933 by Andrey Kolmogorov [42], who gave the basic formulas of probability calculus and its axioms. Although modern statistics is an extremely young branch of mathematics, it is widely used in many fields, ranging from engineering [43] and economics [44] to the scientific aspects of computer science [45]. In navigation, similar to other sciences, it is assumed that the position errors of navigation systems are normally distributed. Major arguments justifying the use of normal distribution in research on positioning in navigation include [46]: intuition and tradition, simplicity of the distribution [47], consistency with the central limit theorem, as well as use as an approximation [48].

Position accuracy in navigation can be defined as a degree of conformance between the estimated or measured position and its true position. Position accuracy can be determined as different types of statistic. These can be calculated related to the true values of coordinates (predictable accuracy), or, if the actual position is not known, the mean position (repeatable accuracy) is often used as an approximation to the actual position. Both position solutions must be based on the same geodetic datum, e.g., the World Geodetic System 1984 (WGS-84) [19]. The most commonly used position accuracy measures in navigation and transport are as follows: CEP (Circular Error Probable, 2D, $p = 0.5$), SEP (Spherical Error Probable, 3D, $p = 0.5$), RMS (1D, $p = 0.632–0.683$), DRMS (2D or 3D, $p = 0.632–0.683$), 2DRMS (2D or 3D, $p = 0.954–0.982$) or 3DRMS (Triple Distance Root Mean Square, 2D or 3D, $p = 0.997$). The description of individual measures is presented in detail in [34]. However, the most common accuracy measure for assessing the positioning accuracy of navigation systems is 2DRMS.

2.2. Reliability Method for Determining the Positioning Accuracy of a Navigation System with 95% Probability

To assess the possibility of using a positioning system in a specific navigation application, a mathematical model based on the general theory of reliability has been proposed. Navigation has long used elements of reliability theory in studies related to the assessment of availability, continuity, integrity and reliability, but so far they have not been applied to assess whether a positioning system meets a certain level of accuracy and to determine the navigation system positioning accuracy [49].

Let us consider a positioning system that determines position as a function of time with an error defined as δ_n. Let us choose a specific type (s) of navigation applications for which we intend to check whether the positioning system meets the application requirements in terms of accuracy and availability. These requirements are presented in [16–24,26]. Let us run a measurement session of the positioning system of a representative length [37] and calculate position errors as a function of time.

Figure 4 (top graph) shows the curve presenting the position error value as a function of time for any positioning system, which should be evaluated in terms of its usability in three exemplary navigation applications. Such applications include: road transport for vehicle identification, with the maximum allowable position error being 1 m with 95% availability, hydrography for orders 1a/1b, with the maximum allowable position error being 5 m with 95% availability and harbour entrance and approach phase for large ships, with the maximum allowable position error being 20 m with 99.7% availability.

Please note that in the presented graph (Figure 4a) the position error value varies as a function of time. Because, near the starting point in this plot, the position error exceeds 20 m, it does not provide the accuracy required for the harbour entrance and approach phase for large ships. This also means that a system with such an error cannot ensure the safe positioning of this process. As a result of exceeding the maximum permissible position error, the system's fitness changes into an unfitness status, as reflected by the "0" values in Figure 4d. After some time, the position error (Figure 4a) decreases to less than 1 m, which means that, for some of the time, the system can be used in all applications. As a result of the reduction in the position error value, for all its applications, the working state of the system changes into a fitness status (Figure 4b–d). The graph (Figure 4a) presents the maximum permissible error values for all three navigation applications (1 m, 5 m and 20 m). If the position error exceeds any of these values, the working state changes. If the position error is smaller than this set value, the system enters the fitness status, and if it is larger, the system is in the unfitness status. Thus, the position coordinate determination can be treated as a two-status stationary renewal process, in which the life and failure times will become random variables, and not the position error as before (a classical approach) [50].

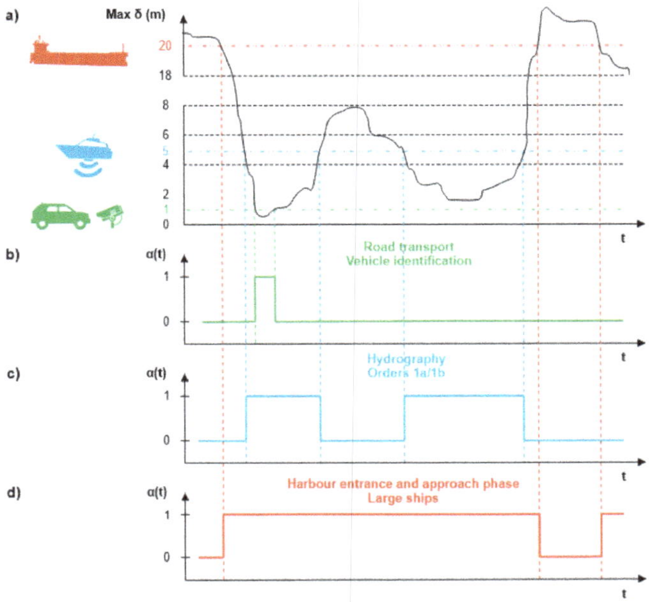

Figure 4. The position error as a function of time (**a**) and three diagrams corresponding to the operational status for: (**b**) vehicle identification; (**c**) orders 1a/1b; (**d**) large ships.

To be able to determine whether the system was fit or not, a U variable was introduced which corresponds to the maximum allowable value of the position error. Let us write it for the three applications under consideration as:

$$U = \begin{cases} 1\ m\ (p = 0.95) \text{ for road transport – vehicle identification} \\ 5\ m\ (p = 0.95) \text{ for hydrography – orders 1a/1b} \\ 20\ m\ (p = 0.997) \text{ for harbour entrance and approach phase – large ships} \end{cases} \quad (3)$$

Assuming that the positioning process varies with time, it can be assigned two states. The first is the life time for which the position error is less than the maximum permissible position error corresponding to the given navigation application ($\delta_n \leq U$ for number of measurements (n) = 1, 2, …). When the inverse relationship occurs ($\delta_n > U$), the system is in a failure time.

Let us assume that X_1, X_2, … correspond to the durations of life times and Y_1, Y_2, … denote the durations of failure times, which are independent and have the same distributions. Changing the durations of life and failure times results in the change of the operational status of a positioning system ($\alpha(t)$). Hence, $Z_n' = X_1 + Y_1 + X_2 + Y_2 + \ldots + Y_{n-1} + X_n$ become the moments of failure, while $Z_n'' = Z_n' + Y_n$ are the moments of life (Figure 5) [26,51].

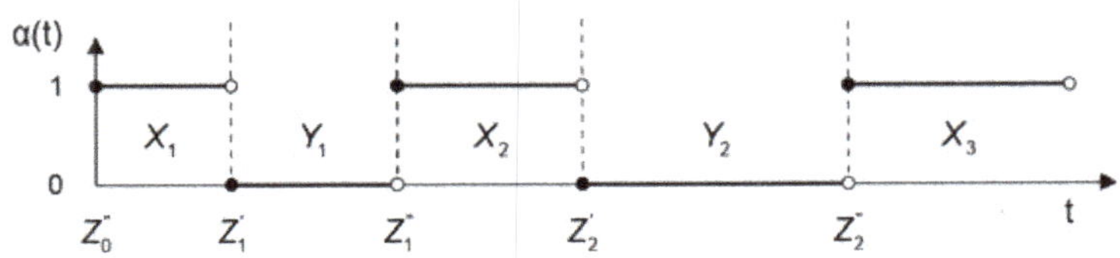

Figure 5. The fitness and unfitness statuses of a positioning system in accordance with the reliability method. Own study based on [51].

For the reliability method, it is necessary to introduce a number of additional assumptions and designations [50]. It should be assumed that the Cumulative Distribution Functions (CDF) of life ($F(x)$) and failure ($G(y)$) times are right-continuous:

$$P(X_i \leq x) = F(x)\,, \tag{4}$$

$$P(Y_i \leq y) = G(y) \;\; \text{for } i = 1,2,\ldots, \tag{5}$$

and that the expected values and variances will take the form:

$$E(X_i) = E(x)\,, \tag{6}$$

$$E(Y_i) = E(y)\,, \tag{7}$$

$$V(X_i) = \sigma_1^2\,, \tag{8}$$

$$V(Y_i) = \sigma_2^2 \;\; \text{for } i = 1,2,\ldots, \tag{9}$$

where: $E(X_i)$—expected value of the life time; $E(Y_i)$—expected value of the failure time; $V(X_i)$—variance of the life time; $V(Y_i)$—variance of the failure time.

Moreover, it should be noted that:

$$\sigma_1^2 + \sigma_2^2 > 0. \tag{10}$$

Based on the above assumptions, it is possible to determine the relationship between the δ_n and U parameters. Thanks to this, the operational status of a positioning system can be assigned as [26,51]:

$$U = \begin{cases} 1 \text{ for } Z''_n \leq t < Z'_{n+1} \\ 0 \text{ for } Z'_{n+1} \leq t < Z''_{n+1} \end{cases} \text{for } n = 0,1,\ldots. \tag{11}$$

Let us define the navigation system positioning availability ($A(t)$) as the probability that at any moment of t, δ_n will not be greater than the value of U [51]:

$$A(t) = P[\delta(t) \leq U], \tag{12}$$

$$A(t) = 1 - F(t) + \int_0^t [1 - F(t-x)]dH_\Phi(x), \tag{13}$$

where:

$$H_\Phi(x) = \sum_{n=1}^{\infty} \Phi_n(x) \tag{14}$$

is a function of the renewal stream made up of the renewal moments of the navigation system complying with a specific operation type, while $\Phi_n(t)$ is a distribution function of the random variable Z''_n.

For the purposes of navigation applications, the distributions of life and failure times are exponential. Therefore, their CDFs and Probability Density Functions (PDF) can be calculated using the following formulas [50]:

$$f(t) = \begin{cases} \lambda \cdot e^{-\lambda \cdot t} \text{ for } t > 0 \\ 0 \text{ for } t \leq 0 \end{cases}, \tag{15}$$

$$g(t) = \begin{cases} \mu \cdot e^{-\mu \cdot t} \text{ for } t > 0 \\ 0 \text{ for } t \leq 0 \end{cases}, \tag{16}$$

$$F(t) = \begin{cases} 1 - e^{-\lambda \cdot t} \text{ for } t > 0 \\ 0 \text{ for } t \leq 0 \end{cases}, \tag{17}$$

$$G(t) = \begin{cases} 1 - e^{-\mu \cdot t} \text{ for } t > 0 \\ 0 \text{ for } t \leq 0 \end{cases}, \tag{18}$$

where: $f(t)$—PDF of the life time; $g(t)$—PDF of the failure time; λ—failure rate; μ—renewal rate.

When these assumptions are adopted, the final form of the availability can be noted as [51]:

$$A_{\exp}(t) = \frac{\mu}{\lambda + \mu} + \frac{\lambda}{\lambda + \mu} \cdot e^{-(\lambda+\mu)\cdot t}, \tag{19}$$

as for the limit value:

$$\lim_{t\to\infty}\left[A_{\exp}(t)\right] = A_{\exp} = \frac{\frac{1}{\lambda}}{\frac{1}{\lambda} + \frac{1}{\mu}} = \frac{\mu}{\mu + \lambda}. \tag{20}$$

2.3. Description of GPS, DGPS and EGNOS Measurement Campaigns

Three different navigation positioning systems, commonly used in world navigation, were used to study the reliability method. They include GPS (measurements from 2013), DGPS (measurements from 2014) and EGNOS (measurements from 2014).

GPS is a space-based radionavigation system owned by the United States Government (USG) and operated by the United States Space Force (USSF). The GPS provides two services, or levels of accuracy: the Precise Positioning Service (PPS) and the Standard Positioning Service (SPS). The PPS is available to authorized users and the SPS is available to all users. SPS is the standard specified level of positioning, velocity, and timing accuracy that is available, without restrictions, to any user on a continuous worldwide basis. It provides a global average predictable positioning accuracy of 8 m (p = 0.95) horizontally and 13 m (p = 0.95) vertically and time transfer accuracy within 30 ns (p = 0.95) of Universal Time Coordinated (UTC) (Figure 6a) [52].

DGPS is an enhancement of the GPS, carried out through the use of differential corrections to the basic satellite measurements performed within the user's receiver. The DGPS is based on accurate knowledge of the geographic location of a reference station, which is used to compute corrections to GPS parameters and the resultant position solution. These differential corrections are then transmitted to DGPS users, who apply the corrections to their received GPS signals or computed position. For a civil user of SPS, differential corrections can improve navigation accuracy to better than 5 m (p = 0.95) (Figure 6b) [53,54].

EGNOS is Europe's regional Satellite-Based Augmentation System (SBAS), which is used to improve the performance of GNSS systems, such as GPS and Galileo. It has been deployed to provide the safety of life navigation services to aviation, maritime and land-based users over most of Europe. According to [55], the positioning accuracy of the Open Service (OS) should be smaller than 3 m (p = 0.95) horizontally and 4 m (p = 0.95) vertically (Figure 6c).

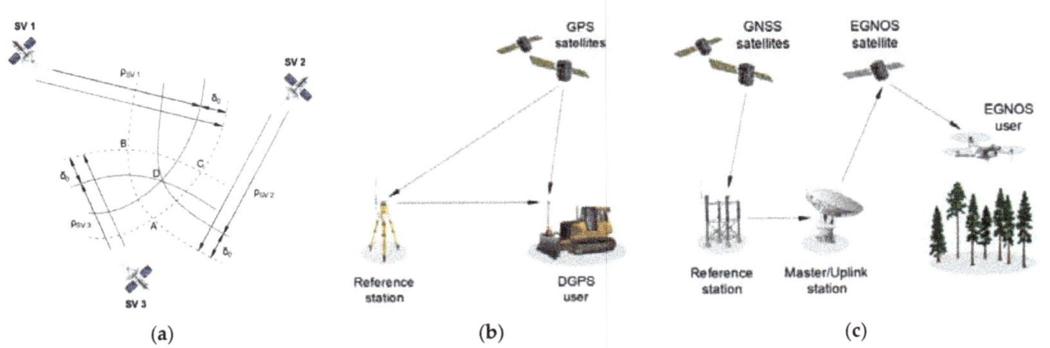

Figure 6. Principles of: (**a**) GPS; (**b**) DGPS; (**c**) EGNOS. Own study based on [56].

The following empirical data were used for the numerical analyses:

- The GPS measurements were carried out at a point with coordinates: $\varphi = 54°32.585029'$ N and $\lambda = 18°32.741505'$ E (Poland). In March 2013, 168′286 fixes were recorded with a recording frequency of 1 Hz. A typical 12-channel GPS code receiver was used in the study;
- The DGPS measurements were carried out at a point with coordinates: $\varphi = 54°31.756087'$ N, $\lambda = 18°33.574138'$ E and $h = 68.070$ m (Poland). In April 2014, 951′698 fixes were recorded with a recording frequency of 1 Hz. 900′000 fixes were used for the analyses, which were the same as for EGNOS. A typical marine DGPS code receiver was used in the study;
- The EGNOS measurements were carried out at a point with coordinates: $\varphi = 54°31.756087'$ N, $\lambda = 18°33.574138'$ E and $h = 68.070$ m (Poland). In April 2014, 927′553 fixes were recorded with a recording frequency of 1 Hz. 900′000 fixes were used for the analyses, which were the same as for DGPS. A typical land EGNOS code receiver was used in the study.

3. Results

The research aims were formulated in the form of questions:

1. Do the empirical (actual) distributions of life and failure times for position errors follow an exponential distribution?
2. Are there distributions other than exponential with a better fit?
3. Depending on the value of the error determining the fitness status (maximum permissible position error for a navigation application), will the statistical distribution of life times change or not?

Research on distributions of life times began with the analysis of GPS position error results. A recording session of 168′286 fixes was studied. This can be considered statistically representative, according to the analyses conducted in [37]. Figure 7 presents the graph of the position error as a function of time. The figure also shows two accuracy values (1 m and 2 m), which correspond to the requirements of road transport for vehicle identification and hydrography for special order.

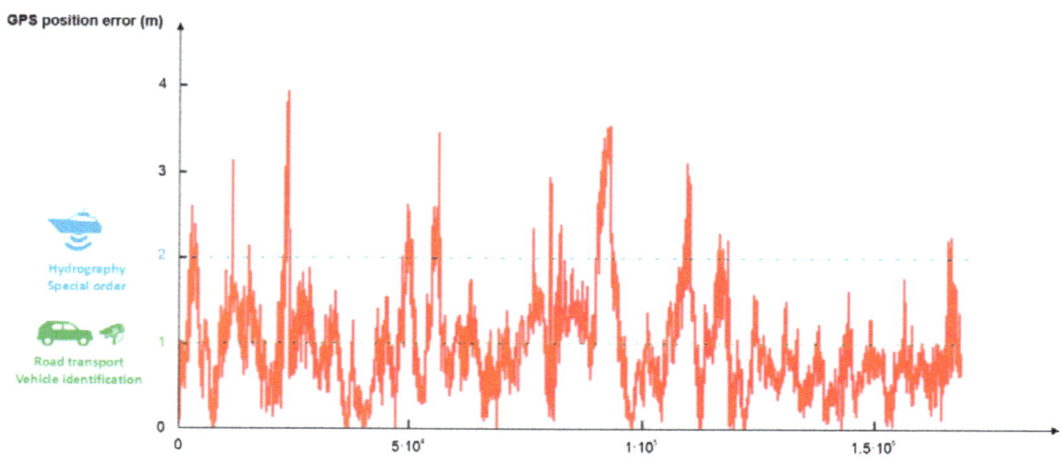

Figure 7. GPS position error as a function of time (168′286 fixes) and two decision thresholds corresponding to the requirements of road transport for vehicle identification and hydrography for special order.

To automate the process of determining life and failure times, the position error values needed to be assigned to one of the operational statuses: 1 (life status, when the temporary position error is below a pre-set limit value) or 0 (failure status, in the opposite case) at any point in time. For this purpose, a calculation sheet was prepared in Mathcad 15 (Figure 8).

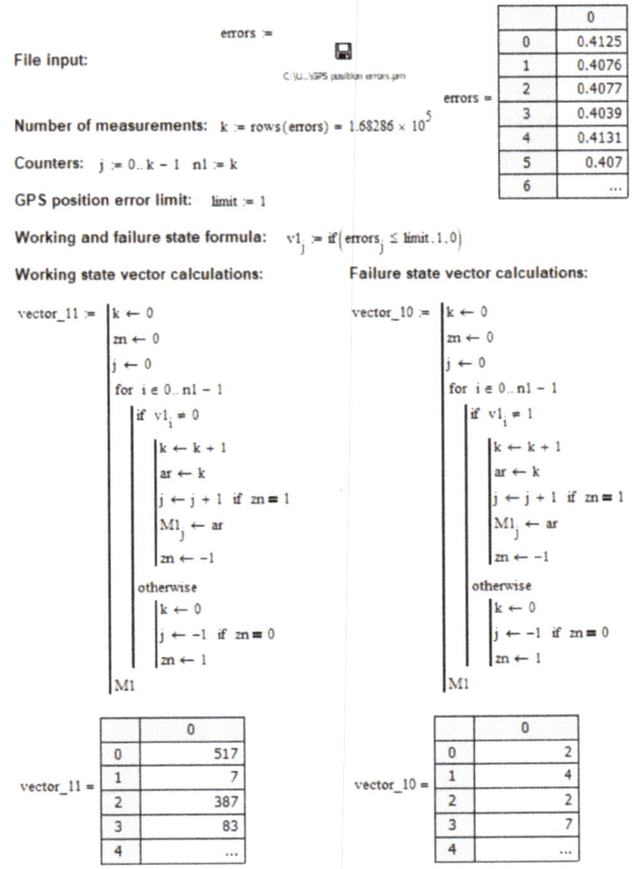

Figure 8. Mathcad worksheet for determining life and failure times based on the GPS 2013 measurement campaign.

In the sheet shown in Figure 8, the calculations began with importing measurement data consisting of position errors. The position error limit was then determined arbitrarily. In the worksheet, it is 1m. Next, two vectors were created consisting of life (vector_11) and failure (vector_10) times.

Both vectors were saved as text files and then uploaded to the EasyFit software, where they were analysed for fit to typical statistical distributions. These distributions included: beta, Cauchy, chi-square, exponential, gamma, Laplace, logistic, lognormal, normal, Pareto, Rayleigh, Student's and Weibull.

For GPS position errors, calculations were carried out for two variants, 1 m and 2 m, corresponding to example navigation applications such as from road transport for vehicle identification (1 m) and hydrography for special order (2 m). In P-P plots, the empirical probability distribution function is plotted against the theoretical distribution. The observations are first sorted in descending order. The i-th observation is then plotted on one axis as $\frac{i}{n}$ (i.e., the value of the observed cumulative distribution) and the other axis as $F(x_i)$, where $F(x_i)$ is the value of the theoretical probability distribution function for respective observation x_i. If the theoretical cumulative distribution is a good approximation of the empirical distribution, then the points on the diagram should be close to the diagonal.

The research began with an analysis of GPS failure times. The results of P-P plot analyses are presented in Table 1. Moreover, the list of theoretical distributions with the best (top) and worst fit can be seen next to each of the graphs.

Table 1. P-P plots of life and failure times for empirical GPS position errors (1 m and 2 m).

GPS

P-P Plot: Life Time for the Position Error Amounted to 1 m

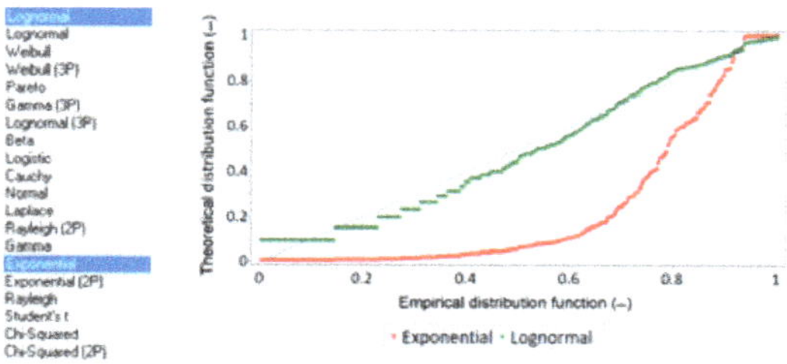

P-P Plot: Failure Time for the Position Error Amounted to 1 m

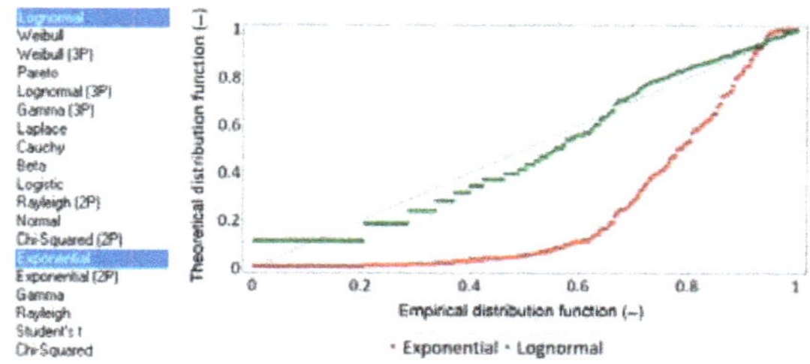

P-P Plot: Life Time for the Position Error Amounted to 2 m

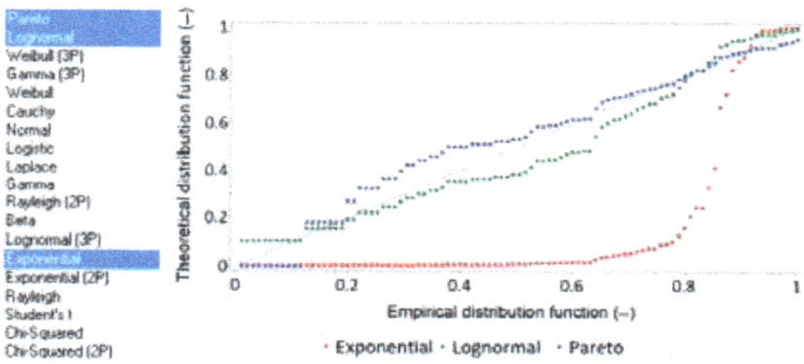

P-P Plot: Failure Time for the Position Error Amounted to 2 m

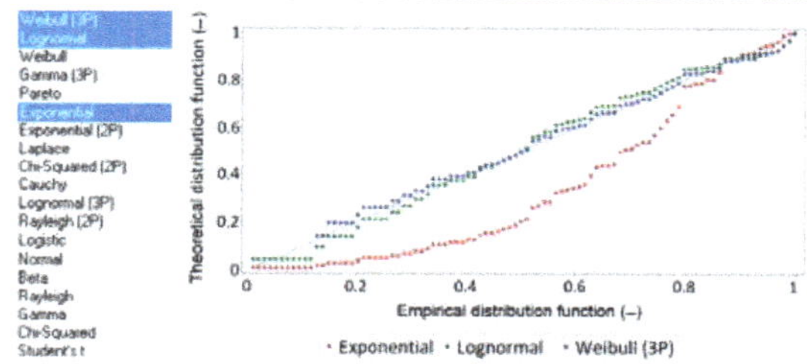

The analyses show that the assumption (very common in technology) that the life and failure times follow exponential distribution is questionable. It is clear that, for distribution function, values below 0.9 (both empirical and theoretical) are poorly fitted. On the contrary, above this value the fit is very good, which may suggest that the failure (λ) and renewal (μ) rates can be calculated very reliably on its basis. It has also been stressed that it is not an exponential distribution but a lognormal distribution that provides a very good approximation of the life and failure time statistics. Furthermore, analyses performed on two different decision thresholds (1 m and 2 m) produced similar conclusions regarding the mismatch between the empirical and theoretical (exponential) distributions for distribution function values below 0.9.

Therefore, it is reasonable to conduct identical analyses for other navigation positioning systems such as DGPS and EGNOS. These analyses have been performed for a very large sample of 900′000 fixes. Table 2 presents the PDFs of life and failure times, together with the best-fit distributions for the DGPS system. Moreover, Table 2 shows the P-P plots for DGPS and EGNOS systems.

The distribution of DGPS system life and failure times is similar to that of GPS, although empirical data fit theoretical exponential distribution much better. On the contrary, for EGNOS it is clear that the empirical distributions of life and failure times deviate significantly from the theoretical distributions. However, as in the case of GPS and DGPS systems, for values above 0.9, the fit is very good.

Table 2. Statistical analysis of life and failure times for empirical DGPS and EGNOS position errors (1 m).

DGPS	
PDF: Life Time for the Position Error Amounted to 1 m	PDF: Failure Time for the Position Error Amounted to 1 m
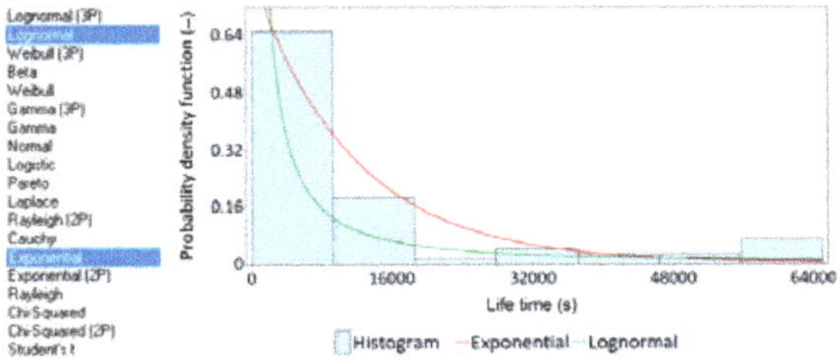	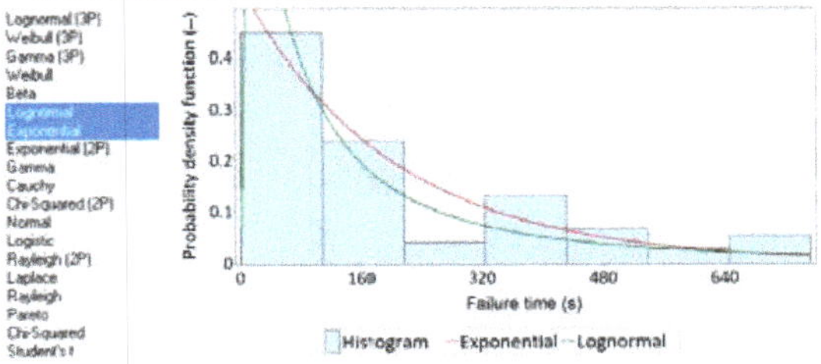
P-P Plot: Life Time for the Position Error Amounted to 1 m	P-P Plot: Failure Time for the Position Error Amounted to 1 m
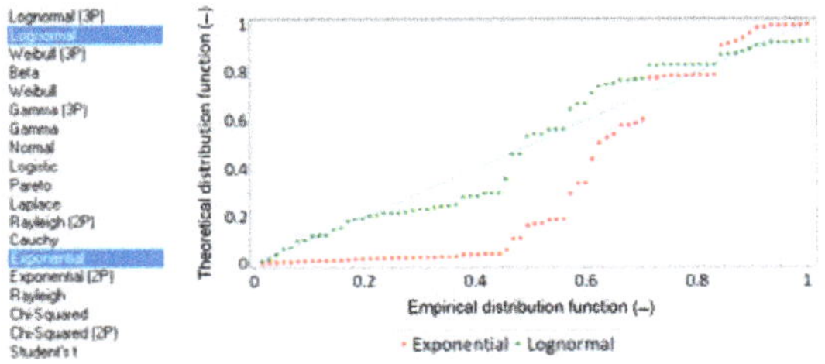	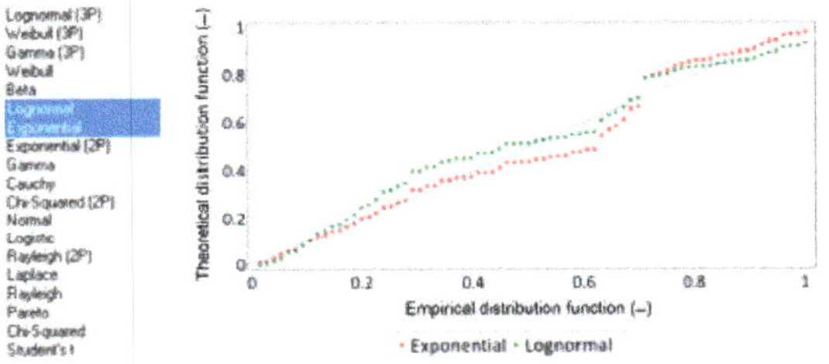
EGNOS	
P-P Plot: Life Time for the Position Error Amounted to 1 m	P-P Plot: Failure Time for the Position Error Amounted to 1 m
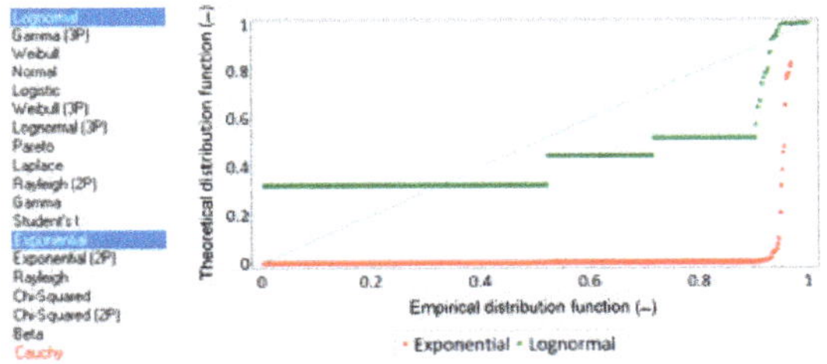	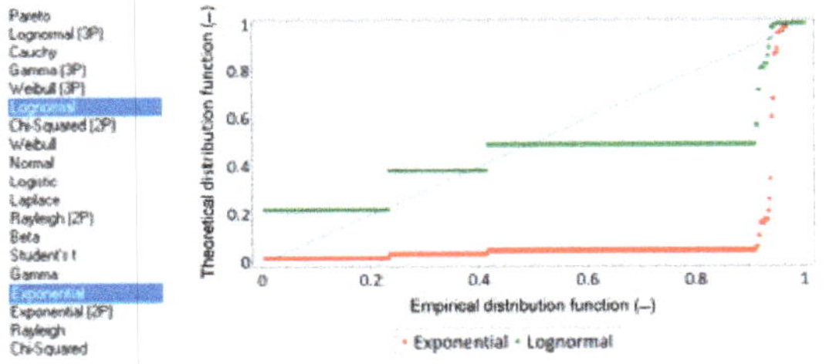

The following conclusions can be drawn from the empirical studies and theoretical analyses carried out:

- The analysis of GPS data indicates that the lognormal distribution reflects the course of the PDF of life and failure times determined for navigation system position errors significantly better than the exponential distribution;
- For values above 0.9, the fit between theoretical and empirical distributions (exponential distribution) is very good in all the analysed cases;
- The results obtained from the GPS system also prove that increasing the decision threshold from 1 m to 2 m causes a previously predictable change in the distributions of life and failure times, which does not explicitly prove that this will affect the final results of positioning accuracy calculations;
- Similarly, as in the case of GPS and DGPS systems, EGNOS exhibits similar properties when it comes to fit between the normal distribution and the empirical data.

The next research stage was to determine the failure and renewal rates of the renewal process. Both of these values made it possible to determine the course of the availability function calculated for an arbitrary position error.

Finally, to present the position error distribution functions of the GPS, DGPS and EGNOS systems, repeated calculations of the probability value corresponding to a given position error value ware performed. It has been assumed that to determine the distribution function, the position error value was increased from 0, every 0.1 m, up to 4 m. The course of distribution functions calculated based on the reliability model is presented in Figure 9.

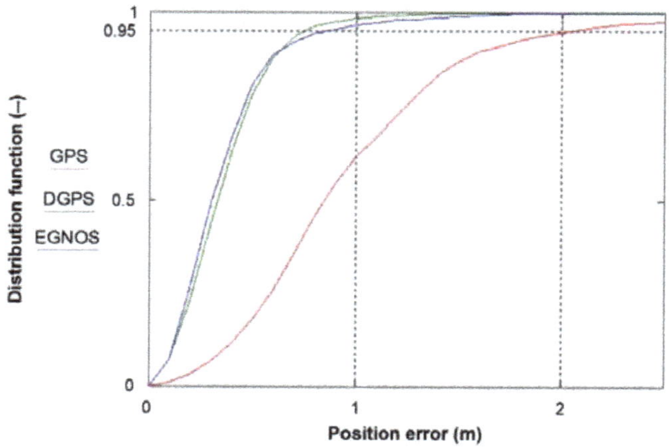

Figure 9. Position error distribution functions of the GPS, DGPS and EGNOS systems calculated using the reliability model.

Please note that the DGPS and EGNOS systems achieved positioning accuracies below 1 m ($p = 0.95$) on a very large and probably representative sample. Therefore, it is concluded that both systems can be successfully used in navigation applications requiring positioning accuracies of 1 m ($p = 0.95$), even though the official system characteristics given in [57] are 3 m ($p = 0.95$) horizontally and 4 m ($p = 0.95$) vertically.

It needs to be particularly emphasised here that it is not the purpose of this article to establish the actual value of the position error of the studied systems, including the EGNOS system. This paper aims to propose a new reliability model which will allow, as an alternative to the classical method using the 2DRMS value, one to determine the navigation system positioning accuracy and the corresponding probability.

4. Discussion

The presented reliability model requires an assessment of its accuracy. This can only be carried out in relation to calculations performed directly on empirical (real) values. The most reliable method of determining the position error value larger than 95% of the population of the remaining errors is to calculate it by sorting the errors from the smallest to the largest. This method of error determination is used in several publications [30,32,33]. There is no doubt that with a very large number of measurements this method produces the most reliable results because it does not assume any statistical distribution of empirical position errors. Therefore, to assess which of the two models (classical or reliability) is closer to the true value, the R95 value obtained by the method of sorting the position errors was used.

There are two curves presented in Figure 10. The first curve is the empirical distribution function of the sorted position errors (red) from the GPS system, which takes the value of 2.039 m ($p = 0.95$). This is an R95 value that can be considered close to true due to a large number of measurements (168′286 fixes). The distribution function calculated from the proposed reliability model (green) reaches a value of 2.044 m ($p = 0.95$). In contrast, the 2DRMS value calculated in the classical way (blue) takes the value of 2.240 m ($p = 0.95$).

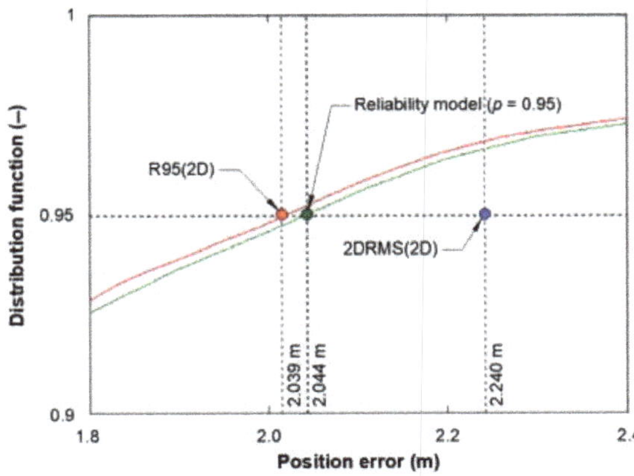

Figure 10. Comparison of three methods for calculating position error values larger than 95% of the population of remaining errors. Analysis of results for the GPS system.

From Figure 10, it follows that a much better approximation of the R95 value was obtained by applying the reliability model than by using the 2DRMS measure.

To verify the accuracy of the reliability method on systems other than GPS, an identical analysis was performed for DGPS and EGNOS systems. It needs to be emphasised that, because the calculations used a very large sample of 900′000 fixes, the results can be considered reliable and representative (Figure 11).

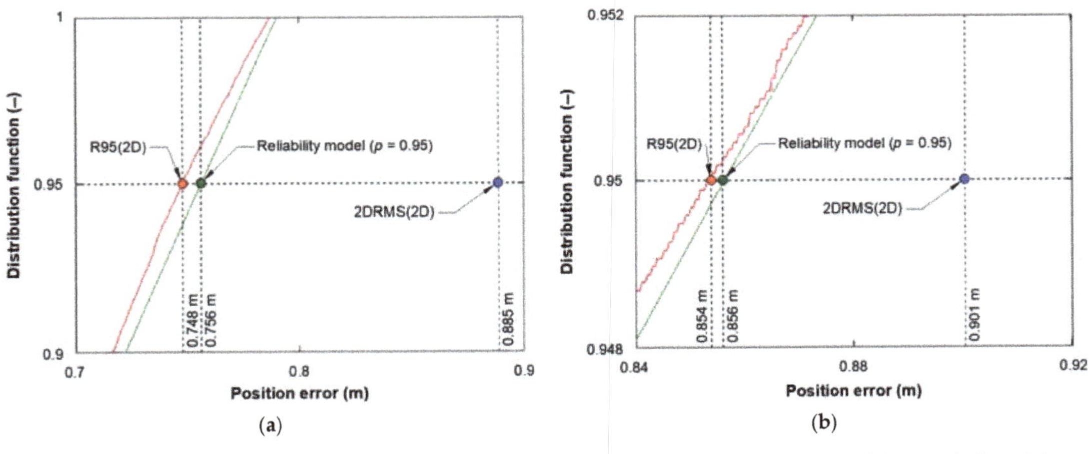

Figure 11. Comparison of three methods for calculating position error value larger than 95% of the population of the remaining errors. Analysis of results for: (**a**) DGPS; (**b**) EGNOS.

Empirical distribution function graphs (obtained by position error sorting) and those obtained based on the reliability model, for DGPS and EGNOS systems, prove that the reliability model provided a better approximation of the true value than the commonly used 2DRMS measure. For the DGPS system, the value considered to be the true R95 was 0.748 m. Calculation using the reliability model yielded 0.756 m (p = 0.95), while the 2DRMS measure was 0.885 m. The same is true for the EGNOS system. The value of the R95 was 0.854 m and the value calculated using the reliability model was 0.856 m, whereas the value of the 2DRMS was 0.901 m. Please note that many authors [29] have already noted the overestimation of the actual values by the 2DRMS measure, which is confirmed by the results presented in Figures 10 and 11. Therefore, it may be concluded that the proposed reliability model calculates the R95 value much more accurately than the classical model.

5. Conclusions

This paper proposes a new method, an alternative to the classical solution based on the 2DRMS measure, for determining navigation system positioning accuracy, which, in its essence, is based on the reliability model. The random variables are life and failure times in the positioning process, and not, as was the case in the classical model, the position errors. This method can be successfully used in assessing the suitability of a positioning system for a specific navigation application. It allows for the calculation of the system's position error with a probability of 95% more accurately than using the classical approach. The method was applied to determine the positioning accuracy of modern navigation systems: GPS (168′286 fixes), DGPS (900′000 fixes) and EGNOS (900′000 fixes). Although empirical distributions of life and failure times differ from the theoretical exponential distribution (for distribution functions with a probability below 0.9), the method provides high accuracy of the final results. An additional advantage of this method lies in the rather simple calculation algorithm.

What was a surprising result of this research was that the lognormal distribution presented a very good fit to the empirical data on life and failure times of all three systems (GPS, DGPS and EGNOS). This requires additional analysis in future research.

Tests conducted on very large measurement samples have proven that the proposed method provides a much more precise determination of positioning accuracy in navigation systems compared to the 2DRMS measure. Thanks to its application, it is possible to determine the position error distribution of the navigation system more precisely, as well as to indicate applications that can be used by this system, ensuring the safety of the navigation process.

It should be noted that the proposed method is not limited only to navigation positioning systems. With minor modifications, it can be successfully applied to the applications listed in Figure 1, which all differ in positioning accuracy and availability. The R95 measure, which was used to compare two models (classical or reliability), is a narrow scope of application. It is intended to determine the position error value for a strictly defined confidence level of 95%.

Funding: This research was funded from the statutory activities of Gdynia Maritime University, grant number WN/2021/PZ/05.

Institutional Review Board Statement: Not applicable.

Informed Consent Statement: Not applicable.

Data Availability Statement: Not applicable.

Conflicts of Interest: The author declares no conflict of interest.

References

1. Brodin, G.; Cooper, J.; Walsh, D.; Stevens, J. The Effect of Helicopter Rotors on GPS Signal Reception. *J. Navig.* **2005**, *58*, 433–450. [CrossRef]
2. Krasuski, K.; Ciećko, A.; Bakuła, M.; Wierzbicki, D. New Strategy for Improving the Accuracy of Aircraft Positioning Based on GPS SPP Solution. *Sensors* **2020**, *20*, 4921. [CrossRef]
3. Krasuski, K.; Savchuk, S. Accuracy Assessment of Aircraft Positioning Using the Dual-frequency GPS Code Observations in Aviation. *Commun. Sci. Lett. Univ. Zilina* **2020**, *22*, 23–30. [CrossRef]
4. Lachapelle, G.; Cannon, M.E.; Qiu, W.; Varner, C. Precise Aircraft Single-point Positioning Using GPS Post-mission Orbits and Satellite Clock Corrections. *J. Geod.* **1996**, *70*, 562–571. [CrossRef]
5. Ochieng, W.Y.; Sauer, K.; Walsh, D.; Brodin, G.; Griffin, S.; Denney, M. GPS Integrity and Potential Impact on Aviation Safety. *J. Navig.* **2003**, *56*, 51–65. [CrossRef]
6. Elhajj, M.; Ochieng, W. Impact of New GPS Signals on Positioning Accuracy for Urban Bus Operations. *J. Navig.* **2020**, *73*, 1284–1305. [CrossRef]
7. MacLean, G. Weak GPS Signal Detection in Animal Tracking. *J. Navig.* **2009**, *62*, 1–21. [CrossRef]

8. Naranjo, J.E.; Jiménez, F.; Aparicio, F.; Zato, J. GPS and Inertial Systems for High Precision Positioning on Motorways. *J. Navig.* **2009**, *62*, 351–363. [CrossRef]
9. Ojeda, L.; Borenstein, J. Non-GPS Navigation for Security Personnel and First Responders. *J. Navig.* **2007**, *60*, 391–407. [CrossRef]
10. Sun, Q.C.; Odolinski, R.; Xia, J.C.; Foster, J.; Falkmer, T.; Lee, H. Validating the Efficacy of GPS Tracking Vehicle Movement for Driving Behaviour Assessment. *Travel Beh. Soc.* **2017**, *6*, 32–43. [CrossRef]
11. Bhatti, J.; Humphreys, T.E. Hostile Control of Ships via False GPS Signals: Demonstration and Detection. *Navig. J. Inst. Navig.* **2017**, *64*, 51–66. [CrossRef]
12. Glomsvoll, O.; Bonenberg, L.K. GNSS Jamming Resilience for Close to Shore Navigation in the Northern Sea. *J. Navig.* **2017**, *70*, 33–48. [CrossRef]
13. Han, J.; Park, J.; Kim, J.; Son, N.-S. GPS-less Coastal Navigation Using Marine Radar for USV Operation. *IFAC-Papers OnLine* **2016**, *49*, 598–603. [CrossRef]
14. Ramesh, R.; Jyothi, V.B.N.; Vedachalam, N.; Ramadass, G.A.; Atmanand, M.A. Development and Performance Validation of a Navigation System for an Underwater Vehicle. *J. Navig.* **2016**, *69*, 1097–1113. [CrossRef]
15. Specht, C. Radio Navigation Systems: Definitions and Classifications. *J. Navig.* **2021**, *74*, 945–954. [CrossRef]
16. EC. *European Radio Navigation Plan*; Version 1.1; EC: Luxembourg, 2018.
17. GLA. *GLA Radio Navigation Plan*; GLA: London, UK, 2007.
18. SMA. *Swedish Radio Navigation Plan, Policy and Plans*; SMA: Norrköping, Sweden, 2009.
19. U. S. DoD. *2019 Federal Radionavigation Plan*; U.S. DoD: Springfield, VA, USA, 2019.
20. GSA. *Report on Rail User Needs and Requirements*; Version 1.0; GSA: Prague, Czech Republic, 2018.
21. IALA. *NAVGUIDE 2018 Marine Aids to Navigation Manual*, 8th ed.; IALA: Saint-Germain-en-Laye, France, 2018.
22. ICAO. *Convention on International Civil Aviation of 7th December 1944*; ICAO: Montreal, QC, Canada, 1944.
23. IHO. *IHO Standards for Hydrographic Surveys; Special Publication No. 44*, 6th ed.; IHO: Monaco, Monaco, 2020.
24. Reid, T.G.R.; Houts, S.E.; Cammarata, R.; Mills, G.; Agarwal, S.; Vora, A.; Pandey, G. Localization Requirements for Autonomous Vehicles. In Proceedings of the WCX SAE World Congress Experience 2019, Detroit, MI, USA, 9–11 April 2019.
25. Specht, C. *System GPS*; Bernardinum Publishing House: Pelplin, Poland, 2007. (In Polish)
26. Specht, M. Method of Evaluating the Positioning System Capability for Complying with the Minimum Accuracy Requirements for the International Hydrographic Organization Orders. *Sensors* **2019**, *19*, 3860. [CrossRef]
27. Bowditch, N. *American Practical Navigator*, 2019 ed.; Paradise Cay Publications: Blue Lake, CA, USA, 2019; Volumes 1–2.
28. U.S. DoD. *Global Positioning System Standard Positioning Service Signal Specification*, 1st ed.; U.S. DoD: Arlington County, VA, USA, 1993.
29. Van Diggelen, F. GPS Accuracy: Lies, Damn Lies, and Statistics. *GPS World* **1998**, *9*, 1–6.
30. Specht, C.; Pawelski, J.; Smolarek, L.; Specht, M.; Dąbrowski, P. Assessment of the Positioning Accuracy of DGPS and EGNOS Systems in the Bay of Gdansk Using Maritime Dynamic Measurements. *J. Navig.* **2019**, *72*, 575–587. [CrossRef]
31. Li, X.; Zhang, X.; Ren, X.; Fritsche, M.; Wickert, J.; Schuh, H. Precise Positioning with Current Multi-constellation Global Navigation Satellite Systems: GPS, GLONASS, Galileo and BeiDou. *Sci. Rep.* **2021**, *5*, 8328. [CrossRef]
32. Specht, C.; Dąbrowski, P.S.; Pawelski, J.; Specht, M.; Szot, T. Comparative Analysis of Positioning Accuracy of GNSS Receivers of Samsung Galaxy Smartphones in Marine Dynamic Measurements. *Adv. Space Res.* **2019**, *63*, 3018–3028. [CrossRef]
33. Szot, T.; Specht, C.; Specht, M.; Dabrowski, P.S. Comparative Analysis of Positioning Accuracy of Samsung Galaxy Smartphones in Stationary Measurements. *PLoS ONE* **2019**, *14*, e0215562. [CrossRef]
34. NovAtel Positioning Leadership. GPS Position Accuracy Measures. Available online: https://www.novatel.com/assets/Documents/Bulletins/apn029.pdf (accessed on 2 November 2021).
35. Specht, M. Consistency Analysis of Global Positioning System Position Errors with Typical Statistical Distributions. *J. Navig.* **2021**, 1–18. [CrossRef]
36. Specht, M. Consistency of the Empirical Distributions of Navigation Positioning System Errors with Theoretical Distributions—Comparative Analysis of the DGPS and EGNOS Systems in the Years 2006 and 2014. *Sensors* **2021**, *21*, 31. [CrossRef]
37. Specht, M. Statistical Distribution Analysis of Navigation Positioning System Errors—Issue of the Empirical Sample Size. *Sensors* **2020**, *20*, 7144. [CrossRef]
38. Cardano, G. *The Book on Games of Chance*; Princeton University Press: Princeton, NJ, USA, 1953.
39. Bellhouse, D. Decoding Cardano's Liber de Ludo Aleae. *Hist. Math.* **2005**, *32*, 180–202. [CrossRef]
40. Huygens, C. De Ratiociniis in Ludo Aleae. In *Francisci a Schooten Exercitationum Mathematicarum libri Quinque*; van Schooten, F., Ed.; Elsevier: Leiden, Netherlands, 1657; pp. 517–524.
41. Laplace, P.S. *Théorie Analytique des Probabilités*; Courcier: Paris, France, 1812. (In French)
42. Kolmogorov, A. Sulla Determinazione Empirica di una Legge di Distribuzione. *Giornale dell'Istituto Italiano degli Attuari* **1933**, *4*, 83–91. (In Italian)
43. Montgomery, D.C.; Runger, G.C. *Applied Statistics and Probability for Engineers*, 5th ed.; John Wiley & Sons, Inc: Hoboken, NJ, USA, 2010.
44. Hong, Y. *Probability and Statistics for Economists*; World Scientific: Singapore, 2017.
45. Baron, M. *Probability and Statistics for Computer Scientists*, 3rd ed.; CRC Press: Boca Raton, FL, USA, 2019.

46. Mertikas, S.P. Error Distributions and Accuracy Measures in Navigation: An Overview. Available online: https://unbscholar.lib.unb.ca/islandora/object/unbscholar%3A8708 (accessed on 2 November 2021).
47. Kendall, M.G.; Buckland, W.R. *A Dictionary of Statistical Terms*, 4th ed.; Longman: Harlow, UK, 1982.
48. Feller, W. *An Introduction to Probability Theory and its Applications*, 3rd ed.; John Wiley & Sons Inc: Hoboken, NJ, USA, 1968; Volume 1.
49. Specht, C.; Rudnicki, J. A Method for the Assessing of Reliability Characteristics Relevant to an Assumed Position-fixing Accuracy in Navigational Positioning Systems. *Pol. Marit. Res.* **2016**, *23*, 20–27. [CrossRef]
50. Barlow, R.E.; Proschan, F. *Statistical Theory of Reliability and Life Testing: Probability Models*; Holt, Rinehart and Winston: New York, NY, USA, 1974.
51. Specht, C. Availability, Reliability and Continuity Model of Differential GPS Transmission. *Annu. Navig.* **2003**, *5*, 1–85.
52. U.S. DoD. *Global Positioning System Standard Positioning Service Performance Standard*, 5th ed.; U.S. DoD: Arlington County, VA, USA, 2020.
53. IALA. *Performance and Monitoring of DGNSS Services in the Frequency Band 283.5-325kHz*; IALA: Saint-Germain-en-Laye, France, 2015.
54. Ketchum, R.L.; Lemmon, J.J.; Hoffman, J.R. *Site Selection Plan and Installation Guidelines for a Nationwide Differential GPS Service*; ITS: Boulder, CO, USA, 1997.
55. GSA. *EGNOS Open Service (OS) Service Definition Document*; Version 2.3; GSA: Prague, Czech Republic, 2017.
56. Keefe, R.F.; Wempe, A.M.; Becker, R.M.; Zimbelman, E.G.; Nagler, E.S.; Gilbert, S.L.; Caudill, C.C. Positioning Methods and the Use of Location and Activity Data in Forests. *Forests* **2019**, *10*, 458. [CrossRef]
57. EC-DG Enterprise and Industry. *EGNOS Safety of Life (SoL) Service Definition Document*; Version 3.3; EC-DG Enterprise and Industry: Brussels, Belgium; Luxembourg, 2019.

remote sensing

MDPI

Article

GNSS-IR Snow Depth Retrieval from Multi-GNSS and Multi-Frequency Data

Jinsheng Tu [1], Haohan Wei [2,*], Rui Zhang [3,4], Lei Yang [5], Jichao Lv [3], Xiaoming Li [6], Shihai Nie [1], Peng Li [1], Yanxia Wang [1] and Nan Li [1]

1 College of Geographic Information and Tourism, Chuzhou University, Chuzhou 239000, China; tujinsheng@chzu.edu.cn (J.T.); nsh1017@chzu.edu.cn (S.N.); lipeng@chzu.edu.cn (P.L.); lidarcenter@chzu.edu.cn (Y.W.); linan@chzu.edu.cn (N.L.)
2 College of Civil Engineering, Nanjing Forestry University, Nanjing 210037, China
3 Faculty of Geosciences and Environmental Engineering, Southwest Jiaotong University, Chengdu 611756, China; zhangrui@swjtu.edu.cn (R.Z.); lvjichao@my.swjtu.edu.cn (J.L.)
4 State-Province Joint Engineering Laboratory of Spatial Information Technology for High-Speed Railway Safety, Southwest Jiaotong University, Chengdu 611756, China
5 College of Information Science and Engineering, Shandong Agricultural University, Tai'an 271018, China; yanglei@sdau.edu.cn
6 College of Surveying and Geoinformatics, Tongji University, Shanghai 200092, China; lxmch@tongji.edu.cn
* Correspondence: weihaohan@njfu.edu.cn; Tel.: +86-139-5185-6853

Citation: Tu, J.; Wei, H.; Zhang, R.; Yang, L.; Lv, J.; Li, X.; Nie, S.; Li, P.; Wang, Y.; Li, N. GNSS-IR Snow Depth Retrieval from Multi-GNSS and Multi-Frequency Data. *Remote Sens.* **2021**, *13*, 4311. https://doi.org/10.3390/rs13214311

Academic Editors: Serdjo Kos, José Fernández and Juan F. Prieto

Received: 3 October 2021
Accepted: 25 October 2021
Published: 26 October 2021

Abstract: Global navigation satellite system interferometric reflectometry (GNSS-IR) represents an extra method to detect snow depth for climate research and water cycle managing. However, using a single frequency of GNSS-IR for snow depth retrieval is often found to be challenging when attempting to achieve a high spatial and temporal sensitivity. To evaluate both the capability of the GNSS-IR snow depth retrieved by the multi-GNSS system and multi-frequency from signal-to-noise ratio (SNR) data, the accuracy of snow depth retrieval by different frequency signals from the multi-GNSS system is analyzed, and a joint retrieval is carried out by combining the multi-GNSS system retrieval results. The SNR data of the global positioning system (GPS), global orbit navigation satellite system (GLONASS), Galileo satellite navigation system (Galileo), and BeiDou navigation satellite system (BDS) from the P387 station of the U.S. Plate Boundary Observatory (PBO) are analyzed. A Lomb–Scargle periodogram (LSP) spectrum analysis is used to compare the difference in reflector height between the snow-free and snow surfaces in order to retrieve the snow depth, which is compared with the PBO snow depth. First, the different frequency retrieval results of the multi-GNSS system are analyzed. Then, the retrieval accuracy of the different GNSS systems is analyzed through multi-frequency mean fusion. Finally, the joint retrieval accuracy of the multi-GNSS system is analyzed through mean fusion. The experimental shows that the retrieval results of different frequencies of the multi-GNSS system have a strong correlation with the PBO snow depth, and that the accuracy is better than 10 cm. The multi-frequency mean fusion of different GNSS systems can effectively improve the retrieval accuracy, which is better than 7 cm. The joint retrieval accuracy of the multi-GNSS system is further improved, with a correlation coefficient (R) between the retrieval snow depth and the PBO snow depth of 0.99, and the accuracy is better than 3 cm. Therefore, using multi-GNSS and multi-frequency data to retrieve the snow depth has a good accuracy and feasibility.

Keywords: GNSS-IR; snow depth; signal-to-noise ratio; multi-GNSS; multi-frequency; mean fusion

1. Introduction

Snow is an important part of the land hydrological cycle and global climate system. Accurate real-time snow depth data are an important reference indicator for water resource management and climate disaster warning [1,2]. Among the existing snow depth detection methods, in situ snow sensor measurement lacks time resolution, and global navigation satellite system interferometric reflectometry (GNSS-IR), as a new microwave sensing technology, has proven to be able to realize snow depth detection [3]. GNSS-IR is a kind of satellite remote sensing technology that uses GNSS signals as the transmitting source to realize the retrieval of the physical parameters of surface targets by receiving and processing the interference effect of GNSS signals formed by direct and surface reflection [4–6]. At present, this technology is mainly employed to retrieve the soil moisture content (SMC), snow depth, and vegetation parameters [7–12].

In recent years, researchers have made remarkable achievements in GNSS-IR snow depth detection. Larson et al. first proposed to extend the application of signal-to-noise ratio (SNR) data observed by the traditional geodetic global positioning system (GPS) receiver in order to detect snow depth [3]. Larson et al. conducted snow depth retrieval using SNR data of a GPS L2C signal at multiple stations of the Plate Boundary Observatory (PBO) and further verified the feasibility of this technology [13]. Later, Larson et al. developed a snow depth retrieval algorithm based on GPS L1 SNR data, compared it with the snow depth results of L2C signal retrieval, and found that the accuracy was improved [14]. Tabibi et al. evaluated the value of GPS L5 SNR data in snow depth detection, compared the results with the L2C signal, and found that there was no detectable deviation in the L5 retrieval results [15]. Tabibi et al. proposed the global orbit navigation satellite system multipath reflectometry (GLONASS-MR) SNR retrieval model, which extended GPS-MR to multiple GNSS, and showed a strong correlation by comparing the retrieval results of GPS L2C and GLONASS R2-coarse acquisition (C/A) signals [16]. Later, Tabibi et al. used simulation and field measurements to evaluate the accuracy of GPS and GLONASS combined multi-GNSS-MR snow depth retrieval. At the same time, the variance factor was used to form the optimal multi-GNSS combined snow depth daily sequence retrieval. Compared with the single signal snow depth retrieval results, the accuracy was significantly improved [17]. Jin et al. used GPS L2P SNR data to retrieve the snow depth, and compared it with GPS L1 C/A code results, which showed a high correlation, indicating that using L2P SNR data can better estimate the snow depth [18]. Zhou et al. used GLONASS L1 SNR data for snow depth detection, and the accuracy reached centimeter-level and showed a strong correlation with the measured data [19]. Zhou et al. also studied the retrieval of different signal combinations and proposed using GPS L1, L2, and L5 signal multipath reflection and SNR combination for the retrieval of snow depth. The results showed that this method can be effectively used for snow depth detection [20].

The above scholars' research on snow depth retrieval is based on GPS and GLONASS observation data and has not been extended to other GNSS systems. To further analyze the potential of other GNSS systems in snow depth detection, Wang et al. used the SNR data of GPS L1 and BeiDou navigation satellite system (BDS) B1I signals to retrieve the snow depth, finally reaching an accuracy of 5 cm in a day [21]. Wang et al. used multi-GNSS system data to retrieve the snow depth and found that the trend of single signal retrieval results of multi-GNSS system constellations was in good agreement, except for the GPS precise code (P-code) signal. Then, the multi-GNSS system combination method based on robust regression was used to combine the signal retrieval between constellations. The results showed that the accuracy, availability, and time sampling of multi-GNSS system combination retrieval were improved [22].

From the current research status, snow depth retrieval is mainly concentrated in single or dual GNSS systems and single frequency SNR data, which is often found to be challenging when attempting to meet the accuracy and time resolution requirements of snow depth detection. Therefore, based on previous studies, this article conducts snow depth retrieval using multi-GNSS and multi-frequency SNR data. For the case that the antenna height of the GNSS receiver is unknown, the mean value of the multi-day Lomb-Scargle periodogram (LSP) spectrum analysis results in the snow-free surface is used as the initial reference reflector height of the multi-GNSS and multi-frequency GNSS-IR in the article. Then, the snow depth retrieval capability of the Galileo satellite navigation system (Galileo) and BDS multi-frequency signal, which are rarely used to retrieve the snow depth, except the GPS and GLONASS signal, are evaluated. A mean fusion of multi-frequency retrieval results of different GNSS systems is proposed to improve the accuracy compared with different frequencies of multi-GNSS system retrieval results. Finally, the multi-GNSS system retrieval results are fused to further evaluate the accuracy of the GNSS combination retrieval of snow depth. In this article, through the above process, the feasibility and accuracy of multi-GNSS and multi-frequency GNSS-IR snow depth retrieval are evaluated.

2. Materials and Methods

2.1. GNSS-IR Snow Depth Retrieval Principle

The snow depth retrieval method used in the article is based on SNR data processing of traditional geodetic GNSS receivers. The SNR is an indicator used to measure the signal strength of global navigation satellites, which is mainly affected by antenna gain, satellite transmission power, and multipath [23–25]. The multipath effect of the SNR decreases with increasing satellite elevation angle. The direct and surface reflected signals will have obvious interference effects at the receiver antenna when the satellite is at low elevation angle. At the same time, the frequency of the reflected signal will also change with the change in antenna height. The snow depth parameter can be obtained by comparing the vertical distance difference between the reflection surface and the receiver antenna phase center under snow-free and snow conditions.

Figure 1 shows that direct signal and reflected signal generate corresponding interference effects at the receiver to form a composite interference signal, which can be expressed as [24]:

$$\mathrm{SNR} = A_{\mathrm{d}}^2 + A_{\mathrm{r}}^2 + 2A_{\mathrm{d}}A_{\mathrm{r}}\cos\ \varphi, \tag{1}$$

where A_{d} and A_{r} are the amplitudes of the direct signal and the reflected signal, and φ is the difference between the phases of the direct signal and the phases of the reflected signal (the unit is rad), which can be expressed as [26]:

$$\varphi = \frac{4\pi h}{\lambda}\sin\ \theta, \tag{2}$$

where λ is the wavelength of the signal; θ is the elevation angle of the satellite; h is the vertical distance from the reflector to the antenna phase center; and the full text is uniformly called the reflector height.

Because the change in snow depth parameters is only related to the reflected signal in the composite SNR data, it is necessary to eliminate the direct signal in the composite SNR to obtain the reflected signal part. In the article, the composite SNR data are fitted by a cubic low-order polynomial, and the composite SNR is linearized before fitting [27]:

$$\mathrm{SNR(volts/volts)} = 10^{\frac{\mathrm{SNR(dB-Hz)}}{20}}, \tag{3}$$

After the trend of the direct signal sequence is fitted, the SNR sequence of the reflected signal that removes the influence of the direct signal can be obtained, which can be expressed as $\mathrm{SNR_r}$ [28]:

$$\mathrm{SNR_r} = A_{\mathrm{r}}\cos(\frac{4\pi h}{\lambda}\sin\ \theta + \varphi), \tag{4}$$

f is the signal frequency of the multipath effect part in the SNR. After simplification, the relationship reflector height and the satellite signal wavelength can be obtained as follows:

$$f = \frac{2h}{\lambda}\ h = \frac{\lambda f}{2}, \tag{5}$$

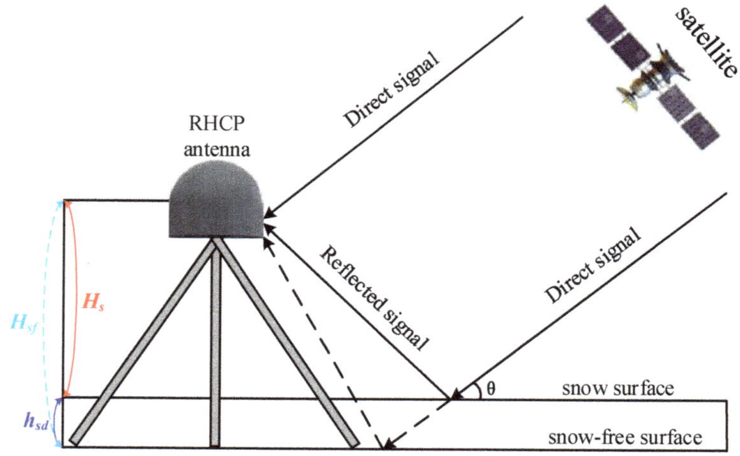

Figure 1. Schematic diagram of GNSS-IR snow depth retrieval. After the satellite sends the signal, the right-handed circular polarized (RHCP) antenna receives the direct signal and the surface reflected signal, and produces interference effect at the receiver. The snow surface reflector height (H_s) and snow-free surface reflector height (H_{sf}) are calculated, respectively, by analyzing the oscillation effect, and the snow depth (h_{sd}) is calculated by comparing the differences between them. θ is the elevation angle of the satellite.

In the article, the LSP method is used to analyze SNR_r, obtain f, and extract h [3,29,30]. As shown in Figure 1, the snow depth parameter is calculated by using the above method by comparing reflector height in the case of snow-free and snow surfaces:

$$h_{sd} = H_{sf} - H_s, \tag{6}$$

where H_{sf} and H_s are the snow-free and snow surface reflector height, and h_{sd} is the snow depth.

Snow depth retrieval is mainly determined by snow and snow-free surface reflector height. Wang et al. mentioned that non-planar reflecting surface and atmospheric refraction in GNSS-IR techniques can lead to errors in reflector height. These two errors are mainly considered in the sea level height retrieval, but rarely in the snow depth, as the snow depth changes slowly, and the reflector height to snow depth is usually smaller than that to the sea level [22]. The purpose of the article is to evaluate the ability of multi-GNSS and multi-frequency GNSS-IR snow depth retrieval, without focusing on the actual error impact caused by these two factors.

2.2. Data Source

2.2.1. Station Location and Surrounding Environment

This article selects GNSS observation data and snow depth data collected at the P387 station (Figure 2) of the PBO, which is located in Sisters, Oregon, the U.S., with an altitude of 963.041 m. The specific location of the station is 44.29675°N and 121.57446°W. The data of the P387 station is collected by SEPTENTRIO (SEPT) POLARX5 receivers and TRM59800.80 antennas pointing toward the zenith with a sampling frequency of 15 s. The terrain around this station is flat without trees, and the signal acquisition conditions are good.

Figure 2 shows that, regarding the P387 site location and surrounding environment, the terrain around P387 is flat. The ability of multi-GNSS and multi-frequency GNSS-IR snow depth retrieval is key to the article. In order to reduce the influence of terrain on the reflector height, the experimental region with small surface fluctuation is selected.

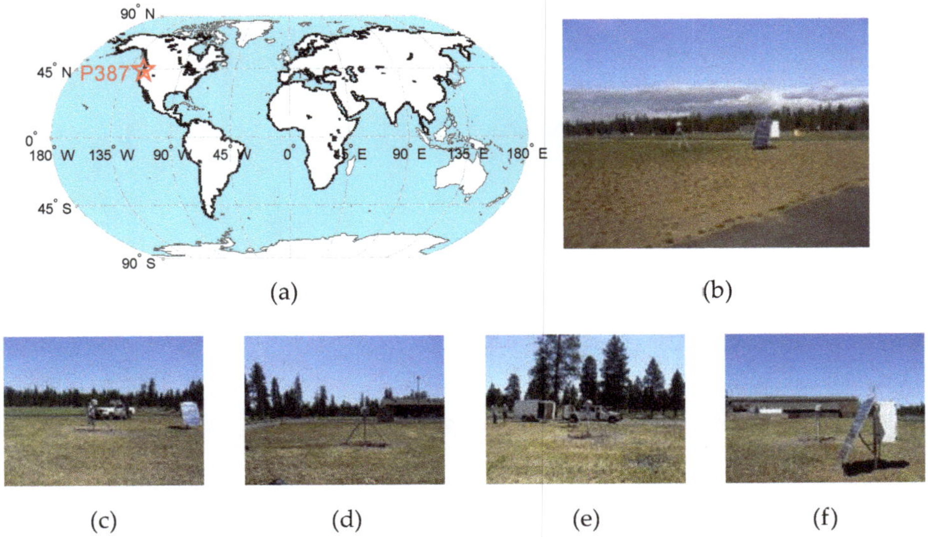

Figure 2. P387 station conditions: (**a**) station location in the world; (**b**) site vision; (**c**) site north; (**d**) site south; (**e**) site east; (**f**) site west.

At the same time, it can be seen that the vegetation around the P387 site is rare, and that the vegetation type is lawn. The surface can be defined as bare soil in winter snow stage. Therefore, with the melting of snow, the surface is gradually exposed, and the signal reflected by the surface is less affected by vegetation attenuation. The surface should be selected to be close to the bare soil in the subsequent analysis of snow-free reflection, so as to reduce the influence of vegetation error caused by the reflector height of snow-free and snow surface.

The roughness of snow surface will also lead to a retrieval error. In the article, the snow surface is regarded as a plane when extracting the reflector height, and it is not corrected temporarily.

For the above description, the calculation results of the reflector height in the snow-free and snow surfaces will not cause significant errors due to the change in the position of the mirror point. Therefore, the rise and fall stages of the satellite can be used as a signal source when selecting the experimental data. Nevertheless, it is necessary to determine whether it is a continuous observation period in advance in order to select the available arc segment.

2.2.2. Selection and Analysis of Experimental Data

Figure 3 shows the PBO snow depth data between days of year (DOYs) 024 and 065 of 2017. The article also selects GNSS observation data at this time, and the observation period selected in the article is when the snow has reached the deepest state, followed by the process of ablation. The feasibility and accuracy of snow depth retrieval using multi-GNSS and multi-frequency SNR data are verified by the change in the snow depth.

During the experiment period, when the GNSS signal is transmitted to the surface receiver, the signal will pass through the atmosphere, and a signal refraction effect will occur when passing through the troposphere. Williams et al. considered that tropospheric delay will cause height error in the obtained vertical reflection distance [31]. Aiming at this problem, this article gives the tropospheric delay information during the experiment, as shown in Figure 4.

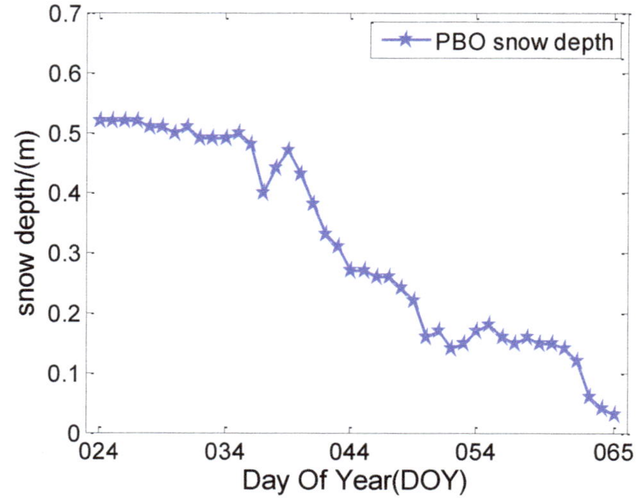

Figure 3. Snow depth of P387 in the experimental period.

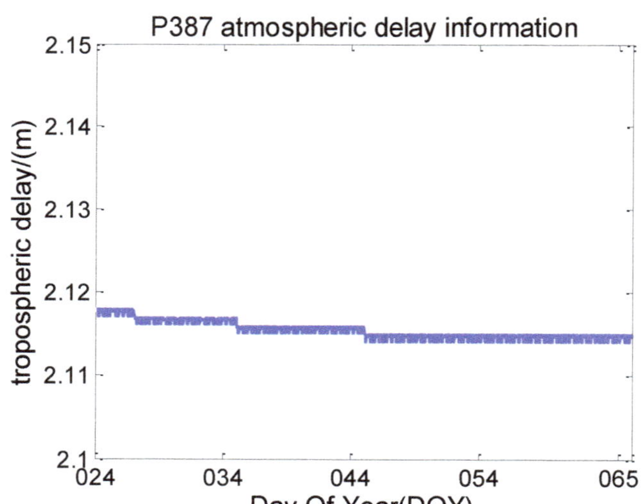

Figure 4. Atmospheric delay information of P387 station during the experiment.

As can be seen from Figure 4, the troposphere delay slowly changes during the 42 days of experiment, and is basically the same. He et al. corrected the tropospheric delay error in the process of retrieving coastal typhoon storm surge by using GNSS-IR signal, and the final accuracy was only improved by approximately 0.5 cm, which can ignore its influence [32]. Therefore, the error caused by tropospheric delay is not especially corrected in the process of snow depth retrieval.

2.2.3. Reflection Region Analysis

The effective reflection region of GNSS signal to the surface can be described by the first Fresnel reflection region, which is a group of ellipses related to the receiver antenna height, satellite azimuth, and satellite elevation angle.

Figure 5a shows the Fresnel reflection region of GPS G10 satellite with DOY of 024 in 2017 at P387 station. Assuming that receiver antenna height is 2 m, different color lines represent the reflection regions with different satellite elevation angles. With the increase in the elevation angle, the Fresnel reflection region will be smaller. The figure shows the Fresnel reflection region map of the satellite elevation angle of 5–25 degrees.

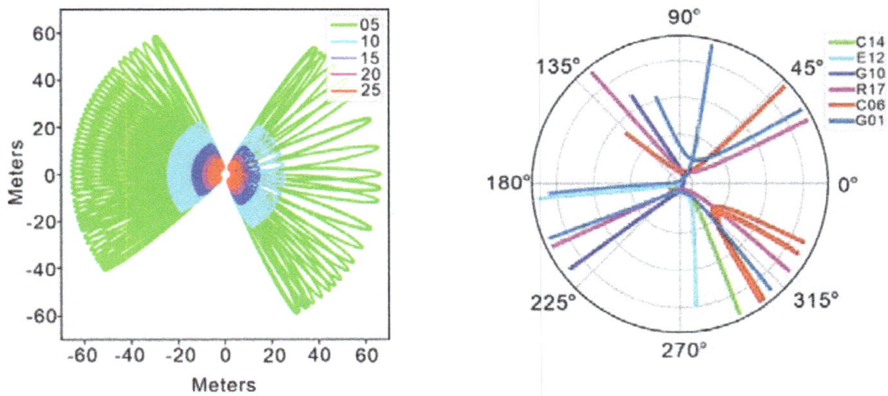

(a) (b)

Figure 5. Reflection region and reflection point track: (**a**) Fresnel reflection region around P387 station; (**b**) ground motion trajectory of reflection points around P387 station.

It can be seen that the effective reflection region is related to the satellite elevation angle. When the satellite elevation angle gradually increases, the effective reflection region will be decrease in size, and will gradually approach the receiver antenna. Large interference will occur when the satellite is at a low elevation angle. The larger the satellite elevation angle, the less affected the satellite is by the multipath of surrounding signals. Therefore, the signal data with low satellite elevation angle should be selected in the process of snow depth retrieval.

Figure 5b shows that the trajectory of reflection points changes with the change in the relative position between GNSS satellite and receiver, showing both different directions and distributions at different arcs and the ground reflection point trajectory formed by some satellites of the four GNSS systems. The combined signal sensing range of the four GNSS systems is significantly expanded, which can provide more data sources and wider sensing region, which is conducive to improving the time resolution of snow depth retrieval.

2.2.4. SNR Types

PBO provides multi-GNSS and multi-frequency SNR data in the observation data of receiver independent exchange format (RINEX) version 3.03. The SNR types of P387 station mainly include S1C, S1W, S2L, S2W, and S5Q for GPS; S1C and S2C for GLONASS; S1C, S5Q, S6C, S7Q, and S8Q for Galileo; S2I, S6I, and S7I for BDS; S1C, S2L, and S5Q for quasi-zenith satellite system (QZSS); and S1C and S5I for satellite-based augmentation system (SBAS). The data sampling interval is 15 s, and the specific information is shown in Table 1.

Table 1 shows that the P387 station provides SNR data of different signal types of six satellite systems. After reading the data, it is found that the QZSS and SBAS have fewer satellites and no available arc segments. Therefore, this article does not consider using QZSS and SBAS data for snow depth retrieval; instead, the SNR data of GPS, GLONASS, Galileo, and BDS are used. In the article, multi-GNSS and multi-frequency SNR data are used to retrieve the snow depth. In addition to GPS, the observation satellites will be extended to other systems, which is of great help in improving the retrieval accuracy of snow depth and expanding the observation range and time resolution.

Table 1. Multi-GNSS and multi-frequency SNR types and description information provided by the P387 station.

Satellite System	Frequency Band/Frequency (MHz)	Channel or Code	Carrier Phase	SNR Types	Yes/No Use
GPS	L1/1575.42	C/A	L1C	S1C	YES
		Z-tracking and similar (AS on)	L1W	S1W	
	L2/1227.60	L2C(L)	L2L	S2L	
		Z-tracking and similar (AS on)	L2W	S2W	
	L5/1176.45	Q	L5Q	S5Q	
GLONASS	G1/(1602 + k*9/16) K = −7 . . . + 12	C/A	L1C	S1C	YES
	G2/(1246 + k*7/16)	C/A	L2C	S2C	
Galileo	E1/1575.42	C	L1C	S1C	YES
	E5a/1176.45	Q	L5Q	S5Q	
	E6/1278.75	C	L6C	S6C	
	E5b/1207.14	Q	L7Q	S7Q	
	E5(E5a + E5b)/1191.795	Q	L8Q	S8Q	
BDS	B1/1561.098	I	L1I	S2I	YES
	B3/1268.52	I	L6I	S6I	
	B2/1207.140	I	L7I	S7I	
QZSS	L1/1575.42	C/A	L1C	S1C	NO
	L2/1227.60	L2C(L)	L2L	S2L	
	L5/1176.45	Q	L5Q	S5Q	
SBAS	L1/1575.42	C/A	L1C	S1C	NO
	L5/1176.45	I	L5I	S5I	

Note: When reading RINEX 3.03, BDS 1I/Q/X and 2I/Q/X can be regarded as the same as 2I/Q/X in the current RINEX standard, and the AS in Table 1 is anti-spoofing.

3. Experiment and Results

3.1. Experimental Technical Scheme

Figure 6 shows the experimental technical scheme of multi-GNSS and multi-frequency GNSS-IR snow depth retrieval. It can be seen that the technical route of the article can be divided into three parts: (1) GNSS-IR data preprocessing is carried out, where the SNR, pseudo-random noise (PRN), satellite elevation angle, azimuth angle, and other data parameters are extracted from the observation (OBS) file and navigation (NAV) file collected by GNSS receivers; (2) the LSP method is used to analyze both the reflector height of snow-free and snow surfaces and the difference between them in order to retrieve the snow depth; (3) the multi-GNSS and multi-frequency GNSS-IR snow depth retrieval results and PBO snow depth data are compared, and then the mean fusion analysis of the multi-GNSS and multi-frequency GNSS-IR snow depth retrieval results is carried out.

Figure 6 shows that RTKLIB software is used for data processing in the article. By reading the OBS file and NAV file in RINEX 3.03 format, the corresponding elevation angle, azimuth angle, SNR, and other data can be extracted. In order to select the satellites in the four GNSS systems that have available observation arc data in 42 days of the experimental stage, this article finally sets the G10 satellite of GPS, R17 satellite of GLONASS, E12 satellite of Galileo, and C14 satellite of BDS as the experimental data source satellites. In the article, the SNR data in the rising stage are mainly selected for processing. When the retrieval results show abnormal values, the SNR data in the falling stage can be selected as a supplement for retrieval. At the same time, the minimum elevation angle threshold of 5 degrees can also be increased accordingly in order to achieve a more practical retrieval value.

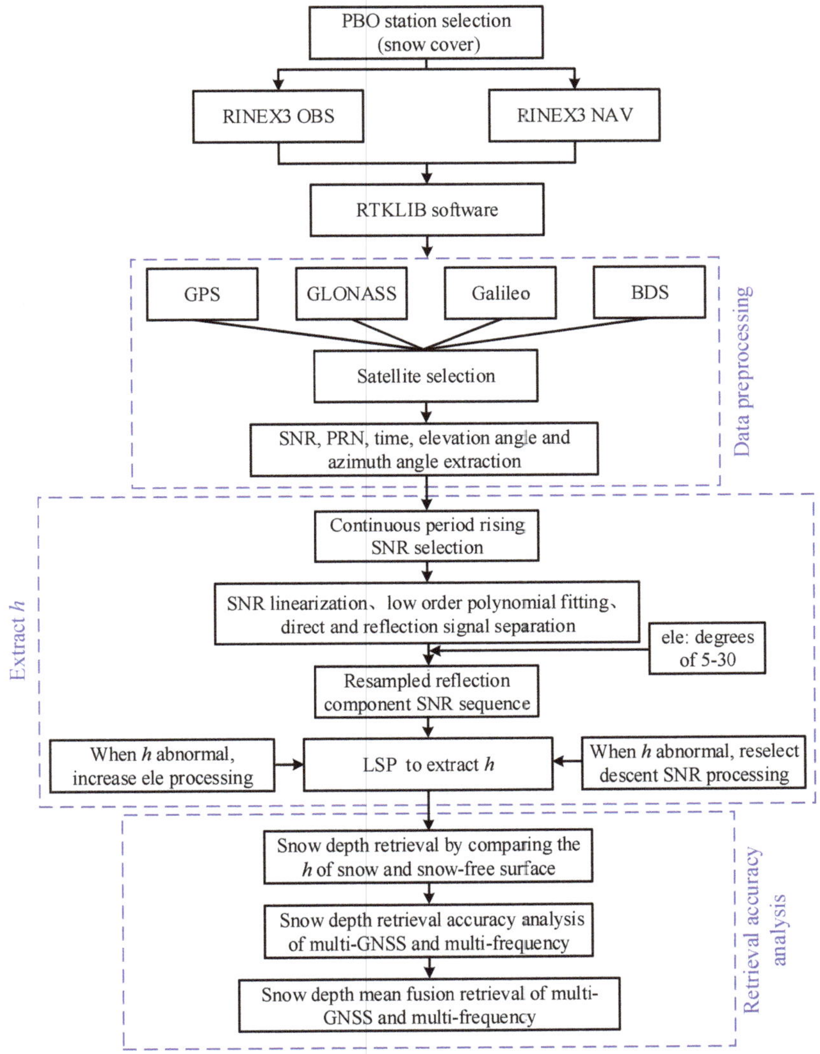

Figure 6. The technical process of multi-GNSS and multi-frequency GNSS-IR snow depth retrieval. Among the ele is the satellite elevation angle.

3.2. Extraction h

3.2.1. Multi-GNSS and Multi-Frequency SNR Sequence Extraction

In the article, multi-GNSS and multi-frequency SNR sequences are extracted by RTKLIB software. Figure 7 shows the rising SNR sequences of G10 (S1C, S2L, and S5Q), R17 (S1C and S2C), E12 (S1C, S5Q, S6C, S7Q, and S8Q), and C14 (S2I, S6I, and S7I) satellites in the four GNSS systems on DOY 024 in 2017. Through data preprocessing, it is found that the SNR values of G10 (S1W and S2W) are the same, and that the SNR difference with other signal frequencies is significant. Therefore, these two types of SNR data are not used for the experimental in the article. Figure 7 shows the SNR sequence change in the elevation angle from low to high.

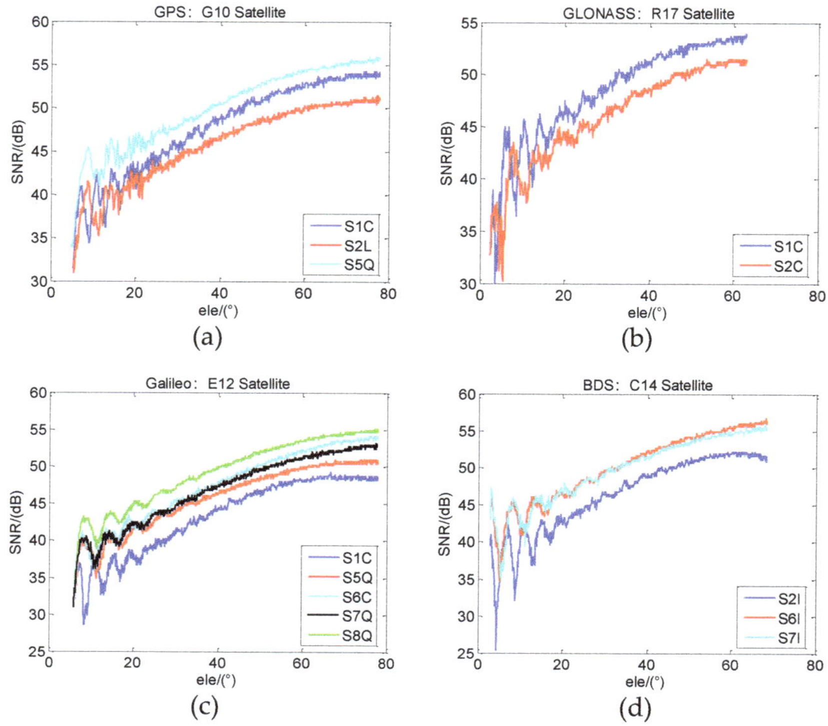

Figure 7. Multi-GNSS and multi-frequency SNR sequences: (**a**) DOY 024: GPS SNR sequence; (**b**) DOY 024: GLONASS SNR sequence; (**c**) DOY 024: Galileo SNR sequence; (**d**) DOY 024: BDS SNR sequence.

Figure 7 shows that the SNR sequence has strong oscillation at a low elevation angle. It is greatly affected by multipath at a low elevation angle, and the interference caused by the direct signal and the reflected signal is obvious. When increasing the elevation angle, the interference degree gradually decreases. Therefore, after removing the direct signal from the SNR sequence, this article mainly selects the SNR sequence of the reflected signal in the range of elevation angles of 5–30 degrees for snow depth retrieval. At the same time, it can be seen that the SNR sequences of the four GNSS systems at different frequencies have specific differences.

3.2.2. SNR Sequence Data Processing

In the article, the composite SNR sequence is linearized first, and then the linearized SNR sequence is fitted by a cubic low-order polynomial to obtain the direct signal part. The SNR sequence of the reflected signal is obtained by subtracting the composite SNR sequence from the direct signal part. Figure 8 shows the processing of GPS S1C SNR data.

3.2.3. LSP Analysis Results of the Snow Surface

Based on the LSP method analysis of the SNR sequence of the multi-GNSS and multi-frequency reflection signal of DOY 024 in 2017, the results are shown in Figure 9.

Figure 9 shows that there are some differences in the results of the multi-GNSS and multi-frequency LSP analysis. Specifically, the S2L and S5Q results in GPS are 1.335 m and 1.330 m, which are consistent and close to 0.1 m compared with the S1C results. The results of S1C and S2C in GLONASS are 1.230 m and 1.285 m, and the difference is 0.055 m. The results of S1C, S5Q, S6C, S7Q, and S8Q in Galileo are 1.355 m, 1.285 m, 1.300 m, 1.315 m, and 1.305 m. The difference between the maximum and the minimum is 0.070 m, and the deviation of S5Q, S6C, S7Q, and S8Q is slight. The results of S6I and S7I in BDS are 1.275 m and 1.310 m, respectively, and the difference is slight. The difference between the results of S6I and S2I is 0.095 m. From the above data, it can be seen that the LSP analysis results of different GNSS systems are different, and that the results of different frequencies in each GNSS system are also different, but the overall consistency is good. In the article, through the LSP method analysis, the reflector height in the multi-GNSS and multi-frequency can be obtained, and the LSP of the snow surface can be obtained.

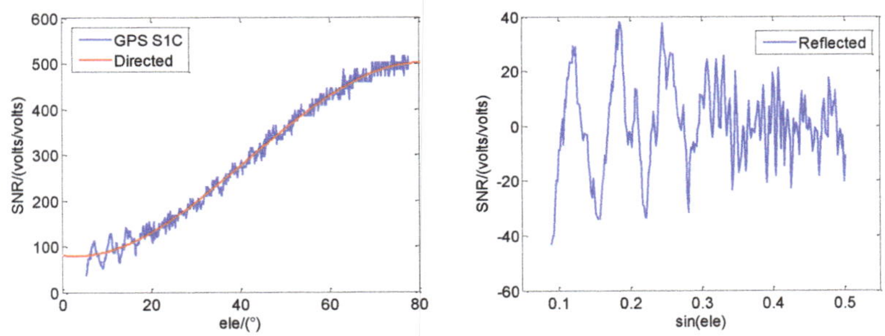

(a) (b)

Figure 8. SNR sequence data processing: (**a**) GPS S1C SNR and direct signal fitting; (**b**) reflected signal extraction.

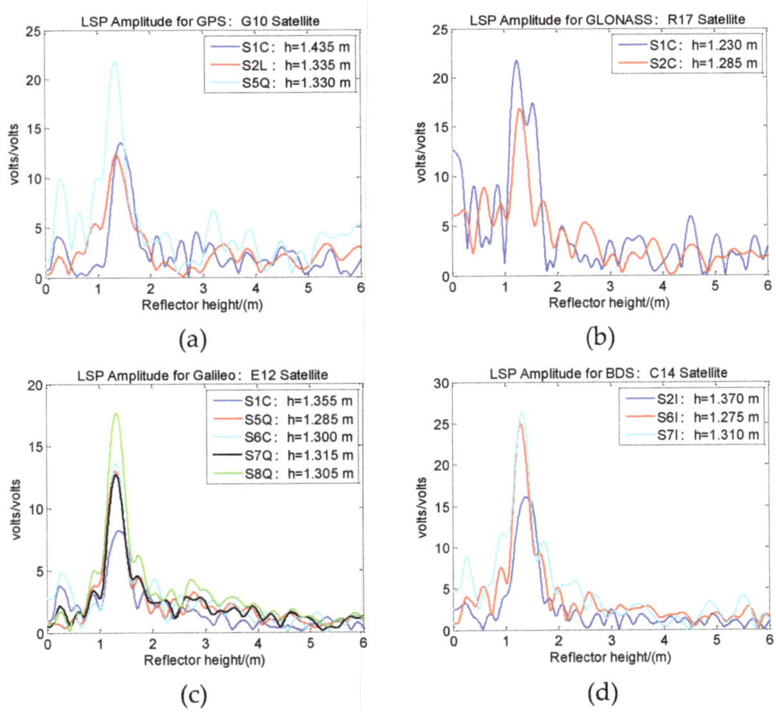

Figure 9. Multi-GNSS and multi-frequency LSP analysis results: (**a**) GPS LSP results; (**b**) GLONASS LSP results; (**c**) Galileo LSP results; (**d**) BDS LSP results.

3.2.4. LSP Analysis Results of the Snow-Free Surface

In order to weaken the difference between the multi-GNSS and multi-frequency LSP results, this article proposes that the snow-free surface reflector height is analyzed by the multi-GNSS and multi-frequency LSP results. Using the above method, the four-day accumulated data (satellites G10, R17, E12, and C14) of DOY 225–228 in 2017 were selected for processing; this period is about October and belongs to the defined bare soil period. This article selects the data of this period to process, which can reduce the influence of surface vegetation on signal propagation. The multi-GNSS and multi-frequency reference values were obtained by averaging the four-day LSP results.

Figure 10 shows that there are some differences in the results of multi-GNSS and multi-frequency LSP. There are some differences between the results of GPS S1C, S2L, and S5Q, but the results of a single frequency at 4 days are basically the same, with slight deviation. The Galileo S5Q, S6C, S7Q, and S8Q results are basically the same, and S1C has a specific difference, but the results of a single frequency at 4 days are basically the same. The deviation of the BDS S6I and S7I results is slight, and the deviation of the S2I results is significant at 225 and 228 on the annual accumulation day. The results of the GLONASS S1C analysis showed a significant variation at 227 accumulated days, and the rest showed good consistency. In the article, the snow depth is obtained through the comparison of the reflector height of different GNSS systems at different frequencies under snow and snow-free surfaces, which can better adapt to the use of multi-GNSS and multi-frequency GNSS-IR technology. The reference value of the reflector height in different GNSS systems at different frequencies is calculated, as shown in Table 2.

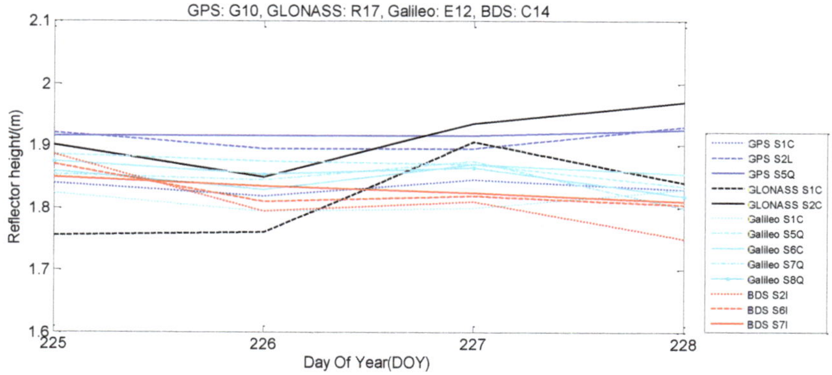

Figure 10. The snow-free surface reflector height reference value of multi-GNSS and multi-frequency LSP analysis results.

Table 2. Multi-GNSS and multi-frequency LSP mean values of the snow-free reflector height.

Satellite System	SNR Types	Mean LSP of 4 Days/m
GPS	S1C	1.834
	S2L	1.910
	S5Q	1.918
GLONASS	S1C	1.815
	S2C	1.914
Galileo	S1C	1.813
	S5Q	1.866
	S6C	1.854
	S7Q	1.844
	S8Q	1.854
BDS	S2I	1.810
	S6I	1.826
	S7I	1.830

Table 2 shows that the results of LSP at different frequencies of the four GNSS systems are more than 1800 m, which is more in line with the actual situation and can be used as the initial reference value of the reflector height of the snow-free surface.

3.3. GNSS-IR Snow Depth Retrieval Results

3.3.1. Multi-GNSS and Multi-Frequency GNSS-IR Snow Depth Retrieval Results

The snow depth is obtained by comparing and analyzing the difference in reflector height under snow-free and snow surfaces. The results of the different frequency retrievals of GPS, GLONASS, Galileo, and BDS are compared with the PBO snow depth data, and the results are shown in Figure 11.

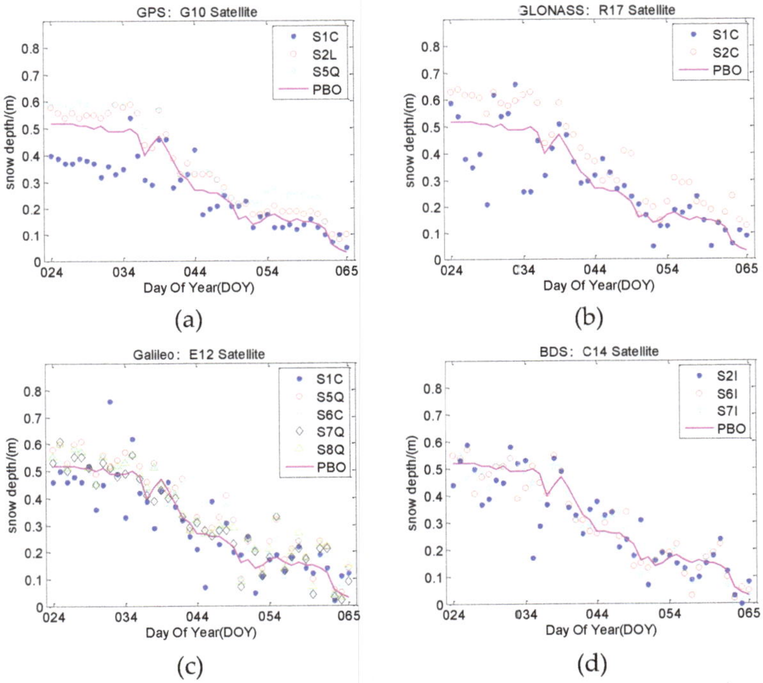

Figure 11. Comparison between multi-GNSS and multi-frequency GNSS-IR snow depth retrieval results and PBO snow depth: (**a**) GPS snow depth retrieval results; (**b**) GLONASS snow depth retrieval results; (**c**) Galileo snow depth retrieval results; (**d**) BDS snow depth retrieval results.

Figure 11 shows that the snow depth retrieval from the multi-GNSS and multi-frequency SNR data has a trend that is similar to that of the PBO snow depth data. Among them, the results of S1C in GPS are more biased than those of S2L and S5Q. The trend of the S1C results in GLONASS is worse than that of S2C. The results of S1C in Galileo are worse than those of S5Q, S6C, S7Q, and S8Q. The trend of the BDS S2I results is worse than that of the S6I and S7I results.

3.3.2. Mean Fusion of Multi-Frequency Retrieval Results in the Four GNSS Systems

The results of the mean fusion of different frequencies in the four GNSS systems are shown in Figure 12.

Figure 12 shows the consistency between the retrieval results of the four GNSS systems and the PBO snow depth, where the trend is more consistent than the previous single frequency retrieval results.

3.3.3. Mean Fusion Retrieval Results of Multi-GNSS System

After the mean fusion of multi-frequency in the four GNSS systems, the retrieval results of highly similar trends are obtained. To further improve the accuracy, the mean fusion of the snow depth retrieval results of the multi-GNSS system is carried out, and the final GNSS system retrieval results are obtained, as shown in Figure 13.

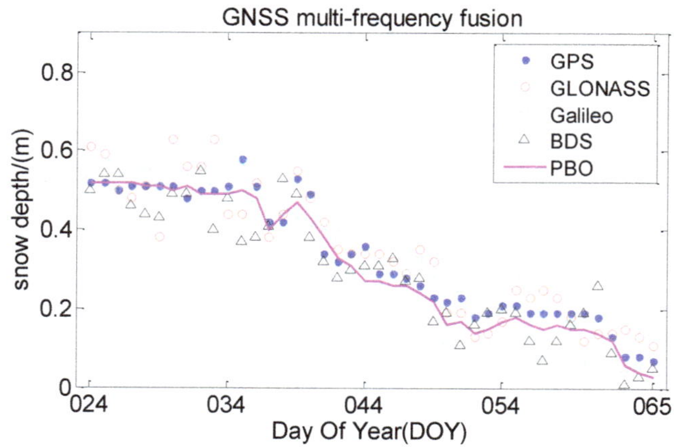

Figure 12. Mean fusion of multi-frequency retrieval results in the four GNSS systems.

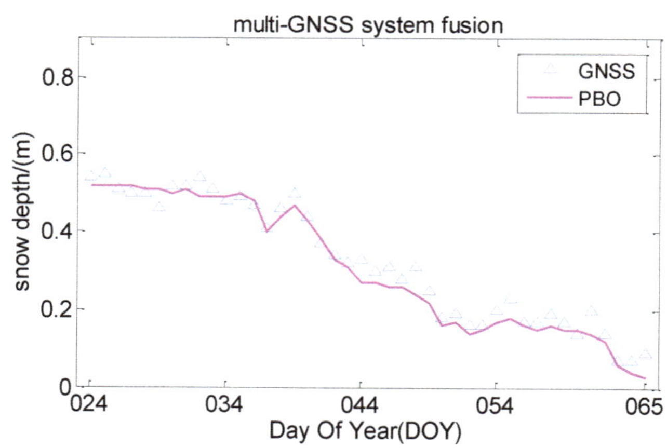

Figure 13. Mean fusion retrieval results of multi-GNSS system.

Figure 13 shows that the mean fusion results between the multi-GNSS system are in good agreement with the PBO snow depth results, and that the trend is basically the same.

4. Discussion

4.1. Accuracy Analysis between Multi-GNSS and Multi-Frequency GNSS-IR Snow Depth Retrieval Results and PBO Snow Depth

The above retrieval results are further analyzed, and the retrieval results of different frequency signals in multi-GNSS system are compared; the correlation coefficient (R) and root mean square error (RMSE) are shown in Figure 14.

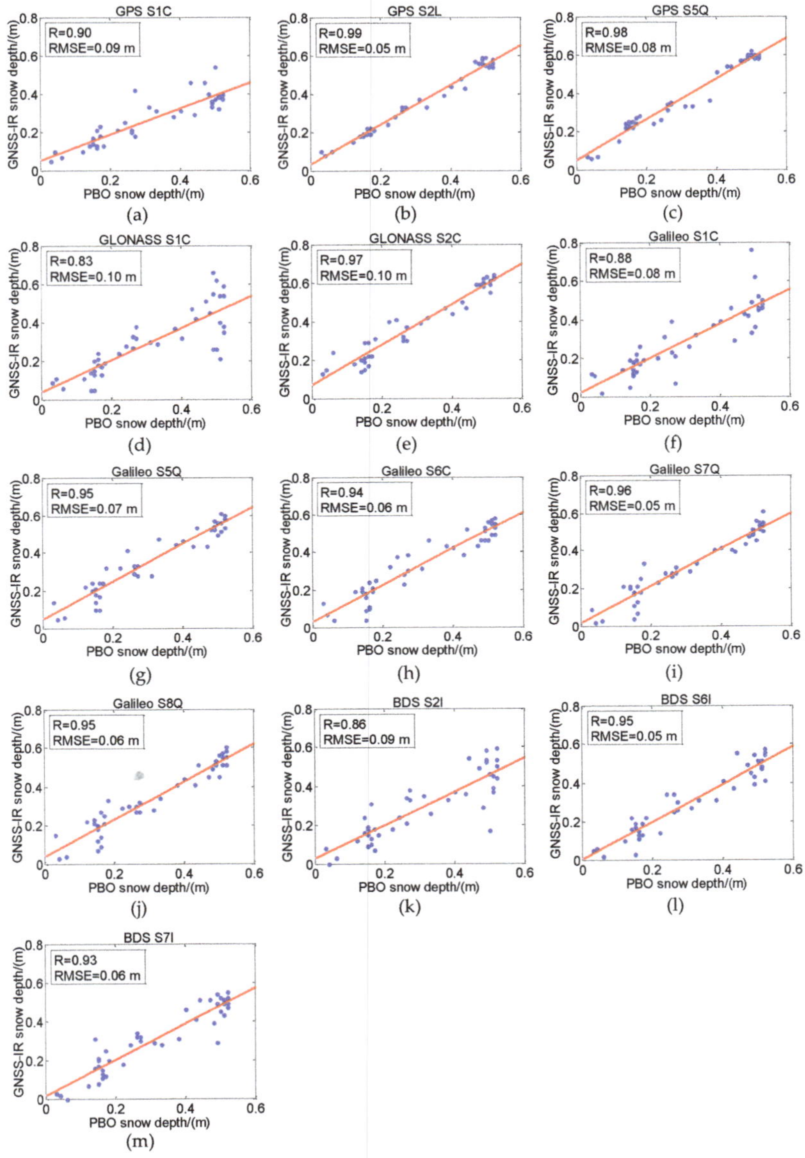

Figure 14. Correlation and RMSE between multi-GNSS and multi-frequency GNSS-IR snow depth retrieval results and PBO snow depth: (**a**) GPS S1C; (**b**) GPS S2L; (**c**) GPS S5Q; (**d**) GLONASS S1C; (**e**) GLONASS S2C; (**f**) Galileo S1C; (**g**) Galileo S5Q; (**h**) Galileo S6C; (**i**) Galileo S7Q; (**j**) Galileo S8Q; (**k**) BDS S2I; (**l**) BDS S6I; (**m**) BDS S7I.

Figure 14 shows the correlation of multi-GNSS and multi-frequency GNSS-IR snow depth retrieval results compared with the PBO snow depth. Table 3 shows the specific R and RMSE.

Table 3. R and RMSE between multi-GNSS and multi-frequency GNSS-IR snow depth retrieval results and PBO snow depth.

Satellite System	SNR Types	R	RMSE/m
GPS	S1C	0.90	0.09
	S2L	0.99	0.05
	S5Q	0.98	0.08
GLONASS	S1C	0.83	0.10
	S2C	0.97	0.10
Galileo	S1C	0.88	0.08
	S5Q	0.95	0.07
	S6C	0.94	0.06
	S7Q	0.96	0.05
	S8Q	0.95	0.06
BDS	S2I	0.86	0.09
	S6I	0.95	0.05
	S7I	0.93	0.06

Figure 14 and Table 3 show that the snow depth results of the GNSS-IR retrieval at different frequencies of the four GNSS systems have a strong correlation with the PBO snow depth. The R of the GPS S2L and S5Q results were 0.99 and 0.98, respectively, which are highly correlated, whereas the R of the S1C was 0.90, which is relatively low. The R of GLONASS S1C and S2C were 0.83 and 0.97, respectively. The correlation of S2C was strong and that of S1C was weak. The R of Galileo S5Q, S6C, S7Q, and S8Q results was approximately 0.95, and that of S1C was 0.88, which is rather weak. The R of BDS S6I and S7I results was 0.95 and 0.93, showing a strong correlation, and that of S2I was 0.86, which was rather weak. At the same time, the RMSE of the results of different frequencies of the four GNSS systems and PBO data were basically in the range of 5 cm to 10 cm, and the error was small. The above data show that the feasibility of retrieving the snow depth using multi-GNSS and multi-frequency GNSS-IR is high. At the same time, for GPS S1C, GLONASS S1C, Galileo S1C, and BDS S2I, the results in the four GNSS systems are rather weak. From the above results, in addition to GPS and GLONASS signals, the Galileo and BDS signals also have a good ability in snow depth retrieval.

4.2. Accuracy Analysis of Multi-Frequency Mean Fusion Results in the GNSS Systems

Mean fusion retrieval is carried out for multi-frequency retrieval results under the four GNSS systems; the accuracy between the retrieval results and PBO snow depth are further analyzed, as shown in Figure 15.

Figure 15 shows that the retrieval results of the four GNSS systems are obtained by averaging the snow depth results of the GNSS-IR retrieval at multi-frequency. The R between the GPS, GLONASS, Galileo, BDS, and PBO results was 0.99, 0.94, 0.97, and 0.95, respectively, showing strong correlations. At the same time, the RMSE of the four GNSS systems have been reduced to a certain extent, to basically in the range of 4 cm to 7 cm, indicating that the mean fusion of different frequency retrieval results in the four GNSS systems has a good effect on the improvement of retrieval accuracy, which can eliminate the weak correlation of GPS S1C, GLONASS S1C, Galileo S1C, and BDS S2I.

Compared with the snow depth of the different frequency signal in the multi-GNSS system retrieval results, the GNSS multi-frequency mean fusion method can effectively improve the retrieval accuracy. The specific improvement accuracy is shown in Table 4.

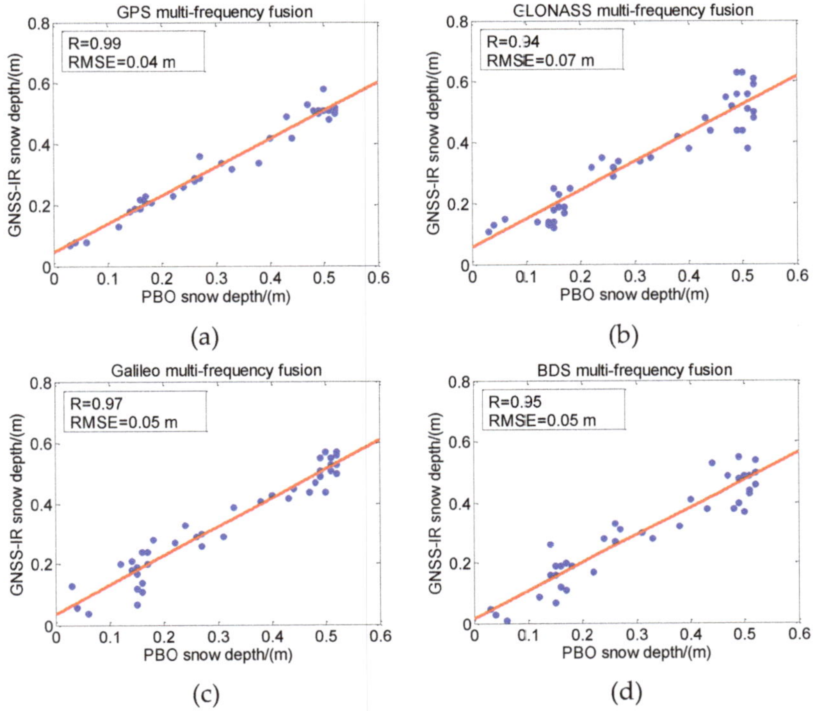

Figure 15. Mean fusion accuracy analysis of multi-frequency retrieval results in the four GNSS systems: (**a**) GPS multi-frequency fusion; (**b**) GLONASS multi-frequency fusion; (**c**) Galileo multi-frequency fusion; (**d**) BDS multi-frequency fusion.

Table 4. Comparison of different frequency signal in the multi-GNSS system retrieval results; multi-frequency mean fusion accuracy increases in R and decreases in the RMSE.

Satellite System	SNR Types	Increases in R	Decreases in the RMSE
GPS	S1C	9.1%	55.6%
	S2L	0.0%	20.0%
	S5Q	1.0%	50.0%
GLONASS	S1C	11.7%	30.0%
	S2C	−3.2%	30.0%
Galileo	S1C	9.3%	37.5%
	S5Q	2.1%	28.5%
	S6C	3.1%	16.7%
	S7Q	1.0%	0.0%
	S8Q	2.1%	16.7%
BDS	S2I	9.5%	44.4%
	S6I	0.0%	0.0%
	S7I	2.1%	16.7%

As can be seen from Table 4, the R of the GNSS multi-frequency mean fusion increases by 11.7%, which is higher than a single frequency in multi-GNSS system retrieval results, and the GLONASS S2C results decrease by 3.2%, but most of the results' correlations are improved. At the same time, the RMSE decreases by 55.6%, and though the retrieval accuracy of Galileo S7Q and BDS S6I is not improved, the other accuracy has been greatly improved. This shows that the mean fusion of the GNSS multi-frequency retrieval results can effectively improve the accuracy of the snow depth retrieval.

4.3. Accuracy Analysis of Mean Fusion Retrieval between Multi-GNSS System

The accuracy of multi-frequency fusion retrieval results has been improved to a certain extent. Furthermore, the fusion accuracy of the retrieval results of the four GNSS systems is analyzed, as shown in Figure 16.

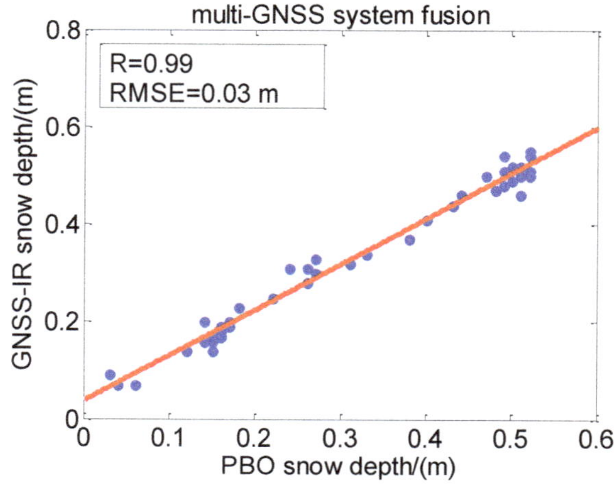

Figure 16. Comparison of the mean fusion results of multi-GNSS system with PBO snow depth.

Figure 16 shows that the R between the mean fusion results of the multi-GNSS system and PBO snow depth was 0.99, showing a strong correlation. At the same time, the RMSE was 3 cm, and the error was also significantly reduced, indicating the effectiveness of this method.

In order to further improve the combined retrieval accuracy of the GNSS system, the GNSS multi-frequency fusion results are further mean fused. The fusion results are compared with the four GNSS systems, and the results are shown in Table 5.

Table 5. Comparison of multi-frequency mean fusion; multi-GNSS system fusion accuracy increases in R and decreases in the RMSE.

Satellite System	Increases in R	Decreases in the RMSE
GPS	0.0%	25.0%
GLONASS	5.1%	57.1%
Galileo	2.0%	40%
BDS	4.0%	40%

It can be seen from Table 5 that the accuracy of the results after multi-GNSS system fusion is further improved. Compared with the results of GNSS multi-frequency fusion, the R of the four GNSS systems are improved, except for GPS. GLONASS has the highest increase of 5.1%. In terms of RMSE, GLONASS increased by 57.1%. The GPS is not improved on R, but it is improved by 25% on RMSE. It can be seen from the above results that the further fusion of the multi-frequency retrieval results of the four GNSS systems can effectively improve the retrieval accuracy. Therefore, the multi-GNSS system combined snow depth retrieval method has strong reliability.

5. Conclusions

Snow is an important indicator to measure the global hydrological cycle and climate. The accurate long-term monitoring of snow depth is helpful for water resource management and climate disaster warning, and has important application prospects. Based on the current GNSS-IR snow depth retrieval method, the temporal resolution is affected by the number of sky satellite arcs, so multi-GNSS and multi-frequency GNSS-IR are introduced as a supplement. The snow depth retrieval is carried out by using the multi-GNSS and multi-frequency SNR data. The snow depth parameters are obtained by comparing the LSP results of snow-free and snow surfaces. At the same time, the results are compared and analyzed in terms of the PBO snow depth, and the correlation and error analysis of the multi-GNSS and multi-frequency GNSS-IR mean fusion results are carried out. The following conclusions are drawn through experimental analysis:

(1) QZSS and SBAS systems in multi-GNSS and multi-frequency SNR data provided by PBO are not suitable for use due to the lack of observation arcs. The GPS S1W and S2W data values are the same, and the other frequency SNR data difference is too large and should not be used;
(2) The LSP results of the snow-free surface can be effectively used as the initial reflector height reference value. The snow depth results of multi-GNSS and multi-frequency GNSS-IR retrieval have a strong correlation with PBO snow depth data, and the RMSE of different frequency retrieval results in the multi-GNSS system is between 5 cm and 10 cm. The correlation between the retrieval results of the GPS L1, GLONASS G1, Galileo E1, and BDS B1 bands in the snow depth retrieval results is rather weak;
(3) The mean fusion of multi-frequency retrieval results in GPS, GLONASS, Galileo, and BDS can effectively improve the accuracy and solve the relatively weak results in some bands. The four GNSS systems retrieval results show a strong correlation, and the RMSE is between 4 cm and 7 cm. Comparing the different frequency signals in the multi-GNSS system retrieval results, the multi-frequency mean fusion increase by 11.7% in R and the RMSE decreases by 55.6%, which is the highest;
(4) The mean fusion accuracy of the retrieval results of the GPS, GLONASS, Galileo, and BDS is significantly improved. The R between the retrieval results and the PBO results is 0.99, and the retrieval accuracy is better than 3 cm, which significantly enhances the accuracy. In the comparison of the multi-frequency mean fusion, the multi-GNSS system fusion increases by 5.1% in R and the RMSE decreases by 57.1%, which is the highest.

Compared with a single GNSS-IR signal, multi-GNSS and multi-frequency GNSS-IR improves the accuracy, continuity, and time resolution of snow depth retrieval. A mean fusion of multi-GNSS and multi-frequency GNSS-IR retrieval results can further enhance the accuracy. With the development of global navigation systems, more types of signals and perfect constellation structures will be provided. Multi-GNSS and multi-frequency GNSS-IR will play a more critical role in the field of snow depth detection.

Author Contributions: Conceptualization, H.W., R.Z. and L.Y.; methodology, J.T.; formal analysis, J.L. and X.L.; data curation, S.N., P.L., Y.W. and N.L.; writing—original draft preparation, J.T.; writing—review and editing, H.W. and R.Z.; supervision, H.W. All authors have read and agreed to the published version of the manuscript.

Funding: This research was funded by the Jiangsu Agriculture Science and Technology Innovation Fund, grant number: CX (21) 3068 and the Anhui Educational Commission Key Project, grant number: KJ2020A00706.

Institutional Review Board Statement: Not applicable.

Informed Consent Statement: Not applicable.

Data Availability Statement: The GNSS site data were provided under the PBO Observation Program of the United States, and the measured snow depth data were obtained from https://data.unavco.org/archive/gnss/products (accessed on 21 October 2021).

Acknowledgments: We thank the PBO Observation Program of the United States for providing the GNSS data, and the University of Colorado for providing data for snow depth comparison analysis. We thank RTKLIB software for providing early SNR extraction. We also thank Roesler, C. and Larson, K.M for providing open access MATLAB code to generate reflector heights from GNSS SNR data (the paper that accompanies this code: Software tools for GNSS interferometric reflectometry (GNSS-IR)).

Conflicts of Interest: The authors declare no conflict of interest.

References

1. Walsh, J.E. Snow cover and atmospheric variability: Changes in the snow covering the earth's surface affect both daily weather and long-term climate. *Am. Sci.* **1984**, *72*, 50–57.
2. Rott, H.; Yueh, S.H.; Cline, D.W.; Duguay, C.; Essery, R.; Haas, C.; Heliere, F.; Kern, M.; Macelloni, G.; Malnes, E.; et al. Cold Regions Hydrology High-Resolution Observatory for Snow and Cold Land Processes. *IEEE Proc.* **2010**, *98*, 752–765. [CrossRef]
3. Larson, K.M.; Gutmann, E.D.; Zavorotny, V.U.; Braun, J.J.; Williams, M.W.; Nievinski, F.G. Can we measure snow depth with GPS receivers? *Geophys. Res. Lett.* **2009**, *36*, 876–880. [CrossRef]
4. Georgiadou, Y.; Kleusberg, A. On Carrier Signal Multipath Effects in Relative GPS Positioning. *Manuscr. Geod.* **1988**, *13*, 172–179.
5. Ge, L.; Han, S.; Rizos, C. Multipath Mitigation of Continuous GPS Measurements Using an Adaptive Filter. *GPS Solut.* **2000**, *4*, 19–30. [CrossRef]
6. Ray, J.K.; Cannon, M.E. Synergy between Global Positioning System Code, Carrier, and Signal-to-Noise Ratio Multipath Errors. *J. Guid. Control Dyn.* **2001**, *24*, 54–63. [CrossRef]
7. Larson, K.M.; Small, E.E.; Gutmann, E.D.; Bilich, A.L.; Braun, J.J.; Zavorotny, V.U. Use of GPS receivers as a soil moisture network for water cycle studies. *Geophys. Res. Lett.* **2008**, *35*, 1–5. [CrossRef]
8. Nievinski, F.G.; Larson, K.M. Inverse Modeling of GPS Multipath for Snow Depth Estimation—Part I: Formulation and Simulations. *IEEE Trans. Geosci. Remote Sens.* **2014**, *52*, 6555–6563. [CrossRef]
9. Nievinski, F.G.; Larson, K.M. Inverse Modeling of GPS Multipath for Snow Depth Estimation—Part II: Application and Validation. *IEEE Trans. Geosci. Remote Sens.* **2014**, *52*, 6564–6573. [CrossRef]
10. Wan, W.; Larson, K.M.; Small, E.E.; Chew, C.C.; Braun, J.J. Using geodetic GPS receivers to measure vegetation water content. *GPS Solut.* **2015**, *19*, 237–248. [CrossRef]
11. Larson, K.M.; Small, E.E. GPS ground networks for water cycle sensing. *IEEE Geosci. Remote Sens. Symp.* **2014**, 3822–3825.
12. Larson, K.M. GPS interferometric reflectometry: Applications to surface soil moisture, snow depth, and vegetation water content in the western United States. *Wiley Interdiscip. Rev. Water* **2016**, *3*, 775–787. [CrossRef]
13. Larson, K.M.; Nievinski, F.G. GPS snow sensing: Results from the EarthScope Plate Boundary Observatory. *GPS Solut.* **2013**, *17*, 41–52. [CrossRef]
14. Larson, K.M.; Small, E.E. Estimation of Snow Depth Using L1 GPS Signal-to-Noise Ratio Data. *IEEE J. Sel. Top. Appl. Earth Observ. Remote Sens.* **2017**, *9*, 4802–4808. [CrossRef]
15. Tabibi, S.; Nievinski, F.G.; Dam, T.V.; Monico, J.F.G. Assessment of modernized GPS L5 SNR for ground-based multipath reflectometry applications. *Adv. Space Res.* **2015**, *55*, 1104–1116. [CrossRef]
16. Tabibi, S.; Nievinski, F.G.; Dam, T.V. Multi-GNSS and multi-frequency SNR multipath reflectometry of snow depth, Trans EOS, G44A-07. In Proceedings of the AGU Fall Meeting Abstract, San Francisco, CA, USA, 17 December 2015.
17. Tabibi, S.; Nievinski, F.G.; Dam, T.V. Statistical Comparison and Combination of GPS, GLONASS, and Multi-GNSS Multipath Reflectometry Applied to Snow Depth Retrieval. *IEEE Trans. Geosci. Remote Sens.* **2017**, *55*, 3773–3785. [CrossRef]
18. Jin, S.; Qian, X.; Kutoglu, H. Snow Depth Variations Estimated from GPS-Reflectometry: A Case Study in Alaska from L2P SNR Data. *Remote Sens.* **2016**, *8*, 63. [CrossRef]
19. Zhou, W.; Liu, L.; Huang, L.; Li, J.; Chen, J.; Chen, F.D.; Xing, Y.; Liu, L. Monitoring snow depth based on the SNR signal of GLONASS satellites. *J. Remote Sens.* **2018**, *22*, 889–899. [CrossRef]
20. Zhou, W.; Liu, L.; Huang, L.; Yao, Y.; Chen, J.; Li, S. A new GPS SNR-Based Combination Approach for Land Surface Snow Depth Monitoring. *Sci. Rep.* **2019**, *9*, 3814. [CrossRef]
21. Wang, Z.; Liu, Z.; An, J.; Lin, G. Snow depth detection and error analysis derived from SNR of GPS and BDS. *Acta Geod. Cartogr. Sin.* **2018**, *47*, 8–16. [CrossRef]
22. Wang, X.; Zhang, S.; Wang, L.; He, X.; Zhang, Q. Analysis and combination of multi-GNSS snow depth retrievals in multipath reflectometry. *GPS Solut.* **2020**, *24*, 1–13. [CrossRef]
23. Bilich, A.; Larson, K.M.; Axelrad, P. Observations of signal-to-noise ratios (SNR) at geodetic GPS site CASA: Implications for phase multipath. *Proc. Cent. Eur. Geodyn. Seismol.* **2004**, *23*, 77–83.
24. Bilich, A.; Larson, K.M. Mapping the GPS multipath environment using the signal-to-noise ratio (SNR). *Radio Sci.* **2007**, *42*, RS6003. [CrossRef]
25. Zhang, S.; Peng, J.; Zhang, C.; Zhang, J.; Liu, Q. GiRsnow: An open-source software for snow depth retrievals using GNSS interferometric reflectometry. *GPS Solut.* **2021**, *25*, 1–8. [CrossRef]
26. Roussel, N.; Ramillien, G.; Frappart, F.; Darrozes, J.; Gay, A.; Biancale, R.; Striebig, N.; Hanquiez, V.; Bertin, X.; Allain, D. Sea level monitoring and sea state estimate using a single geodetic receiver. *Remote Sens. Environ.* **2015**, *171*, 261–277. [CrossRef]
27. Roesler, C.; Larson, K.M. Software tools for GNSS interferometric reflectometry (GNSS-IR). *GPS Solut.* **2018**, *22*, 1–10. [CrossRef]

28. Axelrad, P.; Larson, K.; Jones, B. Use of the correct satellite repeat period to characterize and reduce site-specific multipath errors. In Proceedings of the ION GNSS 2005, Long Beach, CA, USA, 13–16 September 2005; pp. 2638–2648.
29. Lomb, N.R. Least-squares frequency analysis of unequally spaced data. *Astrophys. Space Sci.* **1976**, *39*, 447–462. [CrossRef]
30. Scargle, J.D. Studies in astronomical time series analysis. II—Statistical aspects of spectral analysis of unevenly spaced data. *Astrophys. J.* **1982**, *263*, 835–853. [CrossRef]
31. Williams, S.D.P.; Nievinski, F.G. Tropospheric delays in ground-based GNSS multipath reflectometry—Experimental evidence from coastal sites. *J. Geophys. Res. Solid Earth.* **2017**, *122*, 2310–2327. [CrossRef]
32. He, X.; Wang, J.; Wang, X.; Song, M. Retrieval of coastal typhoon storm surge using multi-GNSS-IR. *Acta Geod. Cartogr. Sin.* **2020**, *49*, 1168–1178. [CrossRef]

remote sensing

MDPI

Article

Fractal Nature of Advanced Ni-Based Superalloys Solidified on Board the International Space Station

Vojislav Mitić [1,2,*], Cristina Serpa [3], Ivana Ilić [4], Markus Mohr [5] and Hans-Jörg Fecht [5]

1 Faculty of Electronic Engineering, University of Nis, 18000 Nis, Serbia
2 Institute Technical Sciences of SASA, 11000 Belgrade, Serbia
3 ISEL, Instituto Superior de Engenharia de Lisboa and CMAFcIO, Center for Mathematics, Fundamental Applications and Operational Research, 1749-016 Lisabon, Portugal; mcserpa@fc.ul.pt
4 Department for Mathematics and Informatics, Faculty of Medical Sciences, University of Nis, 18000 Nis, Serbia; ivana@medfak.ni.ac.rs
5 Institute of Functional Nanosystems, Ulm University, D-89069 Ulm, Germany; markus.mohr@uni-ulm.de (M.M.); hans.fecht@uni-ulm.de (H.-J.F)
* Correspondence: vmitic.d2480@gmail.com; Tel.: +381-63400250

Abstract: Materials science is highly significant in space program investigation, energy production and others. Therefore, designing, improving and predicting advanced material properties is a crucial necessity. The high temperature creep and corrosion resistance of Ni-based superalloys makes them important materials for turbine blades in aircraft engines and land-based power plants. The investment casting process of turbine blades is costly and time consuming, which makes process simulations a necessity. These simulations require fundamental models for the microstructure formation. In this paper, we present advanced analytical techniques in describing the microstructures obtained experimentally and analyzed on different sample's cross-sectional images. The samples have been processed on board the International Space Station using the MSL-EML device based on electromagnetic levitation principles. We applied several aspects of fractal analysis and obtained important results regarding fractals and Hausdorff dimensions related to the surface and structural characteristics of CMSX-10 samples. Using scanning electron microscopy (SEM), Zeiss LEO 1550, we analyzed the microstructure of samples solidified in space and successfully performed the fractal reconstruction of the sample's morphology. We extended the fractal analysis on the microscopic images based on samples solidified on earth and established new frontiers on the advanced structures prediction.

Keywords: Ni-based superalloys; international space station; microstructure; morphology; fractal reconstruction; Fractal Hausdorff dimension

Citation: Mitić, V.; Serpa, C.; Ilić, I.; Mohr, M.; Fecht, H.-J. Fractal Nature of Advanced Ni-Based Superalloys Solidified on Board the International Space Station. *Remote Sens.* **2021**, *13*, 1724. https://doi.org/10.3390/rs13091724

Academic Editor: Serdjo Kos

Received: 31 March 2021
Accepted: 23 April 2021
Published: 29 April 2021

1. Introduction

The superior creep resistance, high-temperature strength, and corrosion resistance make Nickel-based superalloys the material class of choice for high-performance applications, such as turbine blades of aircraft turbines. The precipitation of an ordered cubic gamma' phase within the disordered cubic gamma phase with a similar lattice constant leads to the combination of exceptional mechanical properties. While also Pt- and Pd-based superalloys exist (the platinum-group-metals possess a similar fcc structure), their higher costs make a broad commercial use still unfavorable. Common uses of Nickel-based superalloys are turbine components for aircraft and land-based turbines for power generation.

Manufacturing of turbine blades is usually performed by complex casting procedures, such as investment casting and directional solidification. That route generally achieves polycrystalline morphologies with equiaxed grains or grains aligned along the length of the blade. Furthermore, for advanced performance, single-crystal turbine blades are manufactured, using specially designed alloy compositions, such as CMSX-10. Single-crystalline components exhibit improved performance and lifetime compared with blades of polycrystalline morphology.

1.1. Some Previous Results of Fractal Technique Application on Ceramics Samples

Short Description of the Applied Technique for the Grain Cluster Shape Reconstruction

An image from Figure 1 was imported into a pptx file and a well-definedgrid inserted on it. Equally spaced yellow points were marked on a portion of the contour and the following image was obtained (see Figure 2). Then, an ordered list of yellow points was scaled and registered in a file. After a few simulations on the program Fractal Real Finder [1,2], we find that with $p = 10$ and $L = 2$, a sufficiently good fit is obtained ($p^L = 100$ points) (see the following plot in Figure 3).

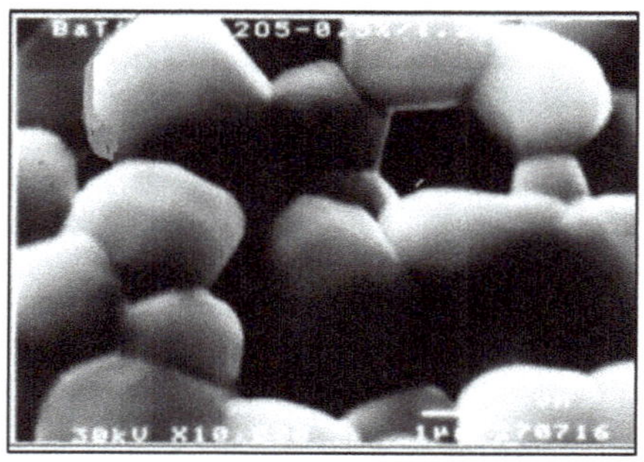

Figure 1. A part of the microstructure morphology of $BaTiO_3$ Ceramics sample.

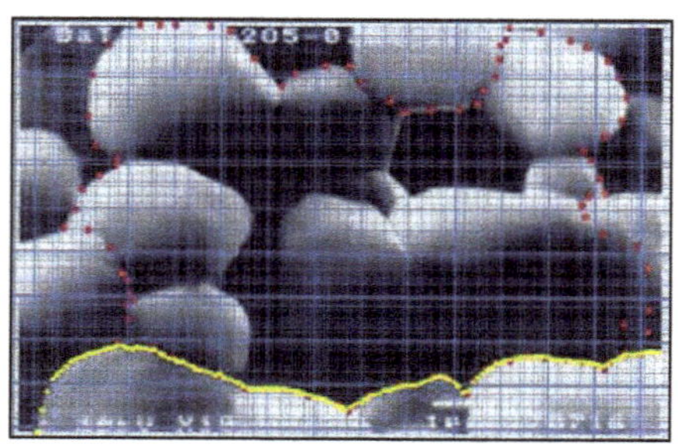

Figure 2. Successful reconstruction of $BaTiO_3$ sample structure by using fractal method.

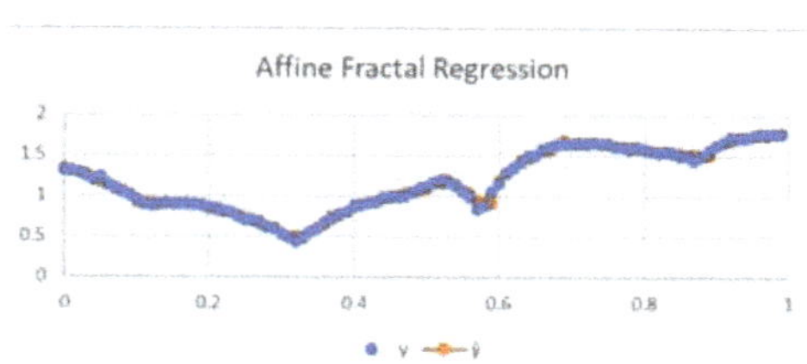

Figure 3. Affine Fractal Regression of the reconstruction of $BaTiO_3$ sample structure by using fractal method.

The statistical Kolmogorov-Smirnov test, as well as the respective plot, strongly confirms the good reconstruction of the original data. From the output of the program, the relevant fractal coefficients are 0.161007796 and 0.188832965.

With these estimated fractal coefficients, an upper estimative for the Hausdorff dimension (which will be denoted with D and explained more firmly in Section 2.1) was computed as the solution of the nonlinear equation:

$$8\left(\frac{1}{10}\right)^{D} + 0.161007796^{D} + 0.188832965^{D} = 1. \tag{1}$$

The calculated Hausdorff dimension is $D = 1.06557$. The result with fractal interpolation has fractal dimension estimated as $D = 1.40792$. Why these two methods give fractal dimensions so different? The difference between interpolation and regression is that the first one finds a function that matches all points, and the second finds a function that approximates the set of points, reducing the error of squared residuals. In practice, the fractal interpolation method introduces between every two points the fractal spikes. This provides a substantial over-estimative for the fractal dimension.

2. Materials and Methods

All samples have been processed and solidified from the liquid state onboard the International Space Station. In Figure 4 we can see the largest space station in the world—the International Space Station (ISS)—which was created as a result of cooperation between Russia, Canada, the United States, Japan, and 11 other member countries of the European Space Agency, including Germany and is in operation since about 20 years. The microstructure of these samples processed in space under microgravity conditions has been analyzed by SEM (Scanning Electron Microscopy—Zeiss LEO 1550).

Figure 4. International Space Station (ISS). Picture taken by a crew member of the space shuttle Atlantis after undocking from the space station (Image source NASA/Crew of STS-132).

The presented structure reconstruction we continued to apply on different micro-superalloy samples, for example, please see Figures 5–8. for the visualization of the space (Figures 5 and 6) and earth (Figures 7 and 8) samples under different magnifications.

An ingot of CMSX-10 was produced from the elemental materials by arc melting. Subsequently, rods were produced by suction casting. Pieces were cut from the rod for EDX investigation and suction casting of spheres of 6.5 mm diameter. The composition and its homogeneity were confirmed by energy-dispersive X-ray spectroscopy (EDX), using an Oxford Instruments Inca X-Sight 7426.

For further information on the EML on ISS, we refer to the literature. The facility, developed and built by airbus defense and space, is centered around a high vacuum experiment chamber that can be operated under high vacuum or in inert gas atmospheres, such as argon or helium. Each sample is stored in an individual sample holder, of which up to 18 are contained in an exchangeable sample chamber with a dedicated sample transport system. For processing, the desired sample can be moved into the experiment chamber. The core of the levitator consists of a coil system (SUPOS coil system) on which two radiofrequency RF generators are connected. One generator is used to produce a quadrupole field, imposing the positioning forces onto the sample. The second RF generator is used to establish a dipole field for heating the sample. The sample, being loosely confined in a wire cage, was placed within the coils during the experiment, leading to the free and extremely stable levitation of the sample.

CMSX-10 is a third generation single-crystal alloy, developed by Cannon Muskegon for a temperature range of 850–950 °C, and is, e.g., used in the Rolls-Royce engine TRENT 800. The composition of the investigated CMSX-10 sample was chosen from Erickson (please see Table 1 for the nominal compositions of the investigated sample).

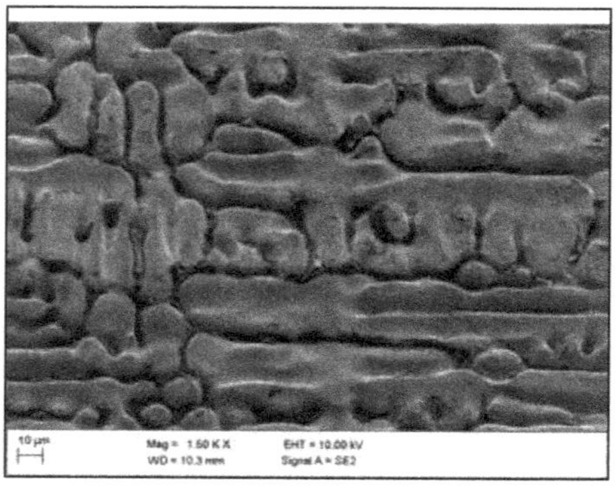

Figure 5. Superalloy with magnification 1.50 KX.

Figure 6. Superalloy with magnification of 200 KX.

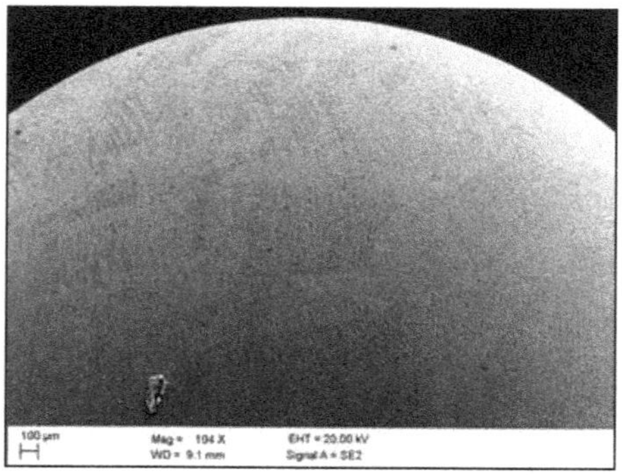

Figure 7. Superalloy with magnification of 104 KX.

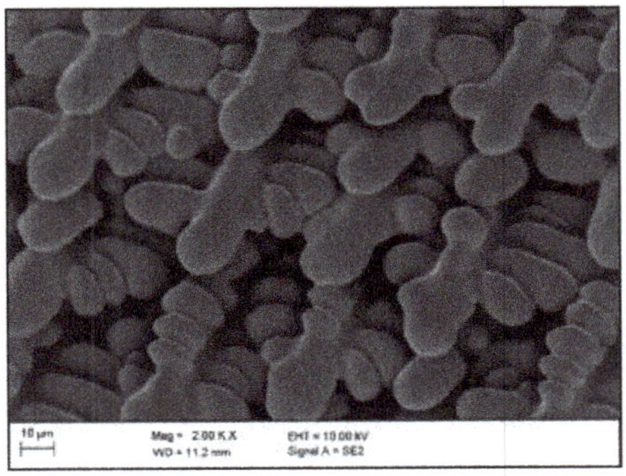

Figure 8. Superalloy with magnification of 2.00 KX.

Table 1. Nominal compositions of the investigated sample.

Composition in wt%	**CMSX-10**
Ni	Bal.
Al	5.7
Cr	2.0
Co	3.0
Mo	0.4
W	5.0
Ti	0.2
Re	6.0
Ta	8.0
Hf	0.03
Nb	0.1

2.1. Short Experimental Review on the Differences in Solidification of CMSX-10 in 1g and 0g

We solidified a CMSX-10 sphere of 6.5 mm diameter in microgravity onboard the international space station. The sample showed high undercooling of about 140 K. Figure 9 shows the corresponding temperature-time diagram recorded for the sample using the EML onboard the ISS. The sample was heated from the solid phase, then molten until fully liquid at the alloys liquidus temperature T_{liq} = 1706 K, then the liquid was further overheated until a maximum temperature of about 1900 K. Subsequently, the sample was cooled freely. This way, the sample undercooled about 140 K below its equilibrium melting point. In comparison, a sample was solidified on ground, while placed on a water cooled copper mold. Due to heterogeneous nucleation on the contact area, this represents the case of minimal undercooling. Figure 9 shows the temperature-time diagram recorded during the relevant melt cycle performed on ISS in microgravity of the 6.5 mm sphere of CMSX-10.

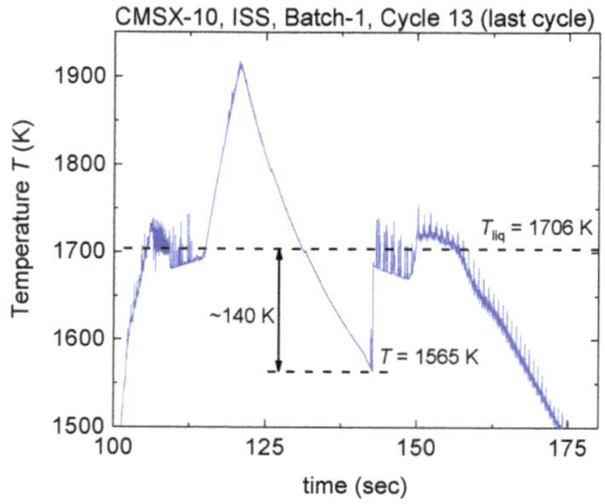

Figure 9. Temperature-Time Diagram of the melt-cycle performed on the CMSX-10 sphere.

2.1.1. SEM Images of the Surface

We have done SEM images on the surface of two samples:

1. CMSX-10–solidified onboard the ISS, "0g-Sample"
2. CMSX-10–solidified on top of a water-cooled copper block, on the ground, in the Arc-Melter, "1g-Sample"

A schematic overview of the situations is given in Figure 10.

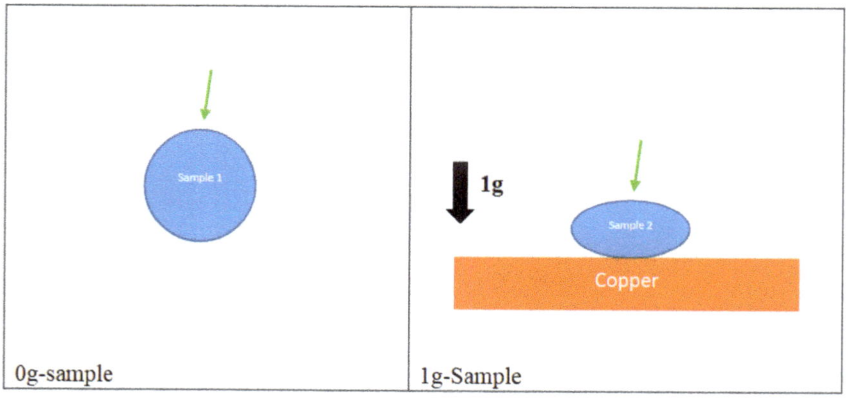

Figure 10. Samples and the solidification conditions.

For each series, 5 magnifications (200, 500, 1000, 1500, 2000) have been used to take the images. The images on the 0g sample were taken at a random position, since the sample appears identical in every single spot. The images of the 1g sample were taken on the top face (green arrow in Figure 10).

Images A, C, E, G, and I in Figure 11 are from the 0-g samples. Images B, D, F, H and J on Figure 11 are from the 1-g samples.

Images A, C, E, G, and I in Figure 12 are from the 0-g samples. Images B, D, F, H and J on Figure 12 are from the 1-g samples.

Images A, C, E, G, and I in Figure 13 are from the 0-g samples. Images B, D, F, H and J on Figure 13 are from the 1-g samples.

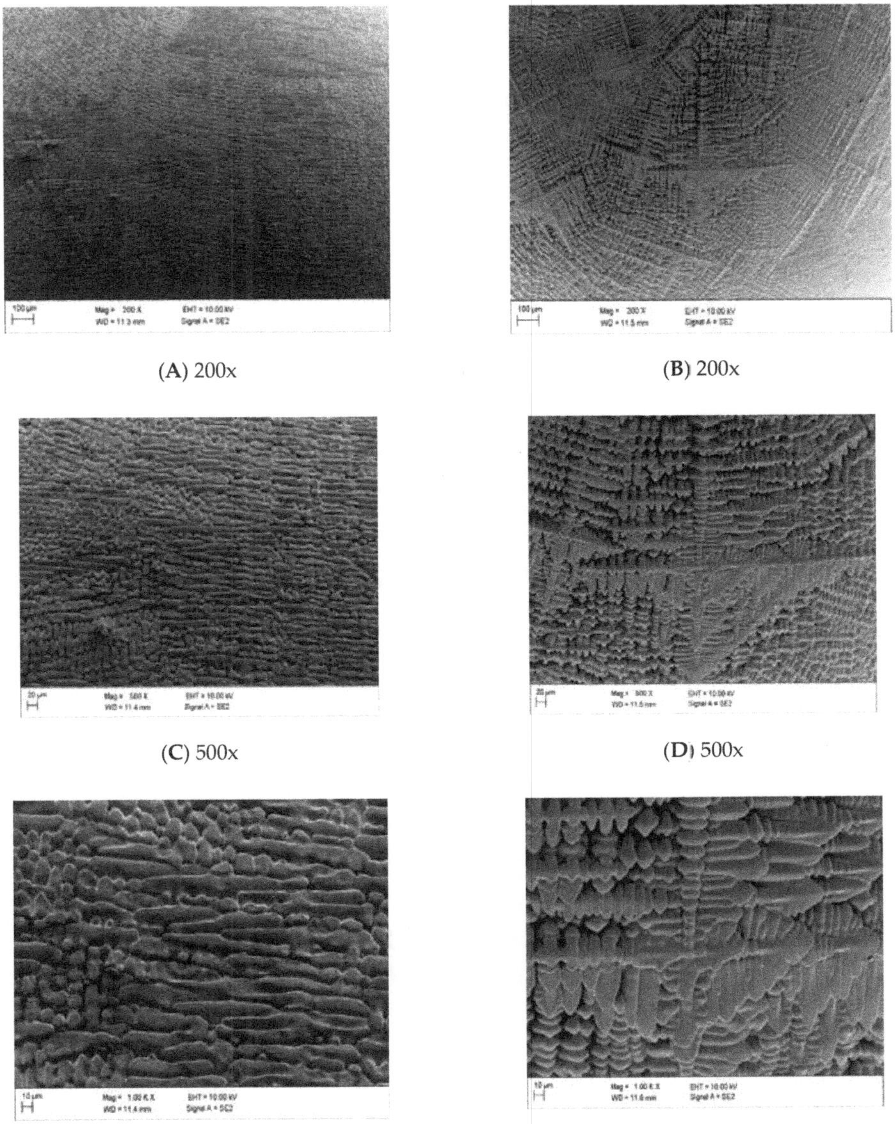

(**A**) 200x (**B**) 200x

(**C**) 500x (**D**) 500x

(**E**) 1000x (**F**) 1000x

Figure 11. *Cont.*

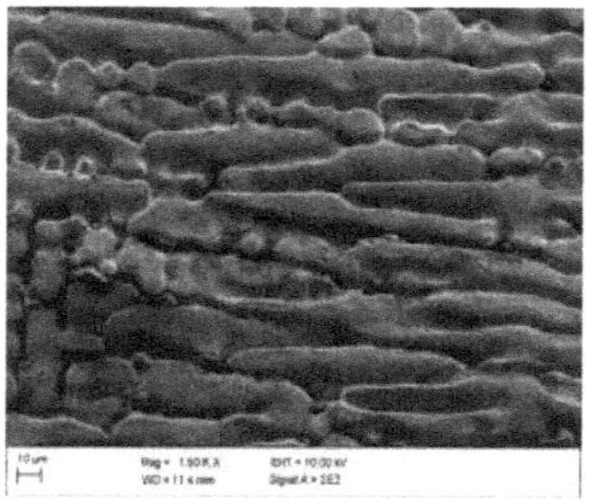

(G) 1500x

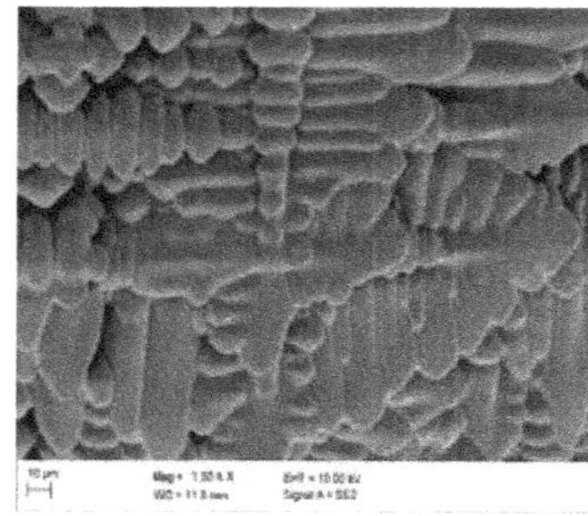

(H) 1500x

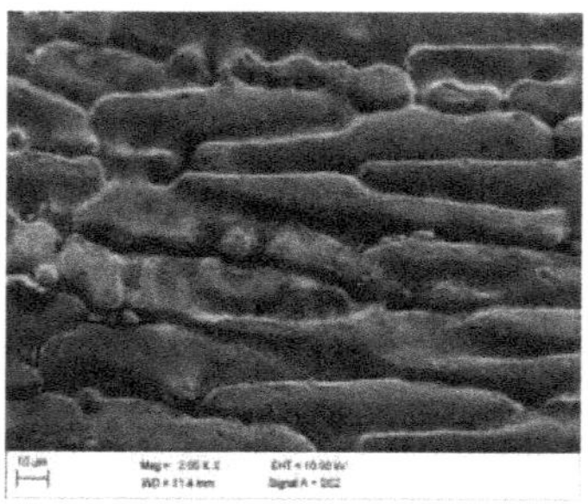

(I) 2000x

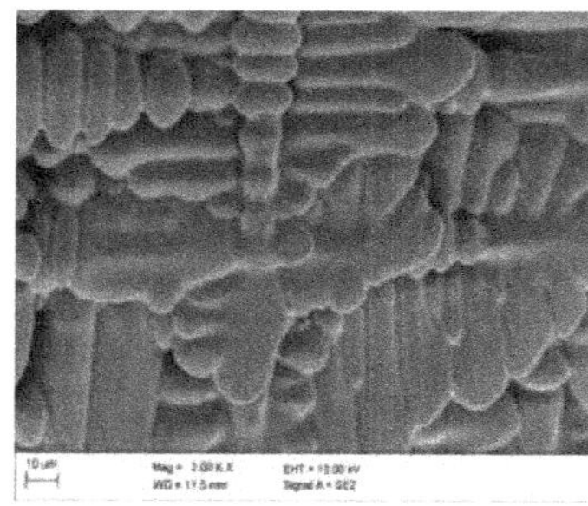

(J) 2000x

Figure 11. Series of 5 SEM images of the CMSX-10 surface of the 0-g sample (**A**,**C**,**E**,**G**,**I**) and the 1-g sample (**B**,**D**,**F**,**H**,**J**).

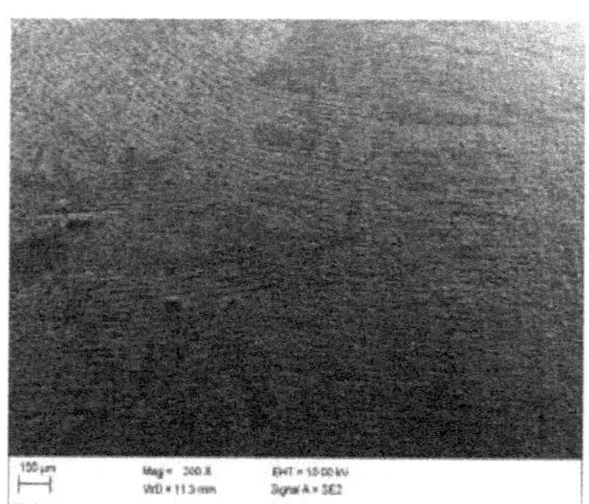

(A) 200x

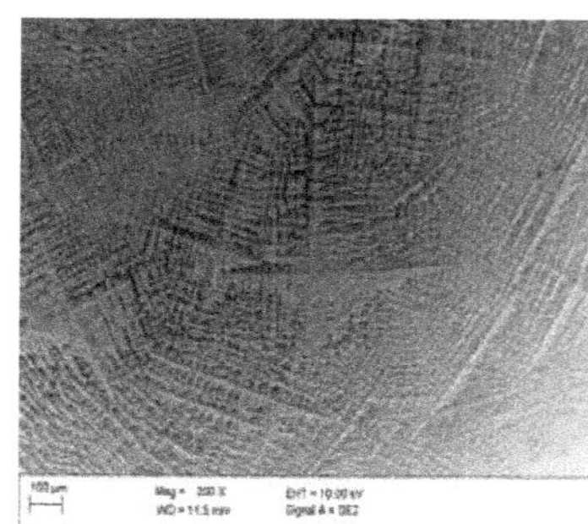

(B) 200x

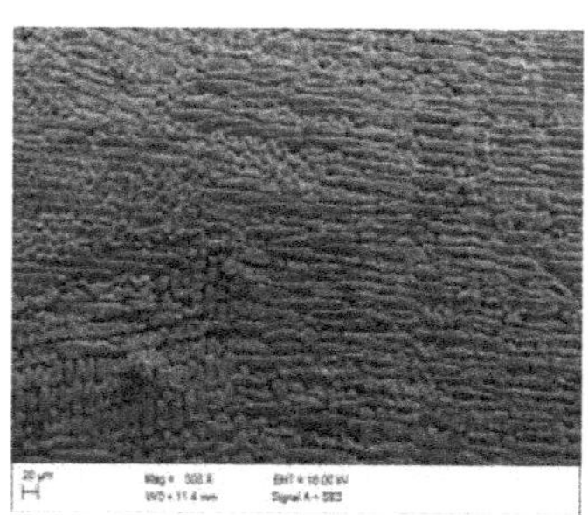

(C) 500x

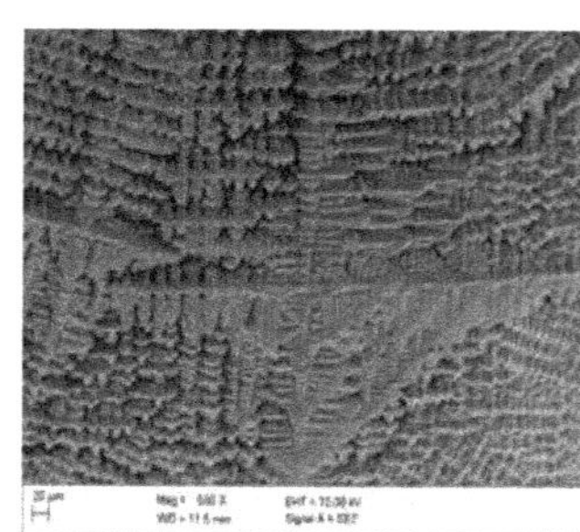

(D) 500x

Figure 12. *Cont.*

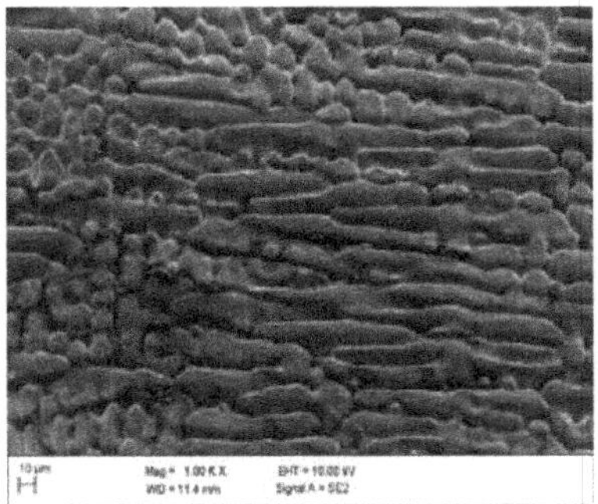

(**E**) 1000x

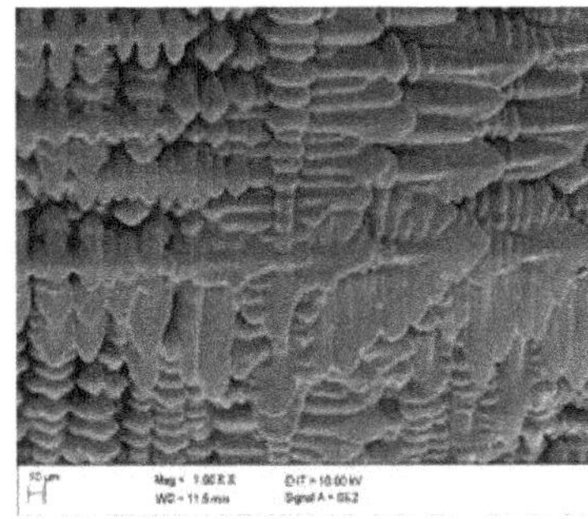

(**F**) 1000x

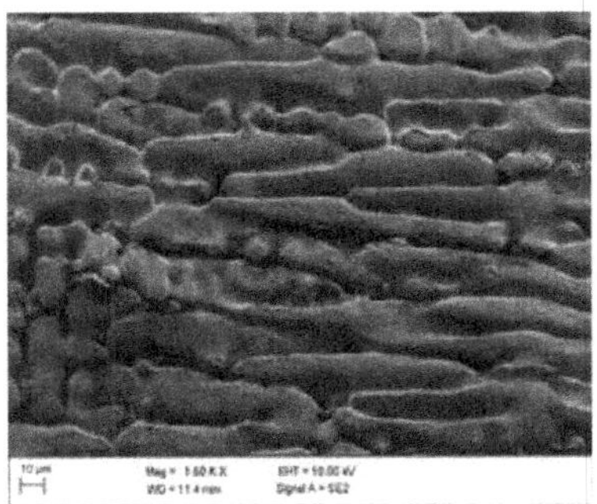

(**G**) 1500x

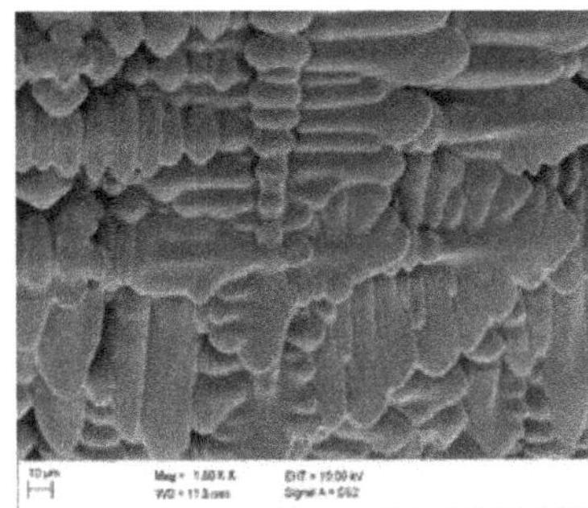

(**H**) 1500x

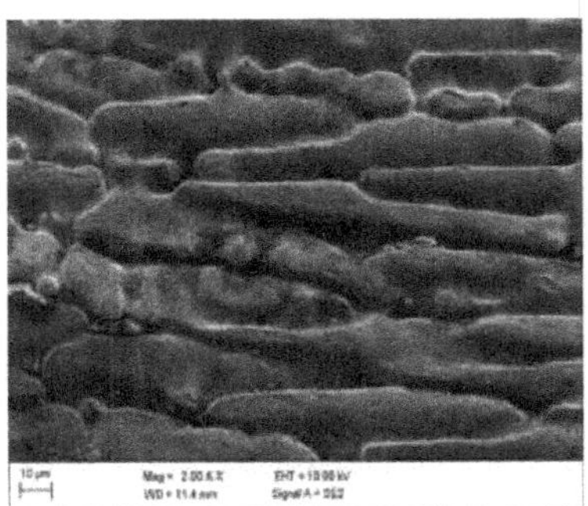

(**I**) 2000x

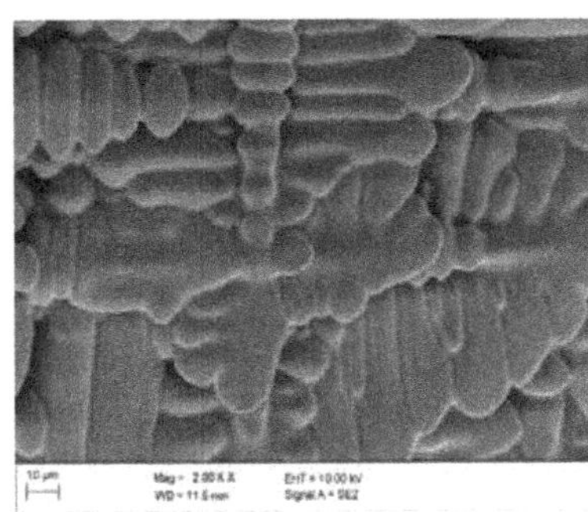

(**J**) 2000x

Figure 12. Series of 5 SEM images of the CMSX-10 surface of the 0-g sample (**A**,**C**,**E**,**G**,**I**) and the 1-g sample (**B**,**D**,**F**,**H**,**J**).

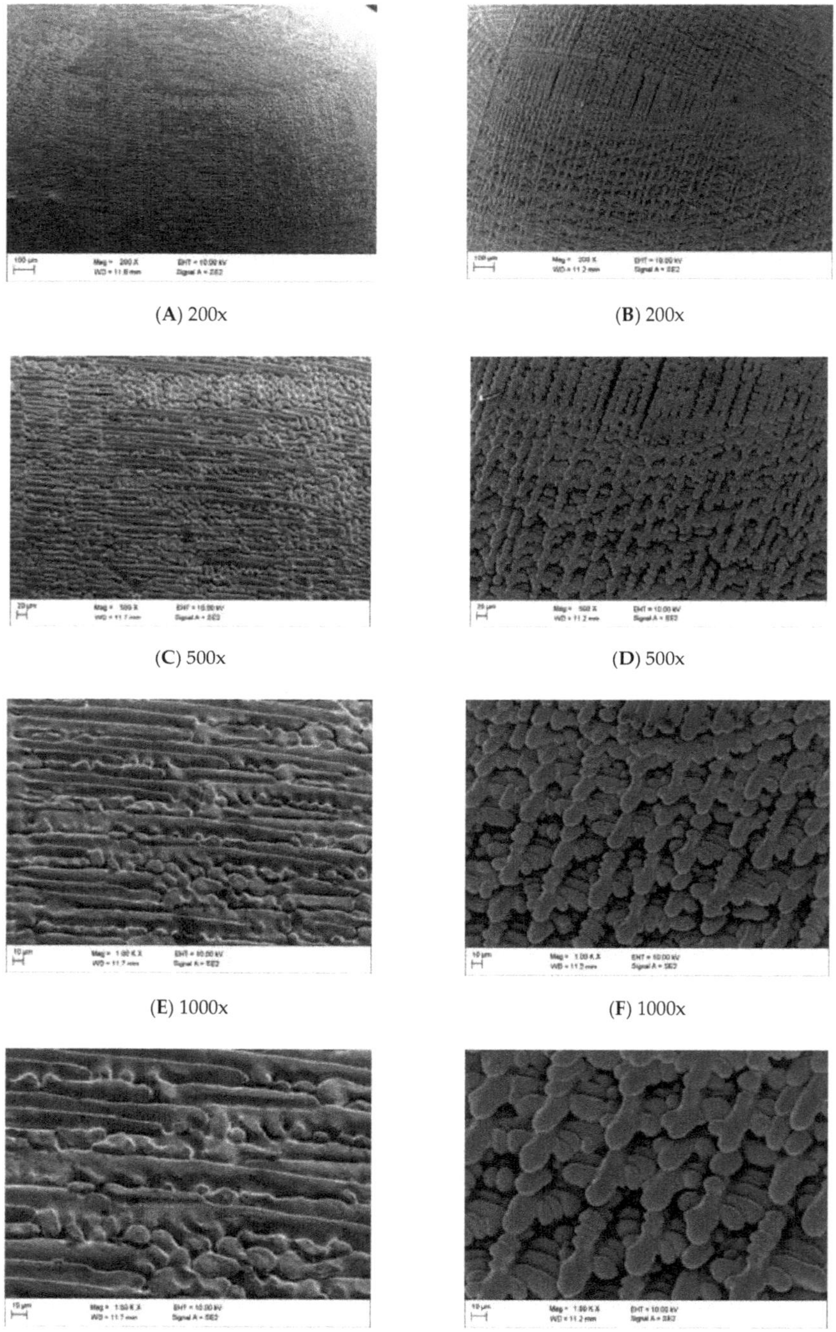

(**A**) 200x (**B**) 200x

(**C**) 500x (**D**) 500x

(**E**) 1000x (**F**) 1000x

(**G**) 1500x (**H**) 1500x

Figure 13. *Cont.*

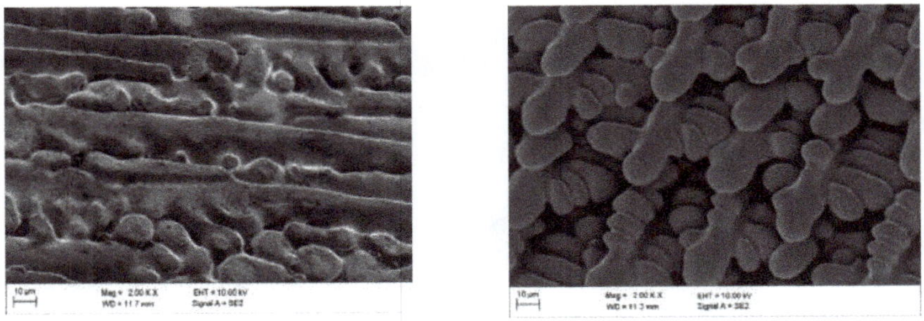

(**I**) 2000x (**J**) 2000x

Figure 13. Series of 5 SEM images of the CMSX-10 surface of the 0-g sample (**A,C,E,G,I**) and the 1-g sample (**B,D,F,H,J**).

2.1.2. Images of Cross-Sections

We have done SEM and optical images on the polished and etched cross-sections of two samples (Figures 14–16):

3. CMSX-10–solidified onboard the ISS, "0g-Sample".
4. CMSX-10–solidified by suction casting, on the ground, in the Arc-Melter, "1g-Sample".

(A) **(B)**

Figure 14. (**A**) 0g sample, (**B**) 1g sample.

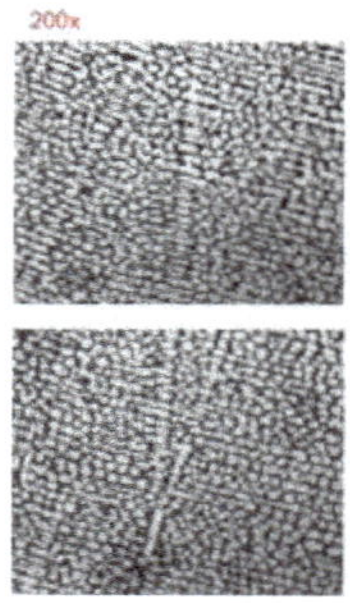

(**A**)

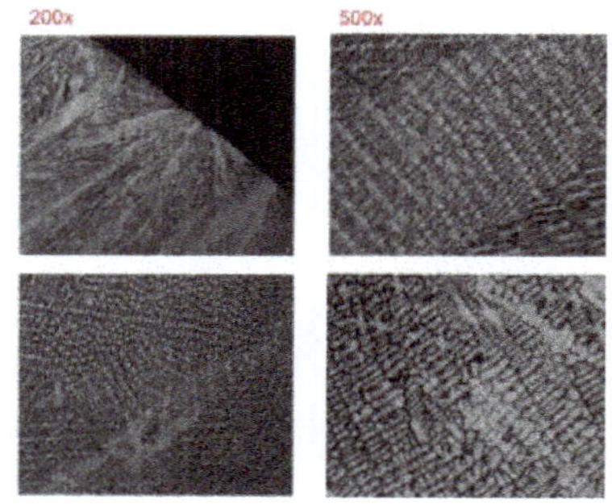

(**B**)

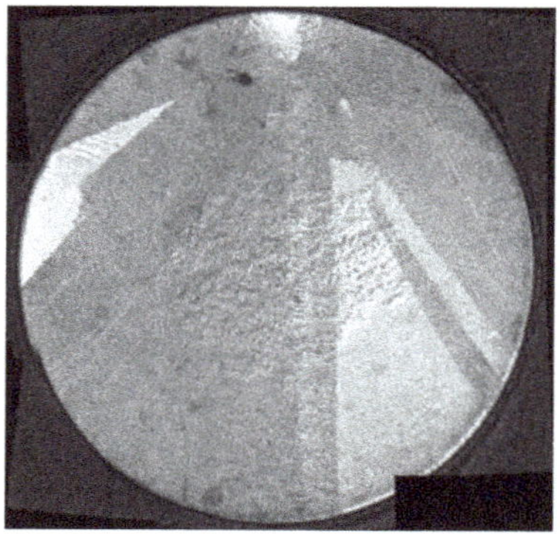

(**C**)

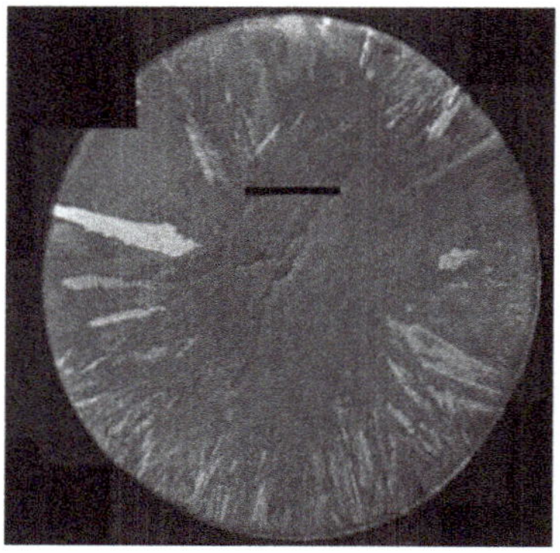

(**D**)

Figure 15. Optical images, 0g sample (**A**,**C**) and 1g sample (**B**,**D**).

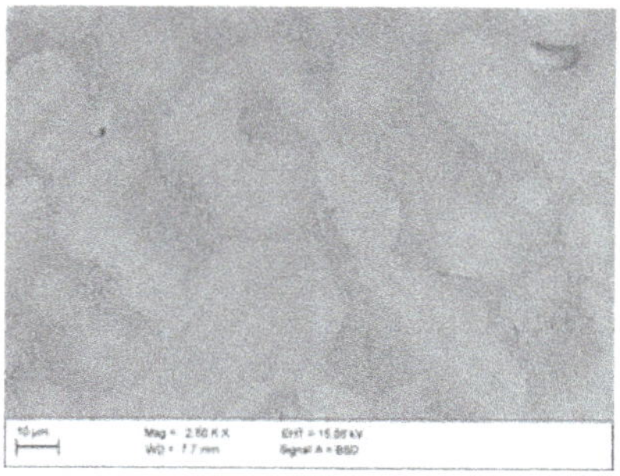

(**A**)

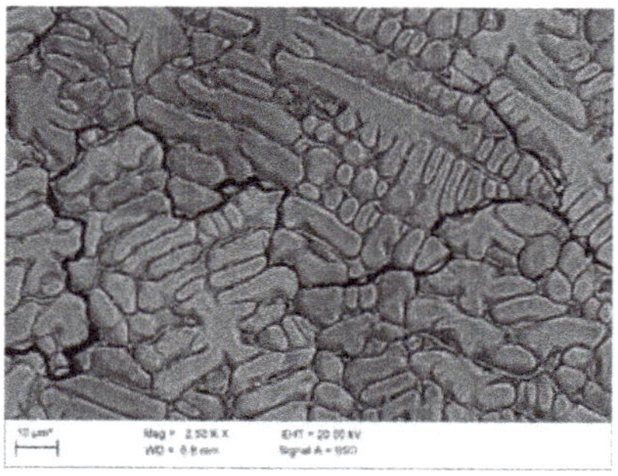

(**B**)

Figure 16. *Cont.*

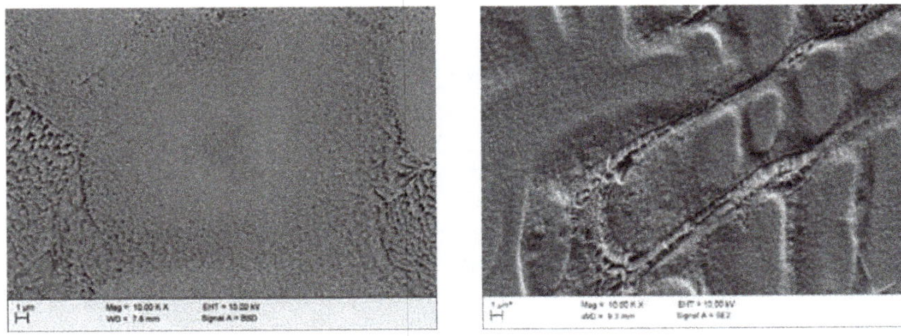

(C) (D)

Figure 16. SEM images, 0g sample (**A**,**C**) and 1g sample (**B**,**D**).

On the Figures 17 and 18 the EDX measurements on the cross section of the 0g-sample are presented showing the partitioning coefficients.

In wt%	Al	Ti	Cr	Co	Ni	Nb	Mo	Hf	Ta	W	Re
Dendrite Core	5.29±0.10	0.14±0.03	1.9±0.06	2.88±0.10	67.60±0.50	0.00±0.07	0.49±0.07	1.38±0.11	5.38±0.24	11.01±0.21	3.95±0.11
Dendrite Shell	5.24±0.06	0.14±0.06	2.02±0.14	2.86±0.01	68.35±0.69	0.00±0.07	0.60±0.01	0.97±0.23	4.45±0.67	11.12±0.06	4.30±0.04
Inter-dentritic region	6.39±0.03	0.28±0.01	1.89±0.06	2.47±0.06	70.59±0.62	0.165±0.04	0.60±0.06	0.72±0.01	7.52±0.68	7.53±0.06	1.86±0.15
Partitioning coefficient	0.83±0.02	0.49±0.11	1.01±0.06	1.17±0.07	0.96±0.02	-	0.82±0.21	1.92±0.10	0.72±0.10	1.46±0.04	2.13±0.23

Partitioning Coefficient = (dendrite composition)/(inter-dendritic composition)

- Enrichment of W, Re, Hf, (Co) in Dendrite
- Enrichment of Al, Ti, Ta, Mo in interdendritic space

Figure 17. EDX measurements on the cross section of the 0g-sample.

EDX measurements on the cross section of the 1g-sample show partitioning coefficients:

In wt%	Al	Ti	Cr	Co	Ni	Nb	Mo	Hf	Ta	W	Re
Dendrite	4.96±0.14	-	2.06±0.13	2.98±0.13	69.61±1.51	-	0.39±0.28	0.70±0.22	5.31±0.56	6.95±0.33	7.06±0.91
Inter-dentritic region	6.67±0.02	-	2.13±0.06	2.57±0.20	70.81±0.40	-	0.44±0.02	0.29±0.09	10.52±0.23	4.49±0.34	2.12±0.18
Partitioning coefficient	0.74±0.02	-	0.97±0.09	1.16±0.14	0.98±0.03	-	0.90±0.69	2.44±1.56	0.50±0.07	1.55±0.19	3.34±0.71

Partitioning Coefficient = (dendrite composition)/(inter-dendritic composition)

- Enrichment of W, Re, Hf, (Co) in Dendrite
- Enrichment of Al, Ta, Mo in interdendritic space

Figure 18. EDX measurements on the cross section of the 1g-sample.

2.1.3. Mathematical Fractal Analysis Technique

In order to describe the precise surface geometry, we introduced the concept of fractals and applied a new method of fractal reconstruction. The fractal analysis of real images was performed by using the technique based on a new affine fractal regression model. This process exploits certain mathematical formulations [2–4] designated for obtaining the coefficients of the equations system that best fit the data. The modeled system is:

$$\varphi\left(\frac{x+j}{p}\right) = a_j\varphi(x) + b_jx + c_j, \tag{2}$$

where $x \in [0,1)$, $0 \leq j \leq p-1$, and a_j, b_j, c_j are the parameters (real numbers) to estimate, with $0 < |a_j| < 1$. The default domain is $[0,1)$.

The solution of this system is a function $\varphi : [0,1) \to \mathbb{R}$ and is called a fractal function [2]. In fact, it is proved [4] that such functions have a mathematical fractal structure, meaning that the plot of their graph is a fractal curve. Theoretical mathematical properties and explicit solutions are provided [2,3]. This model is originated by the system constructed by [5].

The fractal regression method consists of estimating the parameters a_j, b_j, c_j such that they fit the real data. Hence, we consider the form of the explicit solution of the problem that depends on the p-expansion of numbers in $[0,1)$. For $L = 2$, this solution is:

$$\varphi(0) = \frac{c_0}{1-a_0}, \tag{3}$$

$$\varphi\left(\frac{\xi_1}{p}\right) = a_{\xi_1}\frac{c_0}{1-a_0} + c_{\xi_1},\ \xi_1 \neq 0, \tag{4}$$

$$\varphi\left(\frac{\xi_1}{p} + \frac{\xi_2}{p^2}\right) = a_{\xi_1}\left(a_{\xi_2}\frac{c_0}{1-a_0} + c_{\xi_2}\right) + b_{\xi_1}\frac{\xi_2}{p} + c_{\xi_1},\ \xi_2 \neq 0. \tag{5}$$

In order to obtain the best coefficients, the theoretical approach computes the SSR - sum of square residuals in between the formal definition and the real values. Next, it uses the partial derivatives of this SSR and equals to zero, for minimizing this error. The best solution of the problem is given when:

$$\frac{\partial SSR}{\partial a_j} = 0, \frac{\partial SSR}{\partial b_j} = 0, \frac{\partial SSR}{\partial c_j} = 0, \tag{6}$$

for all $j = 0,1,2,\ldots,p-1$. This is a problem with $3p$ parameters to estimate where the equations to solve are nonlinear. The regression method is widely used as linear regression, a much simpler model in data analysis science. For detailed information on this subject see [6].

Parameters a_j are the fractal coefficients and b_j are the directional coefficients. Bigger fractal coefficients mean strong fractal oscillations. Parameter p is the fractal period and L is the fractal level of a curve defined by the system.

The mathematical analytical solution of this partial derivative system (for the fractal regression) is not possible to compute and a numerical approach is needed. By applying the newly available software Fractal Real Finder designed for the numerical computation of the solution, we processed the given samples and estimated the curves and Hausdorff dimension D. With an input of the real data, the program executes simulations and gives an output with a fractal curve as modeled above. With the estimated fractal curves, we may upper estimate the Hausdorff dimension.

Proposition. The Hausdorff dimension of the graph of the function φ solution of the above system is upper bounded by the solution of:

$$\sum_{j=0}^{p-1} {\beta_j}^D = 1, \tag{7}$$

where $\beta_j = max\left\{\frac{1}{p}, |a_j|\right\}$, $0 \leq p \leq p-1$.

The coefficients with fractal relevance are those a_j such that $|a_j| > 1/p$.

All of these additional mathematical calculations with novelty solutions are highly important for better understanding the fractal nature analysis applications in material sciences and specifically for the research data in this paper. The fractal approach is based on the self-similarity of surfaces at different scales. Its superiority is that it is insensitive to the structural details and the structure is specified by a single descriptor, the fractal (noninteger) dimension D. So, the fractal surface analysis was used to describe, by a single parameter, surface roughness over many orders of magnitude. The increasing value of D represents an increasing surface roughness. It gives information on the measure of complexity of different surface topographies. That way, fractal dimension becomes highly suitable for the characterization of various topographies. Particularly, it is crucial in the cases when the land or space microstructures have lots of irregular peaks and valleys that cannot be easily defined and evaluated. Conclusively, fractal dimensions that we obtained from SEM micrographs of surfaces indeed do give us a very good description of the overall topography of the surface, due to the self-similarity.

3. Results

We implement the fractal structures that are originally coming from nature and beunivocally corresponding to chaotic structures in the matter. We considered the microstructures of different images samples. All samples have been processed onboard the International Space Station. In this particular case, the fractal analysis is implemented as the most effective. We applied several aspects of fractal analyses and obtained certain results regarding the Hausdorff dimensions related to the surface and structural characteristics of the CMSX-10 samples.

Several authors in the literature have investigated the mechanisms during the solidification of CMSX-10. It is generally accepted, that a primary solidification takes place, where the gamma-phase dendrites are formed. Subsequently, the interdendritic regions are solidified, together with the precipitation of the gamma′-phase. The solidification sequence of the interdendritic region is a subject of significant research and the assumed solidification paths by different authors do vary [7]. The dendritic formation is common in alloys where solute partitions between the solid and liquid phases. During the growth of the crystal in the melt, solute and heat can accumulate ahead of the growth interface and can lead to an unstable interface and dendritic solidification.

It is commonly observed in Ni-based superalloys, that the heavy elements, such as W and Re are segregated to the gamma-dendrites, the lighter elements, Ti, Ta, and Al are enriched in the remaining liquid [7–9].

3.1. Comparison of the Surface Images

Figures 10–12 (magnification 200 x) show that the 0g-sample shows larger dendrites (shows larger areas of the same repetitive patterns) with the similar orientation, while the 1g-sample shows multiple smaller dendrites that are randomly oriented.

Comparing the images of the 1g-Sample in Figures 11–13 (higher magnifications) one sees that there is only a little or no interdendritic phase. This could be an effect of gravity, which leads to a segregation of the solute phase downwards into the sample. In contrast, for the 0g-sample, there are no visible voids between the dendrite arms. The secondary dendrite arm spacings are similar. It is understandable, that not only the gravity but also the wetting of the solute phase with the solid dendrites influences the segregation of the solute phase. Hence, it might be of interest for further studies to vary the gravitational forces on the sample during solidification in a wide range between 0g and above 1g, in order to get a better understanding of the influence of gravity on this segregation effect.

3.2. Comparison of the Cross-Section Images

Partitioning of the elements in the 1g and 0g-Sample are relatively similar, only Hf and Re have a higher partitioning coefficient in the 1g-sample.

The 1g-Sample shows more small dendrites with finer secondary dendrite arm spacings. The arm spacings are also dependent on the distance to the sample surface.

Voids between the primary dendrite arms can be found in the sample center for the 0g-sample, while this effect is not visible in the 1g-sample.

3.3. Fractal Analysis of the Images Consolidated in Space

A part of an image from the CMSX-10 sample is given in Figure 19, where red points are marked. After a few simulations on the program Fractal Real Finder, we find that with $p = 10$ and $L = 2$, a good fit is obtained ($p^L = 100$ points). The captured result is plotted in Figure 20.

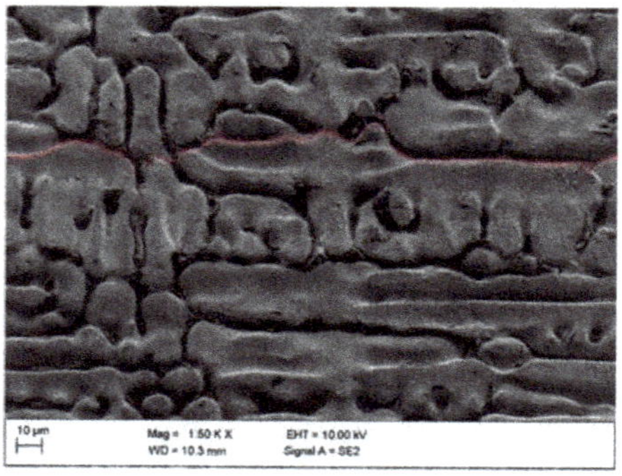

Figure 19. CMSX-10 marked part of the sample.

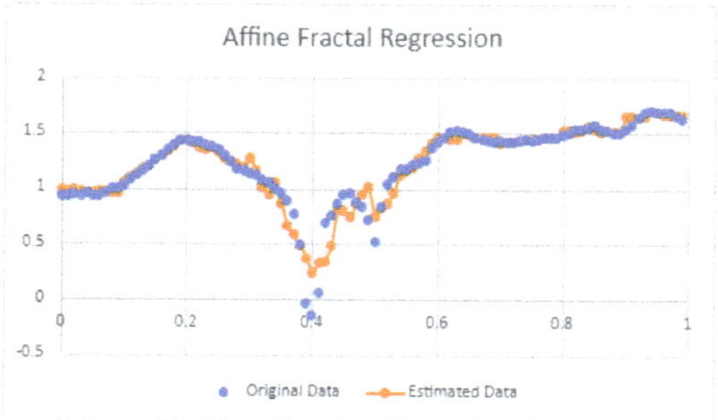

Figure 20. Affine Fractal Regression of the CMSX-10 reconstruction.

The statistical Kolmogorov-Smirnov test, as well as the the respective plot, confirms the good reconstruction of the original data. From the output of the program, the relevant fractal coefficients are −0.175388 and −0.229489. The list of estimated coefficients is presented in Table 2.

Table 2. Estimated coefficients of image from the CMSX-10_ISS_post-flight_007.

	0	1	2	3	4	5	6	7	8	9
a_j	0.018	0.011	−0.046	−0.175	−0.229	−0.073	−0.044	−0.006	−0.038	−0.051
b_j	−0.03	0.401	−0.231	−0.861	1.046	0.778	0.032	0.069	0.008	0.063
c_j	0.967	1.043	1.47	1.437	0.46	0.824	1.494	1.41	1.566	1.699

With these estimated fractal coefficients, an upper estimative for the Hausdorff dimension was computed as the solution of the nonlinear equation:

$$8\left(\frac{1}{10}\right)^D + 0.175388^D + 0.229489^D = 1. \quad (8)$$

The calculated Hausdorff dimension is $D = 1.0906$.

We successfully performed the fractal reconstruction of the sample morphology.

In the next result, the rebuilding of two samples consolidated in the space is analyzed.

The image A is depicted in Figure 21 and the image B is in Figure 22. In each image, red dots have been placed to mark identified boundary lines.

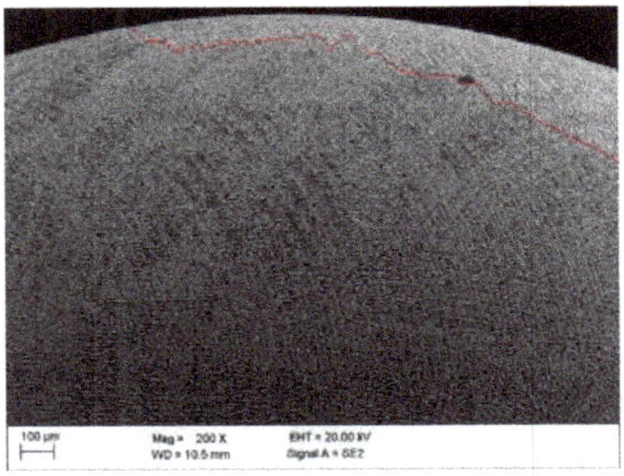

Figure 21. Image A of marked superalloy part.

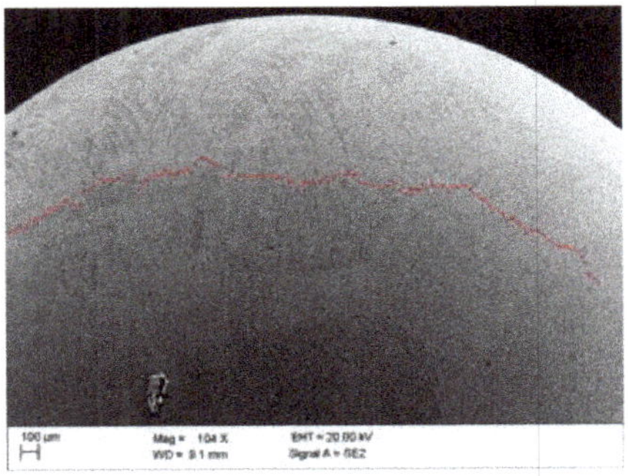

Figure 22. Image B of marked superalloy part.

With the software for numerical computation of the solution, Fractal Real Finder, we obtain the coefficients given in Tables 3 and 4. For the image A, we used 12 fractal periods ($p = 12$) and, for image B, 13 fractal periods ($p = 13$). Both simulations worked with two fractal levels ($L = 2$).

Table 3. Estimated coefficients from Image A.

	0	1	2	3	4	5	6	7	8	9	10	11
a_j	−0.14	0.09	0.059	0.055	0.097	0.268	0.025	−0.098	−0.012	0.043	−0.05	−0.015
b_j	−0.89	0.324	0.318	0.348	0.291	0.941	−0.651	−0.398	−0.49	−0.254	−0.813	−0.607
c_j	5.487	3.788	4.03	4.103	4.072	3.009	4.062	3.987	3.462	2.513	2.513	1.702

Table 4. Estimated coefficients from Image B.

	0	1	2	3	4	5	6	7	8	9	10	11	12
a_j	−0.051	−0.105	0.01	0.141	0.272	0.042	−0.066	−0.018	−0.158	0	0.133	0.028	−0.026
b_j	0.63	0.599	0.262	0.453	0.128	0.004	−0.332	0.269	−0.191	0.155	−0.364	−0.831	−0.632
c_j	1.873	2.685	3.155	2.75	2.862	3.257	3.587	3.226	3.683	0.955	2.713	2.447	1.835

We performed the statistical Kolmogorov-Smirnov test for two samples, and it indicates the goodness of the reconstructions.

The plot of the original data and the estimated curves for the image A and the image B are given respectfully in Figures 23 and 24.

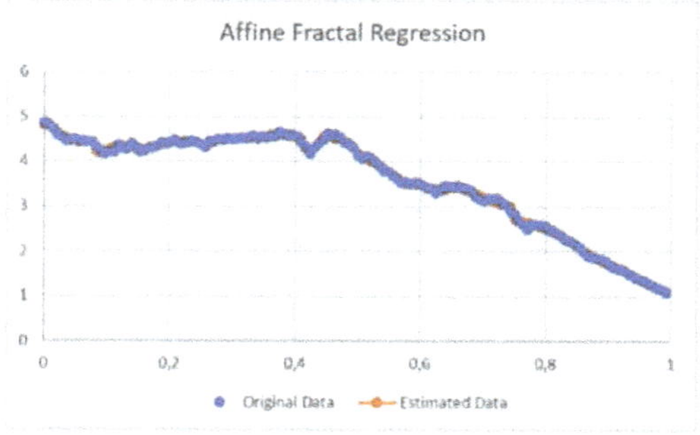

Figure 23. Fractal estimated curve from Image A.

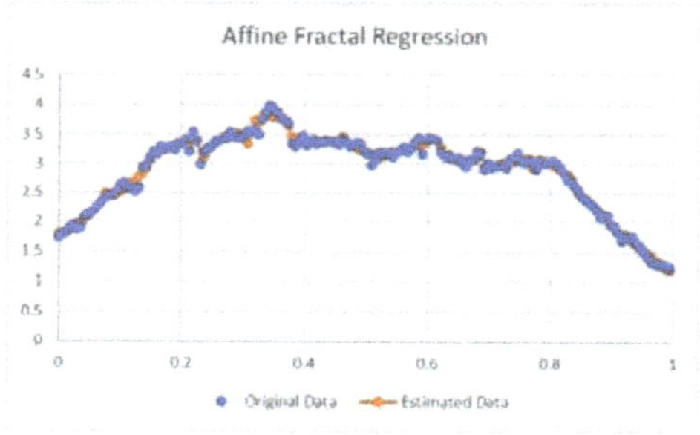

Figure 24. Fractal estimated curve from Image B.

The relevant fractal coefficients, in the case of the image A, are a_0, a_1, a_4, a_5 and a_7 and of image B, a_1, a_3, a_4, a_8 and a_{10}. From the Proposition above, the upper estimators of Hausdorff dimension are respectively D = 1.11368 and D = 1.16975.

3.4. *Fractal Analysis of an Image Consolidated on Earth*

A CMSX-10 spare arc-melter image (Figure 13J) was imported into a pptx file and inserted a grid on it. Then, equally spaced red points were marked on a portion of the contour and the following image was obtained (see Figure 25).

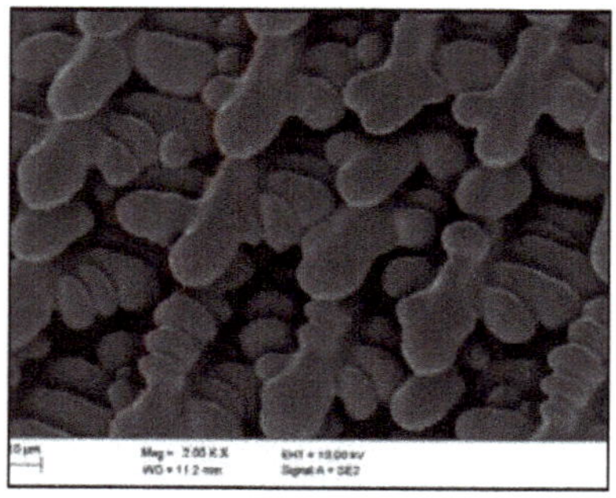

Figure 25. CMSX-10 spare_arc-melter image, with red points on contour.

Then we defined a scale, and we imported the data into a tabulated file in the form of an ordered list of these red points. We tried some options for the number of fractal periods p and fractal levels L and simulated with the software Fractal Real Finder. The result fitted very nicely with $p = 10$, $L = 2$ and $p^L = 100$ points. The estimated coefficients are given in Table 5.

Table 5. Estimated coefficients from the CMSX-10 spare-arc-melter image.

	0	1	2	3	4	5	6	7	8	9
a_j	0.051	−0.025	0.092	−0.07	−0.017	0.037	−0.001	0.025	0.192	−0.026
b_j	−0.58	0.101	0.032	−1.117	0.351	0.561	−0.653	0.334	0.462	−0.374
c_j	3.588	2.859	2.474	2.899	2.111	3.057	1.407	0.818	0.544	0.363

The statistical analysis in the PAST software the Kolmogorov−Smirnov test for the independent samples' comparison of the equality of distributions showed undoubtedly that these distributions could be considered equal. The test showed that the largest difference between corresponding values was 0.06 and from p-value = 0.9921 > 0.05 we concluded that the null hypothesis was not rejected, showing that there is no significant difference between the distribution for the two samples. Please see the respective plot in Figure 26 showing that the reconstruction of the sample morphology is with high accuracy acceptable.

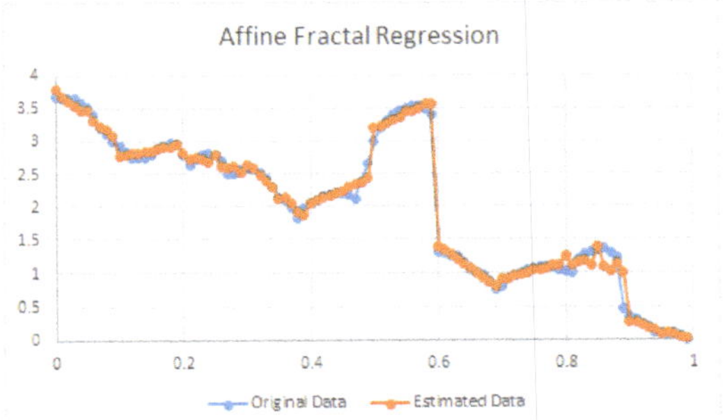

Figure 26. Fractal estimated curve.

The relevant fractal coefficient is $a_8 = 0.1915881$, and, consequently, the upper of the Hausdorff dimension is $D = 1.04007$ (see Proposition above).

From the above results and discussion, it is obvious that fractals help to overbridge the complex structures and processes leading towards controlled disorder and finally the ordered structures [10,11].

Additionally, from the same image (Figure 13J) we selected a circular region and then applied the fractal regression to compare the Hausdorff dimensions of the same sample, but taking into consideration different contour and another fragment of the image. Figure 27 represents this selected region with a polar grid.

We considered the series of the radius (distance between the center point (in blue) and the corresponding red point). We performed the fractal regression, and we obtained the coefficients for the estimated fractal curve (see Table 6). The fit was done for 2 fractal levels and 11 fractal periods.

Figure 27. CMSX-10 spare_arc-melter image, with red points on contour with a polar grid.

Table 6. Estimated coefficients from Image J 2000x.

	0	1	2	3	4	5	6	7	8	9	10
a_j	−0.079	−0.038	0.071	0.06	0.051	0.015	−0.035	−0.023	0.075	−0.033	−0.023
b_j	−0.488	0.733	1.136	−0.534	0.087	−0.904	−0.419	−0.095	0.843	−0.619	0.1
c_j	2.102	1.552	2.247	3.309	2.994	2.979	2.158	1.751	1.742	2.566	1.891

None of the fractal coefficients have sufficient fractal relevance to estimating the Hausdorff dimension bigger than 1, i.e., there is no estimated coefficient above $0.09 = 1/11$. Therefore, the estimated Hausdorff dimension is $D = 1$, which means the estimated fractal oscillations are soft. Reconverting the estimated values back to polar coordinates, we obtained the plot in Figure 28.

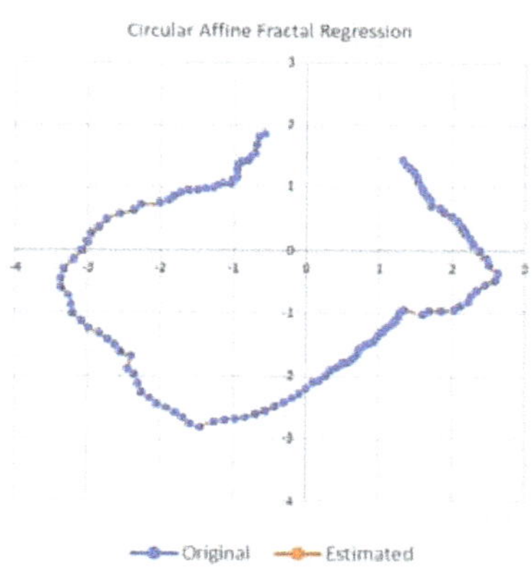

Figure 28. Fractal estimated curve from polar coordinates.

The use of the fractal principle is highly relevant for the evaluation and estimation of dimensional properties of irregular structures in nature. Undoubtedly, they help us in understanding the morphological organization of complex structures that appear in Space and Earth [12–14].

4. Discussion

At the end of this very complex research overview, we obtained different results based on the analysis which included different parameters from surface and cross-sections of the samples studied by SEM and optical microscopy methods. Also, we included some comparative results that inlight the phenomena at the space and land consolidation. From all of these points of view, the fractal analysis provides a better understanding of the consolidation conditions influence on the final morphology of structures. Only fine fractal microstructure analysis enables the comparative differences between the dendrites' sizes, their orientations and voids between the primary dendrite arms.

This is very important for the reason of similarities among the images of space and land samples. Consequently, we can notify a very strong influence of the gravity in the consolidation process, which makes certain differences at microstructures based on different consolidation effects. We obtained certain differences among the land and space Hausdorff dimensions of the samples, but possibly due to different starting conditions in the experiments. However, we must highlight that the comparison of space and land samples was not the target of this paper. The comparison of the basic sample data according to fractal reconstruction could be a direction for some future researches. The aim of this paper is the fractal reconstruction of different surface samples in order to achieve the precise mathematical characterization of their roughness and consequently to predict and design the desired microstructures.

In future research, we plan to analyze more deeply the observed differences between space and land structures and try to reveal potential influences of the internal forces within constituents of samples matter (cohesion and adhesion).

The more biomimetic similarity in our material structures related to nature is a necessity. In our future research, we can also include the questions and relations between entropy and fractals.

5. Conclusions

The importance of these results lies in the fact that we can use these fractal dimensions' characterizations for an additional understanding and insights of microstructures. Since we achieved an improved possibility of getting reconstructed morphology of shapes, we established a completely new perspective and frontiers on the advanced structures' prediction. All this phenomenology is extremely important for the relativization of the scale sizes in the space through fractal nature. In addition, in our further study, we extended fractal analysis on the micro images based on the land-consolidated samples what could be potentially very attractive for the future research in this area.

Finally at the end, when analyzing the data review of all presented experiments we must underline the existence of the dominant internal forces in and between the dendrites. It is necessary to emphasize the real influence of the mentioned forces especially at the space conditions where there is no gravity. Besides, we can also mention the micro capillary and surface tension effects.

Further, it is crucially important to take into consideration that all of these phenomena have been included in the understanding and explanations of the processes in the Space bodies consolidations and even though the whole Space. This fact is not related to any of the Space consolidation theories caused neither by the explosion nor by the high pressure on the micro level. On the other hand, this focuses the roll of the gravity itself, when we have the land consolidation processes and this can potentially provide much thorough approach in the explanation of gravity effect even in the evolution.

Author Contributions: Conceptualization, the idea and supervision: V.M.; Formal analysis, software and writing-original draft preparation: V.M., C.S. and I.I.; Validation, resources data curation and supervision: M.M. and H.-J.F. All authors have read and agreed to the published version of the manuscript.

Funding: V. Mitic and I. Ilic are gratefully thankful to the Ministry for the Education, Science and Technology Development of Serbia regarding the basic science research funds. Cristina Serpa acknowledges partial support from National Funding from FCT - Fundação para a Ciência e a Tecnologia, under the project: UIDB/04561/2020. M. Mohr and H.J. Fecht like to thank DLR (50WM1759) and ESA (AO-099-022) for the generous financial support. M. Mohr and H.J. Fecht acknowledge the access to the ISS-EML, which is a joint undertaking of the European Space Agency (ESA) and the DLR Space Administration. The reported work was conducted in the framework of the ESA project ThermoProp (AO-099-022). We further acknowledge funding from the DLR Space Administration with funds provided by the Federal Ministry for Economic Affairs and Energy (BMWi) under Grant No. 50WM1759.

Institutional Review Board Statement: Not applicable.

Informed Consent Statement: Not applicable.

Data Availability Statement: Not applicable.

Acknowledgments: We would like to thank A. Minkow, R. Wunderlich and Y. Dong for their expert technical support and fruitful discussions.

Conflicts of Interest: The authors declare no conflict of interest.

References

1. Siauwand, T.; Bayen, A.M. *An Introduction to MATLAB Programming and Numerical Methods for Engineers*; Elsevier: Oxford/London, UK; Waltham, MA, USA; San Diego, CA, USA; Amsterdam, The Netherlands, 2015.
2. Serpa, C.; Buescu, J. Constructive Solutions for Systems of Iterative Functional Equations. *Constr. Approx.* **2017**, *45*, 273–299. [CrossRef]
3. Serpa, C.; Buescu, J. Explicitly defined fractal interpolation functions with variable parameters. *Chaos Solitons Fractals* **2015**, *75*, 76–83. [CrossRef]
4. Buescu, J.; Serpa, C. Fractal and Hausdorff dimensions for systems of iterative functional equations. *J. Math. Anal. Appl.* **2019**, *480*, 1–19. [CrossRef]
5. Barnsley, M.F. Fractal functions and interpolation. *Constr. Approx.* **1986**, *2*, 303–329. [CrossRef]
6. Kenney, J.F.; Keeping, E.S. Linear Regression and Correlation, Ch. 15. In *Mathematics of Statistics*, 3rd ed.; Van Nostrand: Princeton, NJ, USA, 1962; pp. 252–285.
7. Terner, M.; Yoon, H.Y.; Hong, H.U.; Seo, S.M.; Gu, J.H.; Lee, J.H. Clear path to the directional solidification of Ni-based superalloy CMSX-10: A peritectic reaction. *Mater. Charact.* **2015**, *105*, 56–63. [CrossRef]
8. Wilson, B.C.; Cutler, E.R.; Fuchs, G.E. Effect of solidification parameters on the microstructures and properties of CMSX-10. *Mater. Sci. Eng. A* **2008**, *479*, 356–364. [CrossRef]
9. Mohr, M.; Furrer, D.; Fecht, H. Thermophysical Properties of Advanced Ni-Based Superalloys in the Liquid State Measured on Board the International Space Station. *Adv. Eng. Mater.* **2020**, *22*, 1901228. [CrossRef]
10. Mitić, V.; Goran, L.; Vesna, P.; Hwu, J.R.; Tsay, S.C.; Perng, T.P.; Sandra, V.; Branislav, V. Ceramic materials and energy-Extended Coble's model and fractal nature. *J. Eur. Ceram. Soc.* **2019**, *39*, 3513–3525. [CrossRef]
11. Mandelbrot, B. *The Fractal Geometry of Nature*; W. H. Freeman: New York, NY, USA, 1977.
12. Arnold, W.; Fischer, H.-H.; Knapmeyer, M.; Krüger, H. Surface Mechanical Properties of Comet 67P. *Jpn. J. Appl. Phys.* **2019**, *58*, SG0801. [CrossRef]
13. Mitić, V.; Kocić, L.J.M.; Mitrović, I.Z. Fractals in Ceramic Structure. In Proceedings of the IX World Round Table Conference on Sintering, Belgrade, Serbia, 1–4 September 1998; Advanced Science and Technology of Sintering. Stojanović, B.D., Skorokhod, V.V., Nikolić, M.V., Eds.; Kluwer Academic/Plenum Publishers: New York, NY, USA, 1999; pp. 397–402.
14. Mitic, V.V.; Lazovic, G.; Mirjanic, D.; Fecht, H.; Vlahovic, B.; Arnold, W. The Fractal Nature as New Frontiers in Microstructure Characterization and Relativisation the Scale Sizes within the Space. *Mod. Phys. Lett. B* **2020**, *34*, 2050421. [CrossRef]

remote sensing

MDPI

Communication

CubeSat Observation of the Radiation Field of the South Atlantic Anomaly

Pavel Kovář [1,*] and Marek Sommer [2,3]

[1] Faculty of Electrical Engineering, Czech Technical University in Prague, Technická 1902/2, 166 27 Prague, Czech Republic
[2] Nuclear Physics Institute of the Czech Academy of Sciences, Husinec-Řež 130, 250 68 Řež, Czech Republic; sommer@ujf.cas.cz
[3] Faculty of Nuclear Sciences and Physical Engineering, Czech Technical University in Prague, V Holešovičkách 2, 180 00 Prague, Czech Republic
* Correspondence: kovar@fel.cvut.cz

Abstract: The movement of the South Atlantic Anomaly has been observed since the end of the last century by many spacecrafts equipped with various types of radiation detectors. All satellites that have observed the drift of the South Atlantic Anomaly have been exclusively large missions with heavy payload equipment. With the recent rapid progression of CubeSats, it can be expected that the routine monitoring of the South Atlantic Anomaly will be taken over by CubeSats in the future. We present one-and-a-half years of observations of the South Atlantic Anomaly radiation field measured by a CubeSat in polar orbit with an elevation of 540 km. The position is calculated by an improved centroid method that takes into account the area of the grid. The dataset consists of eight campaigns measured at different times, each with a length of 22 orbits (~2000 min). The radiation data were combined with GPS position data. We detected westward movement at 0.33°/year and southward movement at 0.25°/year. The position of the fluence maximum featured higher scatter than the centroid position.

Keywords: South Atlantic Anomaly; radiation measurement; CubeSat observation; radiation belts

Citation: Kovář, P.; Sommer, M. CubeSat Observation of the Radiation Field of the South Atlantic Anomaly. *Remote Sens.* **2021**, *13*, 1274. https://doi.org/10.3390/rs13071274

Academic Editor: Serdjo Kos

Received: 8 February 2021
Accepted: 24 March 2021
Published: 26 March 2021

1. Introduction

The Earth has two van Allen radiation belts, i.e., the inner and outer, which consist of trapped high-energy charged particles. Most of the particles within the radiation belts originate from solar winds and galactic cosmic rays [1].

The approximately dipolar magnetic field at low altitudes generated by Earth's core is tilted and shifted from the geographic pole. This creates a region with reduced magnetic field intensity, namely the South Atlantic Anomaly (SAA), which is located approximately on the eastern coast of Brazil. The SAA is a place where the radiation of the inner van Allen belt approaches the closest to Earth [2,3]. One of the manifestations of the SAA is the enhanced count rate of protons and electrons coming from the inner radiation belt [4].

The SAA region presents a threat to low Earth-orbiting satellites due to the high probability of single event upsets, failure of microelectronics and premature ageing, as well as to astronauts and their health [5–7]. Therefore, the region has been thoroughly investigated since its discovery. Many of the published papers have focused on the dynamics of the SAA, mainly the variation in its intensity over time and its drift, to predict its future movement. It has been observed that the SAA moves steadily in a northwest direction. This movement has been registered by measurement of the magnetic field [8,9] and by radiation [2,3,10–21]. Table 1 summarizes the publications that have focused on examining the drift of the radiation center of the SAA. As shown in previous studies, a large spread of drift velocities has been reported. Such large discrepancies could be explained by the dependence of the particle flux on the altitude and on particle energy, as described in [2], as many measurements were performed on different orbits with different instrumentation. Other phenomena that influence the drift of the SAA and its velocity are sun modulation and "geomagnetic jerks". The effect of the solar cycle on drift velocity was described in [3,20,21]. The authors of [19] theorized that the rapid short-term change in the drift of the SAA was caused by a "geomagnetic jerk", which was reported in 2003 by [22]. Geomagnetic jerks occur when the secular acceleration of the magnetic field rapidly changes.

Table 1. Summarized results of measured South Atlantic Anomaly (SAA) drift rates.

Study	Westward Drift (°/Year)	Northward Drift (°/Year)	Altitude (km)	Inclination (°)	Time (Year)
Konradi 1994 [14]	0.32	-	617 450 287 565	28.5 28.5 39 58	1990–1991
Badhwar 1997 [15]	0.28 ± 0.03	0.08 ± 0.03	438 393	50 51.65	1973, 1995
Bühler 2002 [16]	0.06 ± 0.05	0.06 ± 0.05	400	52	1994–1996
Ginet 2006 [17]	0.43 ± 0.13	-	410–1710	69	2000–2006 continuous measurement
Grigorian 2008 [18]	0.1–1.0	0.1	307–393 220 400 500–2500 400 450	65 81.6 51.6 81.3 51.6 51.6	1960–2003
Fürst 2009 [19]	0.248	-	592 in 1996 488 in 2007	23	1996–2007 continuous measurement
Casadio and Arino 2011 [20]	0.24	0.08	782–785 780 800	98.52 98.5 98.55	1991–2010 continuous measurement
Qin 2014 [21]	0.3	0.09	813 833 804 833	98.7 98.6 98.5 98.7	1980–2010 Almost continuous measurement
Schaefer 2016 [10]	0.36 ± 0.06	0.16 ± 0.09	840–860	99	2004–2013 continuous measurement
Jones 2017 [3]	0.20 ± 0.04	−0.11 ± 0.01	400–600		1993–2011 continuous measurement
Ye 2017 [2]	Various (depends on proton energy)	Various (depends on proton energy)	512–687	81.7	1994–2007 continuous measurement
Anderson 2018 [12]	0.277 ± 0.008	0.064 ± 0.008	DMSP F8–F18	DMSP F8–F18	1987–2015 continuous measurement
Aubry 2020 [13]	0.639 0.329 0.256	– – –	715 1336 850	98 66 98	2000–2018 continuous measurement

Another systematic error in the evaluation of SAA drift can be caused by different methods of data processing. One of the most deployed methods for evaluation of the SAA position is Gaussian or Weibull fitting of measured maxima data (flux, dose, and dose rates) over a period of time and calculation of the SAA position based on the fit maximum [3,14–17,21]. In recent years, a new interpolation method based on the calculation of the SAA centroid has been presented [12,13]. The shortcoming of the centroid method used in [12,13] is that it does not consider cosine-latitude effects.

CubeSats have been used for many scientific studies in recent years. Although CubeSat missions cannot fully replace mainstream space missions, they have proven to be a useful and inexpensive platform for small payloads [23,24]. CubeSats have been used in scientific fields of Earth science [25–27], space weather [28–30], and astrophysics [27,31].

Our 1U CubeSat Lucky 7 was launched with the primary objective of testing communication systems and global positioning systems (GPSs). The secondary objective was to study the ionizing radiation field with two devices, i.e., a piDOSE radiation detector and a silicon diode spectrometer. Due to the failure of the spectrometer, only data from piDOSE were obtained.

The objective of this study is to prove that 1U CubeSats can be exploited for the long-term monitoring of the SAA movement, as all measurements introduced in Table 1 were done on professional large satellites. Moreover, this study focuses on the improvement of the centroid method used for the localization of the SAA center.

In our study, we present a modest dataset of processed SAA location data gathered by the radiation detector piDOSE flown on the Lucky 7 satellite. The SAA location was derived from the radiation data by the centroid method normalized for cosine-latitude effects. We show that the centroid method calculates the SAA location with much higher accuracy than the maxima fitting method. Hence, fewer data are needed to evaluate the SAA location when the centroid method is used. Such features might be crucial in the case of 1U CubeSat, which has very limited power, and broadcasting resources and data transfer to Earth might be problematic. We believe that lightweight, cost-effective CubeSats equipped with radiation detectors such as piDOSE can be utilized for the continuous monitoring of SAA drift in the future.

2. Instrumentation and Methods

2.1. CubeSat

The experimental data used for the preparation of this paper were measured by 1U CubeSat Lucky 7 (Figures 1 and 2). Lucky 7 (catalogue no. 19038W) was launched from Vostochny Cosmodrome by the Soyuz-2.1b rocket on 27 June 2019, to a quasi-synchronous orbit of inclination of 97.5008° and altitude of 520 km. Regular scientific data were collected until August 2019 after successful satellite testing. Examples of the points at which the radiation was measured are shown in Figure 3.

Figure 1. Lucky 7 CubeSat.

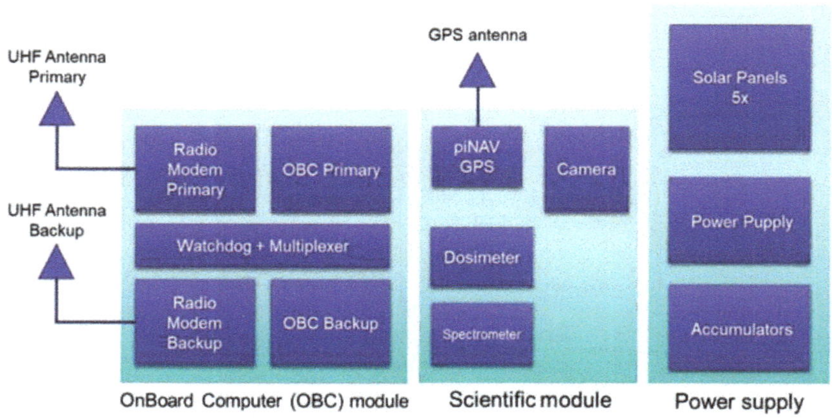

Figure 2. Lucky 7 block diagram.

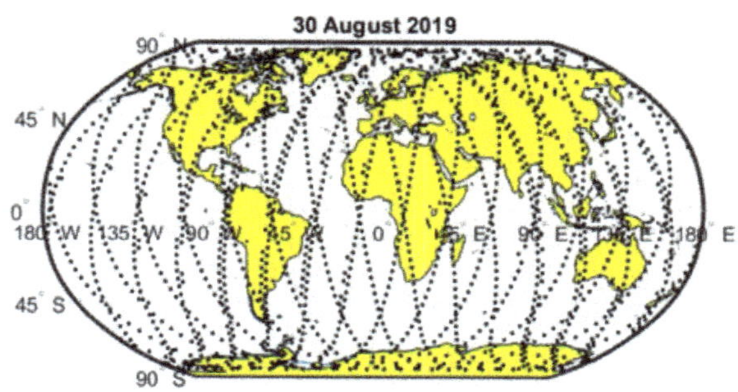

Figure 3. Typical points of radiation measurements.

Lucky 7 CubeSat is a private project of two people, namely Jaroslav Laifr and Pavel Kovar. Mr. Laifr developed and manufactured the satellite hardware including the power supply, radiation detector, and camera. He registered the satellite. Pavel Kovar developed the GPS receiver hardware and software and hardware and the software of the ultrahigh frequency (UHF) satellite modem, including a single-layer communication protocol capable of operating even at a high packet error rate. He is the author of the on-board computer hardware and software. He designed a ground station modem as well as ground station software.

The main component of the satellite is an onboard computer that integrates two independent computers and UHF modems that control the satellite, collect scientific data, and communicate with ground stations. The scientific module (Figure 4) is equipped with the GPS receiver piNAV 2 [32–34], the radiation detector piDOSE [35,36], a spectrometer and a low-resolution camera. The energy for satellite operation is generated by gallium arsenide solar cells that are mounted on the five satellite panels. The radiation hardening power supply is built from bipolar and silicon carbide transistors. Energy is stored in LiFe batteries with a capacity of 4 Wh.

Figure 4. Scientific module.

The Lucky 7 satellite can realize measurements up to 2048 min in duration. The on-board computer saves to memory the number of pulses registered by a radiation detector, position and time of the end of the measurement, and satellite statuses such as power supply voltage, current, currents from the solar cells, the temperature of the detector, and other subsystems.

2.2. PiDOSE

The piDOSE radiation detector is based on a PIN diode X100-7 coupled with a CsI:Tl scintillator with dimensions of 4 × 8 × 8 mm. The detector operates in particle counting mode. The energy deposition is not measured. The integration time is approximately 55 s. The data were converted to counts per minute and corrected for dead time. As shown in [36], the detector can register protons with energies higher than 30 MeV. Since the incident angle of particles and rotation of the satellite could not be determined, the counts were 4 π normalized. Hence, isotropical irradiation is assumed. Although the irradiation in the SAA is strongly directional, the satellite slowly rotates, which helps to mitigate the different directional sensitivities of piDOSE. A typical observed radiation field is shown in Figure 5.

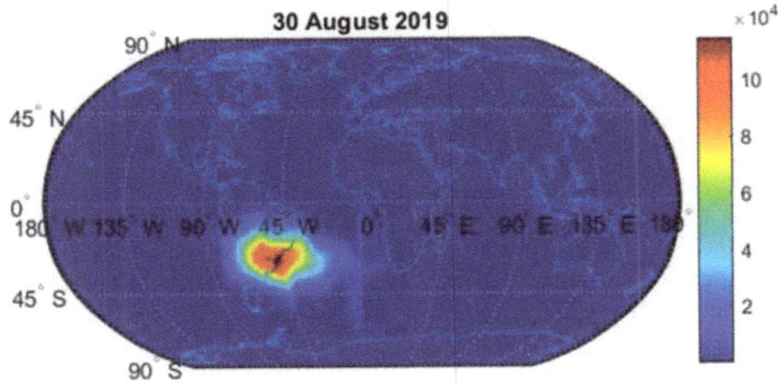

Figure 5. Example of the measurement of the radiation field (number of registered particles per minute (CMP)).

2.3. Position of the Measurement

As the satellite is not equipped with an altitude determination and control system, the satellite slowly rotates. The rotation rate is determined from the current of individual solar panels or from the periodic variation in the carrier to noise (C/N0) indicator of the GPS receiver. The rotation rate is not constant; the typical value is one turn per minute [34]. This rotation causes the GPS position to be unavailable some of the time [34]. The typical availability of PNT (position, navigation, and timing) solutions is 80% of the time. The problem was solved by extrapolation of the satellite position.

The following proposed interpolation algorithm is based on the satellite motion equation and its modification for the Earth-centered Earth-fixed (ECEF) coordinate system [37,38]:

$$
\begin{aligned}
\frac{dx}{dt} &= V_x;\ \frac{dy}{dt} = V_y;\ \frac{dz}{dt} = V_z \\
\frac{dV_x}{dt} &= -\frac{\mu}{r^3}x - \frac{3}{2}J_0^2\frac{\mu.a_e^2}{r^5}x\left(1-\frac{5z^2}{r^2}\right) + \omega^2 x + 2\omega V_y \\
\frac{dV_y}{dt} &= -\frac{\mu}{r^3}y - \frac{3}{2}J_0^2\frac{\mu.a_e^2}{r^5}y\left(1-\frac{5z^2}{r^2}\right) + \omega^2 y + 2\omega V_x \\
\frac{dV_x}{dt} &= -\frac{\mu}{r^3}z - \frac{3}{2}J_0^2\frac{\mu.a_e^2}{r^5}z\left(1-\frac{5z^2}{r^2}\right)
\end{aligned}
\tag{1}
$$

where (x, y, z) and (V_x, V_y, V_z) are the position and velocity vectors, respectively, of the satellite in ECEF coordinates provided by the onboard GPS receiver; $r = \sqrt{x^2+y^2+z^2}$ is a satellite radius; $\mu = 398600.4 \times 10^9\ m^3/s^2$ is the standard gravitation parameter of the Earth; $J_0^2 = 1082625.7 \times 10^{-9}$ is the second-harmonic coefficient of geopotential; $\omega = 7.292115 \times 10^{-5}$ rad/s is the Earth rotation rate; and a_e is the equatorial radius of the reference ellipsoid.

The satellite position is calculated by numerical integration of (1) using the fourth-order Runge–Kutta method [38].

The described algorithm enables the calculations of missing positions of the satellite and also enables the transformation of the position measurement from the end of the measurement cycle to its middle, as the satellite registers the position of the end of the measurement.

The ECEF position is, then, transformed to the geometrical coordinates LLH (longitude, latitude, and height) for further processing [38].

2.4. Position of the South Atlantic Anomaly (SAA)

The position of the SAA can be calculated as a centroid of the radiation field [13]. For this purpose, the data should be transformed to the latitude-longitude grid, and then, the centroid is calculated. The following formulas are presented based on [13]:

$$
\begin{aligned}
Latitude_{centroid} &= \frac{\sum(Interpolated\ Flux \times Flux's\ Latitude)}{\sum Interpolated\ Flux} \\
Longitude_{centroid} &= \frac{\sum(Interpolated\ Flux \times Flux's\ Longitude)}{\sum Interpolated\ Flux}
\end{aligned}
\tag{2}
$$

The problem with Equation (2) is that it does not consider the grid area; therefore, the individual grid points have equal mass. The problem is that the grid area decreases with latitude. Moreover, the method does not take into account that a one-degree grid is a spherical surface, and data are processed in two-dimension (2D). The proposed method is based on centroids in the space of constant curvature defined in [39]. The mass of the individual grids is calculated as a product of the number of registered particles per minute and grid area. The coordinates of the grid are transformed from the geodetic LLA coordinates to the ECEF, then the ECEF centroid coordinates are calculated using (3) and the results are transformed back to the LLH. The adjusted formulas for calculation of the longitudinal ECEF (x_c, y_c, z_c) coordinates of the centroid adjusted to the grid area size are as follows:

$$
\begin{aligned}
x_c &= \frac{\sum(CPM(x,y,z)\cdot cos(\varphi)\cdot x)}{\sum CPM(x,y,z)} \\
y_c &= \frac{\sum(CPM(x,y,z)\cdot cos(\varphi)\cdot y)}{\sum CPM(x,y,z)} \\
z_c &= \frac{\sum(CPM(x,y,z)\cdot cos(\varphi)\cdot z)}{\sum CPM(x,y,z)}
\end{aligned}
\tag{3}
$$

where λ and φ are coordinates of the grid in LLH, (x, y, z) are coordinates of the grid in ECEF, and $CPM(x, y, z)$ is the number of counts per minute in the grid.

2.5. Data Processing

The registered scientific data are downloaded by a ground station for further processing. One dataset contains data from approximately 22 orbits. The data processing can be summarized as follows:

(a) Calculation of the position of the satellite in the middle of the radiation detector, counting time by the method presented in Section 2.3.
(b) Resampling data to a one-degree grid.
(c) Calculation of the position of the radiation maximum as a centroid of the measured data (Section 2.4).
(d) Graphical presentation of the results.

3. Results

The radiation measurement in the SAA region is presented in Figure 6. The figure displays the development of the radiation field expressed as the number of particles registered per minute (CPM counts per minute) over the measurement campaigns, including the positions of the radiation maximum and centroid. The SAA positions are summarized in Table 2.

Table 2. Development of the position of the SAA.

Measurement	Centroid Position Long. Lat. [°]	Max. Position Long. Lat. [°]
(a) 30 August 2019	−25.8637 −48.6520	−27 −49
(b) 30 September 2019	−26.1674 −48.4611	−24 −48
(c) 27 March 2020	−26.2771 −49.1253	−23 −50
(d) 10 October 2020	−26.4505 −48.8551	−29 −55
(e) 1 November 2020	−26.4186 −49.2932	−24 −53
(f) 17 November 2020	−26.5828 −48.7944	−26 −57
(g) 28 December 2020	−26.1477 −48.5793	−29 −53
(h) 2 January 2021	−26.6111 −49.0591	−30 −60

Figure 7 displays the north–south and east–west cross-sections of the SAA radiation field measured in the individual campaigns.

The positions of the SAA radiation field maximum and centroid positions are also displayed in Figure 8. It is evident that the position of the maximum features a much higher scatter than the position of the centroid in which the scatter is considerably lower. The details of the centroid position scatter are shown in Figure 9.

The development of the position of the centroid was interpolated by a first-order polynomial (straight line), as shown in Figure 10. The resulting polynomials for latitude λ_F and longitude φ_F are as follows:

$$\lambda_F = -0.00093d - 26.0229\varphi_F = -0.00062d - 48.65654 \quad (4)$$

where d is the time in days from the first measurement (30 August 2019).

The position of the centroid is, therefore, moving by 0.00093° to the west and 0.00062° to the south per day, which is 0.34° to the west and 0.23° to the south per year.

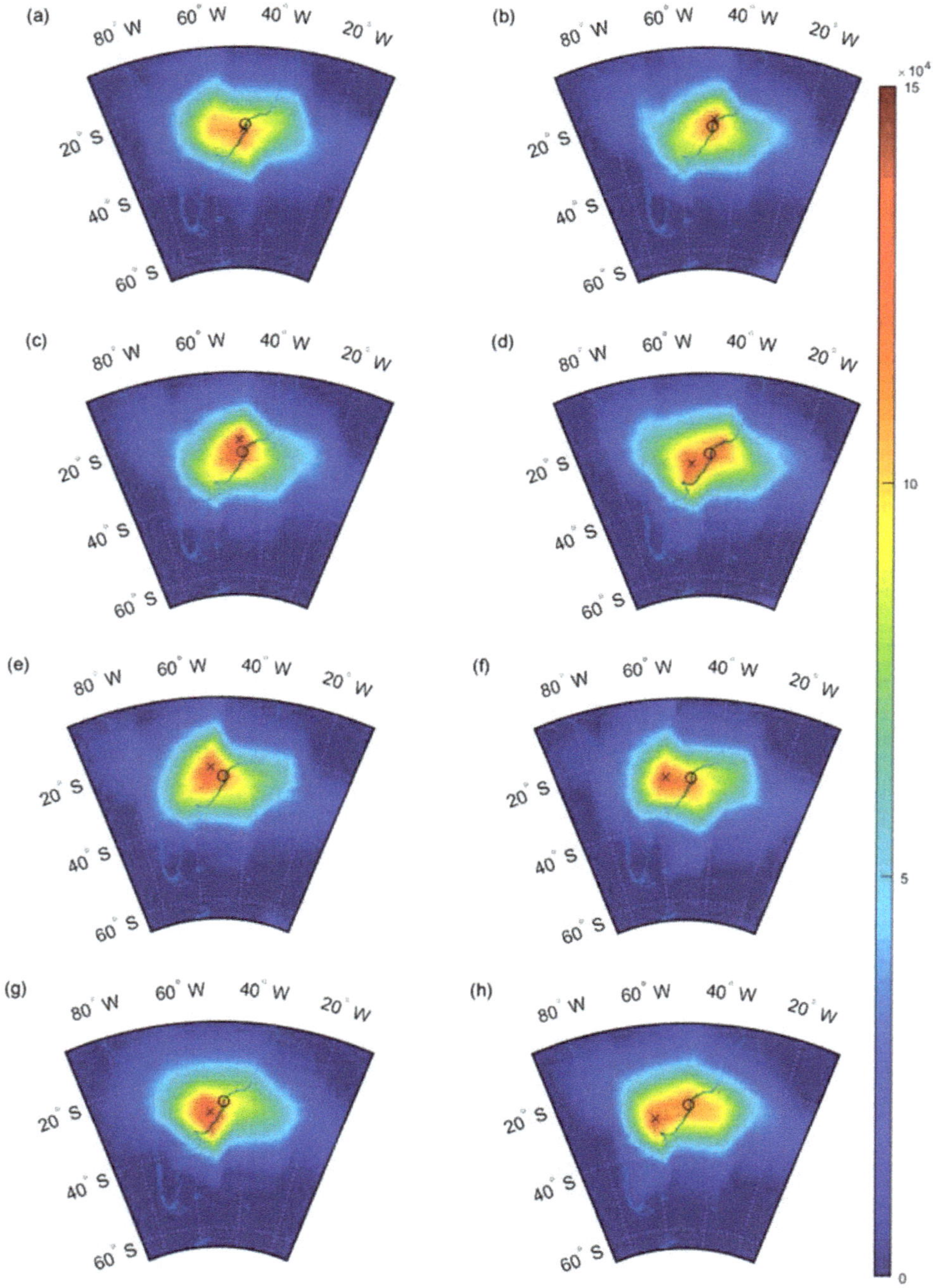

Figure 6. Development of the radiation field (CPM) in the SAA over measurements with the positions of the maximum of radiation (cross) and centroid (circle). (**a**) 30 August 2019; (**b**) 30 September 2019; (**c**) 27 March 2020; (**d**) 10 October 2020; (**e**) 1 November 2020; (**f**) 17 November 2020; (**g**) 28 December 2020; (**h**) 2 January 2021.

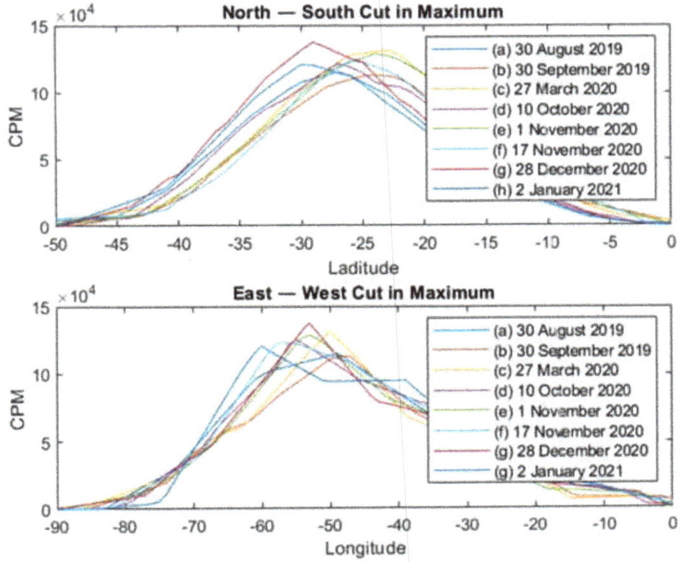

Figure 7. North–south and east–west cross-sections of the SAA radiation field.

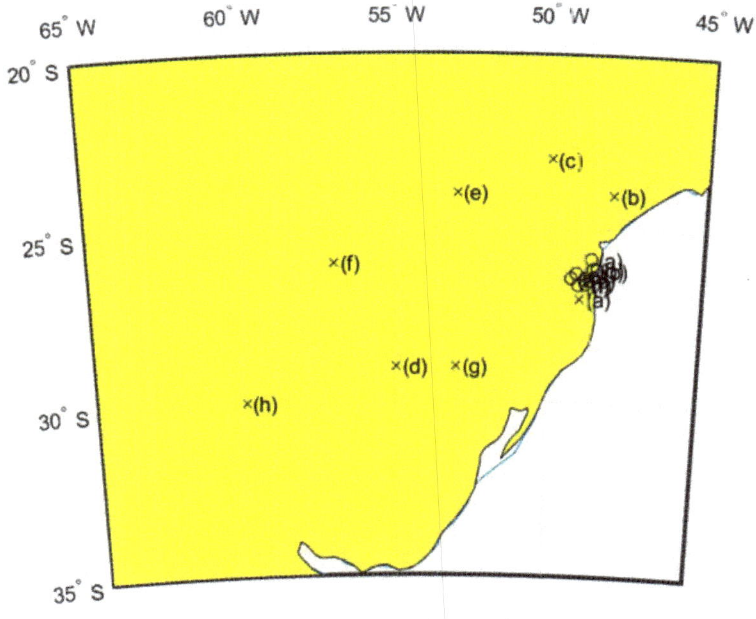

Figure 8. Development of the SAA maximum and centroid.

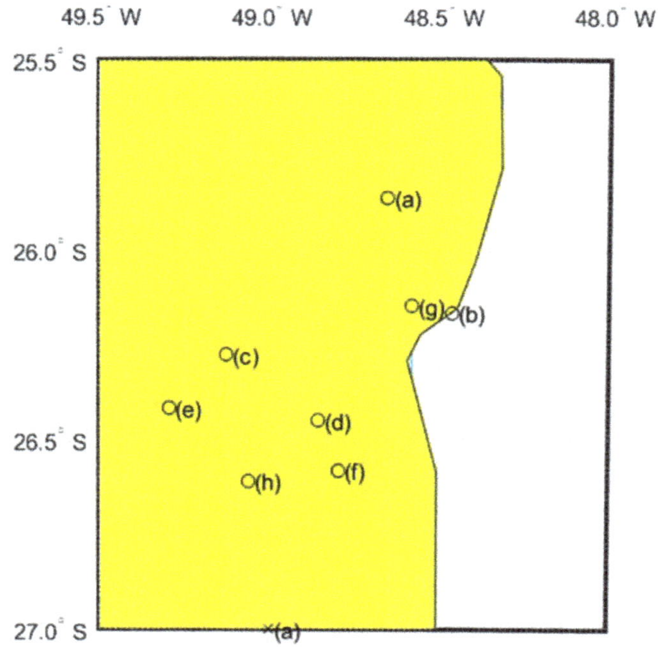

Figure 9. Development of the SAA centroid.

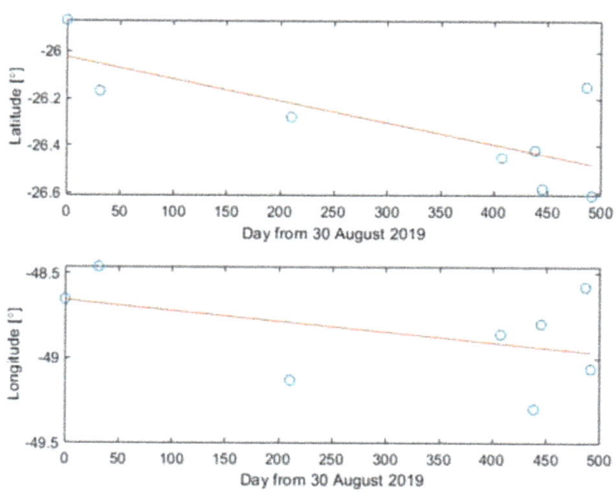

Figure 10. Straight line interpolation of the centroid position.

4. Discussion

The position of the South Atlantic Anomaly can be defined by various methods. The most intuitive is the position of the maximum of the fluence function. Alternatively, the position can be understood as a centroid of this function. In this paper, we applied both methods.

The position of the fluence maximum observed by individual measurement campaigns fluctuates more than the fluence centroid position. The positions of the fluence maxima are more sensitive than the position of the centroid to measurement noise. Although the centroid method is relatively insensitive to the number of measurement points, it would benefit from denser sampling of the SAA region. Denser sampling can be obtained by selecting an orbit with higher inclination or by releasing a swarm of satellites.

Our results show the average westward movement of the SAA by 0.34°/year, which is in good agreement with previously reported measurements summarized in Table 1. The data show an average southward movement rate of 0.23°/year. Even though most of the publications show northward drift, southward drift was also observed in [2,3,21]. The inconsistency in the north–south direction can be explained by applying the improved centroid calculation method that correctly takes into account grid area, in contrast to the application of the less precise centroid position calculation method used in [13].

As shown in [2], the drift of the SAA in latitude is dependent on the energy range of the protons that are measured. Protons within the SAA with lower energy tend to drift towards the north, whereas the most energetic protons of the SAA move towards the south. In [34], piDOSE was estimated to be sensitive to protons with energies larger than 30 MeV due to the shielding of sensitive volume. Moreover, the latitudinal movement of the SAA depends on the solar cycle [21]. Often, there was a shift in direction during the solar cycle minima and maxima. Since our data captured the time of the solar cycle minimum, it can be expected to observe a similar shift in SAA movement. Another effect that influences the measurement of latitudinal movement is a systematic error caused by the fast latitudinal movement of a satellite. The particle counting time is about one minute. In this time, the satellite travels approximately four degrees in orbit which is projected mainly in the latitude. The maximal error in the determination of the latitude position of the fluence maximum within the counting time is two degrees. For the uniform distribution the mean systematic error is about one degree. The error in the east–west direction is considerably lower due to the slower movement of the satellite in longitude.

5. Conclusions

Our study has demonstrated that small, cost-effective 1U CubeSats can be successfully used for the observation of SAA drift. Unlike the measurements introduced in Table 1, which were performed on large and heavy-payload satellites, we used a simply designed detector made of commercial off-the-shelf components that was able to operate for more than 20 months in an increased-radiation environment.

The results show that the method using the fitting of maximum fluxes is suitable for evaluating the drift rate in a large dataset. However, the centroid method normalized for the cosine-latitude effect requires less data to reach the same accuracy. Hence, the latter method is more suitable for CubeSats that have limited performance and data transfer.

We believe that a swarm of 1U CubeSats equipped with similar detectors could be used as a system for continuously monitoring the movement of the SAA. The advantage of such an approach is that individual CubeSats of the swarm can be placed in different orbits and at more suitable inclinations that would allow denser sampling in the SAA region. This would significantly improve the coverage and would allow more complex analysis of SAA drift. Such information would be beneficial for designing more accurate models of the SAA. As shown by [20], on the one hand, the drift of the SAA can change suddenly due to "geomagnetic jerks", which are not incorporated into the models. On the other hand, several studies have shown strong periodicity in SAA movement [13,21] which can be anticipated and foreseen by the models. For example, the ESA's Space Environmental Information System (SPENVIS) [8] models an average drift of the SAA of 0.3° in the westward direction and no movement in latitude, although a number of studies have observed slight movements in the northward or southward direction.

Author Contributions: P.K. developed a satellite UHF communication system ground station, the onboard computer, the satellite software, and the piNAV GPS receiver; he also processed satellite data and drew figures and prepared the parts related to the Lucky 7 satellite; M.S. prepared the introduction and interpreted the experimental results. Both authors have read and agreed to the published version of the manuscript.

Funding: The data processing and preparation of this document were supported by the European Regional Development Fund, Project CRREAT no. CZ.02.1.01/0.0/0.0/15 003/0000481.

Institutional Review Board Statement: Not applicable.

Informed Consent Statement: Not applicable.

Data Availability Statement: The data that support the findings of this study are available from the author upon reasonable request.

Acknowledgments: The authors would like to thank Jaroslav Laifr for building the Lucky 7 CubeSat and developing the piDOSE radiation detector.

Conflicts of Interest: The authors declare no conflict of interest.

References

1. Van Allen, J.A. The geomagnetically trapped corpuscular radiation. *J. Geophys. Res.* **1959**, *64*, 1683–1689. [CrossRef]
2. Ye, Y.; Zou, H.; Zong, Q.; Chen, H.; Wang, Y.; Yu, X.; Shi, W. The Secular Variation of the Center of Geomagnetic South Atlantic Anomaly and Its Effect on the Distribution of Inner Radiation Belt Particles. *Space Weather* **2017**, *15*, 1548–1558. [CrossRef]
3. Jones, A.D.; Kanekal, S.G.; Baker, D.N.; Klecker, B.; Looper, M.D.; Mazur, J.E.; Schiller, Q. SAMPEX observations of the South Atlantic anomaly secular drift during solar cycles 22–24. *Space Weather* **2017**, *15*, 44–52. [CrossRef]
4. Kurnosova, L.V.; Kolobyanina, T.N.; Logachev, V.I.; Razorenov, L.A.; Sirotkin, I.A.; Fradkin, M.I. Discovery of radiation anomalies above the South Atlantic at heights of 310–340 km. *Planet. Space Sci.* **1962**, *9*. [CrossRef]
5. Underwood, C.I. In-orbit radiation effects monitoring on the UoSAT satellites. In Proceedings of the Annual AIAA/Utah State University Conference on Small Satellites, Logan, UT, USA, 27–30 August 1990.
6. Deme, S.; Reitz, G.; Apáthy, I.; Héjja, I.; Láng, E.; Fehér, I. Doses due to the South Atlantic anomaly during the Euromir'95 mission measured by an on-board TLD system. *Radiat. Prot. Dosim.* **1999**, *85*, 301–304. [CrossRef] [PubMed]
7. Heirtzler, J.R. The future of the South Atlantic anomaly and implications for radiation damage in space. *J. Atmos. Solar-Terr. Phys.* **2002**, *64*, 1701–1708. [CrossRef]
8. Heynderickx, D. Comparison between methods to compensate for the secular motion of the South Atlantic Anomaly. *Radiat. Meas.* **1996**, *26*, 369–373. [CrossRef]
9. Hartmann, G.A.; Pacca, I.G. Time evolution of the South Atlantic Magnetic Anomaly. *An. Acad. Bras. Cienc.* **2009**, *81*, 243–255. [CrossRef]
10. Schaefer, R.K.; Paxton, L.J.; Selby, C.; Ogorzalek, B.; Romeo, G.; Wolven, B.; Hsieh, S.Y. Observation and modeling of the South Atlantic Anomaly in low Earth orbit using photometric instrument data. *Space Weather* **2016**, *14*, 330–342. [CrossRef]
11. Lanzerotti, L.J.; Gary, D.E.; Nita, G.M.; Thomson, D.J.; MacLennan, C.G. Noise in wireless systems from solar radio bursts. *Adv. Space Res.* **2005**, *36*, 2253–2257. [CrossRef]
12. Anderson, P.C.; Rich, F.J.; Borisov, S. Mapping the South Atlantic Anomaly continuously over 27 years. *J. Atmos. Solar-Terr. Phys.* **2018**, *177*, 237–246. [CrossRef]
13. Aubry, M.; Costes-Ori, V.; Standarovski, D.; Ecoffet, R. Analysis of the Drift of the South Atlantic Anomaly from ICARE and SEM-2 Flight Data. *IEEE Trans. Nucl. Sci.* **2020**, *67*, 1251–1255. [CrossRef]
14. Konradi, A.; Badhwar, G.D.; Braby, L.A. Recent space shuttle observations of the South Atlantic anomaly and the radiation belt models. *Adv. Space Res.* **1994**, *14*, 911–921. [CrossRef]
15. Badhwar, G.D. Drift rate of the South Atlantic Anomaly. *J. Geophys. Res. Space Phys.* **1997**, *102*, 2343–2349. [CrossRef] [PubMed]
16. Bühler, P.; Zehnder, A.; Kruglanski, M.; Daly, E.; Adams, L. The high-energy proton fluxes in the SAA observed with REM aboard the MIR orbital station. *Radiat. Meas.* **2002**, *35*, 489–497. [CrossRef]
17. Ginet, G.P.; Madden, D.; Dichter, B.K.; Brautigam, D.H. Energetic proton maps for the South Atlantic Anomaly. In Proceedings of the IEEE Radiation Effects Data Workshop, Honolulu, HI, USA, 23–27 July 2007.
18. Grigoryan, O.R.; Romashova, V.V.; Petrov, A.N. SAA drift: Experimental results. *Adv. Space Res.* **2008**, *41*, 76–80. [CrossRef]
19. Fürst, F.; Wilms, J.; Rothschild, R.E.; Pottschmidt, K.; Smith, D.M.; Lingenfelter, R. Temporal variations of strength and location of the South Atlantic Anomaly as measured by RXTE. *Earth Planet. Sci. Lett.* **2009**, *281*, 125–133. [CrossRef]
20. Casadio, S.; Arino, O. Monitoring the South Atlantic Anomaly using ATSR instrument series. *Adv. Space Res.* **2011**, *48*, 1056–1066. [CrossRef]
21. Qin, M.; Zhang, X.; Ni, B.; Song, H.; Zou, H.; Sun, Y. Solar cycle variations of trapped proton flux in the inner radiation belt. *J. Geophys. Res. Space Phys.* **2014**, *119*, 9658–9669. [CrossRef]
22. Olsen, N.; Mandea, M. Investigation of a secular variation impulse using satellite data: The 2003 geomagnetic jerk. *Earth Planet. Sci. Lett.* **2007**, *255*, 94–105. [CrossRef]
23. Toorian, A.; Diaz, K.; Lee, S. The CubeSat approach to space access. In Proceedings of the IEEE Aerospace Conference Proceedings, Big Sky, MT, USA, 1–8 March 2008.
24. Poghosyan, A.; Golkar, A. CubeSat evolution: Analyzing CubeSat capabilities for conducting science missions. *Prog. Aerosp. Sci.* **2017**, *88*, 59–83. [CrossRef]

25. Blackwell, W.J.; Allan, G.; Allen, G.; Burianek, D.; Busse, F.; Elliott, D.; Galbraith, C.; Leslie, R.; Osaretin, I.; Shields, M.; et al. Microwave Radiometer Technology Acceleration Mission (MiRaTA): Advancing Weather Remote Sensing with Nanosatellites. In Proceedings of the 28th Annual AIAA/USU Conference on Small Satellites, Logan, UT, USA, 4–7 August 2014.
26. Westerhoff, J.; Earle, G.; Bishop, R.; Swenson, G.R.; Vadas, S.; Clemmons, J.; Davidson, R.; Fanelli, L.; Fish, C.; Garg, V.; et al. LAICE CubeSat mission for gravity wave studies. *Adv. Space Res.* **2015**, *56*, 1413–1427. [CrossRef]
27. Weiss, W.W.; Rucinski, S.M.; Moffat, A.F.J.; Schwarzenberg-Czerny, A.; Koudelka, O.F.; Grant, C.C.; Zee, R.E.; Kuschnig, R.; Mochnacki, S.; Matthews, J.M.; et al. BRITE-Constellation: Nanosatellites for Precision Photometry of Bright Stars. *Publ. Astron. Soc. Pacific* **2014**, *126*, 573. [CrossRef]
28. Li, X.; Schiller, Q.; Blum, L.; Califf, S.; Zhao, H.; Tu, W.; Turner, D.L.; Gerhardt, D.; Palo, S.; Kanekal, S.; et al. First results from CSSWE CubeSat: Characteristics of relativistic electrons in the near-Earth environment during the October 2012 magnetic storms. *J. Geophys. Res. Space Phys.* **2013**, *118*, 6489–6499. [CrossRef]
29. Desai, M.I.; Allegrini, F.; Ebert, R.W.; Ogasawara, R.; Epperly, M.E.; George, D.E.; Christian, E.R.; Kanekal, S.G.; Murphy, N.; Randol, B. The CubeSat Mission to Study Solar Particles. *IEEE Aerosp. Electron. Syst. Mag.* **2019**, *34*, 16–28. [CrossRef]
30. Gieseler, J.; Oleynik, P.; Hietala, H.; Vainio, R.; Hedman, H.P.; Peltonen, J.; Punkkinen, A.; Punkkinen, R.; Säntti, T.; Hæggström, E.; et al. Radiation monitor RADMON aboard Aalto-1 CubeSat: First results. *Adv. Space Res.* **2020**, *66*, 52–65. [CrossRef]
31. Simms, L.M.; Jernigan, J.G.; Malphrus, B.K.; McNeil, R.; Brown, K.Z.; Rose, T.G.; Lim, H S.; Anderson, S.; Kruth, J.A.; Doty, J.P.; et al. CXBN: A blueprint for an improved measurement of the cosmological x-ray background. In Proceedings of the Hard X-ray, Gamma-ray, and Neutron Detector Physics XIV, San Diego, CA, USA, 24 October 2012; Volume 8507, p. 19.
32. Kovář, P. piNAV L1—GPS receiver for small satellites. *Gyroscopy Navig.* **2017**, *8*, 159–164. [CrossRef]
33. Kovář, P. Start Acceleration of the Space GPS Receiver. *Int. J. Electron. Telecommun.* **2020**, *66*, 745–752.
34. Kovar, P. Experiences with the GPS in Unstabilized CubeSat. *Int. J. Aerosp. Eng.* **2020**, *2020*, 1–20. [CrossRef]
35. Laifr, J.; Kovář, P. High altitude balloon test flight of miniature low power dosimeter for small satellites. *Slabop. Obz.* **2018**, *74*, 1–3.
36. Kovar, P.; Sommer, M.; Matthiae, D.; Reitz, G. Measurement of cosmic radiation in LEO by 1U CubeSat. *Radiat. Meas.* **2020**, *139*, 106471. [CrossRef]
37. Perov, A.; Charisova, V. *GLONASS, Principles, Architecture and Function*; Radiotechnika: Moscow, Russia, 2010; ISBN 978-5-88070-251-0.
38. Subirana, J.S.; Zornoza, J.M.; Hernández-Pajares, M. *GNSS DATA PROCESSING Volume I: Fundamentals and Algorithms*; ESA Communications; ESTEC: Noordwijk, The Netherlands, 2013; ISBN 978-92-9221-886-7.
39. Galperin, G.A. A concept of the mass center of a system of material points in the constant curvature spaces. *Commun. Math. Phys.* **1993**, *154*, 63–84. [CrossRef]

remote sensing

MDPI

Article

Landslide Deformation Prediction Based on a GNSS Time Series Analysis and Recurrent Neural Network Model

Jing Wang [1], Guigen Nie [1,2,*], Shengjun Gao [3], Shuguang Wu [1], Haiyang Li [1] and Xiaobing Ren [1]

1 GNSS Research Center, Wuhan University, Wuhan 430079, China; jingwangwhu@whu.edu.cn (J.W.); shgwu@whu.edu.cn (S.W.); haiyangli@whu.edu.cn (H.L.); xiaobinren@whu.edu.cn (X.R.)

2 Collaborative Innovation Center for Geospatial Information Technology, Wuhan 430072, China

3 Chinese Antarctic Center of Surveying and Mapping, Wuhan 430079, China; sjgao@whu.edu.cn

* Correspondence: ggnie@whu.edu.cn

Abstract: The prediction of landslide displacement is a challenging and essential task. It is thus very important to choose a suitable displacement prediction model. This paper develops a novel Attention Mechanism with Long Short Time Memory Neural Network (AMLSTM NN) model based on Complete Ensemble Empirical Mode Decomposition with Adaptive Noise (CEEMDAN) landslide displacement prediction. The CEEMDAN method is implemented to ingest landslide Global Navigation Satellite System (GNSS) time series. The AMLSTM algorithm is then used to realize prediction work, jointly with multiple impact factors. The Baishuihe landslide is adopted to illustrate the capabilities of the model. The results show that the CEEMDAN-AMLSTM model achieves competitive accuracy and has significant potential for landslide displacement prediction.

Keywords: GNSS time series analysis; landslide displacement prediction; attention mechanism; deep learning

Citation: Wang, J.; Nie, G.; Gao, S.; Wu, S.; Li, H.; Ren, X. Landslide Deformation Prediction Based on a GNSS Time Series Analysis and Recurrent Neural Network Model. *Remote Sens.* **2021**, *13*, 1055. https://doi.org/10.3390/rs13061055

Academic Editors: Serdjo Kos, José Fernández and Juan F. Prieto

Received: 8 February 2021
Accepted: 9 March 2021
Published: 10 March 2021

1. Introduction

Landslide disaster is one of the crucial topics in geological research [1]. The sustainable development of economies and society is seriously threatened as a result of landslide disasters [2]. Reliable early warning systems are a reasonable approach for landslide risk reduction [3,4]. The mechanisms analysis and prediction of landslide movements are the key components of landslide early warning [5–7]. Therefore, it is judicious to carry out landslide displacement prediction.

Landslide displacement prediction models can be divided into two categories: physical models and numerical models [8,9]. Traditional physical models provide a physical explanation for the prediction work according to geological theory [10]. Saito established a three-stage theory of landslide creep failure in 1968 [11,12], and Hoek proposed the extension line method to predict the time-displacement curve of Chilean landslides in 1977 [13]. However, physical models are deficient in their ability to meet the demands of dynamic large landslide prediction [14–16]. With the rapid development of mathematical statistical theory and intelligent algorithms, numerical models have become more popular [5]. Numerical models fully consider the complexity and nonlinearity of the landslide evolution process and have higher prediction accuracy [5,17].

Advances in machine learning provide a powerful tool for numerical landslide model research. Zhou et al. [17] used kernel extreme learning for landslide displacement prediction. Zhu et al. [18] proposed a least squares support vector model and applied it to prediction of the Shuping landslide. Among them, Recurrent Neural Networks (RNNs) have particular advantages in dealing with sequential data [19,20]. Different from other neural networks, RNNs are the deepest algorithms [21], and they can effectively process data information with higher dimensions [22]. As a variant of RNNs, Long Short Term Memory (LSTM) networks perform better at storing and transferring historical information than RNNs [23–26]. The utility of the LSTM in landslide research has been confirmed by many scholars [27–30]. Thus, we choose an LSTM network for landslide displacement prediction in this paper.

The Attention Mechanism (AM) is currently a powerful deep learning toolkit [31]. AM is similar to the human visual observation mechanism that can transfer key information from the input information [32]. AM has been successfully applied in several tasks, such as natural language processing [31], translation [33], and image recognition [34]. Li et al. [35] added the Attention Mechanism to the LSTM model and successfully realized the prediction of personal mobility. Ding et al. [36] proposed a spatio-temporal attention LSTM model for flood forecasting. Thus, we incorporate an Attention Mechanism with an LSTM neural network to capture significant variation and improve the model's performance.

Therefore, a novel model based on time series analysis and Attention Mechanism with Long Short Term Memory (AMLSTM) was proposed to predict landslide displacement. The Baishuihe landslide in China, Hubei province, is utilized for the experiment area. First, we use the Complete Ensemble Empirical Mode Decomposition with Adaptive Noise (CEEMDAN) algorithm to divide the total displacement into the trend term, the periodic term, and the residual term. By analyzing the corresponding relationship between displacement and external factors, a multiple factors AMLSTM model, is applied to predict the displacement, and it is compared with a further four machine learning models. A series of contrastive analyses are conducted to evaluate the performance of all of the models. The results indicate that the proposed CEEMDAN-AMLSTM model performs best in the experiment.

2. GNSS Time Series Analysis

2.1. Landslide Evolution Analysis

The evolution of landslides is the result of the interaction of geological conditions and external factors [37]. The non-linear and non-stationary landslide displacement series are particularly complex and changeable. Therefore, it is necessary to decompose the landslide time series and forecast each component separately. The corresponding time series of the landslide displacement can be expressed by the additive model:

$$y_i = \mathrm{T_i} + S_i + R_i \tag{1}$$

where y_i is the cumulative displacement, $\mathrm{T_i}$ is the trend term, S_i is the period term, and R_i is the residual term.

2.2. Decomposition of Displacement Time Series

Many approaches have been recognized as being powerful tools for decomposing landslide displacement time series, and they include moving average [38], wavelet analysis [39], Variational Mode Decomposition (VMD) [40], and Empirical Mode Decomposition (EMD) [41]. The EMD method is an adaptive method that is used to analyze non-linear signals [42]. However, the model mixing problem constitutes an obstacle when using EMD. To address this problem, the Complete Ensemble Empirical Mode Decomposition with Adaptive Noise (CEEMDAN) method has been proposed in recent years [43]. Compared to the more commonly used EMD method, it has a better separation effect and is noise free. It has many applications in the fields of biological signal processing [44] and engineering [45], but its application in the geological field still needs to be explored.

The CEEMDAN decomposes the complex signal into a finite number of Intrinsic Mode Functions (IMFs). The basic process of the CEEMDAN is as follows [46]:

1. White Gaussian noises is added onto the lines of EEMD. The first IMF can be expressed as:

$$IMF_1 = \sum_{i=1}^{n} \frac{E_1(x + \varepsilon w_i)}{n} \tag{2}$$

where n is the number of decomposition, x is the original signal, ε is a fixed coefficient, w_i is the noise, and E(·) is the decomposition operator.

2. The first residual, r_1, is calculated:

$$r_1 = x - IMF_1 \tag{3}$$

3. For k = 2,3 . . . , K, the IMF_k and the kth residual can be calculated by:

$$IMF_k = \sum_{i=1}^{n} \frac{E_1(r_{k-1} + \varepsilon E_{k-1}(w_i))}{n} \tag{4}$$

$$r_k = r_{k-1} - IMF_k \tag{5}$$

4. The process is calculated until the last residual, R, does not have more than two extrema points; the original signal can be expressed as:

$$x = \sum_{k=1}^{K} IMF_k + R \tag{6}$$

3. Attention Mechanism—LSTM Foresting Framework

3.1. LSTM

Long Short Time Memory (LSTM) was proposed by Hochreiter and Schmidhuber in 1997 [23]. The LSTM can learn information through a well-designed structure called a "gate". The gate can store and control the flow of information so that the state of the previous time step can be transferred to the next time step. The LSTM algorithm has three gates—update gate, forget gate, and output gate—to protect and control the cell state explosion in training [25]. The internal structure of the unit memory is as shown in Figure 1.

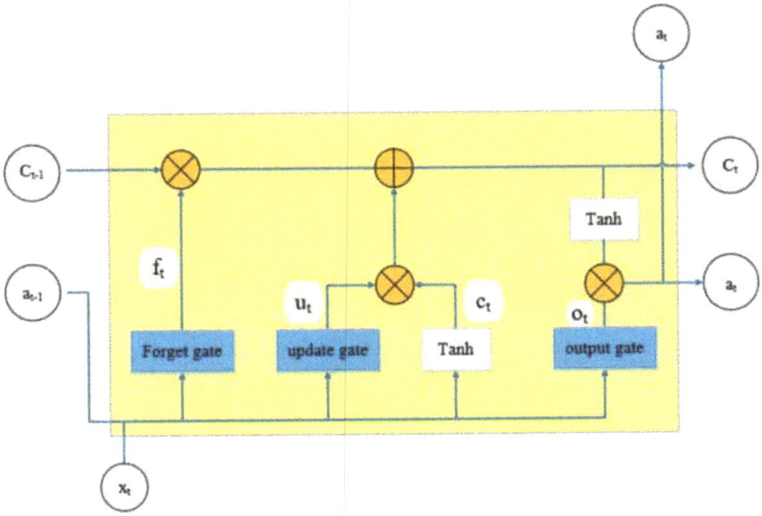

Figure 1. The internal structure of the Long Short Time Memory (LSTM) unit memory.

The ⊗ represents the element-wise product and ⊕ is the element-plus product. The forget gate represents how much of the previous moment unit sate, c_{t-1}, is retained by the current moment, c_t. The input gate determines how much of the current moment input, x_t, is saved in the unit state, c_t. The output gate controls how much of the unit state, c_t, is transferred to the output value, h_t, of the LSTM.

Equations (7)–(12) show the calculation process of LSTM:

$$f_t = \sigma\left(W_f * [a_{t-1}, x_t] + b_f\right) \tag{7}$$

$$u_t = \sigma(W_u * [a_{t-1}, x_t] + b_u) \tag{8}$$

$$\widetilde{c}_t = tanh(W_c * [a_{t-1}, x_t] + b_c) \tag{9}$$

$$c_t = f_t * c_{t-1} + u_t * \widetilde{c}_t \tag{10}$$

$$o_t = \sigma(W_o * [a_{t-1}, x_t] + b_o) \tag{11}$$

$$a_t = o_t * \tanh(c_t) \tag{12}$$

where f_t, u_t, and o_t are gating vectors that respectively store the forgotten, updated, and output information of the storage unit memory; c_t is the vector for the cell state; a_t is the hidden state vector; σ is the sigmoid function; and x_t is the input vector. W_f, W_u, W_c, and W_o are linear transformation matrices whose parameters need to be learned, and b_f, b_u, b_c, and b_o are corresponding bias vectors.

Through the connection of several unit memories, the information flow can be transferred as shown in Figure 2.

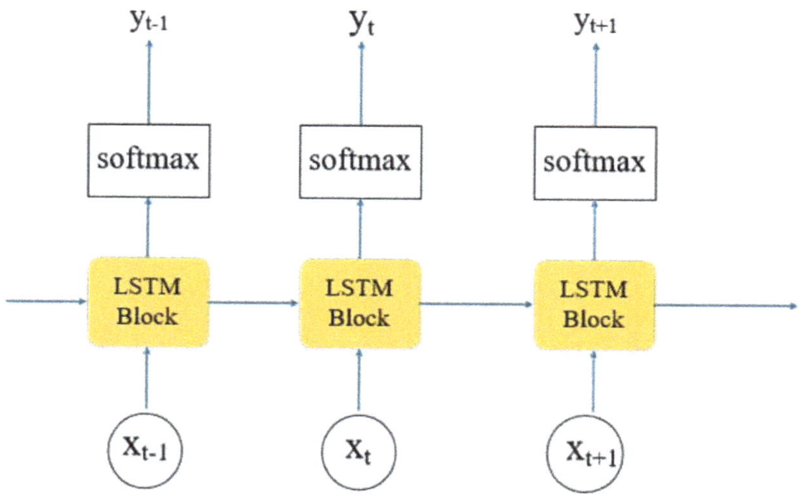

Figure 2. The workflow of LSTM.

3.2. Attention Mechanism

The Attention Mechanism is based on the visual Attention Mechanism found in human observation [32]. This mechanism helps the model focus on the salient information. The schematic of the Attention Mechanism layer is illustrated in Figure 3. The purpose of the attention layer is to enable the model to pay more attention to the significant information. Raffel et al. [47] proposed a reduced Feed-Forward Attention model, which was calculated as follows:

$$score_t = v(a_t) \tag{13}$$

$$w_t = \frac{\exp(score_t)}{\sum_{k=1}^{T} \exp(score_t)} \tag{14}$$

$$\text{s} = \sum_{t=1}^{T} w_t * a_t \tag{15}$$

where the score is the attention score, a is the state vector, v is the learnable function, w is the weight, and s is the context vector.

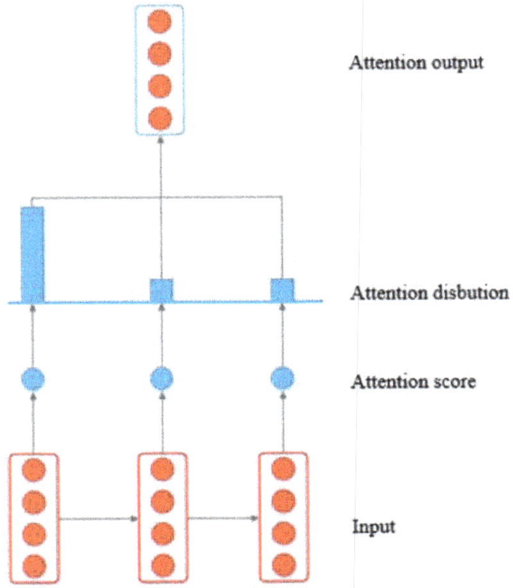

Figure 3. Schematic of the Attention Mechanism.

3.3. *Attention Mechanism—LSTM Model*

Based on the previous discussion, this paper applied the Attention Mechanism with LSTM (AMLSTM) model for landslide displacement prediction. The AMLSTM model includes an input vector, LSTM hidden layers, an attention layer, a fully connected layer, and output predicted values. The architecture of the AMLSTM model is shown in Figure 4.

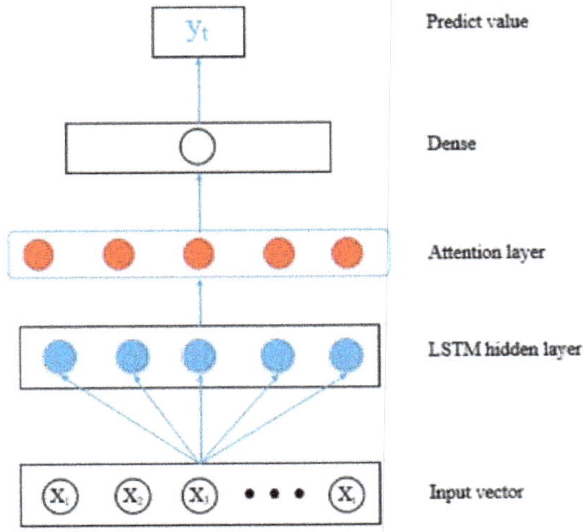

Figure 4. Architecture of the Attention Mechanism with LSTM Neural Network (AMLSTM NN).

3.4. Prediction Process with the Proposed Model

The basic flow of the proposed CEEMDAN-AMLSTM model is shown in Figure 5. Firstly, the landslide cumulative displacement is decomposed into three components: the trend term, the periodic term, and the residual term. The three terms are then predicted separately. The trend displacement is expressed as a monotone increasing function under the influence of internal geological factors. The prediction of the trend term can be carried out by fitting the growth curve with the univariate AMLSTM model. During the construction of the model, the displacement time series is put into the model only. The periodic displacement fluctuates under the influence of two external triggers: rainfall and reservoir water level. Therefore, a multivariable AMLSTM model is established and used to predict the periodic term. Three time series, the historical periodic displacement, rainfall, and reservoir water level are put into the model. Furthermore, the residual displacement affected by random factors shows smooth fluctuation function. The univariate AMLSTM model is adopted for the prediction work.

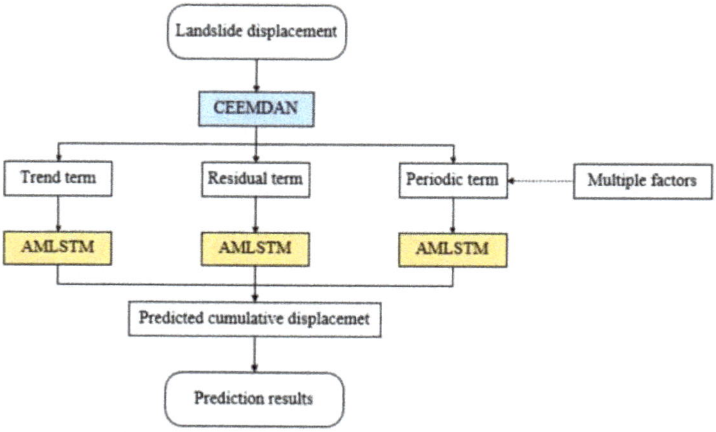

Figure 5. The architecture of the Complete Ensemble Empirical Mode Decomposition with Adaptive Noise (CEEMDAN)-AMLSTM model for landslide displacement prediction.

In the prediction experiments, the majority dataset is used to train the model. The original time series should be normalized and reshaped to meet the requirements of the model. After the AMLSTM model is constructed, the prediction ability is tested and demonstrated with the rest of the dataset.

Ultimately, the cumulative prediction displacement is obtained by adding the trend, the periodic, and the residual prediction displacements. The prediction results should be compared with the actual value to verity the performance.

3.5. Evaluation of Model Accuracy

Quantitative analysis were carried out to access the performance of the model. Three criterions—Root Mean Square Error (RMSE), Mean Absolute Error (MAE), and R^2—were employed to evaluate the prediction work. These metrics are described as follows:

$$\text{RMSE} = \sqrt{\frac{I}{N} * \sum_{i=1}^{N} (y_i - \hat{y}_i)^2} \tag{16}$$

$$\text{MAE} = \frac{1}{N} * \sum_{i=1}^{N} |y_i - \hat{y}_i| \tag{17}$$

$$R^2 = 1 - \frac{\sum_{i=1}^{N}(y_i - \hat{y}_i)^2}{\sum_{i=1}^{N}(y_i - \overline{y})^2} \tag{18}$$

where y_i is the measured value, $\hat{y}_i$ is the prediction value, and $\overline{y}$ is the average value.

4. Experiment and Results

4.1. Study Area

The experimental area is located in Baishuihe, Zigui County, the Three Gorges Reservoir area of the Yangtze River in China. The Baishuihe landslide is located on the south bank of the Yangtze River, with a longitude of 110°32′09″ and a latitude of 31°01′34″ (Figure 6a). The slope is located on the south bank of the Yangtze River, spreading towards the Yangtze River in a ladder shape. The elevation of the back edge of the landslide is 410 m, bounded by the rock-soil boundary, and the front edge is about 70 m. It has been submerged below the reservoir water level. The east and west sides are bounded by bedrock ridges, and the overall slope is about 30°. The length of the north-south direction is 600 m, the width of the east-west direction is 700 m, the average thickness of the sliding body is about 30 m, and the volume is 1.26×10^7 m^3. Six Global Navigation Satellite System (GNSS) deformation monitoring points were installed on the surface of the landslide to form three longitudinal monitoring profiles (Figure 6b). The displacement was monitored once a month. Figure 7 shows the calculated displacement results from December 2006 to December 2012.

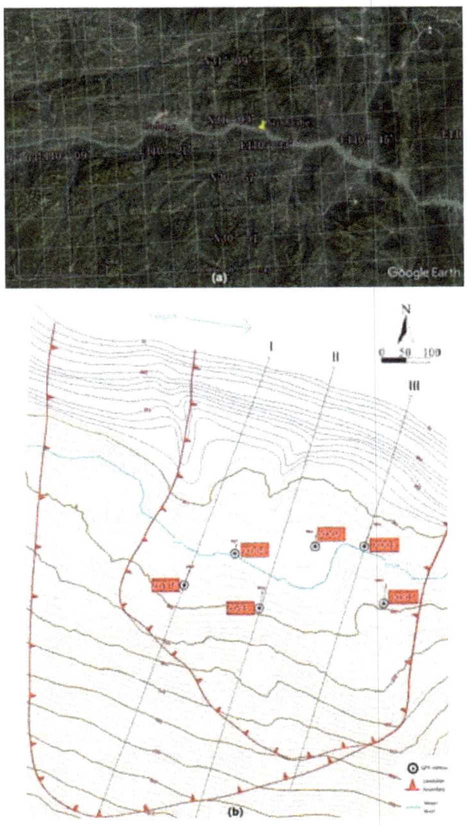

Figure 6. (**a**) Location map from Google Earth and (**b**) Locations of the monitoring Global Navigation Satellite System (GNSS) stations on the landslide.

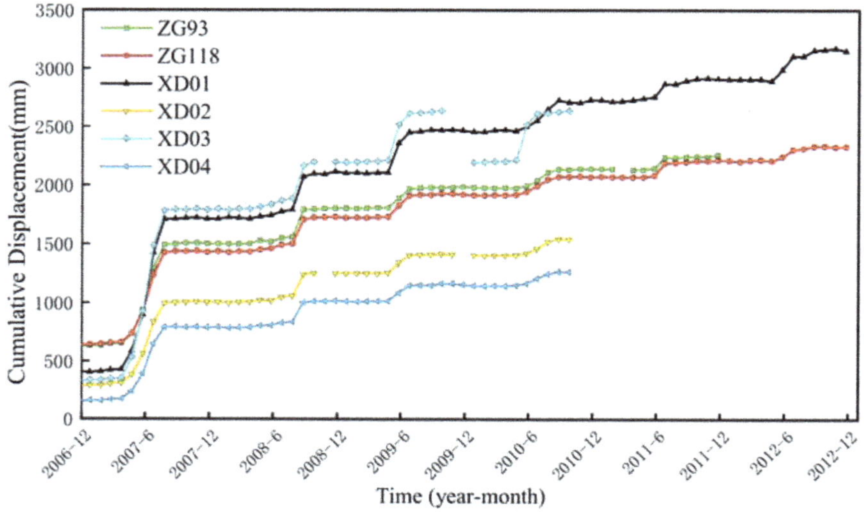

Figure 7. Cumulative displacement monitoring data based on six GNSS points.

It can be seen from Figure 7 that the landslide deformation is characterized by stepwise progressive creep deformation, and the landslide is still in the energy accumulation stage, showing a slow creep deformation state. In this experiment, ZG118 and XD01, the two points with the most abundant dataset, are selected for the prediction work. The measurements from December 2006 to November 2011 are used for training and the measurements obtained from November 2011 to November 2012 are used for testing. Each time interval of the train and test dataset is one month. The cumulative displacements, the reservoir water level, and the rainfall are plotted in Figure 8.

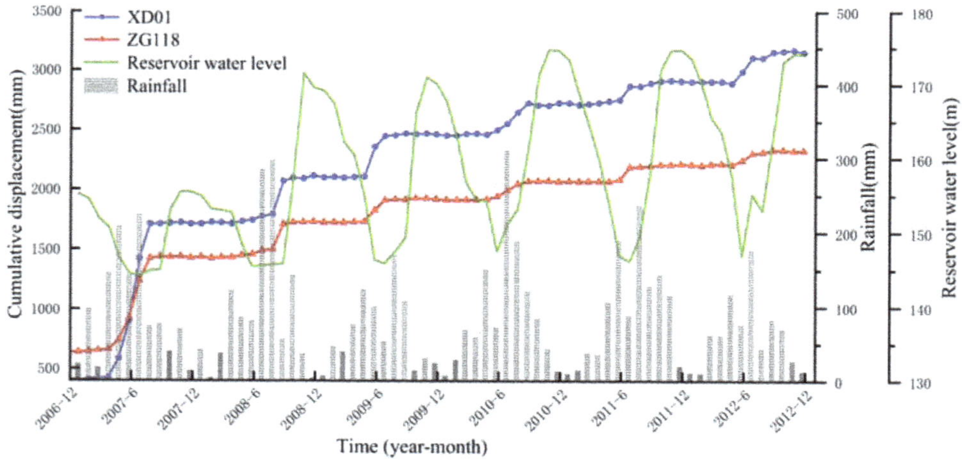

Figure 8. Relationship between rainfall, reservoir water level, and landslide displacement on ZG118 and XD01.

Figure 8 shows that the external periodic rainfall and reservoir water level both have an important influence. The displacement of XD01 and ZG118 increased significantly during a period of drastic decrease of the reservoir water level. For example, from May 2009 to July 2009, the reservoir water level dropped from 160 m to 145 m, and their periodic displacement increased by 200~300 mm, presenting a large step. In addition, heavy rain also had an important effect on landslide displacement fluctuations. For example, from August 2008 to September 2008, the reservoir water level basically did not change but, due to the occurrence of 300 mm of heavy rain during this period, the landslide also showed a large deformation of 200 mm. Therefore, the reservoir water level and rainfall are considered to be the trigger factors of the Baishuihe landslide, leading to the occurrence of the periodic term displacement.

4.2. GNSS Time Series Analysis

According to landslide analysis theory, the cumulative displacement can be decomposed into trend displacement, periodic displacement, and residual displacement using the CEEMDAN algorithm. The results are as follows (Figure 9):

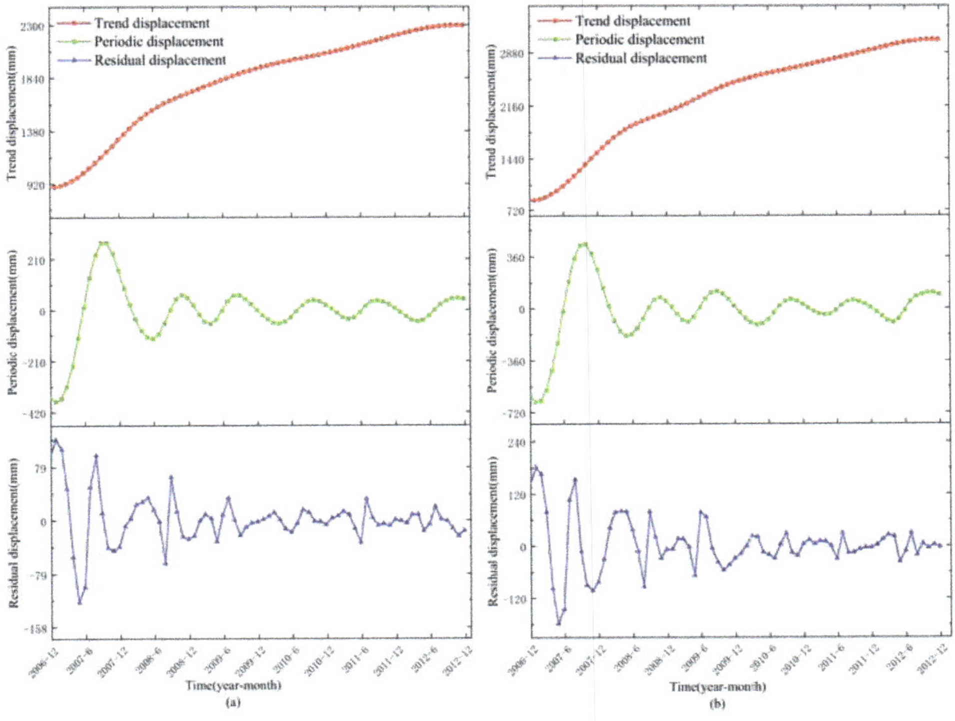

Figure 9. Three decomposed terms of the GNSS time series: (**a**) ZG118, (**b**) XD01.

4.3. Displacement Prediction

4.3.1. Trend Displacement Prediction

Trend displacement is driven by geological conditions. Therefore, the univariate AMLSTM NN model is used to predict the trend displacement. In order to verify the validity of the proposed model, the experiment will be benchmarked with LSTM, Random Forest(RF), RNN, and Support Vector Machine(SVM). The prediction results of the test dataset are shown in Figure 10.

It can be seen in Figure 10 that the trend displacement of the ZG118 and XD01 points represent a smooth monotonically properties. The prediction work by the SVM shows the worst, and the prediction values of the AMLSTM, LSTM, RNN, and RF models show high agreement with the measured true value. The relative error analysis in Table 1 indicates that the AMLSTM, LSTM, and RF have excellent performance in trend term prediction work.

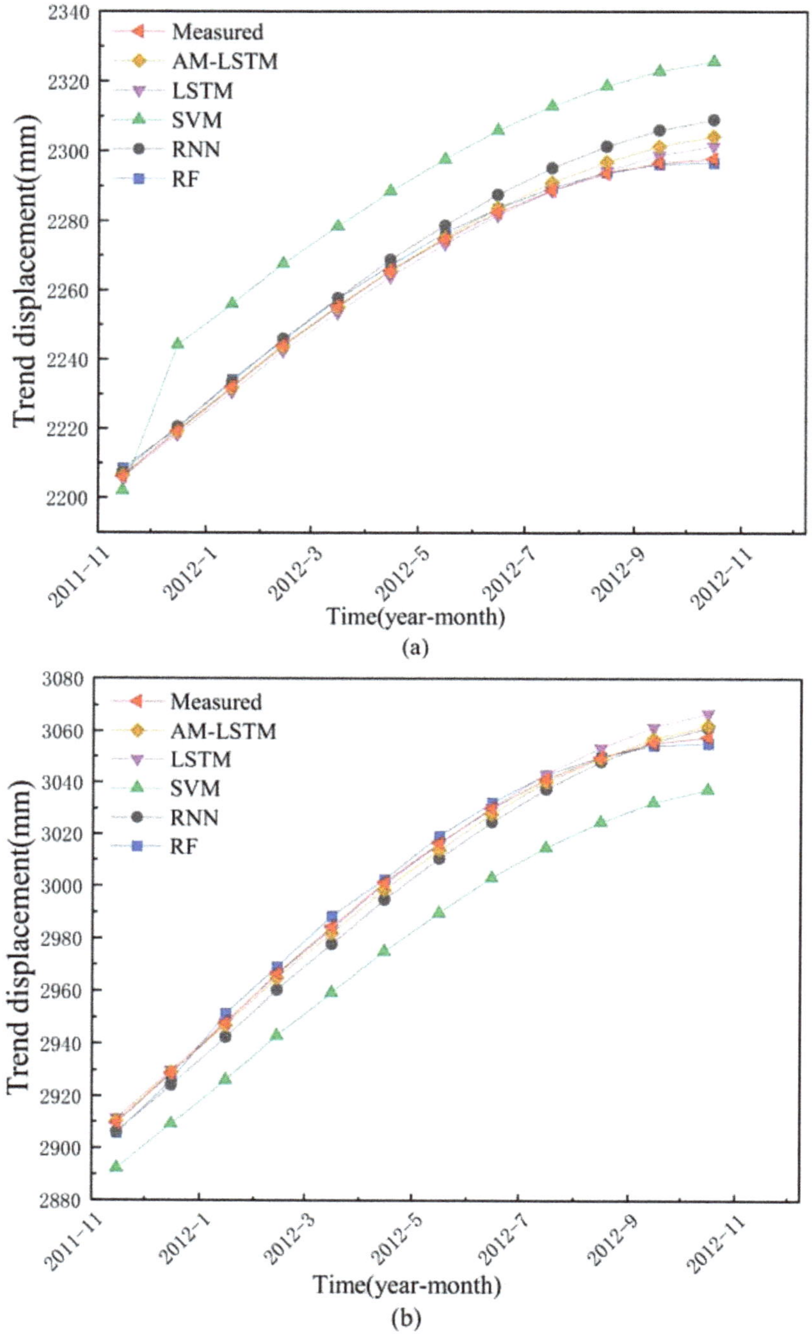

Figure 10. The prediction results of trend displacement by different methods: (**a**) ZG118, (**b**) XD01.

Table 1. The accuracy assessment of trend displacement by different prediction models.

Model	RMSE		MAE		R^2	
	ZG118	XD01	ZG118	XD01	ZG118	XD01
AMLSTM	2.6152	2.1254	1.6785	1.7849	0.9925	0.9981
LSTM	1.6773	3.5006	1.4072	2.2426	0.9969	0.9949
RNN	5.6158	4.8276	4.5776	4.4758	0.9655	0.9904
SVM	23.3985	23.7356	22.6717	23.5418	0.4018	0.7678
RF	1.4897	2.6540	1.3317	2.3943	0.9976	0.9971

4.3.2. Periodic Displacement Prediction

Periodic term is a key component for displacement prediction. According to the analysis in Section 4.1, the external periodic rainfall and reservoir water level both have an important influence. In this section, the periodic displacement will be predicted by the multivariate AMLSTM, and the multivariate LSTM, the SVM, the RF, and the RNN are used as benchmarks. The predictive periodic displacements by the five models are shown in Figure 11 and Table 2.

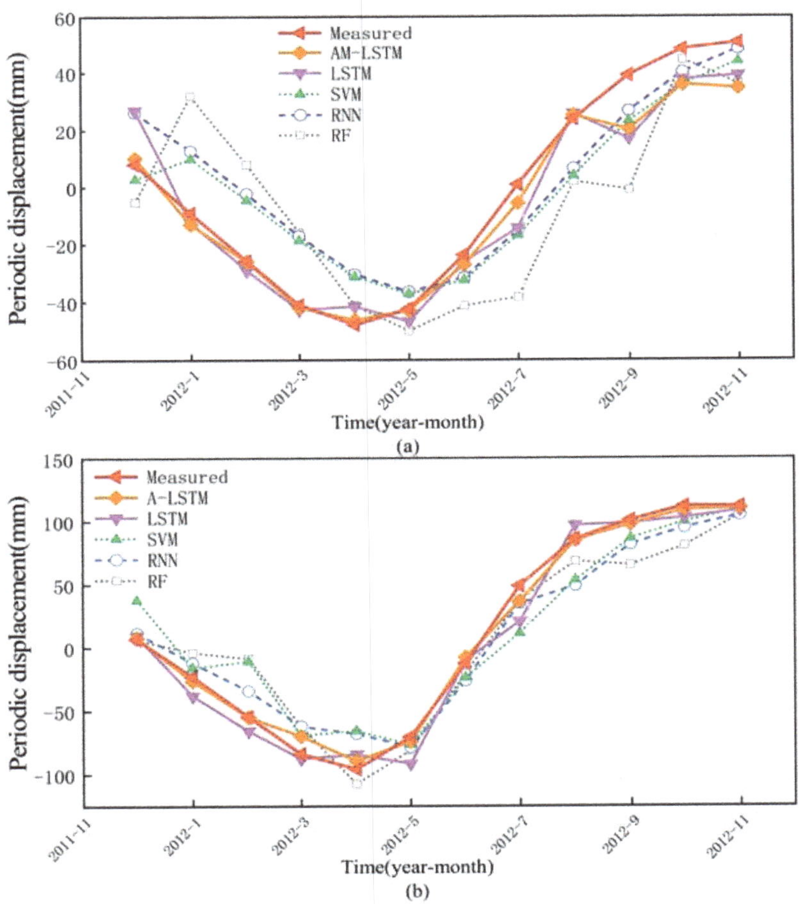

Figure 11. The prediction results of periodic displacement by different methods: (**a**) ZG118, (**b**) XD01.

Table 2. The accuracy assessment of periodic displacement by different prediction models.

Model	RMSE		MAE		R^2	
	ZG118	**XD01**	**ZG118**	**XD01**	**ZG118**	**XD01**
AMLSTM	8.3714	6.1623	5.6456	4.5016	0.9404	0.9933
LSTM	10.9127	12.8428	8.5083	10.1266	0.8987	0.9711
RNN	16.1422	18.9561	14.5892	16.7908	0.7784	0.9371
SVM	15.4854	24.2245	14.236	20.2412	0.796	0.8972
RF	25.6368	22.3304	22.0298	18.298	0.441	0.9126

As shown in Figure 11, the predictions of the AMLSTM and LSTM methods are clearly better than the others, and the quantitative analysis suggest that the AMLSTM achieved the best performance, along with RMSE, MSE, and R^2, in periodic displacement prediction.

4.3.3. Residual Displacement Prediction

Traditionally, the residual term can be regarded as the noise, which is removed during the decomposition procedure. Throughout the test, the residual term does not belong to the white noise. Therefore, the prediction work of this term is necessary. In this experiment, the univariate AMLSTM, LSTM, SVM, RF, and RNN models are used to predict the residual displacement prediction.

Compared with the trend and the periodic term, the residual term is harder to adopt in a model because of its random characteristic. As shown in Figure 12 and Table 3, the AMLSTM offers a better prediction effect than the other four models.

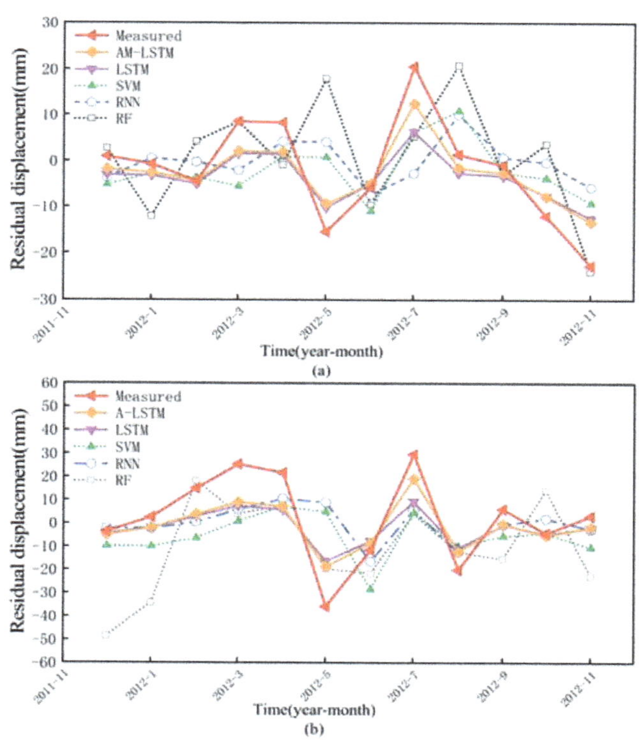

Figure 12. The prediction results of residual displacement by different methods: (**a**) ZG118, (**b**) XD01.

Table 3. The accuracy assessment of residual displacement by different prediction models.

Model	RMSE		MAE		R^2	
	ZG118	**XD01**	**ZG118**	**XD01**	**ZG118**	**XD01**
AMLSTM	5.1002	9.8401	4.2185	8.2213	0.7897	0.7132
LSTM	6.4204	11.9279	5.1768	9.7219	0.6667	0.5785
RNN	11.5546	17.0916	9.0241	12.5840	−0.0796	0.1346
SVM	9.7371	19.2705	8.2718	16.4355	0.2333	−0.1001
RF	13.8302	23.4540	10.1748	20.5233	−0.5467	−0.6296

4.3.4. Total Displacement Prediction

The predicted cumulative displacements can be obtained by taking the sum of the trend, period, and residual displacements. The results are shown in Figure 13 and Table 4.

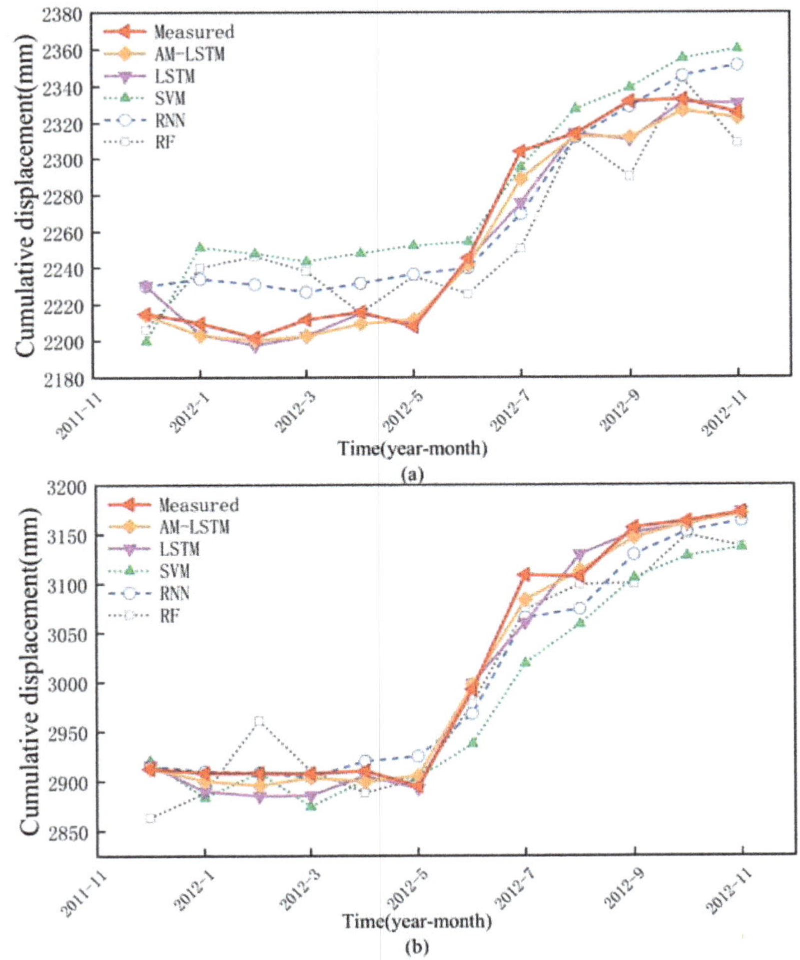

Figure 13. The prediction results of cumulative displacement by different methods: (**a**) ZG118, (**b**) XD01.

Table 4. The accuracy assessment of cumulative displacement by different prediction models.

Model	RMSE		MAE		R^2	
	ZG118	XD01	ZG118	XD01	ZG118	XD01
AMLSTM	8.5514	10.249	6.5395	8.0242	0.9748	0.9918
LSTM	11.7059	18.8873	7.8044	13.3813	0.9528	0.9723
RNN	20.4623	21.4569	17.6515	16.3575	0.8556	0.9634
SVM	29.1695	41.3469	25.6171	33.3799	0.7066	0.8673
RF	28.5883	32.0225	23.5398	26.5033	0.7182	0.9204

The results show that, although some of the prediction values slightly deviate from the real measured data, the AMLSTM model shows the best performance, because this model not only considers multiple external factors, but also optimizes the LSTM algorithm by adding an attention layer. It can better reflect the response relationship between displacement and trigger factors. Moreover, the cumulative displacements are predicted badly by the SVM and RF models.

From a quantitative point of view, the RMSE and MAE of the AMLSTM model are lower than the LSTM, RNN, SVM, and RF models. These results reveal that the AMLSTM shows the most stable prediction performance. Secondly, the R^2 of the AMLSTM are higher than the others. The results indicate that the AMLSTM model has done the best accuracy prediction work. Therefore, the superiority of the AMLSTM can be proved.

5. Conclusions

The traditional landslide prediction model directly deletes the residual items. Moreover, most classic deep learning prediction models do not highlight the impact of important information on the results, so they cannot accurately predict the displacement. This paper used the CEEMDAN and the Attention Mechanism, combined with the LSTM NN to establish a dynamic prediction model for landslide displacement prediction. To corroborate its feasibility and applicability, the proposed model was applied to the Baishuihe landslide area, and joint multiple impact factors were considered here for prediction. By comparing to the prediction effects of other models, the prediction accuracy demonstrated a competitive performance. The results strongly suggest the effectiveness and feasibility of the AMLSTM model in landslide displacement prediction. This novel CEEMDANAM-LSTM strategy can be recommended to other landslide prediction works and has great potential in landslide risk assessment.

Author Contributions: Conceptualization, J.W. and G.N.; methodology, J.W.; software, J.W.; validation, S.W. and X.R.; formal analysis, S.G.; investigation, G.N. and S.G.; writing—original draft preparation, J.W.; writing—review and editing, J.W. and H.L. All authors have read and agreed to the published version of the manuscript.

Funding: This study was financially supported by the National Key Research and Development Scheme Strategic International Cooperation in Science and Technology Innovation Program, grant number: 2018YFE0206500. National Program on Key Basic Research Project (973 Program), grant number: 2013CB733205.

Institutional Review Board Statement: Not applicable.

Informed Consent Statement: Not applicable.

Data Availability Statement: Not applicable.

Acknowledgments: The dataset we used in this paper includes the GNSS time series, rainfall and reservoir water level data set of Baishuihe landslide provided by Chinese National Cryosphere Desert Data Center (http://www.crensed.ac.cn/portal/, accessed on 9 February 2021). The authors acknowledge Google Earth for providing the map and Origin software. Thanks to the editor Aguero Gui and the anonymous reviewers.

Conflicts of Interest: The authors declare no conflict of interest.

References

1. Petley, D. Global patterns of loss of life from landslides. *Geology* **2012**, *40*, 927–930. [CrossRef]
2. Bejar-Pizarro, M.; Notti, D.; Mateos, R.M.; Ezquerro, P.; Centolanza, G.; Herrera, G.; Bru, G ; Sanabria, M.; Solari, L.; Duro, J.; et al. Mapping Vulnerable Urban Areas Affected by Slow-Moving Landslides Using Sentinel-1 InSAR Data. *Remote Sens.* **2017**, *9*, 876. [CrossRef]
3. Dai, F.C.; Lee, C.F.; Ngai, Y.Y. Landslide risk assessment and management: An overview. *Eng. Geol.* **2002**, *64*, 65–87. [CrossRef]
4. Intrieri, E.; Gigli, G.; Casagli, N.; Nadim, F. Brief communication "Landslide Early Warning System: Toolbox and general concepts". *Nat. Hazard. Earth Syst.* **2013**, *13*, 85–90. [CrossRef]
5. Corominas, J.; Moya, J.; Ledesma, A.; Lloret, A.; Gili, J.A. Prediction of ground displacements and velocities from groundwater level changes at the Vallcebre landslide (Eastern Pyrenees, Spain). *Landslides* **2005**, *2*, 83–96. [CrossRef]
6. Huang, F.; Huang, J.; Jiang, S.; Zhou, C. Landslide displacement prediction based on multivariate chaotic model and extreme learning machine. *Eng. Geol.* **2017**, *218*, 173–186. [CrossRef]
7. Casagli, N.; Catani, F.; Del Ventisette, C.; Luzi, G. Monitoring, prediction, and early warning using ground-based radar interferometry. *Landslides* **2010**, *7*, 291–301. [CrossRef]
8. Ren, F.; Wu, X.; Zhang, K.; Niu, R. Application of wavelet analysis and a particle swarm-optimized support vector machine to predict the displacement of the Shuping landslide in the Three Gorges, China. *Environ. Earth Sci.* **2015**, *73*, 4791–4804. [CrossRef]
9. Huang, J.; Griffiths, D.V. Return Mapping Algorithms and Stress Predictors for Failure Analysis in Geomechanics. *J. Eng. Mech.* **2009**, *135*, 276–284. [CrossRef]
10. Mufundirwa, A.; Fujii, Y.; Kodama, J. A new practical method for prediction of geomechanical failure-time. *Int. J. Rock Mech. Min.* **2010**, *47*, 1079–1090. [CrossRef]
11. Saito, M. Forecasting the Time of Occurrence of a Slope Failure. In Proceedings of the 6th International Conference on Soil Mechanics and Foundation Engineering, Montreal, QC, Canada, 8–15 September 1965; Volume 2, pp. 537–541.
12. Saito, M. Forecasting time of slope failure by tertiary creep. In Proceedings of the 7th International Conference on Soil Mechanics and Foundation Engineering, Mexico City, Mexico, 13–16 July 1969; Volume 2, pp. 677–683.
13. Hoek, E.; Bray, J. *Rock Slope Engineering*, Revised 2nd ed.; Publication of Institution of Mining & Metallurgy: London, UK, 1977.
14. Mohammadi, S.; Taiebat, H. Finite element simulation of an excavation-triggered landslide using large deformation theory. *Eng. Geol.* **2016**, *205*, 62–72. [CrossRef]
15. Thiebes, B.; Bell, R.; Glade, T.; Jäger, S.; Mayer, J.; Anderson, M.; Holcombe, L. Integration of a limit-equilibrium model into a landslide early warning system. *Landslides* **2014**, *11*, 859–875. [CrossRef]
16. Ma, J.; Tang, H.; Liu, X.; Hu, X.; Sun, M.; Song, Y. Establishment of a deformation forecasting model for a step-like landslide based on decision tree C5.0 and two-step cluster algorithms: A case study in the Three Gorges Reservoir area, China. *Landslides* **2017**, *14*, 1275–1281. [CrossRef]
17. Zhou, C.; Yin, K.; Cao, Y.; Intrieri, E.; Ahmed, B.; Catani, F. Displacement prediction of step-like landslide by applying a novel kernel extreme learning machine method. *Landslides* **2018**, *15*, 2211–2225. [CrossRef]
18. Zhu, X.; Xu, Q.; Tang, M.; Li, H.; Liu, F. A hybrid machine learning and computing model for forecasting displacement of multifactor-induced landslides. *Neural Comput. Appl.* **2018**, *30*, 3825–3835. [CrossRef]
19. Graves, A.; Mohamed, A.R.; Hinton, G. Speech Recognition with Deep Recurrent Neural Networks. In Proceedings of the 2013 IEEE International Conference on Acoustics, Speech and Signal Processing, Vancouver, BC, Canada, 26–31 May 2013; pp. 6645–6649.
20. Lipton, Z.C.; Berkowitz, J.; Elkan, C. A Critical Review of Recurrent Neural Networks for Sequence Learning. *arXiv* **2015**, arXiv:1506.00019.
21. Schmidhuber, J. Deep learning in neural networks: An overview. *Neural Netw.* **2015**, *61*, 85–117. [CrossRef]
22. Mirikitani, D.T.; Nikolaev, N. Recursive Bayesian Recurrent Neural Networks for Time-Series Modeling. *IEEE Trans. Neural Netw.* **2010**, *21*, 262–274. [CrossRef] [PubMed]
23. Hochreiter, S.; Schmidhuber, J. Long Short-Term Memory. *Neural Comput.* **1997**, *9*, 1735–1780. [CrossRef]
24. Greff, K.; Srivastava, R.K.; Koutnik, J.; Steunebrink, B.R.; Schmidhuber, J. LSTM: A Search Space Odyssey. *IEEE Trans. Neural Netw. Learn. Syst.* **2017**, *28*, 2222–2232. [CrossRef]
25. Gers, F.A.; Schmidhuber, J.; Cummins, F. Learning to Forget: Continual Prediction with LSTM. *Neural Comput.* **2000**, *12*, 2451–2471. [CrossRef]
26. Graves, A. Supervised Sequence Labelling with Recurrent Neural Networks. Available online: https://doi.org/10.1007/978-3-642-24797-2 (accessed on 1 February 2012).
27. Xie, P.; Zhou, A.; Chai, B. The Application of Long Short-Term Memory (LSTM) Method on Displacement Prediction of Multifactor-Induced Landslides. *IEEE Access* **2019**, *7*, 54305–54311. [CrossRef]
28. Xing, Y.; Yue, J.; Chen, C. Interval Estimation of Landslide Displacement Prediction Based on Time Series Decomposition and Long Short-Term Memory Network. *IEEE Access* **2020**, *8*, 3187–3196. [CrossRef]
29. Xing, Y.; Yue, J.; Chen, C.; Cong, K.; Zhu, S.; Bian, Y. Dynamic Displacement Forecasting of Dashuitian Landslide in China Using Variational Mode Decomposition and Stack Long Short-Term Memory Network. *Appl. Sci.* **2019**, *9*, 2951. [CrossRef]
30. Yang, B.; Yin, K.; Lacasse, S.; Liu, Z. Time series analysis and long short-term memory neural network to predict landslide displacement. *Landslides* **2019**, *16*, 677–694. [CrossRef]

31. Parikh, A.; Täckström, O.; Das, D.; Uszkoreit, J. A Decomposable Attention Model for Natural Language Inference. *arXiv* **2016**, arXiv:1606.01933.
32. Bahdanau, D.; Cho, K.; Bengio, Y. Neural Machine Translation by Jointly Learning to Align and Translate. *arXiv* **2014**, arXiv:1409.0473.
33. Luong, M.; Pham, H.; Manning, C. Effective Approaches to Attention-based Neural Machine Translation. *arXiv* **2015**, arXiv:1508.04025.
34. Fu, Q.; Li, S.; Wang, X. MSCNN-AM: A Multi-Scale Convolutional Neural Network with Attention Mechanisms for Retinal Vessel Segmentation. *IEEE Access* **2020**, *8*, 163926–163936. [CrossRef]
35. Li, F.; Gui, Z.; Zhang, Z.; Peng, D.; Tian, S.; Yuan, K.; Sun, Y.; Wu, H.; Gong, J.; Lei, Y. A hierarchical temporal attention-based LSTM encoder-decoder model for individual mobility prediction. *Neurocomputing* **2020**, *403*, 153–166. [CrossRef]
36. Ding, Y.; Zhu, Y.; Feng, J.; Zhang, P.; Cheng, Z. Interpretable spatio-temporal attention LSTM model for flood forecasting. *Neurocomputing* **2020**, *403*, 348–359. [CrossRef]
37. Desai, C.S.; Samtani, N.C.; Vulliet, L. Constitutive modeling and analysis of creeping slopes. *J. Geotech. Eng.* **1995**, *121*, 43–56. [CrossRef]
38. Zhou, C.; Yin, K.; Cao, Y.; Ahmed, B. Application of time series analysis and PSO–SVM model in predicting the Bazimen landslide in the Three Gorges Reservoir, China. *Eng. Geol.* **2016**, *204*, 108–120. [CrossRef]
39. Li, Y.; Sun, R.; Yin, K.; Xu, Y.; Chai, B.; Xiao, L. Forecasting of landslide displacements using a chaos theory based wavelet analysis-Volterra filter model. *Sci. Rep.* **2019**, *9*, 1–9. [CrossRef] [PubMed]
40. Guo, Z.; Chen, L.; Gui, L.; Du, J.; Yin, K.; Do, H.M. Landslide displacement prediction based on variational mode decomposition and WA-GWO-BP model. *Landslides* **2020**, *17*, 567–583. [CrossRef]
41. Xu, S.; Niu, R. Displacement prediction of Baijiabao landslide based on empirical mode decomposition and long short-term memory neural network in Three Gorges area, China. *Comput. Geosci.* **2018**, *111*, 87–96. [CrossRef]
42. Huang, N.E.; Shen, Z.; Long, S.R.; Wu, M.C.; Shih, H.H.; Zheng, Q.; Yen, N.; Tung, C.C.; Liu, H.H. The empirical mode decomposition and the Hilbert spectrum for nonlinear and non-stationary time series analysis. *Proc. R. Soc. A Math. Phys. Eng. Sci.* **1998**, *454*, 903–995. [CrossRef]
43. Torres, M.E.; Colominas, M.A.; Schlotthauer, G.; Flandrin, P. A complete ensemble empirical mode decomposition with adaptive noise. In Proceedings of the 2011 IEEE International Conference on Acoustics, Speech and Signal Processing (ICASSP), Honolulu, HI, USA, 16–20 April 2011; pp. 4144–4147.
44. Colominas, M.A.; Schlotthauer, G.; Torres, M.E. Improved complete ensemble EMD: A suitable tool for biomedical signal processing. *Biomed. Signal Proces.* **2014**, *14*, 19–29. [CrossRef]
45. Zhang, W.; Qu, Z.; Zhang, K.; Mao, W.; Ma, Y.; Fan, X. A combined model based on CEEMDAN and modified flower pollination algorithm for wind speed forecasting. *Energy Convers. Manag.* **2017**, *136*, 439–451. [CrossRef]
46. Han, J.; van der Baan, M. Empirical mode decomposition for seismic time-frequency analysis. *Geophysics* **2013**, *78*, O9–O19. [CrossRef]
47. Raffel, C.; Ellis, D.P.W. Feed-Forward Networks with Attention Can Solve Some Long-Term Memory Problems. *arXiv* **2015**, arXiv:1512.08756.

remote sensing

MDPI

Article

Contribution to the Research of the Effects of Etna Volcano Activity on the Features of the Ionospheric Total Electron Content Behaviour

Ivan Toman [1], David Brčić [2,*] and Serdjo Kos [2]

1 Maritime Department, University of Zadar, Ulica Mihovila Pavlinovića 1, 23000 Zadar, Croatia; itoman@unizd.hr
2 Faculty of Maritime Studies, University of Rijeka, Studentska 2, 51000 Rijeka, Croatia; skos@pfri.hr
* Correspondence: brcic@pfri.hr

Abstract: This research represents a contribution to the theory on the coupling of the volcanic activity and the ionospheric dynamics, represented by total electron content (TEC) patterns and their behaviour. The ionospheric response to the activity of the Etna volcano has been analysed using global navigation satellite system (GNSS)-derived TEC values, employing data from International GNSS Service (IGS) reference station near the volcano and on two distant IGS locations. Volcanic activity has been modelled using volcanic radiative power (VRP) data obtained by the Middle InfraRed Observation of Volcanic Activity (MIROVA) system. The estimated minimal night TEC values have been averaged over defined index days of the VRP increase. During the analysed period of 19 years, the volcano activity was categorised according to pre-defined criteria. The influence of current space weather and short-term solar activity on TEC near the volcano was systematically minimised. The results showed mean/median TEC increases of approximately +3 standard deviations from the overall mean values, with peak values placed approximately 5 days before the VRP increase and followed by general TEC depletion around the time of the actual volcanic activity increase. Additionally, TEC oscillation pattern was found over the volcano site with a half-period of 6.25 days. The main interpretation of results indicates that the volcanic activity has modified the ionospheric dynamics within the nearby ionospheric region before the actual VRP increase, and that the residual impact in the volcano's surrounding area refers to terrestrial endogenous processes and air–earth currents. Those changes can be detected during criteria predefined in the research: during quiet space weather conditions, observing night-time TEC values and within the limits of low short-term solar influence.

Keywords: ionosphere; volcanic activity; total electron content; ground-truth data; seismo-ionospheric coupling

Citation: Toman, I.; Brčić, D.; Kos, S. Contribution to the Research of the Effects of Etna Volcano Activity on the Features of the Ionospheric Total Electron Content Behaviour. *Remote Sens.* **2021**, *13*, 1006. https://doi.org/10.3390/rs13051006

Academic Editors: Magaly Koch and Alessandro Bonforte

Received: 21 January 2021
Accepted: 4 March 2021
Published: 6 March 2021

1. Introduction

The development of the ionospheric total electron content (TEC) determination methods, together with improving the means of volcanic activity monitoring, have allowed for the sensing of the seismic processes and their possible couplings with the ionospheric dynamics. The electromagnetic properties of Earth's upper atmosphere layers are mostly dependent on the solar activity. Compared to solar influences, the seismic causes have a smaller impact on the ionosphere, and details of their coupling mechanisms are still not fully understood. The current body of the scientific literature reflects a substantial amount of research and understanding of the relationships between Earth's seismic activity and resulting changes within the atmospheric layers. Although rarely emphasised, the scientific knowledge on seismic effects on the ionosphere is one of the most intensively researched areas that could lead to a more reliable earthquake and seismic prediction models. An appropriate scientific review on the topic was presented in [1].

Apart from the possibility to employ ionospheric anomalies for earthquake prediction, there are other related and usable effects of seismo-ionospheric coupling, for instance, on satellite communication and radio navigational electromagnetic signals [2]. The ionospheric layers physically modify the path of the satellite positioning signals, consequently introducing a positioning error. The ionospheric delay is subject to various methods of calculation, determination and estimation for high accuracy positioning requirements [3], with TEC as a main indicator of the ionospheric behaviour.

As a dispersive medium feature, the ionospheric refraction depends on the operating signal's frequency. By accurately measuring the effects of the ionospheric delay, sound insights into actual physical state of the ionosphere along the signal ray path can be obtained. Using the data that arise from such examinations, it is possible to investigate a connection between the seismic activity and its effects on the ionization.

The research efforts on relations between earthquakes and ionospheric anomalies are summarised in [4] to a significant extent. The geophysical relationship between seismic activity and associated ionospheric anomalies are generally described by the Lithosphere–Atmosphere–Ionosphere Coupling (LAIC) models [5]. LAIC models explain these relationships as a synergy between different ground surface, atmosphere and ionosphere processes and anomalous variations, the latter being usually recognised as a short-term earthquake precursor [5].

Compared to earthquakes, the effects of volcanic activity on ionospheric anomalies are less investigated. Several authors have proved the high electrification of the volcanic dust, such as in [6], where an electrification mechanism of the volcanic ash was simulated. The authors have found that the dominant source of charging of ash particles occurs as a consequence of the ejection of ions and atomic particles during fracture events of solid silicate particles. Charged volcanic plume is often the reason for occurrence of intense lightning from the volcanic dust clouds, repeatedly present during strong eruptions. In [7], authors have also discussed the theory of ash particles fragmentation as a primary mechanism of volcanic clouds charging.

Volcanic activity has a broad and complex effect on atmospheric electrical properties [8–10]. Apart from the charging of ash particles, the interaction with air ionization is found as the emanation of radioactive radon gas from the ground. Atmospheric conductivity is greatly modified by volcanic eruptions which affect air–earth electric currents and the global atmospheric electrical circuit system. The global electrical circuit (GEC) model is developed to explain atmospheric electric field and currents within Earth's coupled atmosphere–ionosphere–magnetosphere system. There are three major drivers of the atmosphere's electric field and currents [11,12]: thunderstorms, the ionosphere wind dynamo [13] and the solar wind-magnetosphere dynamo [14]. The lithosphere dynamics can locally influence conductivity of near-surface air layer following the radon emanation, changing the current density and modifying the free electron content within the atmosphere [4]. In [15], authors show that boundary layer resistance is decreased following the release of radon, increasing the vertical GEC currents and finally modifying the charge transferred to and from the ionosphere. We consider the possibility that volcanoes are able to exploit the similar mechanism and influence the atmospheric charge to a certain degree.

In [16], the authors found that the ionospheric perturbations, as recorded by the DEMETER satellite network, occurred during intense eruptive events of Etna volcano in 2006. An increase in ion density, particularly O+, was found when a satellite was located at the approximate latitude of the Etna region. At that time, an electric very low frequency (VLF) spectrogram recorded lightning activity in the region. All the findings showed the close relationship between volcanic activity and atmospheric electricity, leading to the possibility of volcanic influence towards ionospheric electrical properties, as discussed and examined further.

Li et al. [17] analysed the coupling relationship between volcanic eruption and TEC anomalies using Global Ionospheric Map (GIM) data before the 15 volcanic eruptions indicated by Volcanic Explosivity Index 4+ (VEI4+). They found anomalies preceding 67%

of the analysed eruption events. Approximately 80% of anomalies occurred between days 6 and 17 before eruptions. The occurrence rate of TEC anomalies before large eruptions was related with the volcanic type and the geographical position. The probability of TEC anomalies' observations before the eruption decreased with latitude; in the equatorial region, 80% of eruptions anomalies were found, but only 60% and 50% of eruptions in mid and high latitudes, respectively, produced preceding TEC anomalies. However, statistical uncertainty was found to be high due to scarce data.

TEC anomalies were analysed in [18] as a response to the Calbuco volcano eruptions. Here, the method of approximation of a spherical wave propagating at a constant velocity from a point source was employed. The authors showed that it is possible, by using dense (30 s) frequency ionospheric GPS data, to calculate the position of the eruptive source within several latitude/longitude degrees. In [19], the relation between volcanic events and ionospheric anomalies was found as well. The total of 269 anomalies were found in relation to 89 eruptions using satellite DEMETER data [20]. The two main types of anomalies were the electrostatic turbulence, and the electromagnetic emissions with a maximum number of those being recorded between 30 and 15 days before the surface volcanic activity. The study on the impact of the 2011 Puyehue-Cordon Caulle volcanic eruption on the GPS ionospheric delay also indicated the presence of ionospheric TEC anomalies several days before the major eruption [21]. In [22], authors analysed global navigation satellite system (GNSS)-derived short-term TEC response to the Merapi 2010 and the Kelud 2014 volcano eruptions. They observed quasi-periodic oscillation of TEC values of a frequency ~4 mHz (period ~250 s), which lasted for ~20 and ~120 min, respectively. Previously [23], it has been shown that these oscillations are related to upward propagation of acoustic waves, causing resonant oscillations in distinct frequencies.

In this paper, we investigated the longer-term relationship of the activity of Etna volcano (Italy), measured by the Middle InfraRed Observation of Volcanic Activity (MIROVA) system, and the ionospheric TEC behaviour calculated using ground-based, dual-frequency GNSS measurements by the International GNSS Service (IGS) network [24]. Almost 20 years' worth of data were analysed and filtered to select significant volcanic activity increases, rather than focusing on volcano eruptions solely. Identified events were further used to investigate the ionospheric TEC dynamics for the same periods, with the primary goal to detect possible anomalous TEC behaviour during days of significant volcanic activity, with an emphasis on days preceding the activity.

In the following chapter, determinants of research fundamentals referring to ionospheric properties, TEC determination and the means of the volcanic activity tracking are briefly provided. The research design and the methodology are defined, presented and explained in the third chapter. The results are presented and discussed afterwards, drawing inferences and conclusions on obtained findings.

2. Background

2.1. TEC Features and the Ionospheric Response to Solar Activity

The ionospheric layers react to the solar activity by changing the temperature and ion production [25,26]. Depending on the user's location on the Earth, the ionosphere can be roughly divided in low-, mid- and high-latitude regions, each characterised by its own processes and responses to the solar forcing [27,28]. There are various methods for evaluation and monitoring of the solar activity.

Certain methods, such as solar radio flux (*sfu*) at 10.7 cm wavelength [29], solar wind [30], or radio noise measurements at centimetre wavelengths [31] are direct, based on instrumental measurements. The others are intermediary, based on the solar activity evaluation effects, implying observations of indices such as sunspot number [32], and planetary geomagnetic *K*, *Kp* and *Dst* indices [33–36].

The total electron content is considered as a basic parameter of the ionospheric electromagnetic properties and their variations. It is defined as a number of free electrons along an equivalent column having a cross-section of one square meter from the satellite to the receiver's antenna [37,38]. It is expressed in TEC units ($TECu$), where [2,39]:

$$1\ TECu = \frac{10^{16}e^{-}}{m^2}. \tag{1}$$

The amount of the GPS ionospheric delay (Δt) can be expressed as [37]:

$$\Delta t = \frac{40.3}{c \cdot f^2} TEC, \tag{2}$$

where f is system operating frequency, and c is electromagnetic signal propagation speed.

The GNSS networks and respective ground-truth observations, data and products are used to make accurate estimates of ionospheric TEC values [40–42]. As the magnitude of the delay effect is proportional to inverse of the square of the frequency (2), it can be used to directly estimate TEC values [43], using a linear combination of code (P) and carrier phase (ϕ) measurements [44,45].

The difference in the signals' reception delay can then be modelled as a sum of the group delays caused by hardware (satellite (Total Group Delay—TGD) and receiver (Receiver Differential Code Bias—rDCB)) biases and the ionospheric TEC. With the receiver calibration and no code smoothing with delay adjustment to code-derived TEC, it is possible to remove sources of error that result from estimation, instead of measurement of the bias, smoothing-induced delay of code TEC, and smoothing-induced ionospheric divergence bias of code TEC [45].

2.2. Specific Parameters of the Volcanic Activity

Used methods for volcanic activity monitoring can be classified as ground-based local, and satellite-based measurements. The former usually consist of developing multidisciplinary surveillance networks that include seismic, positional, tilt, visual and thermal infrared cameras, infrasound, gravity, magnetic and other measurements by local ground-based stations [46–48].

One of the first efforts in volcanic activity remote sensing can be found in [49], where an airborne infrared radiometer was used to measure the activity of Hawaiian volcanoes. First attempts to observe volcanic surfaces from the Earth's orbits are described in [50]. Satellite-derived thermal data within the infrared region between 0.7 and 20 µm are nowadays frequently used as a tool to detect the presence of volcanic eruptions [51]. According to [52], there are 18 satellite-based remote sensing systems which monitor the volcanic activity in terms of radiation measuring, mostly in middle-infrared and thermal-infrared spectrum bands. Notable examples are HOTSAT [53] and MIROVA system [54,55]. MIROVA is a near real-time hot-spot detection system, developed to detect, locate and measure the heat radiation sourced from the volcanic activity.

The main benefit of the system is a combination of the accuracy and ease of exportability of application to different volcanoes. It does not require the analysis of historical data sets but rather produces results of the same quality as the algorithms, specifically calibrated for any volcano target. The system is based on the analysis of the Middle Infrared Radiation (MIR wavelength: 3.44–4.13 µm) extracted from the MODIS Level 1B data [56]. The extraction algorithm is based on spectral and spatial analysis of MIR data, providing detection of heat sources ranging from 1 MW to more than 10 GW of radiated power.

3. Research Methodology

The presented research covered the period from March 2000 to December 2019, containing collected MIROVA data, GNSS observables and selected space weather measurements. MIROVA dataset was obtained courtesy of a collaborative project between the Universities of Turin and Florence (Italy) [57]; GNSS data were retrieved from IGS servers [58], and space weather measurements were obtained from NASA Goddard Space Flight Center

(GSFC) servers [59]. The MIROVA dataset of Etna thermal anomalies contains night-time, daily measured values of volcanic radiative power (VRP) in Watts [60]. We defined criteria for locating days of volcanic activity increase (VAI) as days with at least 25% of VRP higher than the preceding day (ΔVRP), and with the absolute VRP value higher than the threshold value. To analyse the ionospheric TEC dependence on the volcanic influence, i.e., its activity magnitude, three different scenarios were defined depending on VRP threshold value, as shown in Table 1.

Table 1. Volcanic radiative power (VRP) threshold values and number of corresponding days within analysed Middle InfraRed Observation of Volcanic Activity (MIROVA) dataset for three VRP threshold values and 5 different daily increases in VRP.

Criteria	VRP Threshold Value	Number of Days that Satisfy Criteria with Daily VRP Increase				
		5%	15%	25%	35%	45%
A	>1 × 10^8 W	445	402	350	293	226
B	>4 × 10^8 W	260	236	201	167	121
C	>8 × 10^8 W	172	156	131	111	78

The additional criteria requiring a 25% or more day-to-day increase in VRP were defined, aiming to identify either beginning days of eruptions, or days with significant increase in the activity without actual lava eruptions (in these cases, the radiation increase occurred from the thermal energy build-up).

The comprehensive statistical analysis has been conducted evaluating different daily VRP increase values (5, 15, 25, 35 and 45%, respectively), in order to examine the possibility of random affectation and misleading results. The 25% of day-to-day VRP increase was finally selected as a reference value for further calculations, representing a sound compromise between sufficient number of records which satisfy the criteria, and appropriate number of cases of the volcanic activity increase.

The GPS observables have been obtained from the IGS network in the Receiver Independent Exchange Format (RINEX) [61]. The TEC values were estimated using GPS-TEC analysis software [62] following the standard methodology [63]:

$$P^j_{r,4} = P^j_{r,1} - P^j_{r,2} = 40.3\frac{TEC}{f_1^2} - 40.3\frac{TEC}{f_2^2} + (B_{r,1} - B_{r,2}) + \left(B^j_1 - B^j_2\right), \quad (3)$$

$$\phi^j_{r,4} = \phi^j_{r,1} - \phi^j_{r,2} = 40.3\frac{TEC}{f_2^2} - 40.3\frac{TEC}{f_1^2} + (b_{r,1} - b_{r,2}) + \left(b^j_1 - b^j_2\right) - (N_1 - N_2), \quad (4)$$

where j, r are satellite and receiver antenna points, respectively; P, ϕ are pseudorange and carrier phase observations, respectively; TEC is the slant total electron content along the propagation path (in $TECu$) with the ray path zenith angle $\chi_s \neq 0^\circ$; f_1, f_2 are operating frequencies; B_r, B_j are pseudorange code receiver and satellite hardware delay (in m), respectively; b_r, b_j are carrier phase receiver and satellite hardware delay (in m), respectively, and N_1, N_2 are carrier-phase ambiguities (in m).

The simplified representation of the employed TEC determination method is shown in Figure 1. The code difference observables were levelled with phase differences to the correct TEC values, after which they were adjusted with code measurements.

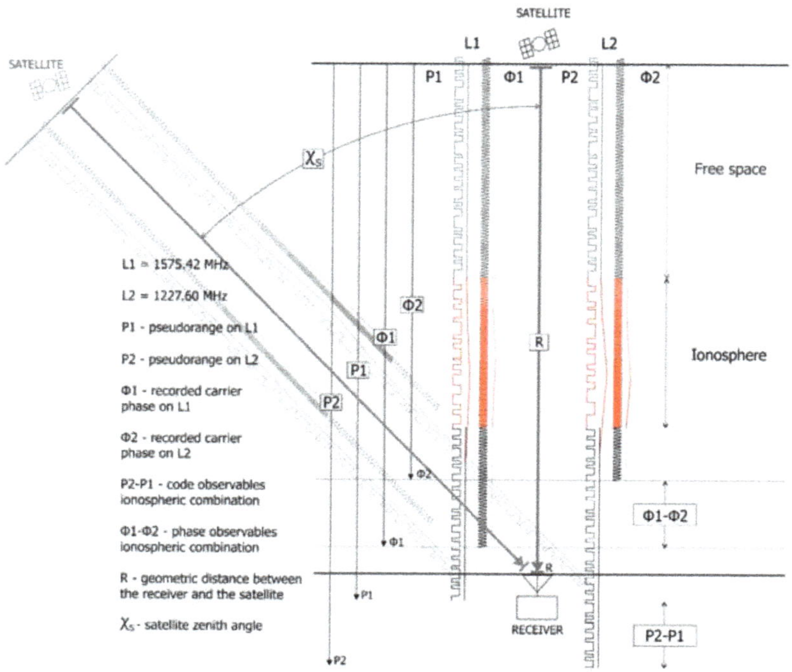

Figure 1. Total electron content (TEC) determination method based on ionospheric GPS dual-frequency combinations. Adapted and modified on the basis of [64,65].

The daily DCB values were retrieved from the Global Ionospheric Maps (GIM) IONEX files [66], employed to refine the ionospheric combinations. Vertical TEC (VTEC) values were obtained by using slant TEC (STEC) to VTEC mapping function [67].

3.1. Mitigation of Short-Term Solar Activity and Other Non-Local Factors' Impact on TEC Behaviour

The daily TEC values are strongly correlated to the solar activity, as shown in Figure 2. The ionospheric TEC values were determined from ground-truth data at IGS Noto, and their mean values were compared to 10.7 cm *sfu* [29], expressed in solar flux units (1sfu = 10^{-22} $Wm^{-2}Hz^{-1}$).

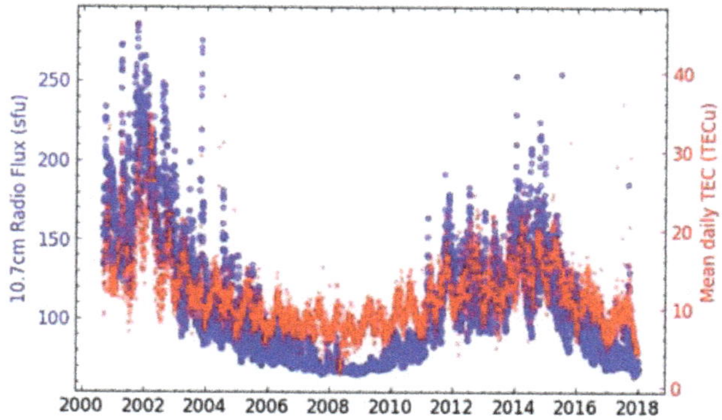

Figure 2. Solar activity expressed as 10.7 cm radio flux (blue) and mean daily ionospheric TEC values as calculated for IGS Noto (red) for the approximate period 2001 to 2018.

The influence of diurnal solar activity represents a challenge in an effort to detect volcanic influences, and appropriate filtering of the solar impact was essential. By examining various TEC subset patterns, we found that minimal night TEC values vary with solar activity in an acceptably small degree (Figure 3). We utilised this observation for detection of volcanic influence on ionospheric TEC along with larger solar effects. Hence, in further analysis minimal night TEC values were considered to be generally unaffected by the solar activity. In order to increase the confidence in practical validity of such an assumption, a series of data analyses were performed to detect possible adverse effects of the remaining solar influence.

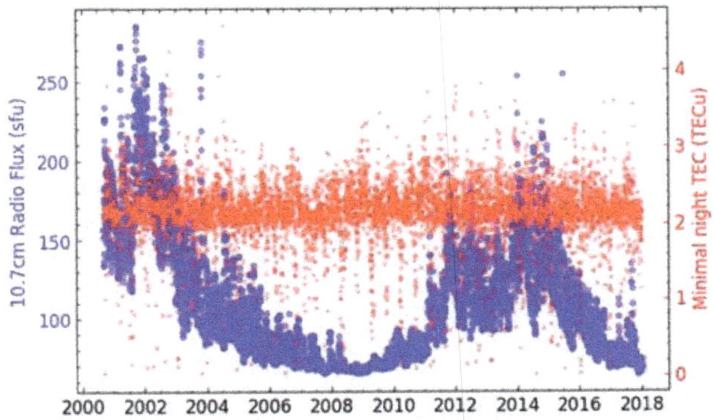

Figure 3. Solar activity expressed as 10.7 cm radio flux (blue) and minimal night ionospheric TEC values at IGS Noto (red).

Due to the night persistence of certain ionospheric layers, the minimal night TEC cannot be considered absolutely unaffected. As an additional control parameter, analyses of *sfu*, solar wind proton density and planetary *Kp* index were performed, thus confirming generally quiet and stable space weather activity during the elaborated periods.

Other causes of TEC anomalies are not directly related to the solar activity. Travelling Ionospheric Disturbances (TIDs) are propagating perturbations in the ionospheric electron density which, if classified as medium scale, have horizontal wavelengths of 100–250 km [68,69].

All mentioned influences have to be considered when examining TEC dynamics in terms of response to the volcanic and tectonic activity, since these affect the ionospheric TEC to a significantly greater extent than the lithosphere dynamics. This is especially important if a single, particular event is analysed.

The last used method to smooth out the mentioned effects was the TEC averaging over the large number of events as per defined criteria. The assumption was that all non-local forcing should equally affect the results over all employed, regional IGS stations. This was done by averaging over a large number of events, rather than affecting one station significantly more than the others, therefore allowing us to consider that the remaining differences can be attributed to local effects with satisfying confidence.

3.2. Determination of Specific Parameters of the Volacno Activity and Ionospheric TEC Features' Behaviour Relation

We have analysed the ionospheric TEC patterns around days of VAI at three IGS reference stations, positioned on similar geographical latitudes: IGS Noto (Italy; 36.88°N, 14.99°E), IGS San Fernando (Spain; 36.46°N, 6.21°W) and IGS Nicosia (Cyprus; 35.141°N, 33.396°E). The IGS Noto was chosen due to its close vicinity to the Etna volcano (distance = 96 km). The other two stations were selected as a control distant stations (1878 km and 1672 km, respectively), geographically considered to be outside of the direct volcano influence (Figure 4).

Figure 4. Geographical positions of IGS Noto, IGS San Fernando and IGS Nicosia in relation to the location of Etna volcano.

In addition, to directly compare TEC patterns during active and quiet volcanic periods, the same calculation was repeated with reverse criteria; by selecting cases where VRP was less than 1×10^6 Watts (total of 493 days), regardless of amount of relative change from the previous day.

As per activity Criteria A, B and C, we extracted minimal night TEC values for defined index days—30 days before and 20 days after the VAI, respectively. The mean value of minimal night TEC was calculated for each index day. The same analysis was performed independently for all locations. In addition, the median TEC values were also calculated to assess the possibility of misleading results regarding the outliers' influence. The mean and median calculations of minimal TEC also increased the confidence in assumption that non-local influences on TEC are statistically averaged in a similar manner for all analysed IGS stations.

The preliminary results referring to specific patterns of TEC time-series dictated further analyses. Pearson correlation coefficients between VRP and TEC time-series have been calculated both for the whole period, and separately for the defined activity criteria periods. The obtained 5-day lag of VRP increase after the TEC peak has been verified using a normalised cross-correlation technique [70]. The observed TEC oscillatory pattern was analysed with the Fast Fourier Transform (FFT) method [71,72], converting a function's time-domain into frequency-domain, with the results confirming the observed periodicity.

4. Structural-Analytical Presentation of the Research Results

The following same-scale figures represent ionospheric TEC patterns through index days according to defined Criteria A, B and C. Horizontal black lines denote mean minimal values. Horizontal, dashed coloured lines represent ± 1 standard deviation from the mean value for the respective criteria. For the Criteria C (for IGS Noto), +2 and +3 standard deviations from the mean value were added as red dotted lines.

The most notable feature within calculated TEC patterns is large and sharp TEC enhancement located at approximately 5 days before the VAI, being most prominent for the Criteria C (Figure 5). This abrupt increase refers to IGS Noto only.

Significantly larger variation of TEC values at IGS Noto can be considered as at least partially attributed to the volcanic activity. However, in this phase other causes cannot be excluded.

The magnitudes of the volcanic activity increase other than 25% were additionally analysed, investigating the possibility of different impacts of VAI threshold values. The pattern which is characterised with the TEC increase 5 days before the VAI is visible in each case (Figure 6), being especially prominent as the VRP increases as well. Since the VRP change did not affected the general pattern, the 25 % day-to-day VRP increase was selected as a valid threshold reference. During the extreme activity increase (45 %—Figure 6e), additional TEC enhancements and depletions were observed, however, due to the small number of days firm conclusions should not be drawn.

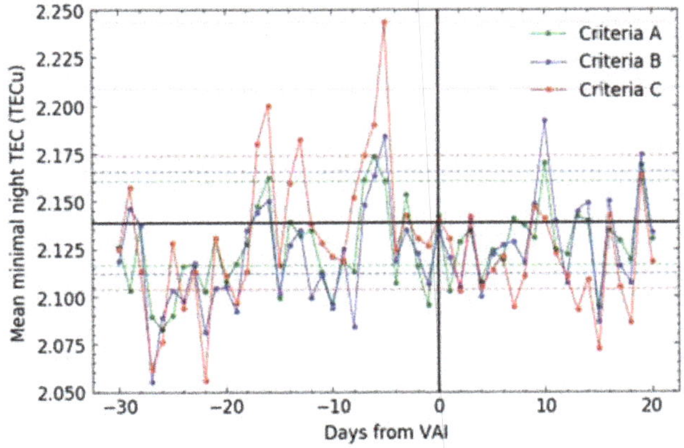

Figure 5. Mean values of minimal night TEC at IGS Noto for Criteria A (green), B (blue) and C (red). The vertical line denotes the day of volcanic activity increase. Mean +/− 1 standard deviation for respective criteria is added with dashed line of the same colour, and for Criteria C, mean +2 and +3 standard deviations are added as red dotted lines.

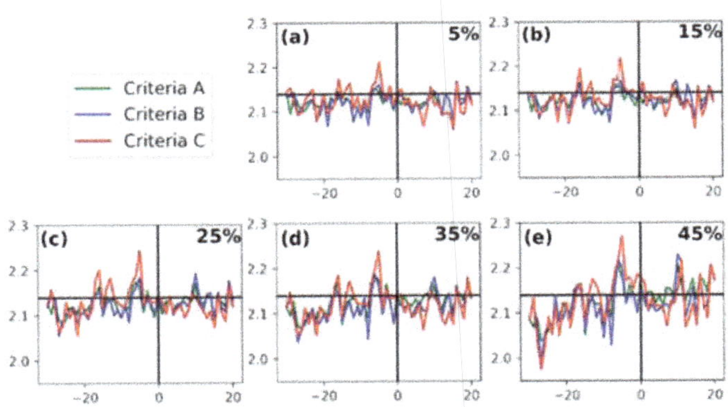

Figure 6. Mean values of minimum night TEC at IGS Noto (y-axis) through index days (x-axis) for different daily increase in the volcanic radiative power (VRP) value: (**a**) 5%, (**b**) 15%, (**c**) 25%, (**d**) 35%, (**e**) 45% of daily VRP increase, respectively. Described TEC pattern is visible in all plots with slight differences, as discussed in text.

The analysis was repeated for IGS San Fernando and IGS Nicosia (Figure 7), revealing significantly different patterns when compared to IGS Noto (Figure 8).

These changes seem to result from noise rather than from a recognizable pattern, especially in the case of IGS Nicosia where the pattern resembles uniform across the whole period.

The median night TEC values were calculated further. According to the similarity of derived time-series (Figure 9), we strengthen our confidence that the possible outlier effects are not a prevalent cause of observed patterns within the analysed datasets.

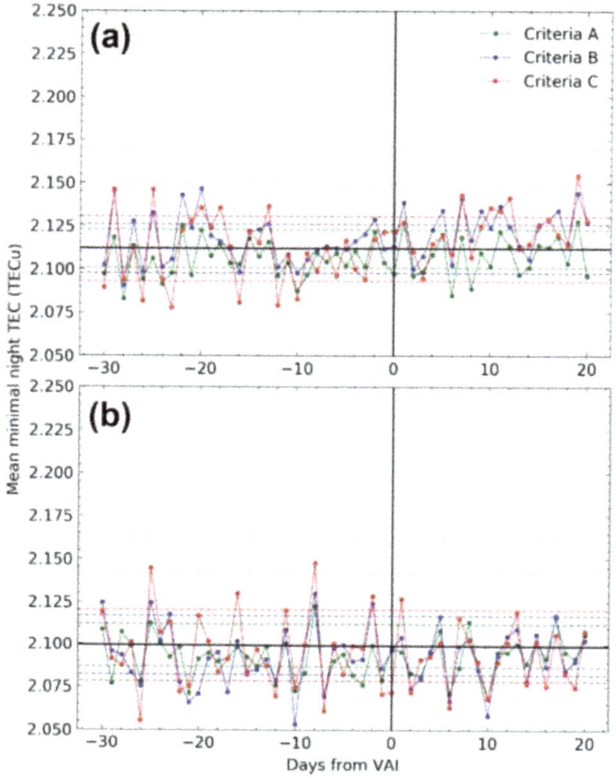

Figure 7. Mean values of minimum night-time TEC at IGS San Fernando (**a**) and Nicosia (**b**).

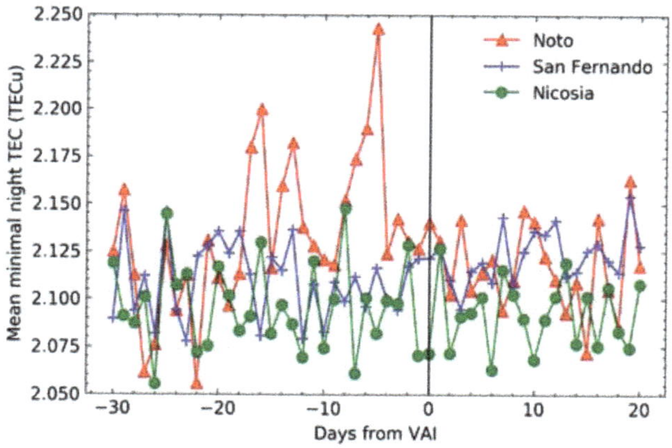

Figure 8. TEC patterns on IGS Noto, IGS San Fernando and IGS Nicosia during index days for the Criteria C. The vertical line denotes the day of defined volcanic activity increase (VAI).

The results of median calculation arguably show similar TEC pattern around the VAI compared to the mean calculation pattern, with the prominent peak more than +3 standard deviations from the median value for the Criteria C.

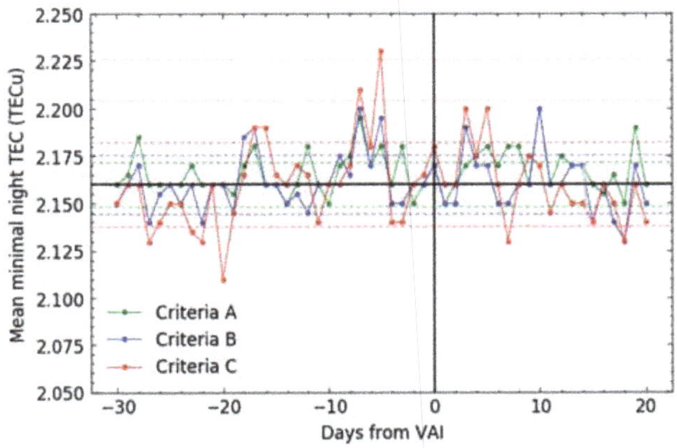

Figure 9. Median values of minimum night TEC at IGS Noto.

Before the observed peak, TEC oscillations (enhancements and depletions during several days before the VAI) at IGS Noto can be also seen. Such a pattern seems to cease with VAI. Applying the FFT method, two prominent peaks were found (Figure 10). The largest oscillation is found to have a period of 12.5 days (half-period of 6.25 days), with the second largest oscillation of 8.5 days (half-period of 4.25 days). This corresponds to the two most prominent peaks in the frequency-domain data. The observed peaks are consistent within all criteria.

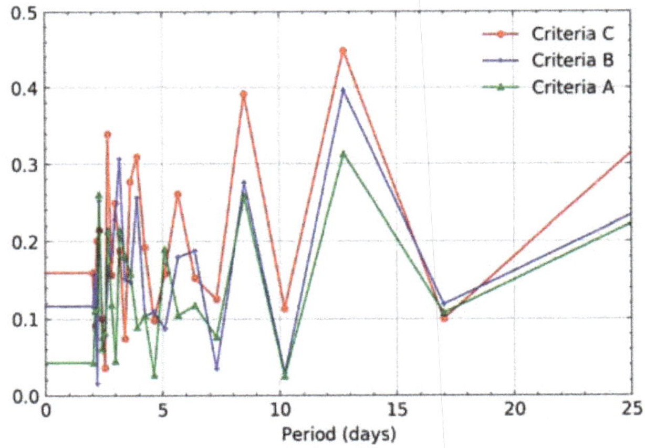

Figure 10. Fast Fourier Transform of mean night TEC values during index days at IGS Noto.

When compared to distant locations, mean TEC values at IGS Noto show increased amplitude and standard deviation. The most prominent anomalous TEC values were found during the activity of Criteria C around the fifth day before the VAI. It can also indicate that the major volcanic activity caused larger deviations in the ionospheric TEC values, when compared to minor- or no-activity periods.

Distribution of minimal night TEC values for IGS Noto (Criteria C) is presented in Figure 11. The majority of data distribution is similar between two datasets: during the day of VAI (Figure 11a) and 5 days before (Figure 11b). The lack of large histogram values additionally suggests that the outlier values have not significantly influenced the datasets.

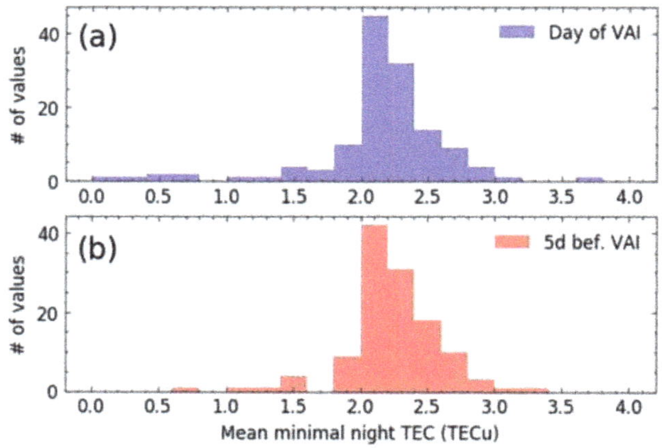

Figure 11. Minimal night TEC values distribution for IGS Noto, criteria C, for the day of VAI (**a**) and 5 days before VAI (**b**).

Basic statistical indicators of minimal night TEC values for all stations are presented in Table 2. Data are shown for the whole dataset, as well as compared to the same indicators for VAI and 5 days before VAI.

Table 2. Statistical indicators of minimal night-time TEC values during all criteria (*Crit*) as calculated for IGS San Fernando (*S*), IGS Nicosia (*NI*) and IGS Noto (*NO*) for the whole dataset (*All*), the day of VAI and 5 days before VAI.

	Crit	Mean			Median			St dev			Max			Min		
Station		S	NI	NO	S	NI	NO	S	NI	NO	S	NI	NO	S	NI	NO
All	-	2.112	2.099	2.139	2.110	2.110	2.160	0.232	0.240	0.441	3.330	3.480	4.580	0.120	0.110	0.020
VAI	A	2.097	2.099	2.142	2.090	2.110	2.170	0.191	0.207	0.492	3.070	2.810	3.700	1.050	1.030	0.020
	B	2.113	2.097	2.134	2.100	2.110	2.170	0.205	0.216	0.503	3.700	2.810	3.700	1.050	1.030	0.020
	C	2.122	2.071	2.140	2.105	2.110	2.180	0.147	0.240	0.501	2.680	2.810	3.700	1.570	1.030	0.020
5 days before VAI	A	2.102	2.094	2.161	2.100	2.100	2.180	0.154	0.247	0.438	2.670	3.120	3.320	1.480	0.230	0.090
	B	2.112	2.100	2.188	2.100	2.120	2.195	0.161	0.293	0.441	2.670	3.120	3.320	1.610	0.230	0.090
	C	2.113	2.082	2.243	2.115	2.120	2.230	0.163	0.280	0.346	2.670	2.770	3.320	1.610	0.230	0.780

The main difference between criteria, in relation to the whole dataset, was the notable increase in mean and median TEC value for Criteria C, 5 days before VAI at IGS Noto solely. The increase is present both compared to Criteria A and B at the same day, as well as compared to Criteria C on VAI. Furthermore, the standard deviation at station Noto is significantly larger when compared to other corresponding values. Absolute minimum values for IGS Noto are significantly smaller than for other two stations. This might indicate the possibility of almost complete temporary removal of free electrons from the atmospheric column during certain stages of volcanic activity.

The space weather indices were calculated employing the same methodology of averaging values over the index days (Figure 12).

There is no visible increase in solar flux and solar wind proton density, as well as in the *Kp* index around the IGS Noto TEC peak days. In Figure 12c, magnitude of geomagnetic disturbance on a 0–9 scale is presented in terms of planetary *Kp* index. *Kp* index is an integer value, but here the mean value of respective integers over index days around VAI is presented, yielding results as a float type. Both 10.7 cm *sfu* and *Kp* index, with the proton density in a lesser amount (Figure 12b), indicate that the days that enter into Criteria B and C, display increased solar activity compared to those of Criteria A. However, there are no visible anomalies that could explain the peak in the night TEC values over IGS Noto.

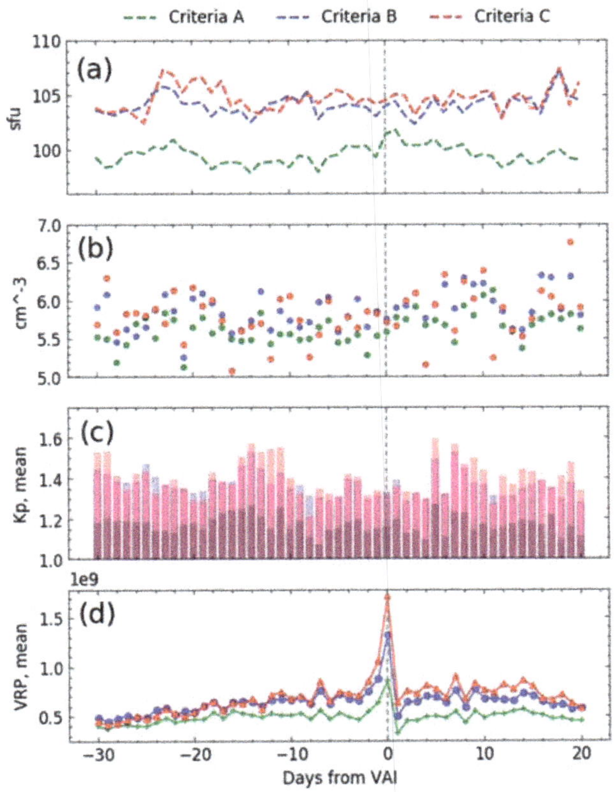

Figure 12. Mean values of 10.7 cm solar flux (**a**), proton density (**b**), *Kp* index (**c**) and VRP (**d**) during index days for Criteria A (green), B (blue) and C (red).

The TEC pattern over IGS San Fernando follows 10.7 cm *sfu* and *Kp* values to a certain extent. Given the dependence of TEC, it was expected to have a possible effect. Therefore, it might be accredited to the remains of the day-time solar influence and geomagnetic processes. Patterns derived from IGS Nicosia show a less similar shape compared to solar parameters. Any remaining similarity between space weather indices and night TEC values is absent in the pattern derived from IGS Noto.

The results of TEC pattern analysis during quiet volcanic periods are presented in Figure 13. The patterns were compared using reverse criterion, i.e., non-active days (NAD), with VRP < $1 \cdot 10^6$ W. The Figure scale was set as previously described.

The TEC patterns are similar, with the IGS Noto mean value being slightly lower than that at San Fernando, and very similar to the case of IGS Nicosia and again, more variable from day to day around NAD than patterns from both control stations. It should be noted that, although these patterns represent generally quieter conditions, the volcano is never absolutely inactive.

The correlation coefficients were calculated for VRP and TEC for all locations. As the most prominent TEC peak was present 5 days before volcanc activity increase, the lagged correlation was performed with TEC time-series, shifted for 5 days before correlating with the VRP (Table 3). The time lag of 5 days VAI increase after the TEC peak was verified using the cross-correlation technique (Figure 14).

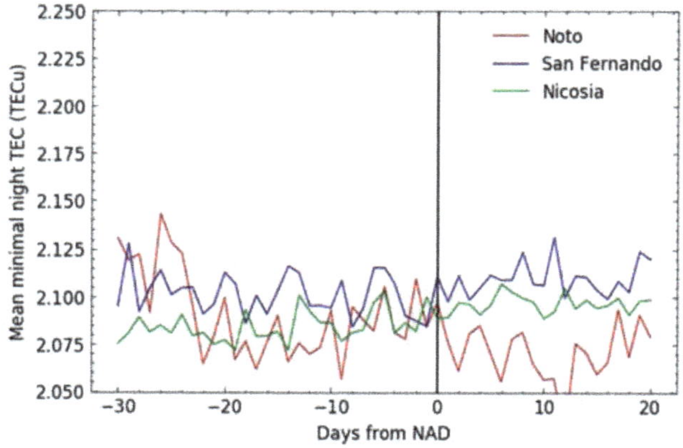

Figure 13. Comparison of minimum night TEC patterns for IGS Noto, IGS San Fernando and IGS Nicosia during non-active days (NAD).

Table 3. Five-day lagged Pearson correlation coefficients between time-series of VRP and TEC for all locations, including whole datasets and defined criteria.

	Station	*Whole Series*	Criteria A	Criteria B	Criteria C
5-day lag	Noto	0.005	0.181	0.231	0.251
	San Fernando	−0.026	−0.029	−0.098	−0.161
	Nicosia	−0.035	−0.037	−0.068	−0.052

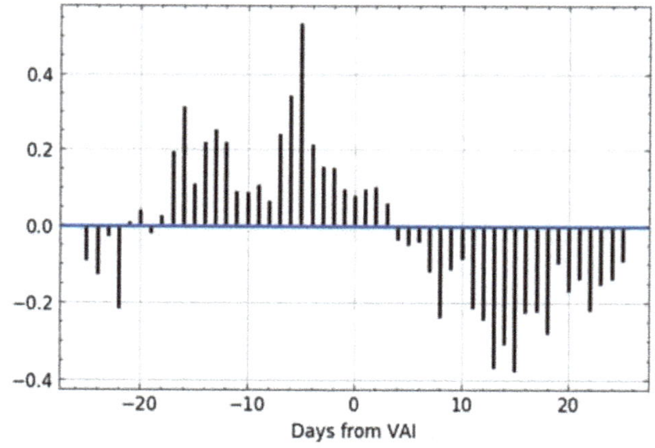

Figure 14. Normalised cross-correlation between VRP and TEC mean values for Criteria C, IGS Noto. The lag with maximum correlation value is equal to 5 days.

The largest correlation took place when a lagged correlation was performed for station Noto, with increasing strength in pace with increasing activity criteria from lower to higher levels. The higher correlation values at IGS Noto strengthen the confidence in the conclusion that the observed TEC peak 5 days before VAI does represent a cause–effect relationship between volcanic activity and the TEC values in nearby ionospheric areas.

5. Brief Overview of the Results' Specifics

The presented results show larger night TEC deviations from the mean dataset value for IGS Noto than those for IGS San Fernando. A general upward and oscillating trend in

TEC pattern with a half-period of 6.25 days is notable until the time reaches the day of VAI, with maximum values generally found around 5 days before. The periodicity analysis also suggests that the pattern oscillation is further modulated with somewhat higher frequency, having a half-period of 4.25 days and a smaller magnitude. Afterwards, a decrease in night TEC values was observed and the pattern loses its wave-like features. The described pattern seems to generally exist for all three activity criteria; however, it is most prominent for the Criteria C, reaching the maximal enhancement at +3 standard deviations from the mean value.

As the Criteria C contains a total of 131 days and is therefore considered to be a representative sample, our argument is that the night TEC increase of +3 standard deviations from the mean value cannot be explained by the random variability. As we could not explain it in this way, or by examining various metrics of solar influence, by comparing with two distant control stations, by considering possible data outlier influence on mean values, or any other way, we conclude that the ionospheric TEC peak located at 5 days before VAI has to be associated with the Etna volcano activity, and moreover, is caused by the same activity. The mentioned enhancement was preceded with oscillatory behaviour of ionospheric TEC; however, the peak was the most prominent observation. As the TEC values increase with tightening criteria definition for the Etna activity level, it follows that the TEC response is also dependent on the amount of the volcanic forcing. Most of the TEC patterns' variations were found within periods of the highest volcano activity.

The results suggest that the observed TEC variations are related to the volcano seismic activity before the active phase of VRP increase. The observed increase in TEC values perhaps can be explained with an increase in the available endogenous energy which firstly produced an increase in air–earth currents. We propose a hypothesis that radon emanation or other seismic-related pre-eruption factors modified the electrical conductivity of the column of air by increasing the charge of boundary layer, leading to a change of vertical electrical air–earth currents and subsequent local modification of GEC properties, detectable as observed oscillations and peaks of the ionospheric TEC. Further investigation of this phenomena, together with an explanation of possible causes and mechanisms of action, remain a matter for future work.

6. Conclusions

The presented results suggest a causal correlation between the Etna volcano activity and the surrounding ionospheric behaviour, providing insights referring to the mutual relation between the two phenomena. Temporal coupling processes and the respective ionospheric total electron content patterns as a consequent dependent variable were investigated. The volcanic influence was found to be present during the peak activity periods, as well as before the actual increase in VRP. The conclusions are based on the analyses of 19 years of ionospheric and volcanic activity data. The specific volcanic activity scenarios were distinguished and correlated with ionospheric TEC values.

The results suggest that 5 days before the commencement of a significant rise in activity, ionospheric TEC within the volcano region sharply increases, followed by the abrupt decline in values towards the day of VAI. The oscillatory pattern of TEC behaviour is found to be present with a main oscillation half-period of 6.25 days. Larger variations in TEC values were observed near the volcano site. A special care was taken in order to exclude possible external influences in terms of the space weather activity, overall solar influence, and possible outliers. Finally, we compared the TEC behaviour during volcano activity with behaviours during non-active periods. The intensity of the TEC anomalous behaviour was found to be positively related to the amount of forcing from the Etna volcano activity and we have not been able to associate observed TEC behaviour prior to volcanic activity increase to the solar forcing.

The proposed approach and presented findings represent a means of generalization of the TEC behaviour pattern during phases before and at the instance of increased volcanic activity and eruptions, with the possibility of separation of the ionospheric activity as a prognostic parameter.

Author Contributions: Conceptualization, I.T., D.B. and S.K.; methodology, I.T. and D.B.; software, I.T.; validation, I.T., D.B. and S.K.; formal analysis, S.K.; investigation, I.T.; resources, I.T. and D.B.; data curation, I.T.; writing—original draft preparation, I.T.; writing—review and editing, D.B.; visualization, I.T. and D.B.; supervision, D.B. and S.K.; project administration, D.B.; funding acquisition, S.K. All authors have read and agreed to the published version of the manuscript.

Funding: This research was funded by the University of Rijeka under the project—uniri-tehnic 18-66: Research of environmental impact on the operation of satellite navigation system in maritime navigation.

Acknowledgments: This work has been fully supported by the University of Rijeka under the project—*uniri-tehnic 18-66*. The authors are grateful for all available MIROVA, GNSS and space weather data and products from the respective organisations, based on which we conducted our research. Finally, we are sincerely grateful to the reviewers for their time, efforts and constructive remarks and observations which significantly led to the improvement of our work.

Conflicts of Interest: The authors declare no conflict of interest.

Abbreviations

TEC	Total Electron Content
TECu	TEC unit
STEC	Slant TEC
sfu	Solar radio flux
GNSS	Global Navigation Satellite System
LAIC	Lithosphere–Atmosphere–Ionosphere Coupling
GEC	Global Electrical Circuit
VEI4+	Volcanic Explosivity Index 4+
MIROVA	Middle InfraRed Observation of Volcanic Activity
rDCB	Receiver Differential Code Bias
TGD	Total Group Delay
VRP	Volcanic Radiative Power
VAI	Volcanic Activity Increase
GIM	Global Ionospheric Maps
IGS	International GNSS Service
NAD	Non-Active Days
FFT	Fast Fourier Transform

References

1. Pulinets, S.; Ouzounov, D. *The Possibility of Earthquake Forecasting*; IOP Publishing: Bristol, UK, 2018.
2. Parkinson, B.W.; Spilker, J.J., Jr.; Axelrad, P.; Enge, P. *Global Positioning System: Theory and Applications*; American Institute of Aeronautics and Astronautics: Washington, DC, USA, 1996; Volume 2.
3. Klobuchar, J.A. Ionospheric Effects on GPS. In *Global Positioning System: Theory and Applications*; Parkinson, B.W., Spilker, J.J., Eds.; American Institute of Aeronautics & Astronautics: Reston, VA, USA, 1996; Volume 1, pp. 485–515.
4. Pulinets, S.; Boyarchuk, K. *Ionospheric Precursors of Earthquakes*; Springer: Berlin, Germany, 2005.
5. Pulinets, S.; Ouzounov, D. Lithosphere–Atmosphere–Ionosphere Coupling (LAIC) model—An unified concept for earthquake precursors validation. *J. Asian Earth Sci.* **2010**, *41*, 371–382. [CrossRef]
6. James, M.R.; Wilson, L.; Lane, S.; Gilbert, J.S.; Mather, T.A.; Harrison, R.G.; Martin, R.S. Electrical Charging of Volcanic Plumes. *Space Sci. Rev.* **2008**, *137*, 399–418. [CrossRef]
7. Mather, T.A.; Harrison, R.G. Electrification of volcanic plumes. *Surv. Geophys.* **2006**, *27*, 387–432. [CrossRef]
8. Kamsali, N.; Prasad, B.S.N.; Datta, J. The Electrical Conductivity as an Index of Air Pollution in the Atmosphere. In *Advanced Air Pollution*; Nejadkoorki, F., Ed.; Tech Open: Shangai, China, 2011; pp. 365–390.
9. Williams, E.R. The global electrical circuit: A review. *Atmos. Res.* **2009**, *91*, 140–152. [CrossRef]
10. Bering, E.A.; Few, A.A.; Benbrook, J.R. The Global Electric Circuit. *Phys. Today* **1998**, *51*, 24–30. [CrossRef]
11. Roble, R.G.; Tzur, I. The Global Atmospheric Electrical Circuit. In *Study in Geophysics: The Earth's Electrical Enviroment*; National Academy Press: Washington, DC, USA, 1986; pp. 206–231.

12. Roble, R.G. On Modeling Component Processes in the Earth's Global Electric Circuit. *J. Atmos. Sol.-Terr. Phys.* **1991**, *53*, 831–847. [CrossRef]
13. Richmond, A.D. Modeling the ionosphere wind dynamo: A review. *Pure Appl. Geophys.* **1989**, *131*, 413–435. [CrossRef]
14. Akasofu, S.I. The solar wind-magnetosphere dynamo and the magnetospheric substorm. *Planet. Space Sci.* **1975**, *23*, 817–823. [CrossRef]
15. Harrison, R.G.; Aplin, K.L.; Rycroft, M.J. Atmospheric Electricity Coupling Between Earthquake Regions and the Ionosphere. *J. Atmos. Sol.-Terr. Phys.* **2010**, *72*, 376–381. [CrossRef]
16. Tramutoli, V.; Marchese, F.; Falconieri, A.; Filizzola, C.; Genzano, N.; Hattori, K.; Lisi, M.; Liu, J.Y.; Ouzounov, D.; Parrot, M.; et al. Tropospheric and Ionospheric Anomalies Induced by Volcanic and Saharan Dust Events as Part of Geosphere Interaction Phenomena. *Geosciences* **2019**, *9*, 177. [CrossRef]
17. Li, W.; Guo, J.; Yue, J.; Shen, Y.; Yang, Y. Total electron content anomalies associated with global VEI4+ volcanic eruptions during 2002–2015. *J. Volcanol. Geotherm. Res.* **2016**, *325*, 98–109. [CrossRef]
18. Shults, K.; Astafyeva, E.; Adourian, S. Ionospheric detection and localization of volcano eruptions on the example of the April 2015 Calbuco events. *J. Geophys. Res. Space Phys.* **2016**, *121*, 10303–10315. [CrossRef]
19. Zlotnicki, J.; Li, F.; Parrot, M. Signals recorded by DEMETER satellite over active volcanoes during the period 2004 August–2007 December. *Geophys. J. Int.* **2010**, *183*, 1332–1347. [CrossRef]
20. Zlotnicki, J.; Li, F.; Parrot, M. Ionospheric Disturbances Recorded by DEMETER Satellite over Active Volcanoes: From August 2004 to December 2010. *Int. J. Geophys.* **2013**, *2013*, 1–18. [CrossRef]
21. Jukić, O.; Sikirica, N.; Rumora, I.; Filić, M. Seven days in Chile: Impact of the 2011 Puyehue-Cordon Caulle volcanic eruption on GPS ionospheric delay. In Proceedings of the International Symposium on Global Navigation Satellite System 2018 (ISGNSS 2018), Bali, Indonesia, 21–23 November 2018; Wijaya, D.D., Ed.; p. 1. Available online: https://www.e3s-conferences.org/articles/e3sconf/abs/2019/20/contents/contents.html (accessed on 12 February 2019).
22. Cahyadi, N.M.; Rahayu, R.W.; Heki, K.; Nakashima, Y. Harmonic ionospheric oscillation by the 2010 eruption of the Merapi volcano, Indonesia, and the relevance of its amplitude to the mass eruption rate. *J. Volcanol. Geotherm. Res.* **2020**, *405*, 1–8. [CrossRef]
23. Tahira, M. Acoustic resonance of the atmospheric at 3.7 mHz. *J. Atmos. Sci.* **1995**, *52*, 2670–2674. [CrossRef]
24. Dow, J.M.; Neilan, R.E.; Rizos, C. The International GNSS Service in a changing landscape of Global Navigation Satellite Systems. *J. Geod.* **2009**, *83*, 191–198. [CrossRef]
25. Liu, L.; Wan, W.; Chen, Y.; Le, H. Solar activity effects of the ionosphere: A brief review. *Chin. Sci. Bull.* **2011**, *56*, 1202–1211. [CrossRef]
26. Kutiev, I.; Tsagouri, I.; Perrone, L.; Pancheva, D.; Mukhtarov, P.; Mikhailov, A.; Lastovicka, J.; Jakowski, N.; Buresova, D.; Blanch, E.; et al. Solar activity impact on the Earth's upper atmosphere. *J. Space Weather Space Clim.* **2013**, *3*, 1–21. [CrossRef]
27. Zolesi, B.; Cander, L.R. *Ionospheric Prediction and Forecasting*; Springer: New York, NY, USA, 2014.
28. Schunk, R.; Nagy, A. *Ionospheres: Physics, Plasma Physics and Chemistry*, 2nd ed.; Cambridge University Press: Cambridge, UK, 2009.
29. Tapping, K.F. The 10.7 cm solar radio flux (F10.7). *Space Weather* **2013**, *11*, 394–406. [CrossRef]
30. Stone, E.; Frandsen, A.; Mewaldt, R.; Christian, E.R.; Margolies, D.; Ormes, J.F.; Snow, F. The Advanced Composition Explorer. *Space Sci. Rev.* **1998**, *86*, 1–22. [CrossRef]
31. Covington, A.E.; Harvey, G.A. Impulsive and long-enduring sudden enhancements of solar radio emission at 10 centimeter wavelength. *J. R. Astron. Soc. Can.* **1958**, *52*, 161–166.
32. Hathaway, D.H. *The Solar Cycle*; Springer Nature: Cham, Switzerland, 2010.
33. Bartels, J.; Heck, N.H.; Johnston, H.F. The three-hour-range index measuring geomagnetic activity. *J. Geophys. Res.* **1939**, *44*, 411–454. [CrossRef]
34. Thomsen, M.F. Why Kp is such a good measure of magnetospheric convection. *Space Weather* **2004**, *2*, 1–10. [CrossRef]
35. Menvielle, M.; Berthelier, A. The K-derived planetary indices: Description and availability. *Rev. Geophys.* **1991**, *29*, 415–432. [CrossRef]
36. Chapman, S.; Bartels, J. *Geomagnetism*; Oxford University Press: Oxford, UK, 1940.
37. Klobuchar, J.A. Ionospheric corrections for timing applications. In Proceedings of the 20th Annual Precise Time and Time Interval (PTTI) Application and Planning Meeting, Vienna, Virginia, 7–9 December 1988; Naval Observatory: Washington, DC, USA, 1988; pp. 193–204.
38. Kintner, P.M.; Ledvina, B. The ionosphere, radio navigation and global navigation satellite systems. *Adv. Space. Res.* **2005**, *35*, 788–811. [CrossRef]
39. Petrovski, I.; Tsujii, T. *Digital Satellite Navigation and Geophysics*; Cambridge University Press: Cambridge, UK, 2012.
40. Liu, J.Y.; Tsai, H.F.; Jung, T.K. The IGS VTEC maps: A reliable source of ionospheric information since 1998. *J. Geod.* **2009**, *83*, 263–275.
41. Liu, J.Y.; Tsai, H.F.; Jung, T.K. Total Electron Content Obtained by Using the Global Positioning System. *Terr. Atmos. Ocean Sci.* **1996**, *7*, 107–117. [CrossRef]
42. Kouba, J. *A Guide to Using International GNSS Service (IGS) Products*; IGS Central Bureau: Pasadena, CA, USA, 2009.
43. Schaer, S. *Mapping and Predicting the Earth's Ionosphere Using the Global Positioning System*; Institut für Geodäsie und Photogrammetrie, Eidg. Technische Hochschule Zürich: Zürich, Switzerland, 1999.

44. Subirana, J.S.; Zornoza, J.M.J.; Hernandez-Pajares, M. *GNSS Data Processing, Volume I: Fundamentals and Algorithms*; ESA Communications: Noordwijk, The Netherlands, 2013.
45. Dyrud, L.; Jovancevic, A.; Brown, A.; Wilson, D.; Ganguly, S. Ionospheric measurement with GPS: Receiver techniques and methods. *Radio Sci.* **2008**, *43*, 1–11. [CrossRef]
46. Bonaccorso, A.; Bonforte, A.; Calvari, S.; Del Negro, C.; Di Grazia, G.; Ganci, G.; Neri, M.; Vicari, A.; Boschi, E. The initial phases of the 2008–2009 Mount Etna eruption: A multidisciplinary approach for hazard assessment. *J. Geophys. Res.* **2011**, *116*, B03203. [CrossRef]
47. Spampinato, L.; Calvari, S.; Oppenheimer, C.; Boschi, E. Volcano surveillance using infrared cameras. *Earth Sci. Rev.* **2011**, *106*, 63–91. [CrossRef]
48. Blackett, M. An overview of infrared remote sensing of volcanic activity. *J. Imaging* **2017**, *3*, 13. [CrossRef]
49. Fisher, W.A.; Mozham, R.M.; Polcyn, F.; Landis, G.H. Infrared surveys of Hawaiian Volcanoes. *Science* **1964**, *146*, 733–742. [CrossRef]
50. Gawarecki, S.J.; Lyon, R.J.P.; Nordberg, W. Infrared spectral returns and imagery of the Earth from space and their application to geological problems: Scientific experiments for manned orbital flight. *Am. Astronaut. Soc. Sci. Technol.* **1965**, *4*, 13–133.
51. Harris, A.J.L. *Thermal Remote Sensing of Active Volcanoes: A User's Manual*; Cambridge University Press: Cambridge, UK, 2013.
52. Coppola, D.; Laiolo, M.; Cigolini, C.; Massimetti, F.; Donne, D.D.; Ripepe, M.; Arias, H.; Barsotti, S.; Parra, C.B.; Centeno, R.G.; et al. Thermal Remote Sensing for Global Volcano Monitoring: Experiences from the MIROVA system. *Front. Earth Sci.* **2020**, *7*, 1–21. [CrossRef]
53. Ganci, G.; Vicari, A.; Fortuna, L.; Del Negro, C. The HOTSAT volcano monitoring system based on a combined use of SEVIRI and MODIS multispectral data. *Ann. Geophys.* **2011**, *54*, 544–550.
54. Coppola, D.; Laiolo, M.; Cigolini, C.; Delle Donne, D.; Ripepe, M. Enhanced volcanic hot-spot detection using MODIS IR data: Results from the MIROVA system. *Geol. Soc. Spec. Publ.* **2016**, *426*, 181–205. [CrossRef]
55. Coppola, D.; Laiolo, M.; Massimetti, F.; Cigolini, C. Monitoring endogenous growth of open-vent volcanoes by balancing thermal and SO2 emissions data derived from space. *Sci. Rep.* **2019**, *9*, 1–12. [CrossRef]
56. Toller, G.N.; Isaacman, A.; Kuyper, J. *MODIS Level 1B Product User's Guide*; Members of the MODIS Characterization Support Team for NASA/Goddard Space Flight Center: Greenbelt, MD, USA, 2009; pp. 1–57.
57. MIROVA. A Collaborative Project between the Universities of Turin and Florence (Italy). 2020. Available online: http://www.mirovaweb.it (accessed on 13 January 2020).
58. International GNSS Service (IGS). Access to Products. 2020. Available online: https://kb.igs.org/hc/en-us/articles/115003935351-Access-to-Products (accessed on 1 July 2020).
59. NASA, Goddard Space Flight Center. Coordinated Data Analysis Web (CDAWeb). 2020. Available online: https://cdaweb.gsfc.nasa.gov/istp_public (accessed on 13 March 2020).
60. Coppola, D.; Laiolo, M.; Piscopo, D.; Cigolini, C. Rheological control on the radiant density of active lava flows and domes. *J. Volcanol. Geotherm. Res.* **2013**, *249*, 39–48. [CrossRef]
61. Gurtner, W.; Estey, L. *RINEX: The Receiver Independent Exchange Format. V3.01*; IGS Central Bureau: Pasadena, CA, USA, 2009; pp. 1–100.
62. Krishna, S.G. GPS-TEC Analysis Software Version 2.9. 2017. Available online: http://seemala.blogspot.in/ (accessed on 21 March 2019).
63. Ren, X.; Zhang, X.; Weiliang, X.; Zhang, K.; Yuan, Y.; Li, X. Global Ionospheric Modelling using Multi-GNSS: BeiDou, Galileo, GLONASS and GPS. *Sci. Rep.* **2016**, *6*, 1–11. [CrossRef]
64. Brčić, D. A Model of Non-Specific Daily Pattern of the Satellite Positioning Signal Ionospheric Delay. Ph.D. Thesis, Faculty of Maritime Studies, University of Rijeka, Rijeka, Croatia, 30 April 2015. Available online: https://repository.pfri.uniri.hr/en/islandora/object/pfri%3A175 (accessed on 13 May 2020).
65. Brčić, D.; Filjar, R.; Kos, S.; Valčić, M. On Global Ionospheric Maps based winter-time ionospheric delay with reference to the Klobuchar model: Case study of the Northern Adriatic. *Pomorstvo* **2019**, *33*, 210–221. [CrossRef]
66. Astronomical Institute of the University of Berne, Center for Orbit Determination in Europe (AIUB CODE). Differential Code Bias IONEX DCB Files. 2020. Available online: Ftp://ftp.aiub.unibe.ch/CODE/ (accessed on 14 February 2019).
67. Xiang, Y.; Yang, G. An Enhanced Mapping Function with Ionospheric Varying Height. *Remote Sens.* **2019**, *11*, 1497. [CrossRef]
68. Otsuka, Y.; Suzuki, K.; Nakagawa, S.; Nishioka, M.; Shiokawa, K.; Tsugawa, T. GPS observations of medium-scale traveling ionospheric disturbances over Europe. *Ann. Geophys.* **2013**, *31*, 163–172. [CrossRef]
69. Hunsucker, R. Atmospheric gravity waves generated in the high-latitude ionosphere: A review. *Rev. Geophys.* **1982**, *20*, 293–315. [CrossRef]
70. Kaso, A. Computation of the normalized cross-correlation by fast Fourier transform. *PLoS ONE* **2018**, *13*, e0203434. [CrossRef] [PubMed]
71. Goldblum, C.E.; Ritter, R.C.; Gillies, G.T. Using the fast Fourier transform to determine the period of a physical oscillator with precision. *Rev. Sci. Instrum.* **1988**, *59*, 778–782. [CrossRef]
72. Brigham, E.O.; Morrow, R.E. The fast Fourier transform. *IEEE Spectr.* **1967**, *4*, 63–70. [CrossRef]

remote sensing

Technical Note

Lithosphere Ionosphere Coupling Associated with Seismic Swarm in the Balkan Peninsula from ROB-TEC and GPS

Lvquan Wei [1,2], Junyu Li [1,2,*], Lilong Liu [1,2], Liangke Huang [1,2], Dunyong Zheng [2,3], Xiangyu Tian [2,4], Ling Huang [1,2], Lv Zhou [1,2], Chao Ren [1,2] and Hongchang He [1,2]

1 College of Geomatics and Geoinformation, Guilin University of Technology, Guilin 541004, China
2 Guangxi Key Laboratory of Spatial Information and Geomatics, Guilin 541004, China
3 National-Local Joint Engineering Laboratory of Geo-Spatial Information Technology, Hunan University of Science and Technology, Xiangtan 411201, China
4 Institute of Surveying and Mapping, Information Engineering University, Zhengzhou 450001, China
* Correspondence: junyu_li@whu.edu.cn; Tel.: +86-13307835838

Abstract: With the rapid development of global navigation satellite systems (GNSS) and their increasingly wide range of applications in atmospheric science, total electron content (TEC) data are widely used in the theoretical study of layer coupling related to seismicity. This study detected and analyzed pre-earthquake ionospheric anomalies (PEIA) by using TEC data from the Royal Observatory of Belgium (ROB), and analyzed coseismic ionospheric disturbance (CID) with vertical TEC (VTEC) from the GPS stations in earthquake preparation areas. The results show that PEIA appear to increase continuously from 08:00–12:00 UT in the 3 days before a seismic swarm of Mw > 5.0. The ionosphere over the seismogenic zones exhibited large-scale anomalies when multiple seismogenic zones of the Balkan Peninsula spatially and temporally overlapped. Moreover, the TEC around the earthquake centers showed a positive anomaly lasting for 7 h. In a single seismogenic zone in Greece, the TEC around the earthquake center reached over +3.42 TECu. In addition, the CID observed from GPS stations shows that with the increase in the number of earthquakes, the ionosphere over the seismogenic area is more obviously disturbed, and after three strong earthquakes, TEC suddenly decreased over the seismogenic area and formed a phenomenon similar to an ionospheric hole. We conclude that a lithosphere–atmosphere–ionosphere coupling mechanism existed before the seismic swarm appeared in the Balkan Peninsula. Earthquake-induced VTEC anomalies occurred more frequently within a 3–10 day window before the earthquake. This phenomenon is particularly evident when multiple seismogenic zones overlap spatiotemporally.

Keywords: TEC; ionosphere; seismic swarm; lithosphere-atmosphere-ionosphere coupling; CID

Citation: Wei, L.; Li, J.; Liu, L.; Huang, L.; Zheng, D.; Tian, X.; Huang, L.; Zhou, L.; Ren, C.; He, H. Lithosphere Ionosphere Coupling Associated with Seismic Swarm in the Balkan Peninsula from ROB-TEC and GPS. *Remote Sens.* **2022**, *14*, 4759. https://doi.org/10.3390/rs14194759

Academic Editors: José Fernández, Juan F. Prieto and Serdjo Kos

Received: 11 August 2022
Accepted: 20 September 2022
Published: 23 September 2022

Publisher's Note: MDPI stays neutral with regard to jurisdictional claims in published maps and institutional affiliations.

1. Introduction

With the development and advancement of space exploration technology, researchers have paid increasing attention to seismic ionosphere detection and lithosphere–ionosphere coupling. However, earthquake prediction has been in the exploratory stage because of the complexities behind the causes of earthquakes. There are many reports regarding pre-earthquake geological structural changes and other pre-earthquake anomalies related to earthquakes; among these, the pre-earthquake ionospheric anomaly is a popular topic of contemporary research [1–3].

Davies (1964) first detected abnormal ionospheric disturbances above the epicenter of a 9.2 magnitude Alaskan earthquake [4]. Pulinets (2004) observed anomalies appearing in electron densities of the ionospheric F region a few days before some strong earthquakes through the analysis of GPS TEC and other ionospheric parameters from different dedicated satellites [5]. The preliminary results from the above research show that the energy released during seismogenesis can propagate upward into the Earth's atmosphere, resulting in ionospheric disturbances. The identification of seismo-ionospheric anomalies was made from the analysis of different GNSS satellite measurements based on different statistical methods. For example, by using GPS TEC, Sanchez (2022) found TEC in the form of traveling ionospheric disturbances (TIDs) within 1 h after the mainshock onset during an earthquake in California [6]. Liu (2001) detected TEC 15 days before an earthquake (Mw 7.7) in Taiwan and found three anomalies that provided empirical evidence of earthquake-executed ionosphere anomalies within five days before the mainshock [7]. Chen (2004) found a significant correlation between anomalies and earthquakes in Taiwan and the anomalous depression of the maximum plasma frequency in the ionospheric F2 layer by analyzing these TEC anomalies [8]. Hobara (2005) found from a dedicated satellite's mission over a seismogenic zone that an abnormal ionospheric disturbance appeared during an 8.3 magnitude earthquake in Hachiko [9]. Ryu (2014) discovered earthquake–ionospheric coupling by detecting ionospheric electron density (IED) data in seismic regions accompanied by PEIA; this report suggested the possible coupling of the lithosphere and ionosphere by the integration of TEC [10]. Similarly, Yao (2016) used a singular spectrum analysis of the TEC time series around the epicenter of the Nepal earthquake and showed positive ionospheric anomalies in the epicenter region before the earthquake [11]. Shah (2020) used global TEC observation data to study ionospheric anomalies before and after earthquakes at different latitudes, and found there were pre-earthquake disturbances and coseismic responses in the temporal and spatial distribution of the TEC data [12]. Tariq (2021) studied the ionospheric anomalies before earthquakes in Pakistan and Islamabad, and the results showed positive anomalies in the ionosphere ten days before the two earthquakes [13]. In general, seismo-ionospheric anomalies appear either a few days to two weeks before large earthquakes or around the earthquake time [14]. Various kinds of great earthquake precursors have been reported so far, but there are few studies on the impact of the seismic swarm on the ionosphere. Therefore, this paper comprehensively describes the ionospheric disturbances of the Balkan–Greece seismic swarm with a case study of both the co- and pre-seismic ionospheric disturbances of the seismic swarm.

2. Experimental Data

To avoid the interference of magnetic storms, earthquakes that occurred under geomagnetic conditions during a quiet period were selected ($Kp < 4$, $F10.7 < 100$ and $Dst > -30$ nT). The Kp and Dst indices were usually lower than 4 and greater than -30 nT, respectively, which indicate that the geomagnetic activity is quiet [15,16]. The seismic data were obtained from the United States Geological Survey (available online: https://earthquake.usgs.gov/earthquakes/search/ (accessed on 10 July 2021)), which provided the reference for calculating the radius of the seismogenic area [11,17]. We also analyzed TEC data from GPS stations operating within the seismogenic zone of the seismic swarm in the Balkan Peninsula (Figure 1). Further details are presented in Table 1.

We retrieved high-quality TEC data from the ROB by using GPS observations, and the products consist of ionospheric vertical TEC maps over Europe, which were estimated in near real-time every 15 min with $0.5^{\circ} \times 0.5^{\circ}$ grids. The maps are available online with a latency of ~3 min in IONEX format at: ftp://gnss.oma.be (accessed on 11 July 2021) and as interactive web pages at: www.gnss.be (accessed on 11 July 2021) [18]. In addition, we calculated the VTEC at seven GPS stations operating within the seismogenic zone of the seismic swarm in the Balkan Peninsula (available online: https://www.epncb.oma.be/ (accessed on 12 July 2021)). The correlation index for the solar activity and geomagnetic activity was obtained from the Goddard Space Flight Center of NASA (available online: https://omniweb.gsfc.nasa.gov/ (accessed on 12 July 2021)).

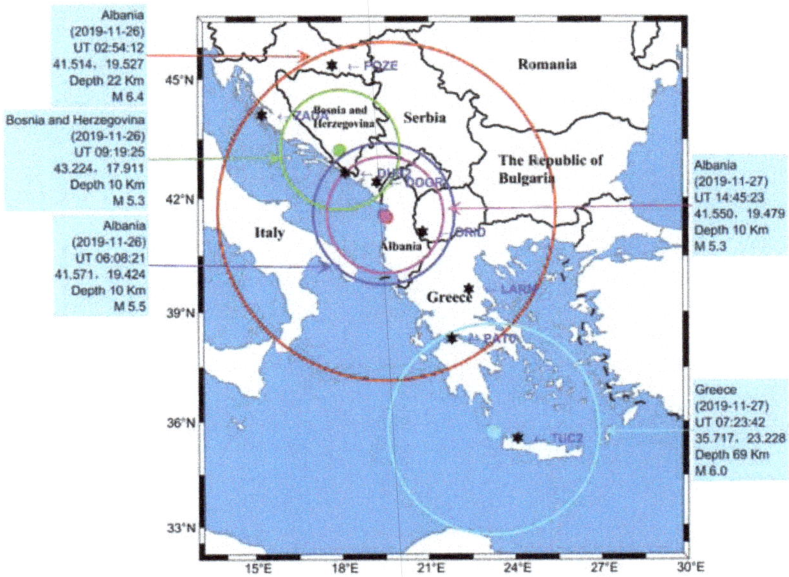

Figure 1. Geographic map of the locations of seismic swarm in the Balkan Peninsula in 2019. Seismogenic zones are shown by circles. Earthquake information is shown with every epicenter (circular). GPS stations are shown by filled hexagons.

Table 1. Details of the earthquakes that occurred in the Balkan Peninsula.

NO.	Date (UT)	Time (UT)	Geog.Lat	Geog.lon	Mag (Mw)	Depth (km)	Regions	Radius of Influence (km)
1	26 November 2019	02:54	41.5138	19.5256	6.4	22	Albania	487.5
2	26 November 2019	06:08	41.5708	19.4242	5.5	10	Albania	204.2
3	26 November 2019	09:19	43.2235	17.9118	5.3	10	Bosnia and Herzegovina	168.3
4	27 November 2019	07:23	35.7174	23.2284	6.0	69	Greece	331.2
5	27 November 2019	14:45	41.5498	19.4787	5.3	10	Albania	168.3

3. Methods

3.1. Bilinear Interpolation

To determine the TEC of the earthquake center, we used the adjacent four grid points to interpolate the TEC using the bilinear interpolation method. Figure 2 shows the ionospheric grid interpolation diagram.

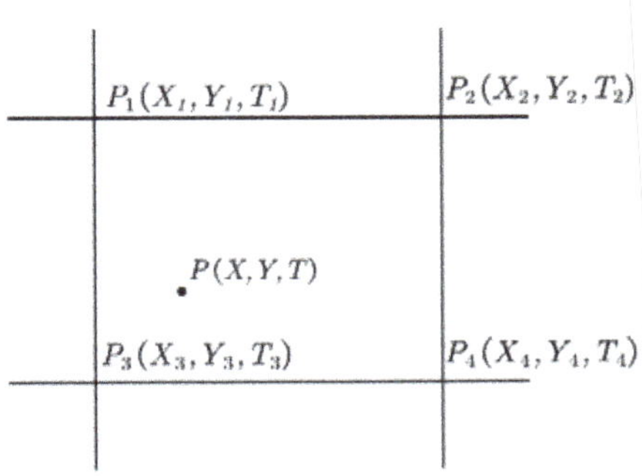

Figure 2. Schematic diagram of the ionospheric grid interpolation algorithm.

According to the Lagrange interpolant, the ionospheric VTEC value of the interpolation point was obtained. The formula is as follows:

$$\begin{array}{c} T = \frac{T_1}{(X_2-X_1)(Y_1-Y_3)}(X-X_1)(Y-Y_3) + \frac{T_2}{(X_2-X_1)(Y_1-Y_3)}(X_2-X)(Y-Y_3) \\ +\frac{T_3}{(X_4-X_3)(Y_4-Y_2)}(X-X_3)(Y_2-Y) + \frac{T_4}{(X_4-X_3)(Y_4-Y_2)}(X_4-X)(Y_2-Y) \end{array} \tag{1}$$

where X and Y are the coordinates of the undetermined interpolation point, the TEC is represented with T, and the grid coordinates are X_i and Y_i, respectively.

3.2. Method for PEIA

Because ionospheric variation is affected by many factors, such as solar activity, earth rotation, geomagnetic conditions, and so on, it has a robust diurnal periodicity. The 15-day moving median (MM) method is a quartile-based statistical analysis technique that is used to express the dispersion of data in a powerful statistical technique proposed by Liu [19,20]. This technique is often used to check data anomalies and identify and strengthen periodic signals. Thus, abnormal disturbances in the TEC time series can be extracted using a quartile-based statistical technique. TEC data that were not disturbed by solar activity and geomagnetic anomalies within 15 days before the earthquake were used. The median (Q_2), upper (Q_1), and lower quartiles (Q_3), and the interquartile range (IQR) were calculated. To further improve the reliability of abnormal signals, the constraint method of 1.5 × IQR is adopted [21]. For instance, in order to generate MM, UB and LB for the 16th day, the TEC values for the first 15 days were utilized. Similarly, 15 days of TEC data from between the 2nd and 16th day were used to generate bounds for the 17th day. If more than one-third of the data (e.g., eight hours are anomalous in a day) were greater or lesser than the UBs and LBs in a day, this day was taken as anomalous [20].

$$\begin{array}{l} TEC_{up} = Q_2 + 1.5 \cdot IQR \\ TEC_{low} = Q_2 - 1.5 \cdot IQR \\ IQR = Q_3 - Q_1 \end{array} \tag{2}$$

where TEC is the median of the first 15 days at the same time, IQR is the quartile of the first 15 days, and equals the IQR multiplied by 1.5 to delimit the upper and lower thresholds. Likewise, TEC_{up} and TEC_{low} are the upper and lower confidence limits of the TEC, respectively, and if the observed TEC falls outside of either of the respective bounds, it is declared that a lower or upper abnormal signal is detected, respectively. If the observed value exceeds the upper threshold, a positive disturbance of the TEC is considered. Meanwhile, when the variations in TEC are below the lower bound, it may be a negative PEIA. The TEC exceptions were calculated as follows:

$$\mathrm{d}TEC = \begin{cases} TEC - TEC_{up} \text{ while } TEC > TEC_{up} \\ 0 \text{ while } TEC_{low} \leqslant TEC \leqslant TEC_{up} \\ TEC - TEC_{low} \text{ while } TEC < TEC_{low} \end{cases} \tag{3}$$

3.3. GNSS TEC Observations and Methods

The TEC from GPS stations operating within the seismogenic zone of the earthquakes was estimated by measuring the phase and amplitude (at a rate of 50 Hz) and code/carrier divergence (at 1 Hz) for each satellite, and the observation interval was 30 s. This calculates the slant TEC (STEC) from the combined frequencies by pseudo-range and carrier-phase measurements using the following equations [22,23]:

$$\begin{array}{l} \text{STEC} = \frac{f_1^2 f_2^2}{40.28(f_1^2 - f_2^2)}(L_1 - L_2 + \lambda_1(N_1 + b_1) - \lambda_2(N_2 + b_2) + \varepsilon) \\ \text{STEC} = \frac{f_1^2 f_2^2}{40.28(f_1^2 - f_2^2)}(P_1 - P_2 - (d_1 - d_2) + \varepsilon) \end{array} \tag{4}$$

where f_1 and f_2 are carrier phase frequencies of GPS signals at the two ends, P and L represent the pseudo-range and carrier phase observation of the delay path of the GPS signal. λ, N, and ε are the wavelength of GPS signals, the ambiguity of the ray path, and random residual of the GNSS signal along ray path, while b and d are the instrumental biases of the carrier phase and pseudo-range of the derived signal. Moreover, the STEC was converted into VTEC using the following equation [24]:

$$\text{VTEC} = \text{STEC} * \cos\left(\arcsin\left(\frac{R \sin z}{R+H}\right)\right) \quad (5)$$

where R and H are the radius of the Earth and the height of the top ionospheric layer in atmospheric altitude, respectively. Z is the elevation angle of the satellite for the ionosphere pierce point.

The ionosphere ranges from 60 km to 1000 km above the Earth's surface; to simplify the ionospheric model, the single-layer ionosphere assumption model is usually introduced. So, in this work, the height of the top layer of the atmospheric altitude was taken as 350 km [25].

4. Analysis and Results

4.1. Geomagnetic Activity Background

The spatial ionospheric variation is closely related to the solar and geomagnetic activity indices [26,27]. Thus, ionospheric anomalies caused by geomagnetism and solar activity should be avoided, and abnormal ionospheric disturbances related to earthquakes were obtained. Figures 3 and 4 show the time series for the F10.7, Dst, and Kp indices from 15 to 29 November 2019. The Dst index does not exceed −30 nT and the Kp index is less than 4, indicating that the geomagnetic field is calm and in a quiet ionospheric period. Therefore, the ionospheric anomalies caused by solar activity and geomagnetic activity can be excluded, and other geophysical signals (e.g., solar and geomagnetic activities) can be distinguished. The data were obtained from NASA's website at: https://omniweb.gsfc.nasa.gov/ (accessed on 15 July 2021).

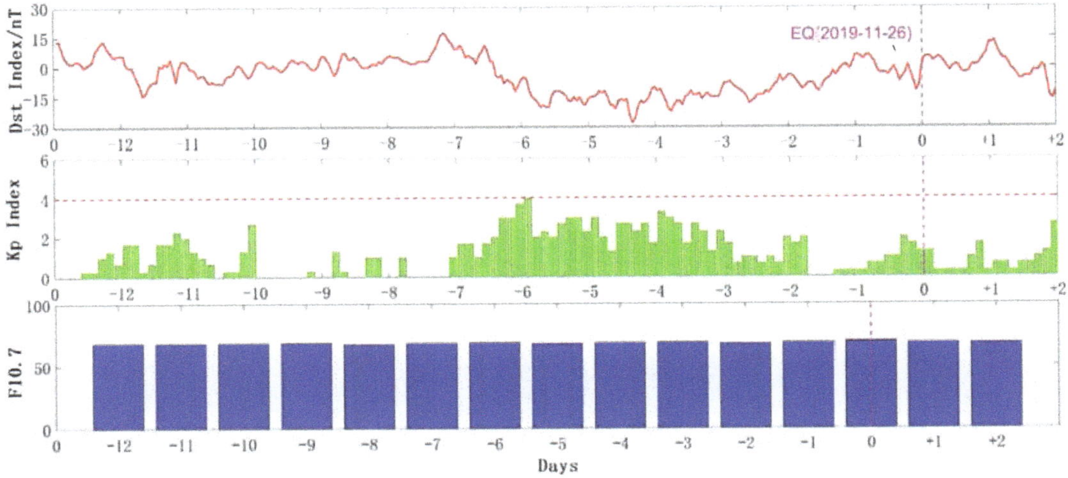

Figure 3. Changes in F10.7, Dst, and Kp indices from 15 November to 29 November 2019.

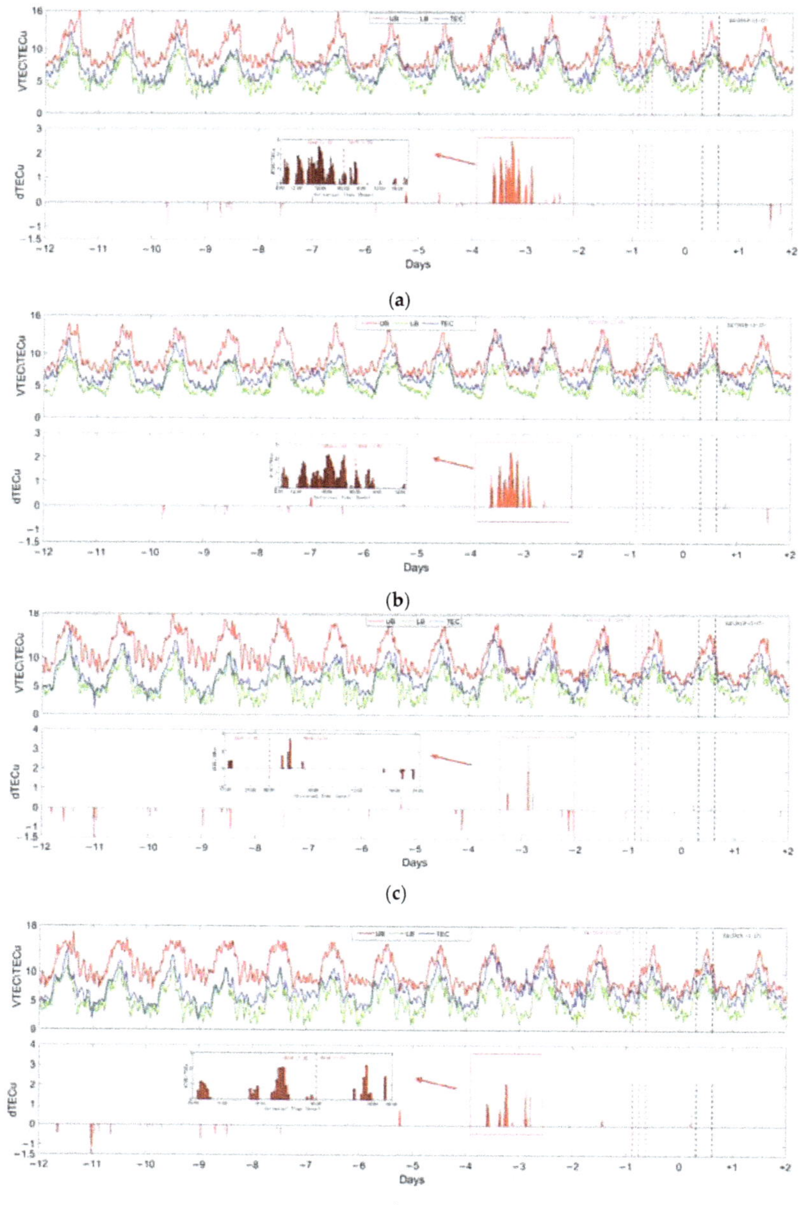

Figure 4. TEC time series diagram of ionospheric anomalies. The red line represents the lower bound (LB), the green line represents the upper bound (UB), the blue line represents the total electron content (TEC), the vertical dashed lines represent the mainshock, dTECu represents the PEIA. (**a**) TEC time series of the Albanian earthquake center; (**b**) TEC time series of earthquake centers in Bosnia and Herzegovina; (**c**) TEC time series of the central grid of the Greek earthquake; (**d**) TEC time series of grid points at the confluence of the Greek and Albanian seismogenic regions.

4.2. Time Series Analysis of Epicenter TEC

To observe the affected hours more closely, TEC data were obtained for 15 days at the intersection of the four epicenters and two seismogenic zones, as shown in Figure 4, where the red curve is the upper quartile threshold, the green curve is the lower quartile threshold, the blue curve is the actual TEC value, and the pink and black vertical bars are the earthquake occurrence times. If the TEC data exceeded the upper and lower quartile thresholds, the TEC deviations were drawn on the red histogram at the bottom of the graph. TEC deviations can be seen beyond the upper limits, 3–4 days before successive earthquakes. We observed slight seismo-ionospheric anomalies 10 days before the earthquake. As shown in Figure 4a, at 07:00 UT on 16 November 2019 (10 days before the earthquake), TEC first appeared as a negative anomaly of −0.92 TECu, and at the same time of the next day (17 November), it also appeared and reached −0.93 TECu. Then, the negative anomaly gradually weakened and became a positive anomaly. The positive anomalies were initiated at 08:45 UT on 22 November (4 days before the earthquake), and these positive anomalies, occurred between 08:45 UT on 22 November and 15:45 UT on 23 November (3 days before the earthquake). The maximum time interval of the anomaly was only 1 h. The peak of the positive anomalies appeared at 17:45 UT on 21 November (5 days before the earthquake), and reached 2.54 TECu. The anomalous characteristics of the epicenters of the Bosnia and Herzegovina earthquakes were consistent with those of the Albania earthquake, as shown in Figure 4b. This phenomenon could be explained because the epicenters of the Bosnia and Herzegovina earthquakes were only 137.8 km apart from the epicenter of the Albania earthquake and the two seismogenic zones coincided. However, as shown in Figure 4c, with a single seismogenic zone at the epicenter, there was a negative anomaly of TEC = −2.91 TECu at 23:30 UT on 15 November (11 days before the earthquake), which is similar to the characteristics of the two previous epicenters, while the intensity of the seismo-ionospheric anomalies had no continuity. Then, positive anomalies appeared at 17:30 UT on 22 November (5 days before the earthquake), with the peak at 03:00 UT on 23 November (3 days before the earthquake), and reached 3.42 TECu. Figure 4d shows the seismic ionospheric anomalies in the area where the Albanian seismic zone meets the Greek seismic zone. The positive anomaly characteristics at the intersection of seismogenic zones were very similar to those in the Albania seismogenic zone, with both showing the characteristics of continuous anomalies. Compared with the single earthquake area, the seismic ionospheric anomalies were significantly different. In addition, earthquake-induced VTEC anomalies occurred more frequently within a 3–10-day window with the approach of the earthquake day. The intersection of seismogenic regions resulted in more frequent ionospheric fluctuations. These results suggest that there is a certain correlation between the intersection of seismogenic regions in time and space, and ionospheric disturbance.

4.3. TEC Changes in the Balkan Peninsula Seismic Swarm

In order to study whether CID are caused by various atmospheric waves (Rayleigh surface waves, and internal gravity waves) generated by the seismic swarm, we investigated the TEC responses to the Balkan Peninsula seismic swarm [28,29]. Figure 5 shows the VTEC time series and the trajectory of the ionospheric penetration point (IPP) observed by the GPS station in the earthquake area, where the orbit color matches the color of the observation station. The pentagram on the orbit is the location of the satellite when the earthquake occurs, and the pentagram color corresponds to the color of the earthquake center (when the red earthquake center occurs, it corresponds to the red five-star position on the IPP path, and the blue earthquake center corresponds to the blue five-star). The red arrow is the direction in which the satellite moves.

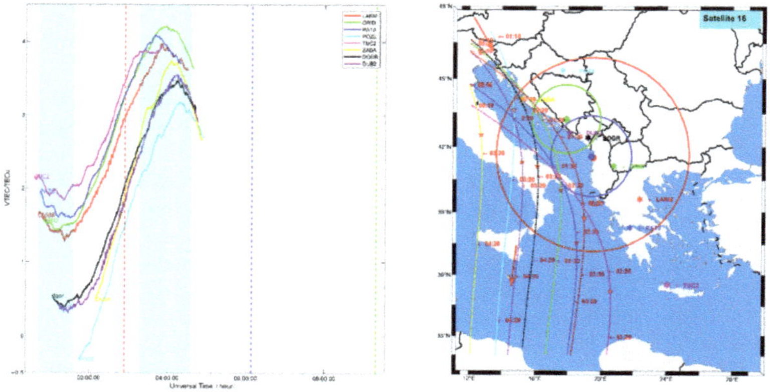

(**a**)

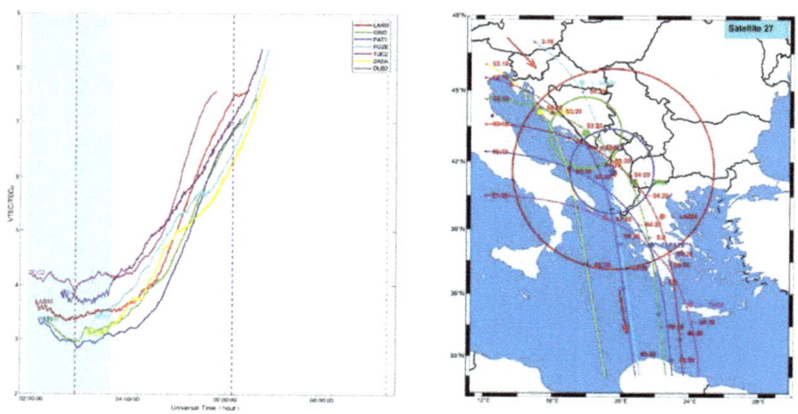

(**b**)

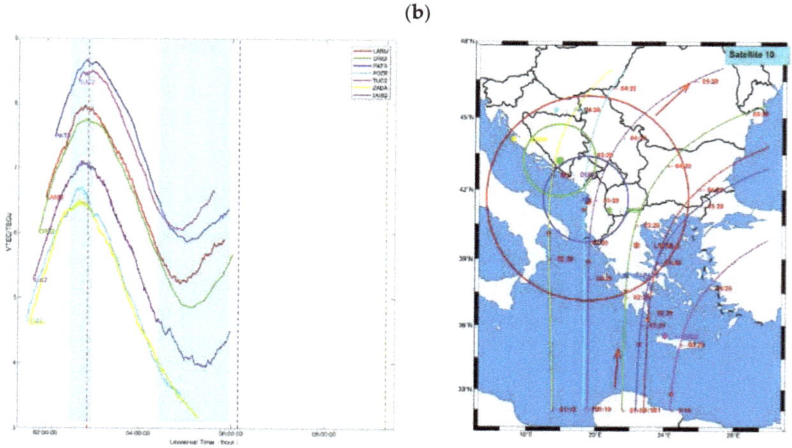

(**c**)

Figure 5. *Cont.*

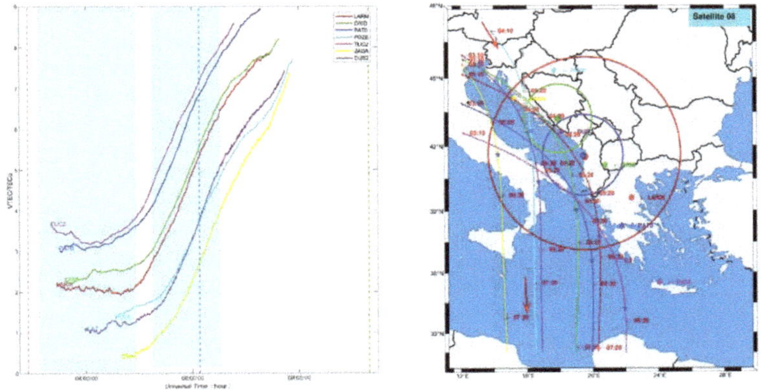

(**d**)

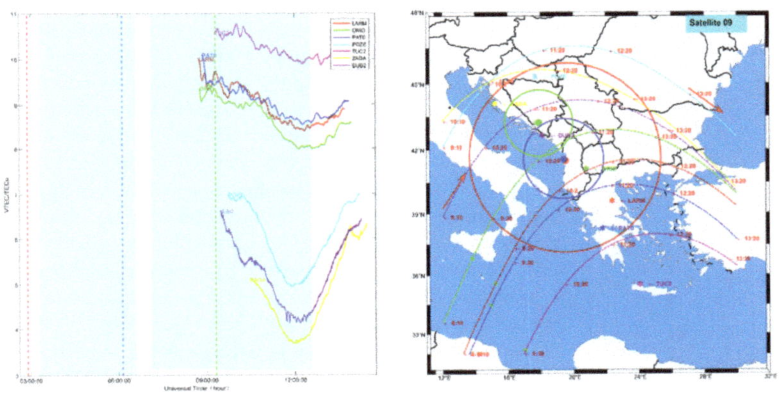

(**e**)

Figure 5. The left panels present the time series of the original VTEC changes observed by different satellites at different stations and the right panels present the corresponding IPP trajectory of the satellite. Additionally, red, blue and green lines indicate the occurrence of an earthquake in the left figure (the color corresponds to the source in the right figure). The small stars on the track represent the IPP position of the satellite when the earthquake occurs (the color corresponds to the source in the figure) and the red thread indicates the moving direction of the satellite in the right figure. (**a**) The time series of the original VTEC changes observed by satellite 16 and the IPP trajectory of the satellite. (**b**) The time series of the original VTEC changes observed by satellite 27 and the IPP trajectory of the satellite. (**c**) The time series of the original VTEC changes observed by satellite 10 and the IPP trajectory of the satellite. (**d**) The time series of the original VTEC changes observed by satellite 08 and the IPP trajectory of the satellite. (**e**) The time series of the original VTEC changes observed by satellite 09 and the IPP trajectory of the satellite.

In the left plane of Figure 5a, we show the VTEC series from 1:00 UT to 5:00 UT recorded by satellite 16 visible from different stations (except for the ZADA station, the IPP longitude range is 11°–13°E) in the seismogenic area. For all stations, CID clearly appear after the earthquake, with time lags of 1–2 h, This phenomenon is consistent with the conclusion of Astafyeva [30,31], and the time needed for acoustic waves to travel from the surface to the IPP. Interestingly, we found that when we compared the VTEC sequence of ZADA station with that of TUC2 station, ZADA station is at the edge of the seismogenic area, IPP also does not pass over the seismogenic area, and the fluctuation amplitude of VTEC sequence is not very obvious. On the contrary, although the TUC2 station is not in the earthquake preparation area, the IPP track is located above the earthquake preparation area and close to the focal center, and it can be clearly seen that it has been strongly disturbed. In addition, it can also be observed that within 2 h before the arrival of the first main earthquake, all stations were disturbed to varying degrees. The perturbation frequency increased as the IPP positions got closer to the source center, with a fluctuation range of 0.1–0.5 TECu, and similar phenomena were observed on other satellites.

The TEC shows rising or falling curvature due to satellite elevation and latitude changes, so it is difficult to observe abnormal conditions. Nevertheless, we observed from the IPP path of satellite 27 (Figure 5b) and satellite 09 (Figure 5e) within 2 h, and its latitude change rate was relatively low. It is obvious that compared with satellite 27 (Figure 5b), when the 6.3 Mw Albanian earthquake struck, CID was observed at all stations on satellite 27. In particular, the ORID station is close to the center of the source, and the IPP is also closest to the center of the source, and its variation in amplitude is large. Figure 5e shows VTEC changes over the seismogenic area after three earthquakes. From the figure on the right, we can see that the IPP of observation satellite 09 (Figure 5e) and the POZE, ZADA and DUB2 stations monitor the VTEC in the upper half of the seismogenic area. In addition, satellite 09 was observed at other stations, and its IPP position was located in the lower half of the seismogenic area. The VTEC observed at the ORID, LARM, and PAT0 stations were disturbed to the same extent, and VTEC change curves showed similar changes. However, at the TUC2 station outside the seismogenic area, it was observed that the VTEC at the edge of the seismogenic area was significantly different from other stations at UT 10:00. Interestingly, the overall trend of VTEC still showed a downward trend, indicating that the decrease in TEC after the earthquake may result from electronic transport. Therefore, it can be inferred that the earthquake swarm will continuously interfere with the ionosphere over the seismogenic area, and this interference will continue to increase with the increase in the number of earthquakes. It is interesting that within 2 h after the earthquake, the VTEC over this area decreased sharply (in the upper half of the seismogenic area). In order to analyze this abnormal phenomenon, the earthquake swarm likely caused the formation of ionospheric holes in the region [32]. In Figure 6b, at UT 9:00–10:00, the VTEC (satellite 22) in the seismogenic area observed by the three stations was lower than 10 tecu. However, after UT 11:00, when satellite 09 moved into the region, VTEC still showed a downward trend, which preliminarily indicated that the ionosphere in this region was abnormal. Figure 6a shows the change in VTEC on the day before the earthquake. It can be seen that its change characteristics align with the daily change characteristics of the ionosphere. With time, VTEC gradually increases.

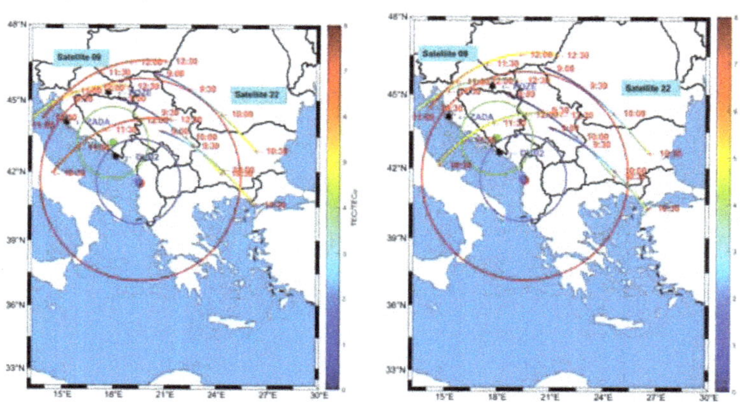

(a) (b)

Figure 6. POZE, ZADA and TUB2 station observed the IPP position and VTEC changes of satellite 09 and satellite 22 at UT 9:00–12:30. The left panels present TEC disturbance (TEC) along the IPP trajectories (trajectories are from left to right) in the vicinity of the epicenter on the previous day on November 25 (**a**) and the right panels show the day of the event on November 26 (**b**).

4.4. Observed Coseismic Ionospheric Disturbances in the Balkan Peninsula Seismic Swarm

To further analyze the CID, we examined TEC data with elevation angles higher than 20°. The TEC disturbance (dTEC) was obtained by employing high pass filtering.

In Figure 7, we compare the IPP trajectories and dTEC of satellite 16, 10, and 09 observed at the GPS stations. Abnormalities can be seen in the dTEC after the onset of the mainshock (26 November 2019) on earthquake day, but no such anomaly can be noted on the previous day or on the day after, respectively. In addition, it can be clearly seen that the anomaly of satellite 16 on the mainshock day is relatively weak, and the anomalies of satellite 10 and satellite 09 are more intense on the mainshock day. The IPPs of satellites 10 and 09 are approaching the source center. After three earthquakes in this area (25 November 2019), we noticed that the variation in the dtec amplitude of satellite 09 was about 0.15 tecu. Therefore, from this spatiotemporal analysis, we concluded that the intensified TEC disturbances in Figure 7 were likely of seismic origin.

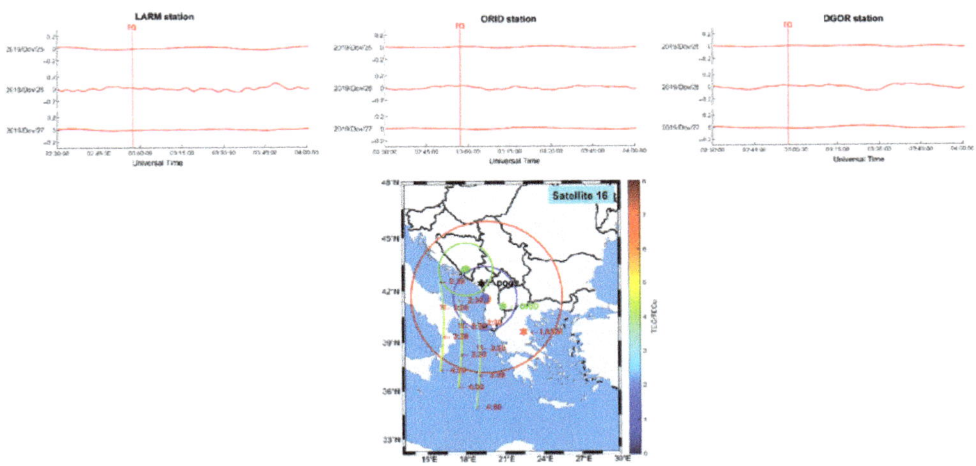

(**a**)

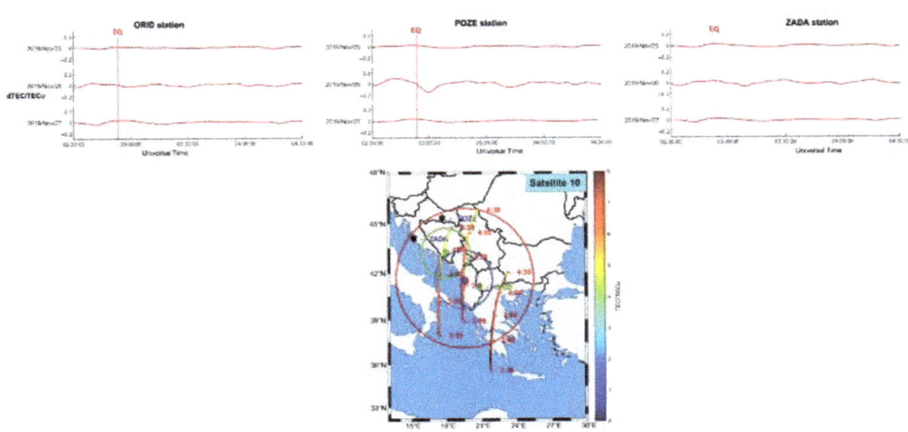

(**b**)

Figure 7. *Cont.*

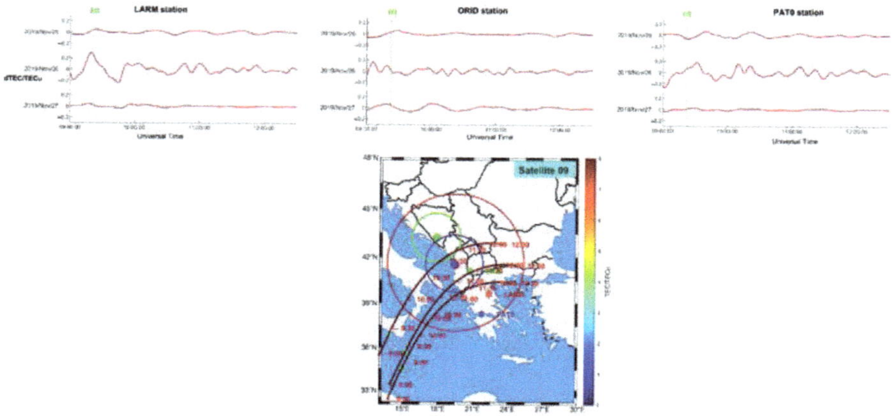

(**c**)

Figure 7. The upper panels present TEC disturbance (dTEC) was observed on three consecutive days (25–27 November 2019, event day on 26 November 2019). The lower panels show their corresponding IPP trajectory of the satellite. (**a**) TEC disturbance (dTEC) on the previous day on 25 November 2019, on the event day on 26 November 2019, and on the day after on 25 November 2019, and the IPP trajectories of the GPS Sat. 16. (**b**) TEC disturbance (dTEC) on the previous day on 25 November 2019, on the event day on 26 November 2019, and on the day after on 25 November 2019, and the IPP trajectories of the GPS Sat. 10. (**c**) TEC disturbance (dTEC) on the previous day on 25 November 2019, on the event day on 26 November 2019, and on the day after on 25 November 2019, and the IPP trajectories of the GPS Sat. 09.

5. Discussions

In this paper, the PEIAs associated with a seismic swarm in Balkan Peninsula were studied. The TEC from ROB-TEC within the seismogenic zone showed significant enhancement before the mainshock day. Ionospheric anomalies caused by solar activity and geomagnetic activity can be excluded, suggesting that earthquakes were the factor in ionospheric anomalies. Temporal TEC from ROB-TEC showed profound variations during the 3–10 days before the mainshock on relatively quiet storm days (Figure 4). Such evidence has already been provided by multiple analyses of temporal TEC over epicentral regions. They (Ulukavak (2020); Nico (2021); Nina (2020)) analyzed global earthquakes with Mw ≥ 6 and possible ionospheric TEC anomalies that occurred before earthquakes. The results of this study contribute to earthquake prediction in the short-term by providing empirical evidence [33–35]. Biagi revealed the INFREP Radio Network variations in connection with six earthquakes (Mw > 5.0) that occurred in the Balkan Peninsula and Adriatic Sea on 26 and 27 November 2019 [36]. All this information increases our knowledge about the origin of different pre-seismic signals associated with the above-mentioned catastrophic earthquakes.

In addition, with the increase in the number of seismogenic zones, the incidence and intensity of ionospheric disturbance has increased, further analyses are needed to quantify the variations with a more statistical and mathematical model and method to convincingly prove such phenomena. However, there are many theories regarding the existence of abnormal TEC from GPS measurements and some have provided enough evidence for the connection of the lithosphere–atmosphere–ionosphere coupling system. For example, Freund has proved the existence of stress-activated electric currents in rocks that sweep through the mineral grains under stress dislocations and cause the peroxy links to break, thus generating flow-down stress gradients, constituting an electric current with attendant magnetic field variations and EM emissions. Further experiments have shown the generation of energetic charge barriers and their thermal conduction as fissure reactions in the seismogenic zone, which could be an intrinsic source of earthquake–ionospheric anomalies [37]. The main findings show that continuous earthquakes have repeatedly triggered the generation of high-energy charge barriers in the seismogenic belt and their heat conduction as crack reactions, which frequently affect the ionosphere.

On the other hand, CID observed by GPS stations (Figure 7) shows that with the increase in the number of earthquakes, the ionosphere over the seismogenic area is more obviously disturbed, which is consistent with the previous findings. For example, the far-field CID following the 2015 Mw 7.8 Nepal earthquake was detected by GPS-TEC [38] as the acoustic gravity wave induced by Rayleigh surface wave propagated into the ionosphere and caused the electronic and plasma density oscillation.

It is worth noting that after three strong earthquakes, the TEC suddenly decreased over the seismogenic area, forming a phenomenon similar to an ionospheric hole (Figure 6b). In further studies, more statistical and mathematical models and methods are required to prove such phenomena. There is evidence of a quantitative relationship between the TEC depression rate of the ionospheric cavity caused by the tsunami earthquake and the intensity of the tsunami earthquake [32,39]. However, more relevant observation data are needed to analyze the ionosphere at different heights in the region and the relevant meteorological data in the region are necessary for the generation of the ionospheric hole. In conclusion, the abovementioned results provide an opportunity to study the correlation between the ionosphere and earthquakes from multiple directions and angles for the earlier detection of specific pre-seismic anomalies related to earthquakes.

6. Conclusions

In this study, we presented research on pre-seismic ionospheric anomalies and CID for a seismic swarm in the Balkan Peninsula in November 2019. The results show that a larger number of positive anomalies appeared in the seismogenic area on 23 November 2019 (3 days before the earthquake), and continuous positive anomaly disturbances appeared in the grid points in the spatially overlapping seismogenic zones. Moreover, there was a small range of negative anomalies in the seismogenic area 6–12 days before the earthquake, although the interval was relatively long. We found that CID could be clearly observed at the stations in the seismogenic area during each earthquake, and the degree of CID and disturbance increased as the IPPs approached the center of the earthquake area. After the three earthquakes, we also found a similar ionospheric hole phenomenon over the seismogenic area, and a large area where the VTEC value plummeted over the seismogenic area. The results showed that with the increase in the number of seismogenic zones, the incidence and intensity of the ionospheric disturbance increased.

These findings suggest that there is synergy between different surface, atmospheric, and ionospheric processes. However, further analysis is required to support the hypothesis regarding lithosphere and ionosphere coupling during the earthquake preparation period as mainshock responses, which will be left for a future work.

Author Contributions: Conceptualization, J.L. and L.L.; methodology, L.W. and J.L.; validation, C.R., L.H. (Liangke Huang) and X.T.; formal analysis, L.W.; writing—original draft preparation, L.W., L.Z. and D.Z.; writing—review and editing, D.Z., J.L., L.H. (Ling Huang) and H.H.; funding acquisition, J.L., D.Z., L.H. (Ling Huang) and L.L. All authors have read and agreed to the published version of the manuscript.

Funding: This work was supported by the Guangxi Natural Science Foundation of China (2020GXNSFBA297145, 2020GXNSFBA159033), the Foundation of Guilin University of Technology (GUTQDJJ661-6032), and the National Natural Science Foundation of China (42004025, 42064002).

Data Availability Statement: The data presented in this study are available on request from the corresponding author.

Conflicts of Interest: The authors declare no conflict of interest.

References

1. Ikuta, R.; Hisada, T.; Karakama, G.; Kuwano, O. Stochastic Evaluation of Pre-Earthquake TEC Enhancements. *J. Geophys. Res. Space Phys.* **2020**, *125*, e2020JA027899. [CrossRef]
2. Thomas, J.N.; Huard, J.; Masci, F. A statistical study of global ionospheric map total electron content changes prior to occurrences of M $\geq$ 6.0 earthquakes during 2000–2014. *J. Geophys. Res. Space Phys.* **2017**, *122*, 2151–2161. [CrossRef]
3. Kakinami, Y.; Saito, H.; Yamamoto, T.; Chen, C.-H.; Yamamoto, M.-Y.; Nakajima, K.; Liu, J.-Y.; Watanabe, S. Onset Altitudes of Co-Seismic Ionospheric Disturbances Determined by Multiple Distributions of GNSS TEC after the Foreshock of the 2011 Tohoku Earthquake on March 9, 2011. *Earth Space Sci.* **2021**, *8*, e2020EA001217. [CrossRef]
4. Davies, K.; Baker, D.M. Ionospheric effects observed around the time of the Alaskan earthquake of March 28, 1964. *J. Geophys. Res. (1896–1977)* **1965**, *70*, 2251–2253. [CrossRef]
5. Pulinets, S. Ionospheric precursors of earthquakes; recent advances in theory and practical applications. *Terr. Atmos. Ocean. Sci.* **2004**, *15*, 413–436. [CrossRef]
6. Sanchez, S.A.; Kherani, E.A.; Astafyeva, E.; de Paula, E.R. Ionospheric Disturbances Observed Following the Ridgecrest Earthquake of 4 July 2019 in California, USA. *Remote Sens.* **2022**, *14*, 188. [CrossRef]
7. Liu, J.; Chen, Y.; Chuo, Y.; Tsai, H. Variations of ionospheric total electron content during the Chi-Chi earthquake. *Geophys. Res. Lett.* **2001**, *28*, 1383–1386. [CrossRef]
8. Chen, Y.-I.; Liu, J.-Y.; Tsai, Y.-B.; Chen, C.-S. Statistical tests for pre-earthquake ionospheric anomaly. *Terr. Atmos. Ocean. Sci.* **2004**, *15*, 385–396. [CrossRef]
9. Hobara, Y.; Parrot, M. Ionospheric perturbations linked to a very powerful seismic event. *J. Atmos. Sol.-Terr. Phys.* **2005**, *67*, 677–685. [CrossRef]
10. Ryu, K.; Lee, E.; Chae, J.; Parrot, M.; Pulinets, S. Seismo-ionospheric coupling appearing as equatorial electron density enhancements observed via DEMETER electron density measurements. *J. Geophys. Res. Space Phys.* **2014**, *119*, 8524–8542. [CrossRef]
11. Chen, P.; Yao, Y.; Yao, W. On the coseismic ionospheric disturbances after the Nepal Mw7. 8 earthquake on April 25, 2015 using GNSS observations. *Adv. Space Res.* **2017**, *59*, 103–113. [CrossRef]
12. Shah, M.; Aibar, A.C.; Tariq, M.A.; Ahmed, J.; Ahmed, A. Possible ionosphere and atmosphere precursory analysis related to Mw> 6.0 earthquakes in Japan. *Remote Sens. Environ.* **2020**, *239*, 111620. [CrossRef]
13. Fernández, J. Geodetic and geophysical effects associated with seismic and volcanic hazards. In *Geodetic and Geophysical Effects Associated with Seismic and Volcanic Hazards*; Springer: Berlin/Heidelberg, Germany, 2004; pp. 1301–1303.
14. Liu, J.-Y.; Chen, Y.; Chuo, Y.; Chen, C.-S. A statistical investigation of preearthquake ionospheric anomaly. *J. Geophys. Res. Space Phys.* **2006**, *111*, A5. [CrossRef]
15. Kumar, S.; Chen, W.; Liu, Z.; Ji, S. Effects of solar and geomagnetic activity on the occurrence of equatorial plasma bubbles over Hong Kong. *J. Geophys. Res. Space Phys.* **2016**, *121*, 9164–9178. [CrossRef]
16. Zheng, D.; Zheng, H.; Wang, Y.; Nie, W.; Li, C.; Ao, M.; Hu, W.; Zhou, W. Variable pixel size ionospheric tomography. *Adv. Space Res.* **2017**, *59*, 2969–2986. [CrossRef]
17. Dobrovolsky, I.; Zubkov, S.; Miachkin, V. Estimation of the size of earthquake preparation zones. *Pure Appl. Geophys.* **1979**, *117*, 1025–1044. [CrossRef]
18. Bergeot, N.; Chevalier, J.-M.; Bruyninx, C.; Pottiaux, E.; Aerts, W.; Baire, Q.; Legrand, J.; Defraigne, P.; Huang, W. Near real-time ionospheric monitoring over Europe at the Royal Observatory of Belgium using GNSS data. *J. Space Weather. Space Clim.* **2014**, *4*, A31. [CrossRef]
19. Liu, J.Y.; Chuo, Y.; Shan, S.; Tsai, Y.; Chen, Y.; Pulinets, S.; Yu, S. Pre-earthquake ionospheric anomalies registered by continuous GPS TEC measurements. In *Annales Geophysicae*; Copernicus GmbH: Göttingen, Germany, 2004; pp. 1585–1593.
20. Liu, J.-Y.; Chen, Y.; Chen, C.-H.; Liu, C.; Chen, C.; Nishihashi, M.; Li, J.; Xia, Y.; Oyama, K.; Hattori, K. Seismoionospheric GPS total electron content anomalies observed before the 12 May 2008 Mw7. 9 Wenchuan earthquake. *J. Geophys. Res. Space Phys.* **2009**, *114*. [CrossRef]
21. Cahyadi, M.N.; Heki, K. Ionospheric disturbances of the 2007 Bengkulu and the 2005 Nias earthquakes, Sumatra, observed with a regional GPS network. *J. Geophys. Res. Space Phys.* **2013**, *118*, 1777–1787. [CrossRef]
22. Dong, Y.; Gao, C.; Long, F.; Yan, Y. Suspected Seismo-Ionospheric Anomalies before Three Major Earthquakes Detected by GIMs and GPS TEC of Permanent Stations. *Remote Sens.* **2022**, *14*, 20. [CrossRef]
23. Zheng, D.; Yao, Y.; Nie, W.; Chu, N.; Lin, D.; Ao, M. A new three-dimensional computerized ionospheric tomography model based on a neural network. *GPS Solut.* **2020**, *25*, 10. [CrossRef]
24. Zheng, D.; Yao, Y.; Nie, W.; Liao, M.; Liang, J.; Ao, M. Ordered subsets-constrained ART algorithm for ionospheric tomography by combining VTEC data. *IEEE Trans. Geosci. Remote Sens.* **2020**, *59*, 7051–7061. [CrossRef]
25. Rao, P.; Ram, S.T.; Krishua, S.; Niranjan, K.; Prasad, D. *Morphological Characteristics of L-Band Scintillations and Their Impact on GPS Signals—A Quantitative Study on the Precursors for the Occurrence of Scintillations*; Andhra University Vishakhapatnam (India) Deptartment of Physics: Visakhapatnam, India, 2006.
26. Filjar, R.; Kos, S.; Krajnovic, S. Dst Index as a Potential Indicator of Approaching GNSS Performance Deterioration. *J. Navig.* **2013**, *66*, 149–160. [CrossRef]
27. Jin, S.; Occhipinti, G.; Jin, R. GNSS ionospheric seismology: Recent observation evidences and characteristics. *Earth-Sci. Rev.* **2015**, *147*, 54–64. [CrossRef]

28. Heki, K.; Enomoto, Y. Preseismic ionospheric electron enhancements revisited. *J. Geophys. Res. Space Phys.* **2013**, *118*, 6618–6626. [CrossRef]
29. Rolland, L.M.; Lognonné, P.; Astafyeva, E.; Kherani, E.A.; Kobayashi, N.; Mann, M.; Munekane, H. The resonant response of the ionosphere imaged after the 2011 off the Pacific coast of Tohoku Earthquake. *Earth Planets Space* **2011**, *63*, 853–857. [CrossRef]
30. Astafyeva, E.; Shalimov, S.; Olshanskaya, E.; Lognonné, P. Ionospheric response to earthquakes of different magnitudes: Larger quakes perturb the ionosphere stronger and longer. *Geophys. Res. Lett.* **2013**, *40*, 1675–1681. [CrossRef]
31. Occhipinti, G.; Kherani, E.A.; Lognonné, P. Geomagnetic dependence of ionospheric disturbances induced by tsunamigenic internal gravity waves. *Geophys. J. Int.* **2008**, *173*, 753–765. [CrossRef]
32. Kakinami, Y.; Kamogawa, M.; Tanioka, Y.; Watanabe, S.; Gusman, A.R.; Liu, J.Y.; Watanabe, Y.; Mogi, T. Tsunamigenic ionospheric hole. *Geophys. Res. Lett.* **2012**, *39*. [CrossRef]
33. Ulukavak, M.; Yalçınkaya, M.; Kayıkçı, E.T.; Öztürk, S.; Kandemir, R.; Karslı, H. Analysis of ionospheric TEC anomalies for global earthquakes during 2000–2019 with respect to earthquake magnitude (Mw $\geq$ 6.0). *J. Geodyn.* **2020**, *135*, 101721. [CrossRef]
34. Nico, G.; Biagi, P.F.; Ermini, A.; Boudjada, M.Y.; Eichelberger, H.U.; Katzis, K.; Contadakis, M.; Skeberis, C.; Moldovan, I.A.; Bezzeghoud, M. Wavelet analysis applied on temporal data sets in order to reveal possible pre-seismic radio anomalies and comparison with the trend of the raw data. In *EGU General Assembly 2021*; Copernicus GmbH: Göttingen, Germany, 2021.
35. Nina, A.; Pulinets, S.; Biagi, P.F.; Nico, G.; Mitrović, S.T.; Radovanović, M.; Popović, L.C. Variation in natural short-period ionospheric noise, and acoustic and gravity waves revealed by the amplitude analysis of a VLF radio signal on the occasion of the Kraljevo earthquake (Mw = 5.4). *Sci. Total Environ.* **2020**, *710*, 136406. [CrossRef]
36. Ulukavak, M.; Yalcinkaya, M. Precursor analysis of ionospheric GPS-TEC variations before the 2010 M 7.2 Baja California earthquake. *Geomat. Nat. Hazards Risk* **2017**, *8*, 295–308. [CrossRef]
37. Freund, F. Toward a unified solid state theory for pre-earthquake signals. *Acta Geophys.* **2010**, *58*, 719–766. [CrossRef]
38. Liu, H.; Zhang, K.; Imtiaz, N.; Song, Q.; Zhang, Y. Relating Far-Field Coseismic Ionospheric Disturbances to Geological Structures. *J. Geophys. Res. Space Phys.* **2021**, *126*, e2021JA029209. [CrossRef]
39. Savastano, G.; Komjathy, A.; Shume, E.; Vergados, P.; Ravanelli, M.; Verkhoglyadova, O.; Meng, X.; Crespi, M. Advantages of geostationary satellites for ionospheric anomaly studies: Ionospheric plasma depletion following a rocket launch. *Remote Sens.* **2019**, *11*, 1734. [CrossRef]

remote sensing MDPI

Technical Note

Multi-Station and Multi-Instrument Observations of F-Region Irregularities in the Taiwan–Philippines Sector

Lung-Chih Tsai [1,2,*], Shin-Yi Su [2], Jun-Xian Lv [1], Terry Bullett [3] and Chao-Han Liu [4]

1 GPS Science and Application Research Center, National Central University (NCU), Taoyuan 320, Taiwan; mark90054@csrsr.ncu.edu.tw
2 Center for Space and Remote Sensing Research, National Central University (NCU), Taoyuan 320, Taiwan; sysu@csrsr.ncu.edu.tw
3 Cooperative Institute for Research in Environmental Sciences, NOAA National Centers for Environmental Information, Boulder, CO 80309, USA; terry.bullett@noaa.gov
4 Academia Sinica, Taipei 115, Taiwan; chliu2@gate.sinica.edu.tw
* Correspondence: lctsai@csrsr.ncu.edu.tw

Abstract: In this study, a multi-station and multi-instrument system, organized and proposed for ionospheric scintillation and equatorial spread-F (ESF) specification and their associated motions in the Taiwan–Philippines sector, is outlined. The issues related to the scintillation and ESF event observed on 26 October 2021, at magnetic quiet conditions are presented and discussed. We first indicate the existence of a plasma bubble in the Taiwan–Philippines sector by using the FormoSat-7/Constellation Observing System for Meteorology, Ionosphere, and Climate-2 (FS7/COSMIC2) GPS/GLONASS radio occultation observations. We verify the latitudinal extent of the tracked plasma bubble using the recorded ionograms from the Vertical Incidence Pulsed Ionospheric Radar located at Hualien, Taiwan. We further discuss the spatial and temporal variabilities of two-dimensional vertical scintillation index VS_4 maps based on the simultaneous GPS L1-band signal measurements from 133 ground-based receivers located in Taiwan and the surrounding islands. We also operate two high-sampling, software-defined GPS receivers and characterize the targeted plasma irregularities by carrying out spectrum analyses of the received signal. As a result, the derived plasma irregularities moved eastward and northward. Furthermore, the smaller the irregularity scale, the higher the spectral index and the stronger the scintillation intensity were at lower latitudes on the aimed irregularity feature.

Keywords: equatorial and low-latitude ionosphere; ionospheric irregularity; scintillation; radio occultation observation; COSMIC; ionogram; GPS/GNSS

Citation: Tsai, L.-C.; Su, S.-Y.; Lv, J.-X.; Bullett, T.; Liu, C.-H. Multi-Station and Multi-Instrument Observations of F-Region Irregularities in the Taiwan–Philippines Sector. *Remote Sens.* **2022**, *14*, 2293. https://doi.org/10.3390/rs14102293

Academic Editors: Serdjo Kos, José Fernández and Juan F. Prieto

Received: 30 March 2022
Accepted: 7 May 2022
Published: 10 May 2022

1. Introduction

Ionospheric scintillation has significant impacts on space-based radio communication, wave propagation, and navigation system performance. The main effects of scintillation on the transionospheric radio system are signal loss and/or phase cycle slips, which cause difficulties in regard to locking the receiver signal. There is no doubt that satellite radio signal scintillation is a consequence of a scattering mechanism, as radio waves are propagating through random electron density (N_e) fluctuations, especially within the F-region ionosphere, where the irregularity layer is sufficiently thick. Many excellent ionospheric scintillation theory and observation reviews have been published [1–4]. Comprehensive studies on the physics and theories of ionospheric irregularities and scintillations can also be found in books by [5,6].

The effects of ionospheric scintillation are most intense in the equatorial region, moderate at high latitudes, and minimum at middle latitudes [7]. At equatorial latitudes and the time near and/or soon after sunset, the zonal neutral wind and conductivity gradient caused by the sunset terminator interact develop an enhanced eastward electric field, called pre-reversal enhancement (PRE), generally at F-region heights [5,8,9]. As a result, the ionosphere moves upward, develops steep density gradients and large-scale plasma depletions in the bottom side F-region, and becomes unstable, triggering the Rayleigh–Taylor (R–T) instability. The plasma depletions, called plasma bubbles, become populated with mesoscale or small-scale irregularities and rise to great heights. These plasma bubbles, extended in altitude, also map out along magnetic field lines to the north and south of the magnetic equator, i.e., higher magnetic latitudes. These structured irregularities generally move eastward by the action of a vertical oriented polarization electric field. Undergoing a non-linear cascade process of electric field driving and wave-wave coupling [10], irregular structures drift and vary from large- to small-scale. The spectral distribution of irregularities includes a broad component associated with waves. Equatorial spread-F (ESF) thus stems from high-frequency radar observations of the "spread" of the ionospheric echoes in ionograms and can be used to describe equatorial and low-latitude F-region instability phenomena. Scintillation and ESF activities attain a maximum value during high sunspot activity periods, especially during equinoxes (March, April, September, and October), owing to the increased value of the background ionization density [11]. In particular, low-latitude scintillation can be dictated by solar transients, such as magnetic storms [11].

Several techniques have been used to observe and study ionospheric irregularities and scintillations. These include sounder, radar backscatter, in-situ measurements onboard rockets or satellites, ground-based satellite beacon signal observations, space-based navigation beacon observations using radio occultation (RO) techniques, etc. Earlier investigations [12] developed an operational system, named the SCIntillation Network Decision Aid (SCINDA), to nowcast and forecast scintillation. At the operator terminal, the SCINDA data were combined with an empirical plasma bubble model to generate three-dimensional maps of irregularity structures and two-dimensional outage maps for the equatorial region in the American sector. In this study, another multi-station and multi-instrument system, developed for ionospheric scintillation and ESF specification in the Taiwan–Philippines sector, is outlined. The issues related to the scintillation and ESF event observed on 26 October 2021, are presented and discussed. We shall first indicate the existence of a plasma bubble in the Taiwan–Philippines sector using the FS7/COSMIC2 Global Positioning System (GPS) or GLObal NAvigation Satellite System (GLONASS) RO observations. We shall verify the latitudinal extent of the tracked plasma bubble using the recorded ionograms from the Vertical Incidence Pulsed Ionospheric Radar (VIPIR) located at Hualien (23.89°N, 121.55°E, dip latitude 17°N), Taiwan. We further discuss the spatial and temporal variabilities of two-dimensional scintillation index maps based on the simultaneous GPS L1-band signal measurements from 133 ground-based receivers located in Taiwan and the surrounding islands. We also operate two high-sampling, software-defined GPS receivers and characterize the targeted plasma irregularities by carrying out spectrum analyses of the received signal. Overall, we summarize the derived plasma irregularity and ESF characteristics and point out a potential precursor for post-sunset scintillation and ESF events.

2. System Description

Since June 2019, the Taiwanese American FS7/COSMIC2 program has been executing active limb sounding of the Earth's neutral atmosphere and ionosphere via GPS/GLONASS RO observations from low-Earth orbiting (LEO) satellites. Similar to the prior mission, FS3/COSMIC, the FS7/COSMIC2 is a six-LEO-satellite constellation mission but orbits at 24° inclination and ~550 km altitudes (~720 km altitudes for parking orbits). It enhances the Global Navigation Satellite System (GNSS) receiver's capability to receive multi-channel (1.5 GHz and 1.2 GHz) GPS and GLONASS satellite signals and can provide more than 5000 RO observations per day within the region between the geographic latitudes of ±40°. Each RO observation can provide a set of limb-viewing measurements on GNSS signal intensity and phase from the LEO satellite altitude to the Earth's surface. Those measurements can be further retrieved into N_e and limb-viewing scintillation index

S_4 profiles in altitudes [13,14]. In this study, we propose to identify a plasma bubble in the Taiwan–Philippines sector by using the FS7/COSMIC2 GPS/GLONASS RO observations. Meanwhile, the Taiwan–Philippines sector is defined as the ionospheric region located in $20° \pm 15°$ N and $120° \pm 15°$ E geographic coordinates, where the geomagnetic latitudes are from $\sim-3°$ to $25°$.

The Hualien VIPIR is a modern ionospheric radar (also termed ionospheric sounder or ionosonde) that fully digitizes complex signal records and uses multiple parallel receiver channels for simultaneous signal measurements from multiple spaced receiving antennas [15]. As a usual ionospheric sounder, the Hualien VIPIR transmits pulsed waveforms in the medium- and high-frequency (MF/HF) bands and measures the envelope group delays, i.e., virtual ranges, of ionospheric echoes to produce ionograms with ionospheric trace h′(f) as a function of the radio carrier frequency. Details of the system can be found in [16].

Other sources of scintillation observations are the 133 ground-based GPS receivers located in Taiwan and the surrounding islands. They are operated and maintained by the Central Weather Bureau (CWB) of Taiwan and routinely provide 1-Hz satellite navigation system data (in the RINEX format). The online processing includes determinations of scintillation index S_4 and the two-dimensional vertical S_4 map in the Taiwan–Philippines sector. Based on earlier investigations [17], the theoretical and experimental analyses show that scintillation index S_4 values become underestimated when a sampling spatial scale is larger than the first Fresnel zone (FFZ). We note that the 1-Hz sampling rate from ground-based GPS receivers is high enough to determine complete S_4 values but not to characterize the signal scintillation spectrum. Thus, we have designed and implemented a software-defined receiver in order to acquire and track GPS (and Satellite-Based Augmentation System, SBAS) L1-band C/A code signals [18,19]. Compared with most commercial GPS receivers, software-defined GPS receivers offer added flexibility and versatility by implementing most functions in software. Another advantage of a software-defined GPS receiver is that it could have a maximum sampling rate of 1000 Hz due to the L1-band Coarse Acquisition (C/A) code duration of 1 millisecond, and the executing sampling rate can be much higher than that of a usual commercial GPS receiver. Spectrum analyses on the received signals of two software-defined GPS receivers located at Chungli (24.97°N, 121.19°E) and Hualien (23.89°N, 121.55°E), Taiwan will be used to characterize the targeted ionospheric irregularities.

3. Results

This section presents the multi-station and multi-instrument observations of a scintillation event that occurred in the Taiwan–Philippines sector on 26 October 2021, which was at magnetic quiet conditions, referring to Kp indexes between 0+ and 2+. Figure 1 shows the geographical geometry of the F-layer irregularity observations obtained during 13:00~15:30 UT, i.e., 21:00~23:30 LT in Taiwan, on 26 October 2021, by the FS7/COSMIC2 GPS/GLONASS RO sounding experiment. Six (#1 to #6 observations) out of fourteen RO observations were recorded and classified as scintillation observations according to their maximum limb-viewing L1-band S_4 values larger than 0.1 [14]. Their recording times, which are the FS7/COSMIC2 LEO satellite orbiting times at the peak altitudes of RO observations, are used to identify the scintillation event period, at least from 13:54 to 14:59 UT. For each RO scintillation observation, a set of RO limb-viewing links at a 15 s sampling rate (between 150 and 450 km altitudes), which connects the occultation points to their conjugate points, is shown in coded colors of L1-band S_4 values to present the possible projection area of ionospheric irregularities, and the trace of perigee points (or tangent points) is also shown. We note that the first three (#1, #2, and #3) RO observations are located near the magnetic equator and experience strong scintillations. The later three (#4, #5, and #6) RO observations are located in the northeast directions of the observations #1, #2, and #3 and could be their latitudinal mapping-out cases; thus, they experience weaker scintillations. In contrast, as shown in Figure 1, the peak tangent point positions of the other eight RO observations without definable scintillations are located outside of the aimed scintillation area covering observations #1 to #6.

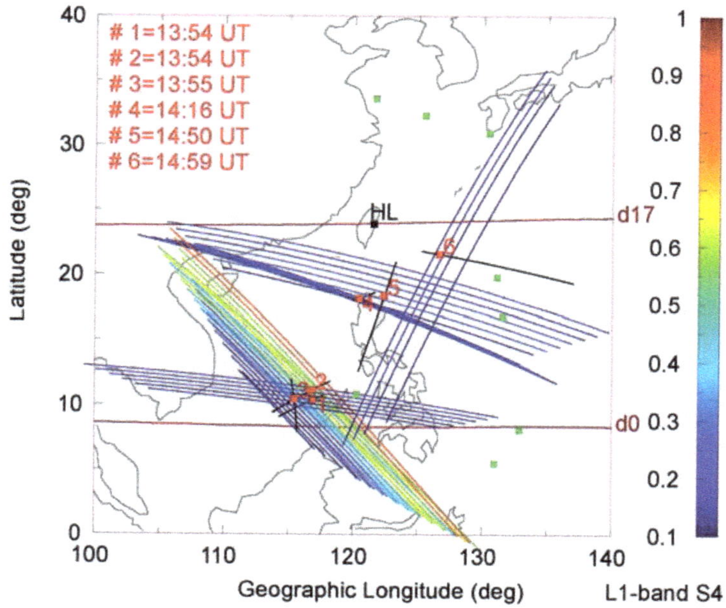

Figure 1. Geographical geometry of the F-layer irregularity observations during 13:00~15:30 UT on 26 October 2021, by the FS7/COSMIC2 GPS/GLONASS RO sounding experiment. Six out of fourteen RO observations were obtained and classified as the scintillation observations #1 to #6. Sets of limb-viewing links (at a de-sampling rate of 15 s and altitudes between 150 to 450 km) connecting the occultation points to their conjugate points are shown in coded colors by L1-band S_4 values to present the possible projection area of ionospheric irregularities. The traces of limb-viewing tangent points are also shown in black. The peak locations of the other eight RO observations without scintillation are shown by green squares, the location of the Hualien VIPIR position is shown by a black square, and two magnetic latitude lines at dip 0° and 17° are shown in brown.

Figure 2 shows the RO scintillation observations #1 and #4 with more information, including the limb-viewing L1-band signal-noise-ratio (SNR) amplitude profiles at both the occulting and calibrating sides, the resulting undersampling S_4 profiles, and the retrieved N_e profile. We note that the L2-band signals are much weaker and thus do not have enough sensitivity to derive reliable S_4 values to be shown in this paper. On the other hand, the derived S_4 values are from "undersampling" measurements because S_4 values become saturated and completed when a sampling spatial scale is less than the FFZ, but otherwise, S_4 values could be underestimated at undersampling conditions [17]. The FFZ is defined by $D_F = \sqrt{\lambda L}$, where λ is the radio wavelength and L is the distance from the irregularity position, which is assumed to be the tangent point position along a limb-viewing GPS/GLONASS-LEO ray, to an LEO satellite position in this study. For those tangent-point altitudes from 400 to 200 km, the FFZ scale sizes for the L1-band signals are between 516 and 643 m. Therefore, the Fresnel frequency can be obtained via $f_F = v/\sqrt{2}\,D_F$, where v is the relative radio-scanning speed to the ionosphere. The derived Fresnel frequency is thus approximately 2.1 (2.6) Hz at 400 (200) km altitude for a frozen ionosphere. We note that the upward drift velocity of plasma irregularities is usually less than 50 m/s [4], which is much lower than the vertical component (~2 km/s) of LEO satellite velocity at F-region altitudes and can be ignored to estimate the corresponding Fresnel frequency. We conclude that the sampling rate of FS7/COSMIC2 RO observations on the ionosphere is 1 Hz and less than the possible Fresnel frequencies of plasma irregularities, and thus the derived S_4 values should be underestimated at undersampling conditions.

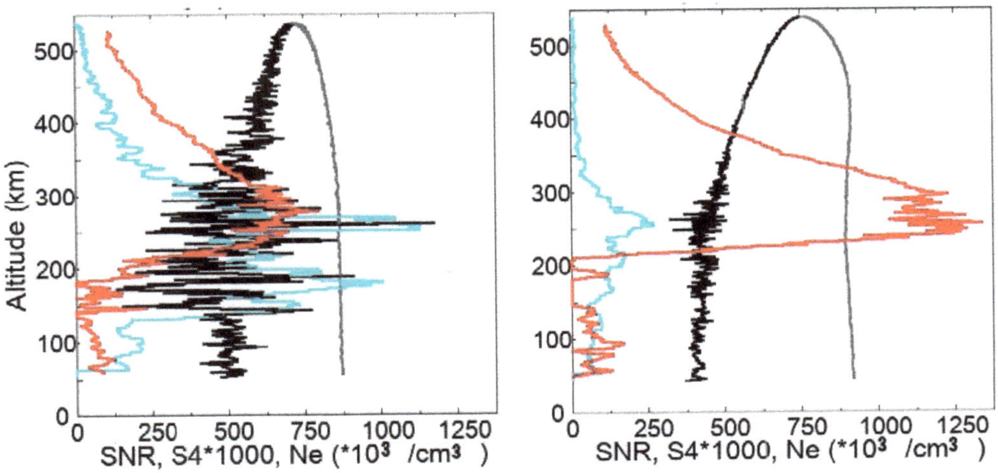

Figure 2. Results of FS7/COSMIC2 RO scintillation observations #1 (**left panel**) and #4 (**right panel**) as in Figure 1. Both figures show the limb-viewing L1-band SNR amplitude profiles at the occulting (calibrating) side in black (grey) and the resulting undersampling S_4 profiles in cyan. The retrieved N_e profiles are shown in red.

As shown in Figure 2, strong and moderate fluctuations occur in the L1-band SNR amplitudes of observations #1 and #4, respectively, from the occulting-side observations but not from the calibrating-side observations. It indicates that the amplitude fluctuations are induced by the ionospheric irregularities, which are located along the limb-viewing links between the occultation points and their conjugate points, as shown in Figure 1. We note that the strong amplitude fluctuations, i.e., large undersampling S_4 values, as well as the N_e fluctuations from observation #1, were distributed over a wider altitude region up to the LEO orbiting altitude than those from observation #4, and the peak scintillations and N_e irregularities happened around the F-layer electron density peaks. This is consistent with the observations from the theoretic descriptions in the Introduction section.

Equatorial spread-F (ESF) features usually accompany equatorial plasma bubbles and have been observed by numerous authors using ionosonde. In this study, we operated the Hualien VIPIR, a modern ionosonde, and observed spread-F features from the ionograms recorded between 13:19~15:04 UT, i.e., 21:19~23:04 LT, on 26 October 2021. The observed ESF event lasted for approximately one hour and forty-five minutes. In Figure 3, the upper ionogram shows two traces of one-hop F-layer echoes, where the upper trace is weaker than the main trace and has higher altitudes with approximately 25 km differences at the start (13:19 UT) of the ESF event. Furthermore, the lower ionogram shows a weak range spread-F, which was recorded during the middle (13:49 UT) of the ESF event. We note that the spread-F frequencies ranged from ~1.7 MHz up to 11.5 MHz, which presents the top frequency of ordinary ionospheric echoes, i.e., *foI*. It is usually difficult to retrieve a critical plasma frequency *foF2* from ionograms with ESF, but not difficult to retrieve *foI*. Multi-trace echoes could be interpreted as being due to the large-scale plasma depletion that happened before an ESF event. Meanwhile, the range-type spread-F indicates the existence of small-scale irregularities on the whole bottom side ionosphere.

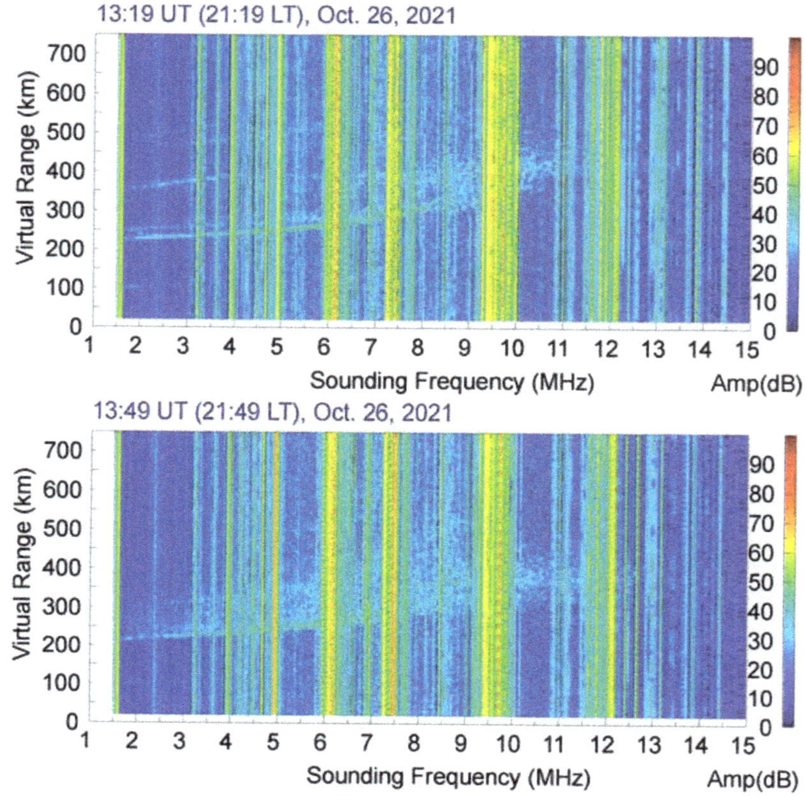

Figure 3. Two typical ionograms showing multi-trace echoes (**upper panel**) and weak range spread-F (**lower panel**) observed at the start (13:19 UT) and mid (13:49 UT) of an ESF event by the Hualien VIPIR on 26 October 2021. Note that the signals throughout all virtual ranges are radio frequency interferences from other radio sources.

Figure 4 shows the geographical geometry of the ionospheric pierce points (IPP) of the simultaneous GPS signal measurements at 14:42 UT on 26 October 2021, using the CWB GPS receiving network. The IPP altitudes were designed to be 300 km in accordance with the statistical peak scintillation or irregularity altitude retrieved from the FS7/COSMIC2 limb-viewing S_4 and/or N_e profiles, as shown in Figure 2. Furthermore, the IPPs were obtained at and limited by a minimum elevation angle of 45° from receivers to avoid multi-path signals. In this study, we define the vertical scintillation index VS_4 as the vertical component of the normalized variance of the signal power intensity I as follows:

$$VS_4 = \sin(\theta) \times sqrt\left(\frac{\left\langle (I - \langle I \rangle)^2 \right\rangle}{\langle I \rangle^2}\right), \quad (1)$$

where θ is the line-of-sight elevation angle from the receiver. We calculate VS_4 values by applying (1) and a 30-s duration window to 1-Hz L1-band signal amplitudes from the CWB GPS data archives. We note that the use of VS_4 to represent a vertical scintillation index is based on an assumption in which irregular N_e distribution is a function of altitude only along a transionospheric radio path. However, equatorial plasma bubbles are plume-like structures from low altitudes [4], and the shapes of bubbles are extended vertically upward and also stretched in the north-south direction along the magnetic field lines. Ref. [20] shows that GPS RO observations with high S_4 values are much more likely to occur when the ray paths are distributed in certain bubbles and more nearly aligned with the magnetic field. The details of two-dimensional and even three-dimensional irregular N_e structures should be figured out in future studies. Meanwhile, we can assume the observing irregularities located at altitudes around 300 km to be the IPP altitude, and the corresponding FFZ scale sizes (D_F) of L1-band signals are between 239 and 275 m for elevation angles from 90° to 45°. We note that GPS satellites have orbiting velocities of approximately 3.9 km/s, and thus the corresponding IPP velocity is ~60 (80) meter/s at an elevation angle of 90° (45°) from the receiver. If we assume a frozen ionosphere, the resulting Fresnel frequencies f_F are between 0.18 and 0.2 Hz. Ref. [11] shows that, using extensive incoherent scatter radar observations from Jicamarca, the daytime westward drifts are significantly smaller than the nighttime eastward drifts, and during solar maximum, the maximum nighttime eastward velocity increases up to about 160 m/s. Even if we include a 160 m/s IPP trace component of plasma irregularity drift velocity, the resulting Fresnel frequencies f_F are approximately 0.66 Hz and are also lower than the 1-Hz sampling rate for the CWB GPS receivers. This means that the derived S_4 values from the CWB GPS data are at complete conditions. As a result, Figure 5 shows three developed structures on a two-dimensional VS_4 map taken at 13:24, 13:49, and 14:42 UT, which are approximately at the start, middle, and end of the scintillation and ESF event. Referring to Figure 4 and the bottom panel of Figure 5 based on the same CWB GPS data at 14:42 UT, we derive a two-dimensional extended VS_4 map deduced through relaxation by applying the red-black smoothing technique [21] on point-distributed VS_4 measurements. We execute one pass through the mesh updating the "red" cells (like the red squares of a checkerboard) and another pass updating the "black" cells (like the black squares) and so forth for two loops. It produces two more cell extents on derived VS_4 maps where the cell (latitudinal or longitudinal) resolution is 0.1°, determined approximately by the medium distance between neighboring GPS stations.

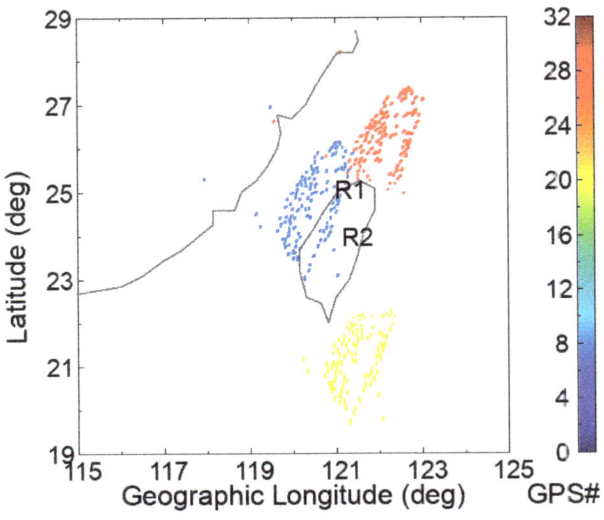

Figure 4. Geographical geometry of the IPPs of the simultaneous GPS signal measurements at 14:42 UT on 26 October 2021, and from 133 ground-based receivers located in Taiwan and the surrounding islands. The dots colored in light blue, yellow, and red show the IPP positions at 300 km altitudes for the lines of sights connecting GPS satellite #8, #21, and #27, respectively, which have a minimum elevation angle of 45° from receivers. The positions of another two software-defined GPS receivers located at Chungli (24.97°N, 121.19°E) and Hualien (23.89°N, 121.55°E), Taiwan, are shown and labeled by "R1" and "R2", respectively.

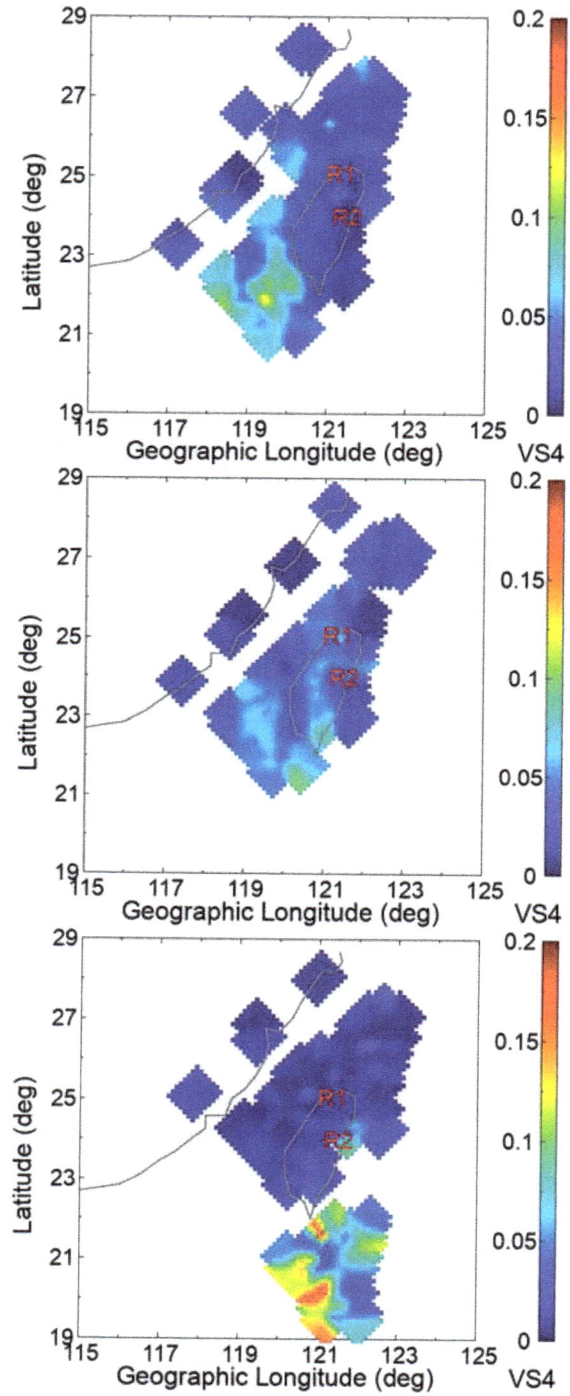

Figure 5. Three developed structures in two-dimensional VS_4 map taken at 13:24 (**upper panel**), 13:49 (**middle panel**), and 14:42 UT (**lower panel**) on 26 October 2021, and derived by the simultaneous GPS signal measurements from the CWB GPS receiving network.

It is now generally accepted that radio scintillation caused by ionospheric plasma irregularities can be characterized by fitting a power-law function with the spectral index p to the derived signal spectrum [2,22–24]. According to the scintillation theory and power spectra analyses [2,25], the intensity spectrum of weak and moderate scintillations and power-law ionospheric irregularities should have more or less a flat portion at low frequencies and start to roll off around the Fresnel frequency in the form f^{1-p}, where p is defined as the spectral index and is near 4 in the irregularity power-law relationship. Furthermore, the higher the spectral index, the stronger the scintillation intensity and the smaller the irregularity scale.

As described in the last section, we also operate two high-sampling, software-defined GPS receivers located at Chungli and Hualien, Taiwan separately to simultaneously receive GPS signals. The locations of the Chungli and Hualien software-defined GPS receivers are shown in Figures 4 and 5 and labeled by "R1" and "R2", respectively. In this study, L1-band C/A code signal amplitudes were recorded at a sampling rate of 50 Hz and for a 70-s duration every five minutes. We obtain three L1-band signal scintillation patches recorded by the Chungli software-defined GPS receiver at different data segments, which are from 13:29 to 13:44 UT, 14:29 to 14:49 UT, and 13:44 to 14:19 UT on 26 October 2021, for GPS satellites #8, #21, and #27 signal observations, respectively. The derived VS_4 values are consistent with CWB GPS observation results and are from 0.19 to 0.25, 0.16 to 0.22, and 0.14 to 0.19 for GPS satellites #8, #21, and #27 signal measurements, respectively. Figure 6 shows the corresponding signal spectrums obtained via a Lomb periodogram algorithm [21]. We note that above approximately 1 Hz, the power spectral densities (PSD) are near the noise level where the minimum frequency of the noise level is defined as the deviation frequency of the signal intensity spectrum [26,27]. The 50-Hz sampling rate is approximately one order higher than the derived deviation frequency. Below 1 Hz, the PSDs are more or less a flat portion at low frequencies and decay from their maxima at a break frequency of f_B toward the noise level in an approximately linear fashion on the log-log scale shown. This indicates and confirms a power-law variation f^{1-p} of the plasma irregularity PSD with the frequency as discussed. We note that the spectrum break frequencies of f_B can be treated as the experimental Fresnel frequencies and are approximately 0.07, 0.15, and 0.12 Hz, and the derived spectral index p values are 3.57, 4.36, and 3.59 for the L1-band signal scintillation patches from GPS satellites #8, #21, and #27, respectively. Comparing with the corresponding IPP positions shown in Figure 4 and the derived VS_4 maps shown in Figure 5, we note that the area with a higher spectral index p has a stronger scintillation intensity. The spectrum analysis results of the L1-band signal scintillation patches recorded by the Hualien software-defined GPS receiver are similar to those from the Chungli system and are not shown in this study.

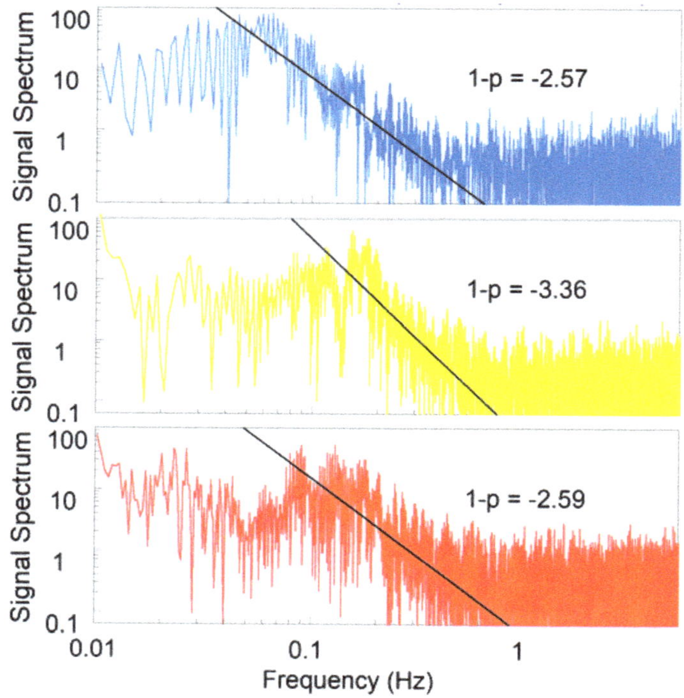

Figure 6. Power spectrums and their spectral index p values of the L1-band signal scintillation patches recorded by the Chungli software-defined GPS receiver from GPS satellites #8 (**upper panel**), #21 (**middle panel**), and #27 (**lower panel**). Three data segments are from 13:29 to 13:44 UT, 14:29 to 14:49 UT, and 13:44 to 14:19 UT on 26 October 2021, for GPS satellites #8, #21, and #27 signal observations, respectively.

4. Discussion

As mentioned in the Introduction section, at the times near and/or soon after sunset, an enhanced eastward electric field, i.e., PRE, causes the F-layer to move upward and develop plasma depletions, triggering the Rayleigh–Taylor instability. As a result, the plasma bubbles develop from the bottom side, and the instabilities cause ionospheric irregularities and radio signal scintillations and intrude into the higher altitudinal and latitudinal ionosphere.

In this study, a multi-station and multi-instrument system is organized and proposed for ionospheric scintillation and ESF specification in the Taiwan–Philippines sector. The FS7/COSMIC2 program can provide more than 5000 GPS/GLONASS RO observations per day within the region between the geographic latitudes of $\pm 40°$, i.e., approximately seven RO observations per hour in the Taiwan–Philippines sector (between $20° \pm 15°$N and $120° \pm 15°$E geographic coordinates). From the FS7/COSMIC2 RO measurements on 26 October 2021, the observed or retrieved limb-viewing SNR, S_4, and N_e profiles have been used to identify an ionospheric irregularity and scintillation event that happened from 13:30 to 15:00 UT, i.e., from 21:30 to 23:00 LT in Taiwan, and at magnetic quiet conditions. The results also show that the N_e irregularities and limb-viewing radio intensity scintillations are stronger and distributed into higher altitudes at the southernmost part of the Taiwan–Philippines sector, i.e., near the geomagnetic equator, compared to those near the sector center (geomagnetic latitude ~10°N). Furthermore, as shown in the right panel of Figure 2, the scintillation altitudes of the RO observation #4 and even observations #5 and #6 were distributed at an altitude range of ~100 km only and around the F-layer N_e peaks. This could indicate that the irregularities at higher latitudes are the latitudinal mapping-out facts from lower latitudes, and that there are stronger irregularities around F-layer peaks because of greater conductivities.

More evidence of the longitudinal extent of plasma irregularities is also shown by the two-dimensional VS_4 maps derived from the CWB GPS data archive. As shown in Figure 5, the significant scintillation event that happened on 26 October 2021, was also observed by the CWB GPS receiving network. The series of VS_4 maps shows plasma irregularities distributed with a stronger intensity at lower latitudes and when moving eastward. This indicates that the FS7/COSMIC2 could provide ionospheric irregularity and scintillation observations scanning in different limb-viewing, i.e., near-vertical, directions, and more than one hundred and thirty ground-based GPS receivers operated by the CWB could do it in horizontal directions.

We note that the ground-based CWB GPS signal observations have a sampling rate of 1 Hz, which is higher than the possible maximum Fresnel frequencies f_F of approximately 0.66 Hz and can thus complete the scintillation index S_4 determination. However, the 1-Hz sampling rate is not enough for irregularity spectral index determination, which needs the rate to be approximately one order larger than its Fresnel frequency. In practice, spectrum analyses applied to the high-sampling, software-defined GPS receiver measurements conform to a power-law variation f^{1-p} of plasma irregularity PSD with frequency. Meanwhile, the derived spectrum break frequencies f_B are more or less 0.1 Hz, which are all less than the corresponding Fresnel frequencies f_F of approximately 0.2 Hz for a frozen ionosphere. This indicates that the targeted plasma irregularities moved northward too and had positive velocity components along the IPP-tracing directions to decrease the relative radio-scanning speed and the experimental Fresnel frequencies obtained by the derived f_B values. In sum, integrating the observations from the FS7/COSMIC2, the CWB GPS receiving network, and two software-defined GPS receivers located in Taiwan, the experimental results show that the targeted plasma irregularities moved eastward and northward. Furthermore, the smaller the irregularity scale, the higher the spectral index and the stronger the scintillation intensity at lower latitudes on the aimed irregularity feature.

As described, ESF features usually accompany equatorial plasma bubbles and can also be observed and scaled from ionograms, as shown in Figure 2. Figure 7 shows the time variations of the virtual ranges of ionospheric echoes at different sounding frequencies *fs* of 1.72, 3.08, 4.72, 5.65, and 7.15 MHz from the Hualien ionograms recorded on 26 October 2021. Figure 7 also shows the corresponding temporal profile of scaled *foI*, which is approximately equal to *foF2* without spread-F and/or sporadic E features but higher than *foF2* with spread-F features. We note that the sunset time of the day was approximately 18:30 LT, i.e., 10:30 UT, at Hualien, Taiwan, and at a 300 km altitude. As shown in Figure 7, the spread-F features happened and were observed between 13:19~15:04 UT, i.e., 21:19~23:04 LT. Furthermore, after 15:04 UT, i.e., the end of the ESF event, the *foI* (*foF2*) values decreased, and the virtual heights of the fixed-frequency ionospheric echoes increased as usual facts of nighttime ionograms. Before 13:19 UT, i.e., the start of the ESF event, the *foI* (*foF2*) values show more or less a flat portion, but the virtual heights of fixed-frequency ionospheric echoes decreased except for those at a sounding frequency of 1.72 MHz, which are actually one-hop and two-hop sporadic E echoes and thus almost invariant in virtual heights. This indicates that after sunset and before the ESF event, the peak ionospheric N_e values were approximately the same but the ionospheric N_e values at the bottom side ionosphere were increased. This could be due to a strong PRE, i.e., eastward electric field enhancement that happened near or after sunset on the day and produced an equatorial fountain effect. As a result, the equatorial ionosphere moved upward, developed steep density gradients and large-scale plasma depletions in the bottom side F-region and became unstable, triggering the R–T instability. Meanwhile, ionospheric plasma moved out along the geomagnetic field line and from the magnetic equator to higher latitudes. In this study, such an equatorial fountain effect was not strong enough to increase the peak N_es and *foF2s* but was strong enough to increase the N_es at the bottom side ionosphere in the low-latitude region, e.g., the Taiwan area. We found that a post-sunset decrement of the virtual heights of fixed-frequency ionospheric echoes could be a good precursor for post-sunset scintillation and ESF events.

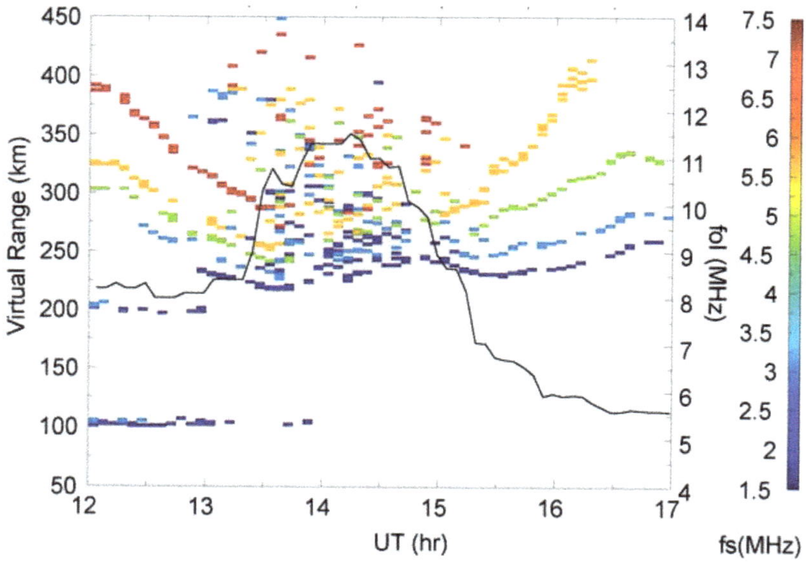

Figure 7. Time variations of virtual ranges (referring to the left *y*-axis) of ionospheric echoes at different sounding frequencies *fs* of 1.72 (dark blue), 3.08 (light blue), 4.72 (green), 5.65 (orange), and 7.15 MHz (red) from the Hualien ionograms recorded on 26 October 2021. Another temporal profile of scaled *foI* is also shown and referring to the right text. Note that the spread-F features were happened between 13:19~15:04 UT, i.e., 21:19~23:04 LT in Taiwan.

5. Conclusions

Ionospheric irregularities and scintillations and their associated motions in the Taiwan–Philippine sector have been observed and specified by the FS7/COSMIC2 data, the VIPIR located at Hualien, Taiwan, 133 ground-based GPS receivers located in Taiwan and the surrounding islands, and two high-sampling, software-defined GPS receivers. The integrated system has the potential to provide scintillation intensities, zonal drift measurements, and even three-dimensional irregularity structures. We also suggest that a post-sunset decrement of the virtual heights of fixed-frequency ionospheric echoes could be a good precursor for post-sunset scintillation and ESF events. In the future, we expect to identify the ionospheric conditions in the Taiwan–Philippines sector that led to the onset of plasma/R–T instabilities and to forecast the growth and the timing/duration of each instability. An examination of these instabilities will form the basis for the forecast of the timing and severity of (GNSS) radio scintillations.

Author Contributions: Conceptualization, L.-C.T., S.-Y.S. and C.-H.L.; Data curation, L.-C.T., J.-X.L. and T.B.; Formal analysis, L.-C.T.; Funding acquisition, L.-C.T. and S.-Y.S.; Investigation, L.-C.T., J.-X.L. and T.B.; Methodology, L.-C.T., S.-Y.S. and C.-H.L.; Project administration, L.-C.T. and S.-Y.S.; Resources, L.-C.T., J.-X.L. and T.B.; Software, L.-C.T. and J.-X.L.; Supervision, L.-C.T., S.-Y.S. and C.-H.L.; Validation, L.-C.T.; Visualization, L.-C.T.; Writing—original draft, L.-C.T.; Writing—review & editing, S.-Y.S., T.B. and C.-H.L. All authors have read and agreed to the published version of the manuscript.

Funding: This research was funded by the Office of Naval Research, U.S.A. [grant number N00014-21-1-2486] and in part by the Ministry of Science and Technology, Taiwan [grant number MOST 110-2119-M-008-003].

Data Availability Statement: The FS7/COSMIC2 "ionPhs" data can be downloaded from the COSMIC Data Analysis and Archive Center (CDAAC, http://cdaac-www.cosmic.ucar.edu/ (accessed on 1 December 2021)) and the Taiwan Analysis Center for COSMIC (TACC, http://tacc.cwb.gov.tw/cdaac/ (accessed on 1 December 2021)).

Acknowledgments: The authors would also like to thank UCAR's CDAAC, NSPO Satellite Operations Control Center (SOCC), and the CWB, Taiwan, for providing FS7/COSMIC2 data and GPS data.

Conflicts of Interest: The authors declare no conflict of interest.

References

1. Aarons, J. Global morphology of ionospheric scintillations. *Proc. IEEE* **1982**, *70*, 360–378. [CrossRef]
2. Yeh, K.C.; Liu, C.H. Radio wave scintillations in the ionosphere. *Proc. IEEE* **1982**, *70*, 324–360. [CrossRef]
3. Basu, S.; Basu, S. Equatorial scintillations: Advances since ISEA-6. *J. Atmos. Terr. Phys.* **1985**, *47*, 753–768. [CrossRef]
4. Basu, S.; Groves, K.M.; Basu, S.; Sultan, P.J. Specification and forecasting of scintillations in communication/navigation links: Current status and future plans. *J. Atmos. Sol.-Terr. Phys.* **2002**, *64*, 1745–1754. [CrossRef]
5. Kelley, M.C. *The Earth Ionosphere: Plasma Physics and Electrodynamics*, 2nd ed.; Elsevier: Amsterdam, The Netherlands; Academic Press: Boston, MA, USA, 2009.
6. Rino, C.L. *The Theory of Scintillation with Applications in Remote Sensing*; John Wiley & Sons, Inc.: Hoboken, NJ, USA, 2011.
7. Basu, S.; MacKenzie, E.; Basu, S. Ionospheric constraints on VHF/UHF communications links during solar maximum and minimum periods. *Radio Sci.* **1988**, *23*, 363–378. [CrossRef]
8. Farley, D.T.; Bonelli, E.; Fejer, B.G.; Larsen, M.F. The pre-reversal enhancement of the zonal electric field in the equatorial ionosphere. *J. Geophys. Res.* **1986**, *91*, 13723. [CrossRef]
9. Crain, D.J.; Heelis, R.A.; Bailey, G.J.; Richmond, A.D. Low-latitude plasma drifts from a simulation of the global atmospheric dynamo. *J. Geophys. Res.* **1993**, *98*, 6039. [CrossRef]
10. Kelley, M.C.; Franz, T.L.; Prasad, G. On the turbulent spectrum of equatorial spread F: A comparison between laboratory and space results. *J. Geophys. Res.* **2002**, *107*, 1432. [CrossRef]
11. Fejer, B.G.; de Paula, E.R.; Gonzalez, S.A.; Woodman, R.F. Average vertical and zonal F region plasma drifts over Jicamarca. *J. Geophys. Res.* **1991**, *96*, 13901. [CrossRef]
12. Groves, K.M.; Basu, S.; Weber, E.J.; Smitham, M.; Kuenzler, H.; Valladares, C.E.; Sheehan, R.; MacKenzie, E.; Secan, J.A.; Ning, P.; et al. Equatorial scintillation and systems support. *Radio Sci.* **1997**, *32*, 5. [CrossRef]
13. Tsai, L.-C.; Chang, K.K.; Liu, C.H. GPS radio occultation measurements on ionospheric electron density from low Earth orbit. *J. Geod.* **2011**, *85*, 941–948. [CrossRef]
14. Tsai, L.-C.; Su, S.-Y.; Liu, C.-H. Global morphology of ionospheric F-layer scintillations using FS3/COSMIC GPS radio occultation observations. *GPS Solut.* **2017**, *21*, 1037–1048. [CrossRef]
15. Grubb, R.N.; Livingston, R.; Bullett, T.W. *A New General Purpose High Performance HF Radar*; XXIX URSI General Assembly: Chicago, IL, USA, 2008.
16. Tsai, L.-C.; Tien, M.-H.; Chen, G.-H.; Zhang, Y. HF radio angle-of-arrival measurements and ionosonde positioning. *Terr. Atmos. Ocean. Sci.* **2014**, *25*, 401–413. [CrossRef]
17. Tsai, L.-C.; Su, S.-Y.; Liu, C.-H.; Schuh, H.; Wickert, J.; Alizadeh, M.M. Diagnostics of Es layer scintillation observations using FS3/COSMIC data: Dependence on sampling spatial scale. *Remote Sens.* **2021**, *13*, 3732. [CrossRef]
18. Tsui, J.B.Y. *Fundamentals of Global Positioning System Receivers: A Software Approach*; John Wiley & Sons, Inc.: New York, NY, USA, 2000.
19. Tsai, L.-C.; Su, S.-Y.; Chien, H.; Liu, C.-H.; Schuh, H.; Wickert, J.; Alizadeh, M.M. Coastal sea-surface wave measurements using software-based GPS reflectometers in Lanyu, Taiwans. *GPS Solut.* **2021**, *25*, 133. [CrossRef]
20. Anderson, P.C.; Straus, P.R. Magnetic field orientation control of GPS occultation observations of equatorial scintillation. *Geophys. Res. Lett.* **2005**, *32*, L21107. [CrossRef]
21. Press, W.H.; Teukolsky, S.A.; Vetterling, W.T.; Flannery, B.P. *Numerical Recipes in C: The Art of Scientific Computing*, 2nd ed.; Cambridge University Press: New York, NY, USA, 1992.
22. Crane, R.K. Spectra of ionospheric scintillation. *J. Geophys. Res.* **1976**, *81*, 2041–2050. [CrossRef]
23. Rino, C.L. A power law phase screen model for ionospheric scintillation, 1. Weak scatter. *Radio Sci.* **1979**, *14*, 1135–1145. [CrossRef]
24. Rino, C.L. A power law phase screen model for ionospheric scintillation, 2. Strong scatter. *Radio Sci.* **1979**, *14*, 1147–1155. [CrossRef]
25. Singleton, D.G. Power spectra of ionospheric scintillations. *J. Atmos. Terr. Phys.* **1974**, *36*, 113–133. [CrossRef]
26. McCaffrey, A.M.; Jayachandran, P.T. Spectral characteristics of auroral region scintillation using 100 Hz sampling. *GPS Solut.* **2017**, *21*, 1883–1894. [CrossRef]
27. Demyanov, V.; Danilchuk, E.; Yasyukevich, Y.; Sergeeva, M. Experimental estimation of deviation frequency within the spectrum of scintillations of the carrier phase of GNSS signals. *Remote Sens.* **2021**, *13*, 5017. [CrossRef]

MDPI
St. Alban-Anlage 66
4052 Basel
Switzerland
Tel. +41 61 683 77 34
Fax +41 61 302 89 18
www.mdpi.com

Remote Sensing Editorial Office
E-mail: remotesensing@mdpi.com
www.mdpi.com/journal/remotesensing

www.ingramcontent.com/pod-product-compliance
Lightning Source LLC
Chambersburg PA
CBHW061615220926
51850CB00005B/514
9783036575940